MATHEMATIQ
&
APPLICATIONS

Directeurs de la collection:
J. M. Ghidaglia et X. Guyon

26

Springer
*Paris
Berlin
Heidelberg
New York
Barcelone
Budapest
Hong Kong
Londres
Milan
Santa Clara
Singapour
Tokyo*

Lars Hörmander

Lectures on Nonlinear Hyperbolic Differential Equations

Springer

Lars Hörmander
Department of Mathematics
University of Lund
Box 118, S-221 00 Lund, Sweden

Mathematics Subject Classification:

35L05, 35L10, 35L15, 35L45, 35L60, 35L65, 35L67, 35L70, 35B65, 35S50, 34A12

ISBN 3-540-62921-1 Springer-Verlag Berlin Heidelberg New York

SPIN: 11735205   46/3111 - 5 4 3 2 1 - Imprimé sur papier non acide

PREFACE

This book is a somewhat revised version of notes from lectures given at the University of Lund during three semesters 1986–87. The aim of those lectures was to present the main results then known about global existence or "blowup" of solutions of nonlinear hyperbolic differential equations and propagation of singularities for the solutions. Unfortunately lack of time made it impossible to cover such topics as the equations of fluid dynamics, the methods based on conformal transformations for the study of nonlinear perturbations of the wave equation or the Yang-Mills equation, and the propagation of conormal singularities for solutions of nonlinear equations.

When I was offered to publish the notes in the series "Mathématiques & Applications" my first reaction was that now, almost ten years later, the list of missing topics would be much longer. However, since the notes have been rather frequently quoted in the literature I finally accepted the offer to publish them after a minor revision which might make them more useful but still allows identification of references already in the literature. In spite of the revision this book still has the character of lecture notes; it is far less systematic than a monograph and generality is often sacrificed in the interest of simplicity.

The following are the major changes:

1. A short exposition of the use of conformal transformations to study the solutions of nonlinear perturbations of the wave equation has been added as Section 6.7, and it is supported by a new appendix and an addition to Section 6.2. This material has been taken over from notes of lectures given at the University of California at San Diego in the Winter Quarter of 1990.

2. In the original lecture notes several questions on the lifespan of solutions of nonlinear perturbations of the Klein-Gordon equation with one or two space dimensions were raised and motivated by heuristic arguments. Since affirmative answers are now available, this part of Section 7.5 has been replaced by a new Section 7.8 presenting work of Shatah which was mentioned but not covered in the original notes. It is the basis for the recent work in the area just mentioned, which is also sketched briefly.

3. Since the earlier version of Chapter IX was published with small changes as reference Hörmander [9] this chapter has been rather extensively rewritten to take into account improvements published in reference Hörmander [10].

4. Some references to related more recent work have been added but not in a systematic way aiming for completeness.

The book assumes some familiarity with basic distribution theory, measure theory and functional analysis. From Chapter VIII on it is useful to have some prior knowledge of basic microlocal analysis, but references are given to such results where the exposition is not selfcontained. Only the last section requires somewhat more advanced pseudo-differential techniques.

I am grateful to Hans Lindblad for information on recent work concerning nonlinear Klein-Gordon equations which was very valuable to me, and to the editors of this series and Springer Verlag for encouraging me to revive my notes.

Lund in December 1996

Lars Hörmander

CONTENTS

CHAPTER I

ORDINARY DIFFERENTIAL EQUATIONS

1.1. Introduction. Ordinary differential equations are always hyperbolic. One should therefore have a clear understanding of their properties with respect to existence and uniqueness of solutions before considering partial differential equations. Moreover, in many cases ordinary differential equations play an essential role in the study of partial differential equations. We shall not content ourselves here by just recalling the basic facts on ordinary differential equations but shall in fact discuss the proofs in some detail. The reason for that is that they give prototypes for arguments used in more general situations, and the ideas are more transparent in a simple case.

1.2. Local existence and uniqueness for the Cauchy problem. Let f be a continuous function in a neighborhood of $(t_0, x_0) \in \mathbf{R}^{1+n}$, with values in $\mathbf{R}^n$. We want to solve the Cauchy problem

$$(1.2.1) \qquad dx(t)/dt = f(t, x(t)); \quad x(t_0) = x_0;$$

for t in a neighborhood of t_0. To do so we choose $a > 0$, $b > 0$, and M such that f is defined in $R = \{(t, x) \in \mathbf{R}^{1+n}; |t - t_0| \le a, |x - x_0| \le b\}$ and

$$(1.2.2) \qquad |f(t, x)| \le M \quad \text{when } (t, x) \in R.$$

If x is a C^1 solution of (1.2.1) with $|x(t) - x_0| \le b$ for $|t - t_0| \le T$, it follows from (1.2.1) that

$$(1.2.3) \qquad |x(t) - x_0| = \left| \int_{t_0}^{t} f(s, x(s)) \, ds \right| \le M |t - t_0|, \quad |t - t_0| \le T.$$

The right-hand side is $< b$ if $M|t - t_0| < b$.

Proposition 1.2.1. *If $T \le \min(a, b/M)$ and $x(t)$ is any solution of (1.2.1) for $|t - t_0| \le T$, it follows that $|x(t) - x_0| \le b$ then.*

Proof. Let F be the set of all $T' \in [0, T]$ such that $|x(t) - x_0| \le b$ for $|t - t_0| \le T'$. This set is closed. If the supremum T_1 of F is less than T, we have just proved that $|x(t) - x_0| \le T_1 M < b$ for $|t - t_0| \le T_1$, so a neighborhood of $\pm T_1$ is in F since $x(t)$ is continuous. Hence $T_1 = T$ which proves the statement.

The preceding "continuous induction argument" will be used very frequently to take care of nonlinear effects without running into circular arguments.

Theorem 1.2.2. *Assume in addition to (1.2.2) the Lipschitz condition*

$$(1.2.4) \qquad |f(t, x) - f(t, y)| \le C|x - y| \quad \text{if } |t - t_0| \le a, |x - x_0| \le b, |y - x_0| \le b.$$

Then (1.2.1) has a unique C^1 solution for $|t - t_0| \le T$ if $T \le \min(a, b/M)$.

Proof. We shall solve (1.2.1) by successive approximation starting from $x_0(t) \equiv x_0$ and defining $x_k(t)$ for $k > 0$ by

$$(1.2.1)' \qquad dx_k(t)/dt = f(t, x_{k-1}(t)); \quad x_k(t_0) = x_0.$$

Thus

$$x_k(t) = x_0 + \int_{t_0}^t f(s, x_{k-1}(s))ds, \quad |t - t_0| \leq T.$$

By induction we conclude as in (1.2.3) that $|x_k(t) - x_0| \leq b$ when $|t - t_0| \leq T$, and for $k > 1$ we have

$$|x_k(t) - x_{k-1}(t)| = \left| \int_{t_0}^t (f(s, x_{k-1}(s)) - f(s, x_{k-2}(s)))ds \right|$$

$$\leq C \left| \int_{t_0}^t |x_{k-1}(s) - x_{k-2}(s)|ds \right|.$$

We have $|x_1(t) - x_0(t)| \leq M|t - t_0|$ and conclude inductively that

$$|x_k(t) - x_{k-1}(t)| \leq MC^{k-1}|t - t_0|^k/k! \leq MC^{-1}(CT)^k/k!.$$

Since $\sum (CT)^k/k! < \infty$, it follows that $x_k(t)$ converges uniformly to a limit $x(t)$ for $|t-t_0| \leq T$, that $|x(t) - x_0| \leq b$, and that

$$x(t) = x_0 + \int_{t_0}^t f(s, x(s))ds, \quad |t - t_0| \leq T.$$

Hence $x \in C^1$, and (1.2.1) is fulfilled.

If $\tilde{x}(t)$ is another solution then

$$x(t) - \tilde{x}(t) = \int_{t_0}^t (f(s, x(s)) - f(s, \tilde{x}(s)))ds,$$

and since $|x(t) - \tilde{x}(t)| \leq 2M|t - t_0|$, we obtain as before

$$|x(t) - \tilde{x}(t)| \leq 2MC^{-1}(CT)^k/k!$$

for any k. When $k \to \infty$ we conclude that $x(t) = \tilde{x}(t)$ identically.

Another more illuminating proof of the uniqueness is given in the proof of the following:

Theorem 1.2.3. *Assume* (1.2.2) *and* (1.2.4). *Then the Cauchy problem*

(1.2.5) $$dx/dt = f(t, x); \quad x(t_0) = y;$$

has a unique solution $x = x(t, y)$ *for* $|t - t_0|M + |y - x_0| \leq b$, $|t - t_0| \leq a$, *and*

(1.2.6) $$|x(t, y) - x(t, z)| \leq e^{C|t-t_0|}|y - z|,$$

if $|t - t_0| \leq a$ *and* $|t - t_0|M + \max(|y - x_0|, |z - x_0|) \leq b$.

Proof. The existence statement follows from Theorem 1.2.2 and the triangle inequality. Let $x(t, y)$ be any solution with initial data y. Since

$$d(x(t, y) - x(t, z))/dt = f(t, x(t, y)) - f(t, x(t, z))$$

we obtain if $R(t) = |x(t, y) - x(t, z)|$ (Euclidean norm!) that

$$|dR(t)/dt| \leq CR(t)$$

when $R(t) > 0$, for

$$\tfrac{1}{2}dR(t)^2/dt = \langle x(t, y) - x(t, z), d(x(t, y) - x(t, z))/dt \rangle$$
$$= \langle x(t, y) - x(t, z), f(t, x(t, y)) - f(t, x(t, z)) \rangle$$

can be estimated in absolute value by $CR(t)^2$. Hence

$$R(t) \leq e^{C|t-t_0|}R(t_0),$$

which proves (1.2.6) and, when $y = z$, also the uniqueness.

The proof gives in fact the uniqueness statement in a stronger result:

Theorem 1.2.3′. *Theorem 1.2.3 remains valid for $t \geq t_0$ if (1.2.4) is weakened to*

$$(1.2.4)' \qquad\qquad \langle y - z, f(t, y) - f(t, z) \rangle \leq C|y - z|^2.$$

The existence statement will be included in Theorem 1.2.6. In the scalar case the condition $(1.2.4)'$ means that

$$f(t, y) - f(t, z) \leq C(y - z) \quad \text{if } z \leq y,$$

so it is a one-sided Lipschitz condition. Let us also note that if this condition is fulfilled and

$$dx(t)/dt \geq f(t, x(t)), \quad dy(t)/dt \leq f(t, y(t)), \quad x(t_0) \geq y(t_0)$$

then $x(t) \geq y(t)$ for $t \geq t_0$. In fact, if $R(t) = y(t) - x(t) \geq 0$ then

$$dR(t)/dt \leq f(t, y(t)) - f(t, x(t)) \leq CR(t).$$

If $R(t) \geq 0$ for $t_1 < t < t_2$ and $R(t_1) = 0$ it follows that $R = 0$ in $[t_1, t_2]$ since $R(t)e^{-Ct}$ is nonnegative and decreasing. Since the left end point t_1 of a maximal interval I where $R(t) > 0$ would be a zero of R, it follows that $R = 0$ in I, so there is no such interval.

Before the next theorem we apply Theorem 1.2.3 to a linear differential equation.

Proposition 1.2.4. *Let $A(t)$ be a continuous function for $|t - t_0| \leq a$ with values in $n \times n$ matrices. Then the Cauchy problem*

$$dx(t)/dt = A(t)x(t), \quad x(t_0) = x_0 \in \mathbf{R}^n,$$

has a unique solution with values in $\mathbf{R}^n$ for $|t - t_0| \leq a$, and $x(t) = F(t)x_0$ where F is the unique $n \times n$ matrix valued function satisfying

$$(1.2.7) \qquad\qquad dF(t)/dt = A(t)F(t), \quad F(t_0) = \mathrm{Id},$$

where Id is the unit matrix. If $\|A(t)\| \leq M$ for $|t - t_0| \leq a$, then

$$(1.2.8) \qquad\qquad \|F(t)\| \leq e^{|t - t_0|M}, \quad \|F(t)^{-1}\| \leq e^{|t - t_0|M}, \quad |t - t_0| \leq a,$$

$$(1.2.9) \qquad\qquad \det F(t) = \exp\left(\int_{t_0}^t \mathrm{Tr}\, A(s)ds\right), \quad |t - t_0| \leq a.$$

Proof. We apply Theorem 1.2.3 with $b = 1$ and $z = 0$ and conclude that for $|t - t_0| \leq \min(a, 1/2M)$ and $|y| < 1/2$ there is a solution with $|x(t, y)| \leq e^{M|t - t_0|}|y|$. Since $x(t, y) = \varepsilon^{-1}x(t, \varepsilon y)$, the condition $|y| < 1/2$ can be dropped, and the solution is extended in at most $1 + 4Ma$ steps to the whole interval where $|t - t_0| \leq a$. By the uniqueness and linearity we have $x(t) = F(t)y$ where $\|F(t)\| \leq e^{|t - t_0|M}$, for this is true when $M|t - t_0| < 1/2$, and $F \in C^1$ since $F(t)y$ is in C^1 for fixed y. For fixed y we have

$$d(F(t)y)/dt = A(t)F(t)y$$

so (1.2.7) holds, and (1.2.7) of course determines F uniquely. The second part of (1.2.8) is obtained if we write $F(t) = F(t, t_0)$ and note that $F(t, t_0)F(t_0, t) = \mathrm{Id}$. To prove (1.2.9) we first observe that

$$F(t) = \mathrm{Id} + (t - t_0)A(t_0) + o(|t - t_0|),$$

which implies
$$\det F(t) = 1 + (t - t_0) \operatorname{Tr} A(t_0) + o(|t - t_0|).$$

Hence
$$\frac{d}{dt} \det F(t) = \operatorname{Tr} A(t_0) \quad \text{when } t = t_0.$$

By the group property observed above,
$$\det F(t + s)/\det F(t) = \det F(t + s, t),$$

which proves that
$$\left(\frac{d}{dt} \det F(t)\right)/\det F(t) = \operatorname{Tr} A(t).$$

Integration gives (1.2.9).

The following theorem is the first where we compare a general nonlinear differential equation with its *linearisation* at a solution:

Theorem 1.2.5. *Assume that (1.2.2) holds and that $\partial f/\partial x$ exists and is continuous in R. Then the function $x(t, y)$ defined in Theorem 1.2.3 is in C^1 when $|t - t_0|M + |y - x_0| \le b$, $|t - t_0| \le a$, and the Jacobian matrix $J = \partial x/\partial y$ satisfies*

$$(1.2.10) \qquad dJ(t, y)/dt = f_x'(t, x(t, y))J(t, y), \quad J(t_0, y) = \operatorname{Id}.$$

In particular,

$$(1.2.11) \qquad \det J(t, y) = \exp\left(\int_{t_0}^{t} \operatorname{Tr} f_x'(s, x(s, y))\, ds\right).$$

Proof. Fix z and set for $|t - t_0|M + \max(|y - x_0|, |z - x_0|) \le b$, $|t - t_0| \le a$
$$\psi(t) = x(t, y) - x(t, z), \quad \text{thus } \psi(t_0) = y - z.$$

Then
$$d\psi(t)/dt = f(t, x(t, y)) - f(t, x(t, z)) = f_x'(t, x(t, z))\psi(t) + \varrho_y(t)|y - z|$$

where by (1.2.6) and Taylor's formula for any $\varepsilon > 0$ we have $|\varrho_y(t)| \le \varepsilon$ if $|y - z|$ is small enough. Let $F(t)$ be the C^1 matrix function with
$$dF(t)/dt = f_x'(t, x(t, z))F(t), \quad F(t_0) = \operatorname{Id},$$

and set $\psi(t) = F(t)\varphi(t)$. For $F(t)$ and $F(t)^{-1}$ we have the estimates (1.2.8), so the equation
$$F(t)d\varphi(t)/dt = d\psi(t)/dt - (dF(t)/dt)\varphi(t) = \varrho_y(t)|y - z|$$

implies $|\varphi(t) - \varphi(t_0)| \le C\varepsilon|y - z|$, hence
$$|\psi(t) - F(t)(y - z)| = |F(t)(\varphi(t) - \varphi(t_0))| \le C\varepsilon|y - z|$$

for another C. This implies that $x(t, y)$ is differentiable with respect to y, and that $\partial x(t, y)/\partial y = F(t)$. Hence we have proved (1.2.10), and (1.2.11) follows from Theorem 1.2.4. That $J(t, y)$ is a continuous function of y follows from the uniqueness of the solution of (1.2.10) and the equicontinuity of $J(t, y)$ as a function of t.

If we drop the Lipschitz condition in Theorem 1.2.2, then only the uniqueness statement is lost:

Theorem 1.2.6 (Peano's theorem). *If f is continuous in R and satisfies (1.2.2) then (1.2.1) has a C^1 solution for $|t - t_0| \leq T$ if $T \leq a$ and $T \leq b/M$.*

Proof. We can extend f to $[t_0 - a, t_0 + a] \times \mathbf{R}^n$ by defining

$$f(t, x) = f(t, x_0 + (x - x_0)b/|x - x_0|) \quad \text{if } |x - x_0| > b.$$

By regularization with respect to x we then obtain a sequence of functions $f_\nu(t, x)$ which are continuous functions of t with values in $C^\infty(\mathbf{R}^n)$, such that $f_\nu \to f$ uniformly in R when $\nu \to \infty$ and $|f_\nu| \leq M$ for all ν. By Theorem 1.2.2 the Cauchy problem

$$dx_\nu(t)/dt = f_\nu(t, x_\nu(t)), \quad x_\nu(t_0) = x_0,$$

has a unique solution for $|t - t_0| \leq T$, and $|x_\nu(t) - x_0| \leq b$, $|dx_\nu(t)/dt| \leq M$ (by the differential equation). Thus we can find a uniformly convergent subsequence. Changing notation we may assume that $x_\nu(t)$ converges uniformly to $x(t)$, which implies that

$$dx_\nu(t)/dt = f_\nu(t, x_\nu(t)) \to f(t, x(t))$$

uniformly for $|t - t_0| \leq T$. Hence $x(t)$ satisfies (1.2.1).

Every solution of (1.2.1) can be obtained in the manner just described:

Proposition 1.2.7. *With f and T as in Theorem 1.2.6 let $x(t)$ be any solution of (1.2.1) for $|t - t_0| \leq T$. Then one can find $f_\nu(t, x)$ such that*

(i) *(1.2.1) holds with f replaced by f_ν also;*
(ii) *f_ν is a continuous function of t with values in $C^\infty(\mathbf{R}^n)$;*
(iii) *$f_\nu \to f$ uniformly in R as $\nu \to \infty$.*

Proof. If we define f_ν as in the proof of Theorem 1.2.6 we have the properties (ii) and (iii) but not necessarily (i). However,

$$dx(t)/dt = f(t, x(t)) = f_\nu(t, x(t)) + f(t, x(t)) - f_\nu(t, x(t)).$$

Replacing $f_\nu(t, x)$ by $f_\nu(t, x) + f(t, x(t)) - f_\nu(t, x(t))$ we have all the properties (i)–(iii).

This proposition gives at once:

Theorem 1.2.8. *Let the hypotheses of Theorem 1.2.6 be fulfilled, and let*

$$X = \{(t, x(t)); |t - t_0| \leq T, x(t) \text{ satisfies } (1.2.1)\}.$$

Then X is compact, and the slice $X_t = \{x; (t, x) \in X\}$ is connected for each t.

Proof. The compactness of X follows from the compactness in the uniform topology of the set of solutions of (1.2.1). It suffices to prove the connectedness when $T < b/M$. If X_s is not connected, we can write $X_s = F_0 \cup F_1$ with F_j compact, disjoint and nonempty. Let the distance between F_0 and F_1 be $2\delta > 0$, and choose solutions $x^{(j)}(t)$ of (1.2.1) with $x^{(j)}(s) \in F_j$ for $j = 0, 1$. Using Proposition 1.2.7 we choose $f_\nu^{(j)}$ satisfying (ii) and (iii) there for $j = 0, 1$, and satisfying (i) for the solution $x^{(j)}$. For $0 \leq \lambda \leq 1$ we let $x_{\nu,\lambda}(t)$ be the solution of

$$dx(t)/dt = (1 - \lambda)f_\nu^{(0)}(t, x(t)) + \lambda f_\nu^{(1)}(t, x(t)), \quad x(t_0) = x_0.$$

Since $(1-\lambda)f_\nu^{(0)}(t, x) + \lambda f_\nu^{(1)}(t, x) \to f(t, x)$ uniformly in $R \times [0, 1]$ as $\nu \to \infty$ and $T < b/M$, the solution exists for large ν. We have $x_{\nu,\lambda}(t) = x^{(\lambda)}(t)$ for $\lambda = 0, 1$, and $x_{\nu,\lambda}(t)$ is a C^1

function of (t, λ) by Theorem 1.2.5. (We may add the differential equation $d\lambda/dt = 0$ to express that λ is a parameter.) Hence there is some $\lambda = \lambda_\nu \in (0, 1)$ for which the distance of $x_{\nu,\lambda_\nu}(s)$ to $X_s = F_0 \cup F_1$ is $\geq \delta$. As in the proof of Theorem 1.2.6 we can find a uniform limit $x(t)$ of $x_{\nu,\lambda_\nu}(t)$, and $x(t)$ satisfies (1.2.1). Since the distance from $x(s)$ to X_s is $\geq \delta$ we have a contradiction proving the theorem.

When $n = 1$ the result is obvious, for a solution of $dx(t)/dt = f(t, x)$ from s backwards to t_0 starting between two solution curves of (1.2.1) must hit one of them before or when arriving at t_0, and from there on we can let it continue to t_0 as one of these curves. An example with $n = 1$ is the differential equation

$$dx(t)/dt = \tfrac{3}{2}x^{1/3}$$

which does not satisfy the Lipschitz condition at $x = 0$. For $t \geq 0$ we have the two solutions $x(t) = t^{3/2}$ and $x(t) = 0$. All solutions starting at $(0, 0)$ lie between these two solutions, and the solutions

$$x_c(t) = 0 \quad \text{for } t \leq c, \quad x(t) = (t - c)^{3/2} \quad \text{for } t \geq c$$

fill out the funnel between them.

When $n = 1$ the boundary of X always consists of solution curves:

Theorem 1.2.9. *With the hypotheses and notation of Theorem 1.2.8 it follows when $n = 1$ that*

$$X(t) = \{(t, x); x^{(0)}(t) \leq x \leq x^{(1)}(t)\}$$

where $x^{(j)}(t)$ are solutions of (1.2.1) called minimal and maximal respectively. The set of all solutions is closed under the lattice operations max and min.

Proof. If $x(t)$ and $\tilde{x}(t)$ are solutions then $\hat{x}(t) = \max(x(t), \tilde{x}(t))$ is also a solution. This is clear where $x(t) \neq \tilde{x}(t)$. When $x(t) = \tilde{x}(t)$ we have

$$x(t + s) = x(t) + sf(t, x(t)) + o(s), \quad \tilde{x}(t + s) = x(t) + sf(t, x(t)) + o(s),$$

hence $\hat{x}(t+s) = x(t)+sf(t, x(t))+o(s)$ so $\hat{x}$ is differentiable at t with derivative $f(t, x(t))$. This proves the last statement; the minimum is treated in the same way.

By the compactness we can for any s choose a solution x_s with $x_s(s) = x^{(1)}(s)$. If $s_1, s_2, \ldots$ is a countable dense set, it follows that

$$X_N(t) = \max_{j \leq N} x_{s_j}(t)$$

is a solution for $|t - t_0| \leq T$ which is equal to $x^{(1)}$ at s_j when $j \leq N$. By the equicontinuity it follows that $X_N \to x^{(1)}$ uniformly as $N \to \infty$, so $x^{(1)}$ is a solution. In the same way it follows that $x^{(0)}$ is a solution.

1.3. Existence of solutions in the large. The simple example of the Cauchy problem

(1.3.1) $$dx(t)/dt = x(t)^2, \quad x(0) = x_0,$$

shows that one cannot expect in general to have global solutions of a nonlinear differential equation. In fact, the solution of (1.3.1) is

(1.3.2) $$x(t) = x_0/(1 - x_0 t)$$

so a solution exists in the interval where $x_0 t < 1$ but it becomes unbounded as $x_0 t \to 1$. We shall come back to this example later on. However, we shall first show that the domain of existence is lower semicontinuous for perturbations of the data.

Theorem 1.3.1. *Let $f(t,x)$ and $\partial f(t,x)/\partial x$ be continuous in an open set $\Omega \subset \mathbf{R}^{1+n}$, and assume that the Cauchy problem (1.2.1) has a solution with graph in Ω for $t_1 \leq t \leq t_2$ where $t_1 < t_0 < t_2$. Then there is a neighborhood U of x_0 in $\mathbf{R}^n$ such that for every $y \in U$ the Cauchy problem (1.2.1) with x_0 replaced by y has a unique solution $x(t,y)$ for $t_1 \leq t \leq t_2$. The solution is in $C^1([t_1, t_2] \times U)$.*

Proof. If $x_0(t)$ is the solution with $x_0(t_0) = x_0$, we can take t, $x - x_0(t)$ as new coordinates to reduce to the case where $x_0 = 0$ and $x_0(t) \equiv 0$. Thus we assume that $f(t, 0) = 0$. Choose $b > 0$ and M, C so that f is bounded by M and has Lipschitz constant C with respect to x in $[t_1, t_2] \times \{y; |y| \leq b\}$. By Theorem 1.2.3 the solution with initial data y at t_0 exists in $[t_1, t_2]$ as long as $|t - t_0|M + |y| \leq b$, and we have

$$|x(t,y)| \leq e^{C|t-t_0|}|y|$$

then. Repeated use of this result shows that a solution with $|x(t, y)| \leq b$ exists in $[t_1, t_2]$ provided that $e^{C|t_j - t_0|}|y| < b$ for $j = 1, 2$. By Theorem 1.2.5 it is a C^1 function of y and t.

Note that we could also allow a small perturbation of f in the theorem, for we could just replace f by $f + \varepsilon g$ where ε is small, adding the equation $d\varepsilon/dt = 0$.

We shall now return to the simple example at the beginning of the section. When doing so we shall study existence for positive time of the solution of the Cauchy problem. Then we have global existence if $x_0 \leq 0$ but blowup at a finite time if $x_0 > 0$. In the following two lemmas we allow perturbations of the Cauchy problem (1.3.1) but one should think of the coefficients a_1 and a_2 as being small.

Lemma 1.3.2. *Let $x(t)$ be a solution in $[0, T]$ of the ordinary differential equation*

$$(1.3.3) \qquad dx(t)/dt = a_0(t)x(t)^2 + a_1(t)x(t) + a_2(t)$$

with all a_j continuous and $a_0 \geq 0$. Let

$$(1.3.4) \qquad K = \int_0^T |a_2(t)|dt \exp\left(\int_0^T |a_1(t)|dt\right).$$

If $x(0) > K$ it follows that

$$(1.3.5) \qquad \int_0^T a_0(t)dt \exp\left(-\int_0^T |a_1(t)|dt\right) < (x(0) - K)^{-1}.$$

Proof. Let us first assume that $a_1 = 0$, and introduce

$$x_2(t) = \int_0^t |a_2(s)|ds.$$

Then $x_2(0) = 0$ and $x_2(T) = K$. Let x_1 be the solution of the Cauchy problem

$$dx_1(t)/dt = a_0(t)(x_1(t) - K)^2; \quad x_1(0) = x(0).$$

Thus

$$(x_1(t) - K)^{-1} - (x_1(0) - K)^{-1} = -\int_0^t a_0(s)ds,$$

so x_1 is increasing, and if x_1 exists in $[0, T]$ then

$$(1.3.5)' \qquad \int_0^T a_0(s)ds < (x_1(0) - K)^{-1}.$$

Since

$$d(x_1(t) - x_2(t))/dt = a_0(t)(x_1(t) - K)^2 - |a_2(t)| \le a_0(t)(x_1(t) - x_2(t))^2 + a_2(t)$$

and $x_1(t) - x_2(t) = x(t)$ when $t = 0$, we obtain $x_1(t) - x_2(t) \le x(t)$ in $[0, T]$ as long as $x_1(t)$ exists. Thus x_1 cannot become infinite in $[0, T]$ which proves that $(1.3.5)'$ holds. For a general a_1 we just set

$$x(t) = X(t) \exp\left(\int_0^t a_1(s)ds\right).$$

This reduces (1.3.3) to

$$dX(t)/dt = a_0(t) \exp\left(\int_0^t a_1(s)ds\right) X(t)^2 + a_2(t) \exp\left(-\int_0^t a_1(s)ds\right),$$

and we just have to apply the special case of the lemma already proved.

Lemma 1.3.3. *Let a_j be continuous functions in $[0, T]$, set $a_0^+ = \max(a_0, 0)$, and define K by (1.3.4). If $x_0 \ge 0$ and*

$$(1.3.6) \qquad \int_0^T a_0^+(t)dt \exp\left(\int_0^T |a_1(t)|dt\right) < (x_0 + K)^{-1},$$

$$(1.3.7) \qquad \int_0^T |a_0(t)|dt \exp\left(\int_0^T |a_1(t)|dt\right) < K^{-1},$$

then (1.3.3) has a solution in $[0, T]$ with $x(0) = x_0$, and

$$(1.3.8) \quad x(T)^{-1} \ge (x_0 + K)^{-1} - \int_0^T a_0^+(t)dt \exp\left(\int_0^T |a_1(t)|dt\right) \quad \text{if } x(T) \ge 0,$$

$$(1.3.9) \quad |x(T)|^{-1} \ge K^{-1} - \int_0^T |a_0(t)|dt \exp\left(\int_0^T |a_1(t)|dt\right) \quad \text{if } x(T) < 0.$$

Proof. Assume first that $a_1 \equiv 0$ and let x_2 again be the integral of $|a_2|$ with $x_2(0) = 0$. Thus $x_2(T) = K$. Now let x_1 be the solution of

$$dx_1(t)/dt = a_0^+(t)(x_1(t) + K)^2, \quad x_1(0) = x_0,$$

that is,

$$(x_1(t) + K)^{-1} = (x_0 + K)^{-1} - \int_0^t a_0^+(s)ds.$$

By (1.3.6) the increasing function x_1 exists in $[0, T]$. Since

$$d(x_1(t) + x_2(t))/dt = a_0^+(t)(x_1(t) + K)^2 + |a_2(t)| \ge a_0(t)(x_1(t) + x_2(t))^2 + a_2(t)$$

and $x_1 + x_2 = x$ at 0, we obtain $x \le x_1 + x_2 \le x_1 + K$ in $[0, T]$ if x exists, hence

$$x(T)^{-1} \ge (x_1(T) + K)^{-1} = (x_0 + K)^{-1} - \int_0^T a_0^+(t)dt$$

if $x(T) > 0$, which proves (1.3.8).

If on the other hand x has a zero in $[0, T]$ then we can apply (1.3.8) to $-x$, with x_0 replaced by 0 and to an interval starting at the zero of x. This gives (1.3.9). Now if we do not assume a priori that x exists in $[0, T]$, it follows that (1.3.8), (1.3.9) hold with T replaced by any smaller t such that a solution exists in $[0, t]$. Hence we have a fixed upper bound in any such interval. It follows at once that a solution does exist in $[0, T]$, for the considered set of t values is both open and closed. (Compare with the proof of Proposition 1.2.1.) Finally when a_1 is not identically 0 we can reduce to the case already studied just as in the proof of Lemma 1.3.2. The proof is complete.

1.4. Generalized solutions. The proof of the Peano existence theorem (Theorem 1.2.6) can be applied under much more general conditions on f. This leads to various classes of generalized solutions. In this spirit we shall discuss briefly the results of Filippov [1] since they throw interesting light on the solutions of conservation laws in one space dimension.

Let $f(t, x)$ be a real valued measurable function defined in

$$R = \{(t, x) \in \mathbf{R}^{1+n}; |t - t_0| \leq a, |x - x_0| \leq b\}.$$

We wish to discuss generalized solutions of the Cauchy problem

$$(1.4.1) \qquad dx(t)/dt = f(t, x(t)); \quad x(t_0) = x_0,$$

for t in a neighborhood of t_0. In doing so we assume that

$$(1.4.2) \qquad |f(t, x)| \leq M(t),$$

where $M \in L^1$. It is convenient to extend f to $\mathbf{R}^{1+n}$ by defining $f = 0$ outside R; the hypotheses just made are then fulfilled in the whole space, so we let $a = \infty$ from now on. To simplify notation we assume that $t_0 = 0$.

Let f_ε be regularizations of f,

$$f_\varepsilon(t, x) = \int f(t, x - \varepsilon y)\chi(y)dy$$

where $0 \leq \chi \in C_0^\infty(\mathbf{R}^n)$, $\int \chi(y)dy = 1$. Then f_ε also satisfies (1.4.2), $f_\varepsilon(t, x(t))$ is an integrable function of t for every continuous $x(t)$, by the Fubini theorem, and

$$|\partial f_\varepsilon(t, x)/\partial x| \leq CM(t)/\varepsilon, \quad C = \int |\partial \chi/\partial y|dy.$$

This allows us to use the proof of Theorem 1.2.2 to solve the integral equation

$$(1.4.3) \qquad x(t) = x_0 + \int_0^t f_\varepsilon(s, x(s))ds, \quad |t| \leq T,$$

if $C \int_{-T}^T M(t)dt < \varepsilon$. Indeed, for the successive approximations

$$x_k(t) = x_0 + \int_0^t f_\varepsilon(s, x_{k-1}(s))ds, \quad k = 1, 2, \ldots; \quad x_0(t) \equiv x_0,$$

we obtain

$$|x_k(t) - x_{k-1}(t)| \leq \varepsilon/C \left(C \int_{-T}^T M(t)dt/\varepsilon \right)^k,$$

so $x_k(t)$ has a uniform limit $x(t)$ satisfying (1.4.3) as $k \to \infty$. Hence it follows that $x(t)$ is absolutely continuous for $|t| \leq T$ and that (1.4.1) holds almost everywhere with f replaced by f_ε. Repeating the argument we get a solution $x_\varepsilon(t)$ for $|t| \leq a$, with

$$|x_\varepsilon(t) - x_0| \leq \int_{-a}^{a} M(t)dt; \quad |dx_\varepsilon(t)/dt| \leq M(t) \text{ almost everywhere.}$$

Hence x_ε is uniformly bounded and equicontinuous. For any limit $x(t)$ when $\varepsilon \to 0$ we have, with dx/dt defined in the sense of distribution theory,

$$|\langle dx/dt, \psi \rangle| \leq \int M|\psi|dt, \quad \psi \in C_0^\infty, \quad \text{for } |\langle dx_\varepsilon/dt, \psi \rangle| \leq \int M|\psi|\, dt,$$

so dx/dt is an L^1 function $\leq M$ in norm almost everywhere, which means that $x(t)$ is absolutely continuous. In what sense does $x(t)$ satisfy the differential equation (1.4.1)?

To answer this question we observe that $f_\varepsilon(t, x)$ is for fixed t an average of the values of f in a neighborhood of x. Thus values taken on a set of measure 0 play no role. Let us introduce the essential supporting function of the values taken by f at x through

$$H(t, x, \xi) = \lim_{\delta \to 0} \text{ess sup}_{|x-y|<\delta} \langle f(t, y), \xi \rangle,$$

which is clearly semi-continuous from above in x for fixed t and convex, positively homogeneous in ξ. Thus H is the supporting function of a closed convex set $F(t, x)$ which is an upper semi-continuous function of x for fixed t. It is the smallest closed convex set such that any neighborhood contains the values of $f(t, y)$ for almost all y in some neighborhood of x.

For x_ε we have when $-a \leq t_1 \leq t_2 \leq a$

$$\langle x_\varepsilon(t_2) - x_\varepsilon(t_1), \xi \rangle = \int_{t_1}^{t_2} \langle f_\varepsilon(t, x_\varepsilon(t)), \xi \rangle dt$$

$$\leq \int_{t_1}^{t_2} \text{ess sup}_{|x(t)-y|<\delta} \langle f(t, y), \xi \rangle dt$$

if $\varepsilon + |x_\varepsilon(t) - x(t)| < \delta$ for $|t| < a$. The essential supremum is a measurable function of t since it is the logarithm of the limit as $N \to \infty$ of

$$\left(\int_{|x(t)-y|<\delta} \exp(N\langle f(t, y), \xi \rangle)\, dy \right)^{1/N}.$$

If $x(t)$ is the limit of a subsequence, it follows that

$$\langle x(t_2) - x(t_1), \xi \rangle \leq \int_{t_1}^{t_2} H(t, x, \xi)dt.$$

Hence we have at every differentiable point

$$\langle x'(t), \xi \rangle \leq H(t, x, \xi),$$

that is, $x'(t) \in F(t, x)$ almost everywhere.

Theorem 1.4.1. *Under the assumption (1.4.2) the Cauchy problem (1.4.1) has a solution in the sense that x is absolutely continuous and for almost all $t \in [t_0 - a, t_0 + a]$*

$$x'(t) \in F(t, x(t))$$

where $F(t, x)$ is the smallest closed convex set such that every neighborhood contains the values of $f(t, y)$ for almost all y in some neighborhood of x.

When f is continuous with respect to x we recover the generalized solutions of Carathéodory. Let us also consider an example where $f(t, x)$ is in C^1 on the closure of each side of a C^1 hypersurface S where a jump takes place. Assume that for the boundary values f_+ and f_- on the two sides of the surface the vector fields $(1, f_\pm)$ are transversal to S. If they point to the same side of S then the integral curves just pass through the surface. If they point to the surface from the sides where they are defined then an integral curve arriving at S must stay in S as an integral curve of the equation

$$dx/dt = \lambda f_+ + (1 - \lambda) f_-$$

with $\lambda \in (0, 1)$ chosen so that $(1, \lambda f_+ + (1 - \lambda) f_-)$ is tangential to S. If they point away from S then the integral curve may remain in S as a solution of this equation up to an arbitrary point where it leaves on either side as a standard integral curve.

The proof of Theorem 1.4.1 shows easily that the solutions of (1.4.1) in the generalized sense form a compact set. In fact, it follows at once from (1.4.2) that they are uniformly bounded and equicontinuous. The discussion preceding the statement of Theorem 1.4.1 also gives that uniform limits of solutions are solutions. Proposition 1.2.7 remains valid with (ii) and (iii) weakened. To achieve (i) we can choose f_ε as in the proof of Theorem 1.4.1 and then modify f_ε to

$$\tilde{f}_\varepsilon(t, x) = (1 - \psi((x - x(t))/\varepsilon) f_\varepsilon(t, x) + \psi((x - x(t))/\varepsilon) f(t, x(t)),$$

noting that $x'(t) = f(t, x(t))$ almost everywhere. Here $\psi \in C_0^\infty$, $0 \le \psi \le 1$, and $\psi(0) = 1$. This changes $f_\varepsilon(t, x)$ only for x close to $x(t)$, and the modified values will then lie in any neighborhood of $F(t, x(t))$ for small ε. Hence Theorems 1.2.8 and 1.2.9 remain valid for the generalized solutions, with only minor changes of the proof. We leave it as an exercise to check these statements.

The improved version of the uniqueness statement in Theorem 1.2.3 given in Theorem 1.2.3' remains valid for the generalized solutions. It allows f to have negative jumps when x is a scalar variable, which agrees well with the example discussed above.

Theorem 1.4.2. *Assume in addition to the hypotheses in Theorem 1.4.1 that*

$$(1.4.4) \qquad \langle y - z, f(t, y) - f(t, z) \rangle \le C|y - z|^2, \quad y, z \in \mathbf{R}^n, |t - t_0| \le a.$$

Then (1.4.1) has a unique solution for $t_0 \le t \le t_0 + a$, and we have for the solution $x(t, y)$ with initial data y

$$(1.4.5) \qquad |x(t, y) - x(t, z)| \le e^{C(t - t_0)}|y - z| \quad \text{if } t_0 \le t \le t_0 + a.$$

Proof. If $|y' - y| < \delta$ and $|z' - z| < \delta$ then we have by (1.4.4)

$$\langle y' - z', f(t, y') - f(t, z') \rangle \le C|y' - z'|^2 \le 2C(|y - z|^2 + 4\delta^2),$$

hence

$$\langle y - z, f(t, y') - f(t, z') \rangle \le 2C(|y - z|^2 + 4\delta^2) + 4\delta M(t).$$

When $\delta \to 0$ we conclude that for almost all t

$$\langle y - z, Y - Z \rangle \leq 2C|y - z|^2 \quad \text{if } Y \in F(t, y) \text{ and } Z \in F(t, z).$$

Let $x(t, y)$ denote any solution with initial data y; if we prove (1.4.5) then the uniqueness follows when $y = z$ and the theorem will be proved. Let $R(t) = |x(t, y) - x(t, z)|$ with Euclidean norm. This is an absolutely continuous function, and from the proof of Theorem 1.2.3 we conclude that for almost all t we have

$$R(t)dR(t)/dt = \langle x(t, y) - x(t, z), Y - Z \rangle$$

where $Y \in F(t, x(t, y))$ and $Z \in F(t, x(t, z))$. We have just seen that this implies

$$R(t)dR(t)/dt \leq 2CR(t)^2.$$

The constant 2 obtained in the preceding proof can be replaced by any constant > 1 if we use that $(a + b)^2 \leq a^2(1 + \varepsilon) + b^2(1 + 1/\varepsilon)$ for every $\varepsilon > 0$. Hence

$$dR(t)/dt \leq CR(t)$$

almost everywhere when $R(t) \neq 0$. This implies

$$R(t) \leq R(t_0)e^{C(t-t_0)} \quad \text{when } 0 \leq t - t_0 \leq a,$$

which completes the proof.

Remark. The Lipschitz continuity assumed in Theorem 1.2.2 can be replaced by a somewhat weaker modulus of continuity in x by a classical theorem of Osgood. It was extended by Chemin and Lerner [1] to Banach space valued functions with such a condition in x and only an integrability condition in t.

CHAPTER II

SCALAR FIRST ORDER EQUATIONS WITH ONE SPACE VARIABLE

2.1. Introduction. In Section 2.2 we shall briefly discuss the linear case, which is essentially another way of looking at ordinary differential equations. However, in spite of the simplicity this leads naturally to the introduction of the method of characteristics which we shall use in the quasilinear case. Next we discuss "Burgers' inviscid equation"

$$\partial u/\partial t + u\partial u/\partial x = 0$$

at some length in Sections 2.3 and 2.4. It has the advantage that it can be studied quite explicitly. All the same it illustrates two main points: the blowup of classical solutions at a finite time and the existence and uniqueness of generalized solutions satisfying an admissibility (entropy) condition. In fact, in Section 2.5 we extend the results on Burgers' equation to a conservation law

$$\partial u/\partial t + \partial f(u)/\partial x = 0$$

with strictly convex f. The much more complex situation for general f is postponed to Chapter III where we allow several space variables also.

The inspiration for the study of discontinuous solutions comes from the equations of gas dynamics. However, the restriction to *scalar* equations with *one* space variable does not cover any realistic models. We shall come closer to the equations of gas dynamics in Chapter IV, but the situation is far from completely understood.

2.2. The linear case. In this section we shall discuss the Cauchy problem

$$(2.2.1) \qquad \partial u/\partial t + a(t,x)\partial u/\partial x = b(t,x)u; \quad u(0,x) = u_0(x);$$

where $x \in \mathbf{R}$. In fact, nothing essential has to be changed if $x \in \mathbf{R}^n$ and $a(t,x)\partial u/\partial x$ is interpreted as a scalar product. We assume that $a(t,x)$, $\partial a(t,x)/\partial x$ and $b(t,x)$ are continuous functions for $t \geq 0$.

From Section 1.2 we know that the ordinary Cauchy problem

$$(2.2.2) \qquad dx(t)/dt = a(t,x(t)), \quad x(0) = y$$

defining the characteristics has a solution $x(t,y)$ defined and C^1 in a neighborhood of $\{0\} \times \mathbf{R}$. If a is bounded, say, the solution exists in $\mathbf{R} \times \mathbf{R}$. If we take (t,y) as new variables instead of (t,x) and set

$$U(t,y) = u(t,x(t,y)), \quad B(t,y) = b(t,x(t,y)),$$

then (2.2.1) is changed to

$$(2.2.3) \qquad \partial U/\partial t = B(t,y)U; \quad U(0,y) = u_0(y),$$

if we use that $a = \partial x/\partial t$. This change of variables is legitimate if u is a continuous function satisfying (2.2.1) in the sense of distribution theory. The solution of (2.2.3) is obvious,

$$(2.2.4) \qquad U(t,y) = u_0(y) \exp \left(\int_0^t B(s,y)\,ds \right),$$

and returning to the (t, x) variables we obtain a unique continuous solution of (2.2.1) in the set which can be reached by the characteristics (2.2.2); it is open by Theorem 1.3.1 and equal to the upper half plane if a is bounded, for example.

We shall now prove a uniqueness theorem due to Olejnik [1] which will be important in the nonlinear case. We shall use the notation

$$S_T = \{(t,x); 0 < t < T\}.$$

Theorem 2.2.1. Let $u \in L^\infty(S_T)$ satisfy the Cauchy problem

$$(2.2.5) \qquad \partial u/\partial t + \partial(au)/\partial x = bu, \quad u = 0 \text{ for } t = 0,$$

in the weak sense, that is,

$$(2.2.5)' \qquad \iint u(\partial\varphi/\partial t + a\partial\varphi/\partial x + b\varphi)\,dx\,dt = 0 \quad \text{for } \varphi \in \overline{C}_0^\infty(S_T).$$

(This means that φ is the restriction to S_T of a function in $C_0^\infty(\mathbf{R}^2)$ vanishing in a neighborhood of $\{(t,x); t \geq T\}$.) Assume that $a, b \in L^\infty$, which makes this condition well defined, and that we have a one-sided Lipschitz condition

$$(2.2.6) \qquad a(t,x) - a(t,y) \leq C(t)(x-y), \quad 0 < t < T,\ x > y,$$

where $C(t)$ is a finite decreasing function for $t > 0$ (which may tend to ∞ when $t \to 0$). Then it follows that $u = 0$ almost everywhere.

Note that φ is not assumed to vanish at $t = 0$, which is the reason why (2.2.5)' also includes the Cauchy condition.

Proof. Let a_ε, b_ε be C^∞ functions with the same bounds as a and b, including (2.2.6), and converging to a and b almost everywhere as $\varepsilon \to 0$. We also require that $b_\varepsilon(t, x) = 0$ for $t < \varepsilon$. With $\psi \in C_0^\infty(S_T)$ we solve the adjoint equation

$$(2.2.7) \qquad \partial\varphi_{\varepsilon\delta}/\partial t + a_\varepsilon \partial\varphi_{\varepsilon\delta}/\partial x + b_\delta\varphi_{\varepsilon\delta} = \psi; \quad \varphi_{\varepsilon\delta}(T, x) = 0.$$

We shall later on let $\varepsilon \to 0$ first and $\delta \to 0$ afterwards. The solution exists and is in C^∞ by the first part of this section. Application of (2.2.5)' to $\varphi_{\varepsilon\delta}$ gives

$$(2.2.8) \qquad \iint u\psi\,dx\,dt = \iint u((a_\varepsilon - a)\partial\varphi_{\varepsilon\delta}/\partial x + (b_\delta - b)\varphi_{\varepsilon\delta})\,dx\,dt.$$

We shall give uniform estimates of $\varphi_{\varepsilon\delta}$ which imply that the right-hand side tends to 0 when first $\varepsilon \to 0$ and then $\delta \to 0$. This will prove that the left-hand side is always 0 and prove the theorem.

Let $x_\varepsilon(t; s, y)$ be the solution of the characteristic equation starting at (s, y):

$$dx_\varepsilon(t; s, y)/dt = a_\varepsilon(t, x_\varepsilon(t; s, y)); \quad x_\varepsilon(s; s, y) = y.$$

Then the solution of (2.2.7) is given by

$$(2.2.9) \qquad \varphi_{\varepsilon\delta}(t,x) = \int_T^t \psi(s; x_\varepsilon(s;t,x)) \exp\left(\int_t^s b_\delta(\tau, x_\varepsilon(\tau;t,x))\,d\tau\right) ds.$$

It vanishes for t near T, and if $\psi(t,x) = 0$ and $b_\delta(t,x) = 0$ for $t < \delta$ then the integrals stop at δ when $t < \delta$. Now it follows from the uniform bound of a_ε that the supports of all $\varphi_{\varepsilon\delta}$ lie in a fixed compact set, and by Theorem 1.2.3′ it follows from (2.2.6) (for a_ε) that

$$|\partial x_\varepsilon(s;t,x)/\partial x| \le C_t' \quad \text{for } t \le s \le T.$$

Hence (2.2.9) gives a uniform bound

$$(2.2.10) \qquad |\partial\varphi_{\varepsilon\delta}(t,x)/\partial x| \le C_{\tau\delta}'', \quad 0 < \tau \le t \le T,$$

which is uniform with respect to ε. Hence

$$(2.2.11) \qquad \iint_{t>\tau} |u(a_\varepsilon - a)\partial\varphi_{\varepsilon\delta}/\partial x|\,dx\,dt \to 0 \quad \text{as } \varepsilon \to 0 \text{ if } \tau > 0.$$

From (2.2.10) with $t = \delta$ and the fact that for $t < \delta$

$$\varphi_{\varepsilon\delta}(t,x) = \varphi_{\varepsilon\delta}(\delta, x_\varepsilon(\delta;t,x))$$

we see that for $t < \delta$

$$(2.2.12) \qquad \int |\partial\varphi_{\varepsilon\delta}(t,x)/\partial x|\,dx = \int |\partial\varphi_{\varepsilon\delta}(\delta,x)/\partial x|\,dx \le C_\delta'''.$$

Now (2.2.9) implies a uniform bound for $\varphi_{\varepsilon\delta}$, so (2.2.8), (2.2.11) and (2.2.12) give when $\varepsilon \to 0$

$$\left|\iint u\psi\,dx\,dt\right| \le M \iint_{|x|<M,|t|<T} |b_\delta - b|\,dx\,dt + MC_\delta'''\tau$$

for every $\tau > 0$. Letting first $\tau \to 0$ and then $\delta \to 0$ we now conclude that $\iint u\psi\,dx\,dt = 0$ which completes the proof.

2.3. Classical solutions of Burgers' equation. We shall now discuss the Cauchy problem

$$(2.3.1) \qquad \partial u/\partial t + u\partial u/\partial x = 0 \quad \text{for } t \ge 0, \quad u(0,x) = u_0(x).$$

This equation is often called Burgers' (inviscid) equation, occasionally also Hopf's equation. However, it was already studied by Poisson, Airy, Challis and Stokes one hundred years earlier (see Stokes [1]). (It is more appropriate to use the term Burgers' equation for the corresponding heat equation discussed in Section 2.4.) From Section 2.2 we know that if $u \in C^1$ then u is constant on the characteristics defined by

$$dx/dt = u(t,x).$$

Hence they are straight lines, so

$$(2.3.2) \qquad u(t,x) = u_0(y) \quad \text{when } x = y + tu_0(y).$$

Since $\partial x/\partial y = 1 + t u_0'(y)$ we see that if $u_0 \in C^1$ and u_0 and u_0' are bounded, then the equation $x = y + t u_0(y)$ determines y as a C^1 function of (t, x) when $0 \le t < T$ if

$$1/T = \sup -u_0'.$$

Hence we have a unique solution in C^1 for $0 \le t < T$. However, $\partial u/\partial x = u_0'(y)/(\partial x/\partial y)$ is unbounded as $t \to T$ if $-u_0'$ attains a positive maximum at y, so there is no C^1 solution beyond the strip $0 \le t < T$. The solution can be thought of geometrically as follows. Draw the graph of $y \mapsto u_0(y)$ and refer it to the y axis and the axis $y + tu = 0$ instead of the u axis, for a fixed t. Then we obtain the graph of $u(t, \cdot)$, which thus gets a steeper and steeper negative slope for increasing t until it gets a vertical tangent as $t \to T$. However, we should note that for $0 \le t < T$

$$(2.3.3) \qquad \sup |u(t, x)| = \sup |u_0(x)|; \qquad \int |\partial u(t, x)/\partial x|\, dx = \int |u_0'(x)|\, dx.$$

This makes it natural to try to construct a bounded solution of bounded total variation beyond the development of singularities.

The blowup of the first order derivatives as $t \to T$ can also be seen by differentiating Burgers' equation with respect to x. With $w = \partial u/\partial x$ we obtain, if $u \in C^2$,

$$Lw = -w^2; \quad L = \partial/\partial t + u\partial/\partial x.$$

Thus L denotes differentiation with respect to t along the characteristics. This is the main example from Section 1.3 of an ordinary differential equation with solutions blowing up at a finite time; we have

$$w(t) = w(0)/(1 + w(0)t)$$

on the bicharacteristic, which gives back the earlier result.

Thus the solution of the Cauchy problem (2.3.1) has a finite *lifespan* T unless u_0 is increasing, which never happens if u_0 has compact support and is not identically 0. Note that it does not matter if the Cauchy data are small; if u_0 is replaced by εu_0 then T is replaced by $T_\varepsilon = T/\varepsilon$. In fact, the solution $u(t, x)$ is replaced by $\varepsilon u(\varepsilon t, x)$, so the time variable just slows down.

We have the same homogeneity property for the Cauchy problem

$$(2.3.4) \qquad \partial u/\partial t + au\partial u/\partial x = bu^2 \quad \text{for } t \ge 0; \ u(0, x) = u_0(x),$$

with constant a and b. If $a \ne 0$ then $1/T = \sup(bu_0 - au_0')$. In fact, on the characteristics we have

$$dx/dt = au, \quad du/dt = bu^2.$$

On the characteristic starting at $(0, y)$ we obtain by integrating these equations

$$(2.3.5) \qquad u = u_0(y)/(1 - u_0(y)bt),$$

$$(2.3.6) \qquad x = y - ab^{-1} \log(1 - u_0(y)bt),$$

$$(2.3.7) \qquad \partial u/\partial y = (u/u_0)^2 u_0' = (1 - u_0 bt)^{-2} u_0',$$

$$(2.3.8) \qquad \partial u/\partial x = u_0'(1 - u_0 bt)^{-1}(1 + u_0' at - u_0 bt)^{-1},$$

where (2.3.6) should be read as $x = y + au_0(y)t$ if $b = 0$. A classical solution will exist until u or $\partial u/\partial x$ blows up. If $a \ne 0$ then u is still bounded when $\partial u/\partial x$ blows up, provided that $u_0 \in C_0^1$:

Lemma 2.3.1. *Assume that $a \neq 0$, $0 \not\equiv u_0 \in C_0^1$. Then*

$$M = \sup(bu_0 - au_0') > 0,$$

and $M_0 = \sup bu_0 < M$. Thus $u_0' \neq 0$ when M is attained.

Proof. We may assume that $b \geq 0$ for otherwise we can just change the sign of u_0, and that $a > 0$, for otherwise we just have to change the sign of y. If $bu_0 - au_0' \leq 0$ then $u_0 \exp(-bx/a)$ is increasing, which is impossible since the support is compact. Hence $M > 0$. If $b > 0$ and M_0 is attained at y_0, then $u_0'(y_0) = 0$, so $M \geq M_0$. To exclude the equality we consider the nonpositive function

$$x \mapsto (u_0(x) - u_0(y_0))e^{-bx/a}.$$

The derivative is

$$x \mapsto a^{-1}(au_0'(x) - bu_0(x) + M_0)e^{-bx/a},$$

and since u_0 has compact support it is negative for some x. At such a point we have $bu_0(x) - au_0'(x) > M_0$, which proves the lemma.

2.4. Weak solutions of Burgers' equation. If we write Burgers' equation in the form

$$(2.4.1) \qquad \partial u/\partial t + \tfrac{1}{2}\partial u^2/\partial x = 0,$$

then it is clear what it means to say that a function in L^∞_{loc} is a weak (distribution) solution. If $\tilde{u}(t, x) = u(t, x)$ when $t \geq 0$ and $\tilde{u}(t, x) = 0$ for $t < 0$, we can interpret the Cauchy problem (2.3.1) as the equation

$$\partial \tilde{u}/\partial t + \tfrac{1}{2}\partial \tilde{u}^2/\partial x = u_0 \otimes \delta(t)$$

in the sense of distribution theory, or explicitly

$$(2.4.1)' \qquad -\iint_{t>0}(u\partial\varphi/\partial t + \tfrac{1}{2}u^2\partial\varphi/\partial x)\,dx\,dt = \int u_0(x)\varphi(0, x)\,dx, \quad \varphi \in C_0^\infty(\mathbf{R}^2).$$

In particular, the equation (2.4.1) in the weak sense means that if u is in C^1 on each side of a C^1 curve with equation $x = X(t)$, then

$$(2.4.2) \qquad X'(t)[u] = \tfrac{1}{2}[u^2] \quad \textit{(the Rankine-Hugoniot condition)}$$

where $[\,\cdot\,]$ denotes the jump across the curve at $(t, X(t))$ for increasing x. To see this we can integrate by parts in $(2.4.1)'$ on each side of the curve or note that

$$u = u_+H(x - X(t)) + u_-H(X(t) - x), \quad u^2 = u_+^2 H(x - X(t)) + u_-^2 H(X(t) - x),$$

where H is the Heaviside function and $u_\pm$ are smooth extensions of the restrictions of u to the two sides of the jump. Thus

$$X'(t) = \tfrac{1}{2}(u_+ + u_-)$$

is the average of the characteristic speeds on the two sides.

However, one should keep in mind that the differential equation itself does not uniquely determine a notion of weak solution. Indeed, suppose that we introduce a new variable v by $u = f(v)$ where $f'(v) > 0$. Then the differential equation becomes

$$\partial v/\partial t + f(v)\partial v/\partial x = 0$$

after cancellation of a factor $f'(v)$, and we would call v a weak solution if

$$\partial v/\partial t + \partial F(v)/\partial x = 0$$

where $F' = f$. Now the jump condition becomes

$$X'(t)[v] = [F(v)], \text{ that is, } X'(t) = (F(v_+) - F(v_-))/(v_+ - v_-) = \int_{v_-}^{v_+} f(v)\, dv/(v_+ - v_-)$$

where $v_\pm$ are the values of v on the two sides of the discontinuity. This should be compared with the condition (2.4.2),

$$X'(t) = (f(v_+) + f(v_-))/2$$

which is just the trapezoidal approximation for the average of f over (v_-, v_+). Thus the two jump conditions only agree up to terms of second order in the "shock strength" $v_+ - v_-$.

The ambiguity just discussed does not arise in the equations of physics since they are presented from the beginning as conservation laws, that is, in weak form. We shall therefore assume that our equation is stated as a conservation law from the beginning and shall stick to the given dependent variable.

As a first example of weak solutions we shall give a solution of Burgers' equation with initial condition

$$u_0(x) = u_\pm \quad \text{for } \pm x > 0,$$

where $u_\pm$ are constants. A weak solution is given by

$$u(t, x) = u_\pm \quad \text{for } \pm (x - ct) > 0,$$

if the constant speed is $c = \frac{1}{2}(u_+ + u_-)$ as required by (2.4.2). However, when $u_+ > u_-$ we can find a more natural solution if u_0 is approximated by increasing functions, for which we know that a global smooth solution always exists. This gives the solution

$$u(t, x) = \begin{cases} u_- \text{ for } x < u_- t \\ x/t \text{ for } u_- t \leq x \leq u_+ t \\ u_+ \text{ for } x > u_+ t. \end{cases}$$

This example suggests that to get uniqueness one should only allow the unavoidable *negative* jumps for fixed t, that is, require that $u(x - 0, t) \geq u(x + 0, t)$ for the solution.

A more rational guide to the choice of an "admissible" solution, due to Burgers [1] and Hopf [1], is to consider the Cauchy problem for the corresponding heat equation

$$(2.4.3) \qquad \partial u/\partial t + u\partial u/\partial x = \mu\partial^2 u/\partial x^2; \quad t \geq 0; \quad u(0, x) = u_0(x);$$

where $\mu > 0$ is thought of as viscosity. Letting $\mu \to 0$ one should obtain a physically natural solution of (2.3.1). The equation (2.4.3) can be written

$$\partial u/\partial t = \partial(\mu\partial u/\partial x - \tfrac{1}{2}u^2)/\partial x.$$

If we set $\varphi = \exp(-1/2\mu \int u\, dx)$ then

$$-2\mu\varphi'_x = \varphi u, \quad -2\mu^2\varphi''_{xx} = \varphi(\mu u'_x - \tfrac{1}{2}u^2), \quad u = -2\mu\partial_x \log \varphi = -2\mu\varphi'_x/\varphi,$$

so (2.4.3) becomes

$$\frac{\partial}{\partial x}(-2\mu\varphi'_t/\varphi + 2\mu^2\varphi''_{xx}/\varphi) = 0.$$

We can choose the integration constant (depending on t) in the definition of φ so that

$$\varphi'_t = \mu\varphi''_{xx},$$

for replacing $\varphi(t, x)$ by $\varphi(t, x)e^{\gamma(t)}$ adds $-\gamma'(t)$ to $(-\varphi'_t + \mu\varphi''_{xx})/\varphi$. If we just know that

$$(2.4.4) \qquad\qquad \int_0^x u_0(\xi)\, d\xi = o(x^2),$$

then

$$\varphi_0(x) = \exp\left(-1/2\mu \int_0^x u(\xi)\, d\xi\right) = O(e^{\varepsilon x^2}), \quad \varepsilon > 0,$$

so we have a positive solution of the Cauchy problem given by

$$\varphi(t, x) = (4\pi\mu t)^{-1/2} \int_{-\infty}^{\infty} \varphi_0(y) \exp(-(x - y)^2/4\mu t)\, dy.$$

(See e.g. Hörmander [4, Theorem 3.3.3]. Positive solutions are unique by a theorem of Widder.) Returning to u we have obtained the solution

$$(2.4.5) \qquad u_\mu(t, x) = \int_{-\infty}^{\infty} (x - y)/t\, e^{-F(t,x,y)/2\mu}\, dy \bigg/ \int_{-\infty}^{\infty} e^{-F(t,x,y)/2\mu}\, dy$$

where

$$(2.4.6) \qquad F(t, x, y) = (x - y)^2/2t + \int_0^y u_0(\eta)\, d\eta.$$

We shall prove that u_μ converges as $\mu \to 0$ to a solution of (2.4.1)$'$ satisfying the jump condition suggested above, in a quantitative form which will be shown to characterize it uniquely. When discussing the limit of (2.4.5) we note that the main contributions to the integrals occur where $F(t, x, y)$ achieves its minimum. Since $F(t, x, y) \to \infty$ as $|y| \to \infty$, the minimum is achieved in a compact set. We denote the smallest interval containing this set by $[y_-(t, x), y_+(t, x)]$ where $y_-(t, x) \le y_+(t, x)$. Since

$$F(t, x, y) \ge F(t, x, y_\pm), \quad F(t, x, y) \ge |y|^2/3t \text{ for large } |y|,$$

a routine argument left as an exercise shows that

$$(2.4.7) \qquad (x - y_+(t, x))/t \le \varliminf_{\mu \to 0} u_\mu(s, y) \le \varlimsup_{\mu \to 0} u_\mu(s, y) \le (x - y_-(t, x))/t,$$

where the limits are taken for $(\mu, s, y) \to (0, t, x)$. The essential point in the proof is that the contributions to the integrals from the complement of a neighborhood of $[y_-, y_+]$ are negligible.

Lemma 2.4.1. *If $x_1 < x_2$ then $y_+(t, x_1) \leq y_-(t, x_2)$. In particular, $y_\pm$ are increasing, continuous to the right resp. left, equal and hence continuous except at countably many points, for fixed t. If z is any minimum point of $F(t, x, \cdot)$ then*

$$y_\pm(\lambda t, \lambda x + (1 - \lambda)z) = z \quad \text{for } 0 < \lambda < 1.$$

Proof. Set $y_+ = y_+(t, x_1)$. Then

$$F(t, x_1, y) \geq F(t, x_1, y_+)$$

for all y, and we claim that

$$F(t, x_2, y) > F(t, x_2, y_+) \quad \text{if } y < y_+.$$

In fact, if $y < y_+$ and $x_2 > x_1$, then

$$F(t, x_2, y) - F(t, x_2, y_+) \geq F(t, x_2, y) - F(t, x_1, y) + F(t, x_1, y_+) - F(t, x_2, y_+)$$
$$= ((x_2 - y)^2 - (x_1 - y)^2 + (x_1 - y_+)^2 - (x_2 - y_+)^2)/2t = (y - y_+)(x_1 - x_2)/t > 0.$$

This proves the first statement which implies the monotonicity stated, in particular that $y_-(t, x - 0) \leq y_-(t, x)$. Equality follows since the minimum $m(t, x)$ of $F(t, x, y)$ is a continuous function of (t, x), so $F(t, x, y)$ assumes its minimum at $y_-(t, x - 0)$. To prove the last statement we note that since for all y

$$F(t, x, y) - F(t, x, z) \geq 0$$

we have

$$F(\lambda t, \lambda x + (1 - \lambda)z, y) - F(\lambda t, \lambda x + (1 - \lambda)z, z) \geq$$
$$(|\lambda x + (1 - \lambda)z - y|^2 - |\lambda x + (1 - \lambda)z - z|^2)/2\lambda t - (|x - y|^2 - |x - z|^2)/2t$$
$$= (z - y)(2\lambda x + 2(1 - \lambda)z - z - y - (2x - y - z)\lambda)/2\lambda t =$$
$$= (z - y)^2(1 - \lambda)/2\lambda t > 0 \quad \text{if } z \neq y,$$

which completes the proof. Note that $F(\lambda t, \lambda x + (1 - \lambda)z, \cdot)$ has a unique strict minimum point, which is a stable property for perturbations of (t, x) if $u_0 \in C^1$.

Theorem 2.4.2. *Assume that u_0 satisfies $(2.4.4)$ and that $u_0 \in L_{loc}^\infty$. Then the solution u_μ to $(2.4.3)$ converges as $\mu \to 0$ for almost all (t, x) to a solution $u(t, x)$ of $(2.4.1)'$, in fact,*

$$(2.4.7)' \qquad\qquad u_\mu(t, x) \to u(t, x) = (x - y(t, x))/t$$

for all (t, x) such that $F(t, x, y)$ has a unique minimum point $y(t, x)$. This is true except in a set which is countable for fixed t, and u is continuous in (t, x) where the minimum point is unique. In general

$$(2.4.7)'' \qquad\qquad u(t, x \pm 0) = (x - y_\pm(t, x))/t$$

if $y_\pm(t, x)$ is the largest (smallest) minimum point. If $x_1 \leq x_2$ then

$$(2.4.8) \qquad\qquad u(t, x_2 + 0) - u(t, x_1 - 0) \leq (x_2 - x_1)/t.$$

If $|u_0| \leq M$ then $|u| \leq M$. The condition (2.4.8) then characterizes u among all solutions of (2.4.1)′; it is called the Hopf solution.

Proof. The monotonicity properties in Lemma 2.4.1 show that for fixed t we have $y_+(t, x) = y_-(t, x)$ except for countably many values of x, and $u_\mu(t, x) \to (x - y_+(t, x))/t$ then by (2.4.7). We define $u(t, x) = (x - y_+(t, x))/t$ in general. From (2.4.7) it follows that u is then continuous at (t, x) if $y_+(t, x) = y_-(t, x)$, and the continuity of $y_\pm$ to the right (left) gives (2.4.7)″. Since

$$\partial u_\mu/\partial t + \partial(u_\mu^2/2)/\partial x = \mu \partial^2 u_\mu/\partial x^2$$

and $u_\mu \to u$, $u_\mu^2 \to u^2$ weakly for $t > 0$ as $\mu \to 0$, it follows that $\partial u/\partial t + \partial(u^2/2)/\partial x = 0$ for $t > 0$. If $u_0 \in C_0^1$ then it is clear that for small t we have the solution discussed in Section 2.3, so (2.4.1)′ holds then. For the general u allowed in the theorem the fact that

$$(y - x)^2/2t + \int_x^y u_0(\eta)\, d\eta \leq 0$$

when y is a minimum point of $F(t, x, \cdot)$ shows that when (t, x) belongs to a compact set then y has a uniform bound. Thus we can estimate $|u_0|$ by a constant A in the integrand and conclude that $|y - x|/t \leq 2A$, so u is locally bounded when $t \geq 0$. Now we can choose a sequence $u_0^j \in C_0^1$ such that $u_0^j \to u_0$ in L^1_{loc} and (2.4.4) holds uniformly in j. Then the corresponding solutions u^j converge to u locally boundedly for $t \geq 0$. Since (2.4.1)′ holds for u^j it follows by dominated convergence for u as $j \to \infty$. (The condition that u_0 is locally bounded can easily be eliminated using the results of Theorem 2.4.3 below to get enough control of u^2 to justify the passage to the limit. One can also show directly that the L^2 norm with respect to x over a compact set is $O(t^{-1/2})$ which suffices to justify the passage to the limit in (2.4.1)′ for any $u_0 \in L^1_{loc}$ satisfying (2.4.4).)

The upper Lipschitz continuity is obvious for if $x_1 \leq x_2$ then

$$t(u(t, x_2 + 0) - u(t, x_1 - 0)) = x_2 - y_+(t, x_2 + 0) - (x_1 - y_-(t, x_1 - 0)) \leq x_2 - x_1.$$

If $y_\pm(t, x)$ are equal, then Lemma 2.4.1 shows that $u = u(t, x) = (x - y(t, x))/t$ on the open line segment between (t, x) and $(0, y_+(t, x))$. If $u_0 \in C^1$ this enters the set near the x axis where we know that $|u| \leq M$, so we conclude that $|u| \leq M$ then. This follows in general by approximation as above. The uniqueness statement is a consequence of Theorem 2.2.1. In fact, if we have another solution v with the same initial data, then $w = u - v$ is a weak solution of

$$\partial w/\partial t + \partial(aw)/\partial x = 0, \quad a = (u + v)/2,$$

with initial data 0, and (2.2.6) is valid with $C(t) = 1/t$ by (2.4.8). The proof is complete.

As observed in the preceding proof, every (t, x) with $t > 0$ is the vertex of a triangle $T_{t,x}$ with the other vertices at $(0, y_\pm(t, x))$, such that u is continuous, equal to the speed $(x - y_\pm(t, x))/t$ on the open sides to (t, x). For every $t' > t$ there is a unique x' such that $T_{t',x'} \supset T_{t,x}$. For the proof we observe that either $T_{t,x} \subset T_{t',x'}$ or else these triangles are disjoint, for two open sides can only intersect if they have the same speeds, and then one triangle is contained in the other. If (t, x) lies on an open side of $T_{t',x'}$, then $T_{t,x}$ is a part of that side. Let x' be the infimum of all x such that $T_{t',x'}$ lies to the right of (t, x). Then the right side of $T_{t',x'}$ lies to the right of (t, x) by the right continuity of $y_+(t, x)$, but the left side cannot lie strictly to the right of (t, x) by the left continuity of $y_-(t', x')$ and the minimal property of x'. This proves that $T_{t',x'} \supset T_{t,x}$. The uniqueness follows from the fact that two such triangles $T_{t',x'}$ would have intersecting boundaries in the upper half plane since they have a point in common. Hence we see that through every point there is a well defined curve in the direction of increasing t such that $T_{t,x}$ is increasing; it is geometrically evident that it is Lipschitz continuous. Let $x = x(t)$ be such a curve. We claim that

$$(2.4.9) \quad \lim_{t' \to t+0, t'' \to t+0} (x(t'') - x(t'))/(t'' - t') = (u(t, x - 0) + u(t, x + 0))/2, \quad x = x(t).$$

For the proof we let $x' = x(t')$, $x'' = x(t'')$ and assume that $t'' > t' > t$. Then

$$y_-(t'', x'') \le y_-(t', x') \le y_-(t, x) \le y_+(t, x) \le y_+(t', x') \le y_+(t'', x''),$$

and $y_\pm(t'', x'') \to y_\pm(t, x)$ as $t'' \to t$, for by the proof of right (left) continuity of $y_\pm$ in Lemma 2.4.1 it is clear that $T_{t,x(t)}$ is a right continuous function of t. If $y_-(t', x') = y_+(t', x')$ for some $t' > t$ then (2.4.9) is trivial, so we exclude this case. Set

$$I'(\eta) = u_0(\eta) + (\eta - x')/t', \quad I''(\eta) = u_0(\eta) + (\eta - x'')/t'',$$
$$y' = y_-(t', x'), \quad y'' = y_+(t'', x'').$$

By the minimizing property of y'' resp. y' we have

$$\int_{y'}^{y''} (I''(\eta) - I'(\eta)) \, d\eta = -\int_{y''}^{y'} I''(\eta) \, d\eta - \int_{y'}^{y''} I'(\eta) \, d\eta \le 0.$$

Hence

$$(y'' - y')(x'/t' - x''/t'') + (1/t'' - 1/t')(y''^2 - y'^2)/2 \le 0,$$

that is,

$$x'/t' - x''/t'' \le (1/t' - 1/t'')(y'' + y')/2,$$
$$x't'' - x''t' \le (t'' - t')(y_+(t'', x'') + y_-(t', x'))/2$$

Similarly,

$$x''t' - x't'' \le (t' - t'')(y_-(t'', x'') + y_+(t', x'))/2$$

and since $(x''t' - x't'')/(t' - t'') = x' - t'(x'' - x')/(t'' - t')$ these inequalities together prove (2.4.9).

It follows from (2.4.9) that the curve $x = x(t)$ is an integral curve of the differential equation

$$dx(t)/dt = u(t, x)$$

in the sense of Theorem 1.4.1. In view of Theorem 1.4.2 and (2.4.8) such a curve is uniquely determined in the future by its starting point. Going backward in time starting from (t, x), however, the two sides of $T_{t,x}$ are integral curves. In fact, they are the maximal and minimal integral curves. To prove this we note that for arbitrary $y_- < y_-(t, x)$, $y_+ > y_+(t, x)$ the monotonicity and continuity properties listed in Theorem 2.4.2 show that

$$(x' - y_+)/t' < u(t', x') < (x' - y_-)/t'$$

for all points of continuity (t', x') in a neighborhood of $T_{t,x}$. Hence the integral curves going backwards from (t, x) must remain in the triangle with vertices at (t, x) and $(0, y_\pm)$.

If $u_0 \in C^k$ for some $k \ge 1$ and $F(t, x, y)$ has a unique minimum point $y(t, x)$ where $F''_{yy} > 0$, then this is true in a full neighborhood and $u \in C^k$ there. In fact, the equation

$\partial F(t', x', y)/\partial y = 0$ has a unique solution $y(t', x')$ near y if (t', x') is close to (t, x), and it is a C^k function of (t', x') by the implicit function theorem. Similarly, if $F(t, x, y)$ has such a strong minimum at $y_+(t, x)$ and at $y_-(t, x)$ and no other minimum points, then the equation $\partial F(t', x', y)/\partial y = 0$ has unique C^k solutions $y = Y_\pm(t', x')$ equal to $y_\pm(t, x)$ at (t, x). We have $u \in C^k$ except where

$$F(t', x', Y_+(t', x')) = F(t', x', Y_-(t', x')).$$

Since

$$\partial\big(F(t', x', Y_+(t', x')) - F(t', x', Y_-(t', x'))\big)/\partial x' = (Y_-(t', x') - Y_+(t', x'))/t' \neq 0$$

this equation determines x' as a C^{k+1} function of t. Thus the shock curve has one additional derivative.

Note also that if u_0 is just continuous, then $u_0(y) + (y - x)/t$ vanishes at $y_\pm(t, x)$. If $u_0 \in C^1$ it follows that $u_0'(y) + 1/t$ has a zero between $y_-(t, x)$ and $y_+(t, x)$ if these are different. If u_0 is chosen so that u_0' does not take the same nonzero value more than a finite number of times, it follows that there are only finitely many discontinuities for fixed t. Thus u is then in C^k outside some curves of discontinuities satisfying the Rankine-Hugoniot condition. When two such curves collide, the interaction produces one shock curve for larger t. Schaeffer [1] has shown that for generic u_0 the solution is piecewise C^k except at the shock generation points where singularities are a bit more complicated. To discuss a typical example we choose

$$u_0(x) = -x + x^3$$

in a neighborhood of 0, which gives a shock generation point at $(1, 0)$. Since u_0 is odd, it follows that u is odd in x (by the uniqueness), so the shock curve is defined by $x = 0$. On the two sides we have $u(t, x) = -y + y^3$ where $x = y + t(-y + y^3)$ with y having the sign of x. When $t = 1$ we obtain the equation $y^3 = x$, hence

$$u(1, x) = -x^{1/3} + x,$$

and when $t = 1 + s$, $s > 0$, we have the equation $(1 + s)y^3 - sy = x$. For $x > 0$ it has one positive and two negative or complex roots. Putting $y = z(s/(1 + s))^{1/2}$ we obtain

$$z^3 - z = x((1 + s)/s^3)^{1/2} = \xi,$$

which gives z as a C^∞ function of $\xi \geq 0$, with $z(0) = 1$ and $z(\xi) = \xi^{1/3}+$ lower order terms in $\xi^{1/3}$ as $\xi \to \infty$. We have $u(1 + s, x) = (s/(1 + s)^3)^{1/2}(s\xi - z(\xi))$.

One can pose other conditions on the weak solutions which are equivalent to (2.4.8) but are applicable in more general situations. To do so we multiply Burgers' viscous equation

$$\partial u_\mu/\partial t + u_\mu \partial u_\mu/\partial x = \mu \partial^2 u_\mu/\partial x^2$$

by $\Phi'(u_\mu)$, where $\Phi \in C^2$, and rewrite it in the form

$$\partial \Phi(u_\mu)/\partial t + \partial Y(u_\mu)/\partial x = \mu \partial^2 \Phi(u_\mu)/\partial x^2 - \mu \Phi''(u_\mu)(\partial u_\mu/\partial x)^2.$$

Here Y is defined by

$$Y'(s) = s\Phi'(s).$$

If Φ is chosen convex, that is, $\Phi'' \geq 0$, it follows that

$$\partial\Phi(u_\mu)/\partial t + \partial Y(u_\mu)/\partial x \leq \mu \partial^2 \Phi(u_\mu)/\partial x^2.$$

Since u_μ is locally bounded as $\mu \to 0$, it follows that in the sense of distribution theory

$$(2.4.10) \qquad\qquad \partial\Phi(u)/\partial t + \partial Y(u)/\partial x \leq 0,$$

when $t \in \mathbf{R_+}$, that is, $t > 0$. Explicitly this means that

$$(2.4.10)' \qquad \iint (\Phi(u)\partial\chi/\partial t + Y(u)\partial\chi/\partial x)\, dt\, dx \geq 0 \quad \text{if } 0 \leq \chi \in C_0^\infty(\mathbf{R_+} \times \mathbf{R}).$$

Here Φ is any convex C^2 function. Taking

$$\Phi(s) = ((s-k)^2 + \varepsilon^2)^{1/2}$$

with $k \in \mathbf{R}$ and $\varepsilon > 0$, we conclude when $\varepsilon \to 0$ that we can take

$$\Phi(s) = |s-k|, \quad Y(s) = \tfrac{1}{2}(s^2 - k^2)\operatorname{sgn}(s-k) = \tfrac{1}{2}|s-k|(s+k).$$

Thus $(2.4.10)$ implies that for every $k \in \mathbf{R}$

$$(2.4.10)'' \qquad \iint |u-k|(\partial\chi/\partial t + \tfrac{1}{2}(u+k)\partial\chi/\partial x)\, dt\, dx \geq 0 \quad \text{if } 0 \leq \chi \in C_0^\infty(\mathbf{R_+} \times \mathbf{R}).$$

Conversely, $(2.4.10)'$ follows from $(2.4.10)''$ since all convex functions can be represented on any compact set as a superposition of functions of the form $|x - k|$ and linear functions. Note that when we take k very large positive or negative respectively, it follows from $(2.4.10)''$ that

$$\mp \iint (u\partial\chi/\partial t + \tfrac{1}{2}u^2\partial\chi/\partial x)\, dt\, dx \geq 0,$$

so $(2.4.10)''$ implies that u is a weak solution of Burgers' equation.

Theorem 2.4.3. *Let u and v satisfy $(2.4.10)''$ for all $k \in \mathbf{R}$ and be bounded in absolute value by M for $0 < t < T$. Then it follows that*

$$\int_{a+Mt}^{b-Mt} |u(t,x) - v(t,x)|\, dx$$

is a decreasing function of t as long as $b - a > 2Mt$ and $t < T$. In particular,

$$\int_{-\infty}^{\infty} |u(t,x) - v(t,x)|\, dx$$

is a decreasing function of t. If u and v are the solutions given by Theorem 2.4.2 for the initial data u_0 and v_0, then the limit as $t \to 0$ is

$$\int_a^b |u_0(x) - v_0(x)|\, dx.$$

Proof. We apply $(2.4.10)''$ with $k = v(s,y)$. This gives

$$\iint |u(t,x) - v(s,y)|(\partial\chi(t,x)/\partial t + \tfrac{1}{2}(u(t,x) + v(s,y))\partial\chi(t,x)/\partial x)\, dt\, dx \geq 0$$

if $0 \le \chi \in C_0^\infty((0, T) \times \mathbf{R})$. If $\chi \in C_0^\infty(\{(t, x, s, y); 0 < t < T, 0 < s < T\})$ we can use $\chi(\cdot, \cdot, s, y)$ here and integrate with respect to s and y also. Adding the analogous equation where u and v have been interchanged, we obtain

$$\iiiint |u(t, x) - v(s, y)|(\partial\chi/\partial t + \partial\chi/\partial s$$
$$+ \tfrac{1}{2}(u(t, x) + v(s, y))(\partial\chi/\partial x + \partial\chi/\partial y)) \, dt \, ds \, dx \, dy \ge 0.$$

Replacing t by $t + s$ and x by $x + y$ we conclude with $\Psi(t, x, s, y) = \chi(t + s, x + y, s, y)$ that

$$\iiiint |u(t + s, x + y) - v(s, y)|(\partial\Psi/\partial s$$
$$+ \tfrac{1}{2}(u(t + s, x + y) + v(s, y))\partial\Psi/\partial y) \, dt \, ds \, dx \, dy \ge 0,$$

if $\Psi \in C_0^\infty(\mathbf{R}^4)$ has support in the set where $0 < s < T$, $0 < s + t < T$. Choose $\Psi(t, x, s, y) = \psi_1(s, y)\psi_2(t, x)$ and note that

$$\iint |u(t + s, x + y) - v(s, y)|(\partial\psi_1/\partial s + \tfrac{1}{2}(u(t + s, x + y) + v(s, y))\partial\psi_1/\partial y) \, ds \, dy$$

is a continuous function of (t, x) because $|u - v|$ and $u|u - v|$ are Lipschitz continuous functions of (u, v) when $|u| \le M$, $|v| \le M$, and $(s, y) \mapsto u(t + s, x + y)$ is a continuous function of (t, x) with values in L_{loc}^1. Hence it follows that this is nonnegative for fixed (t, x), and in particular for $(t, x) = (0, 0)$ we obtain for $0 \le \psi_1 \in C_0^\infty((0, T) \times \mathbf{R})$

$$(2.4.11) \quad \iint |u(s, y) - v(s, y)|(\partial\psi_1(s, y)/\partial s + \tfrac{1}{2}(u(s, y) + v(s, y))\partial\psi_1(s, y)/\partial y) \, ds \, dy \ge 0.$$

Now choose

$$\psi_1(s, y) = \varphi(s)c(s, y)$$

where $0 \le \varphi \in C_0^\infty(0, T)$ and $c(s, y)$ is nonnegative and vanishes for large y,

$$(2.4.12) \qquad \qquad \partial c/\partial s + M|\partial c/\partial y| \le 0.$$

Then it follows that

$$\iint |u(s, y) - v(s, y)|c(s, y)\varphi'(s) \, ds \, dy \ge 0.$$

We can let c converge to the characteristic function of the set

$$\{(s, y); a + Ms < y < b - Ms\} = \{(s, y); |y - y_0| + Ms < \tfrac{1}{2}(b - a)\}, \quad y_0 = \tfrac{1}{2}(a + b),$$

by a standard regularisation, for any decreasing function of $|y - y_0| + Ms$ has the desired property (2.4.12). Hence it follows that if $b - a \ge 2MT$ and $0 \le \varphi \in C_0^\infty(0, T)$, then

$$\int \varphi'(s) \, ds \int_{a+Ms}^{b-Ms} |u(s, y) - v(s, y)| \, dy \ge 0,$$

which proves that the inner integral is decreasing.

Let u be the Hopf solution of Burgers' equation with initial data $u_0 \in L^\infty$. Choose a sequence $u_{0j} \in C_0^\infty$ with $u_{0j} \to u_0$ in L^1_{loc}, and let u_j be the corresponding Hopf solution of Burgers' equation. Then

$$\int_{a+Mt}^{b-Mt} |u_j(t,x) - u_k(t,x)| dx \leq \int_a^b |u_{0j}(x) - u_{0k}(x)| dx,$$

and letting $k \to \infty$ we find that

$$\int_{a+Mt}^{b-Mt} |u_j(t,x) - u(t,x)| dx \leq \int_a^b |u_{0j}(x) - u_0(x)| dx.$$

Since $u_j(t, \cdot) \to u_{0j}$ in L^1 as $t \to 0$, we obtain

$$\overline{\lim_{t \to 0}} \int_{a+Mt}^{b-Mt} |u_0(x) - u(t,x)| dx \leq 2 \int_a^b |u_{0j}(x) - u_0(x)| dx.$$

Hence $u(t, \cdot) \to u_0$ in L^1_{loc} as $t \to 0$. The proof is complete.

The contraction property in Theorem 2.4.3 is of course much stronger than a uniqueness theorem. Thus (2.4.8) and (2.4.10)″ are two equivalent ways of characterizing the admissible solution of Burgers' equation. At a simple jump discontinuity it is clear that they agree, for (2.4.8) just means that the jump must be negative while (2.4.10)″ means that, if the speed of the shock is s, then

$$|u_+ - k|(s - \tfrac{1}{2}(u_+ + k)) \geq |u_- - k|(s - \tfrac{1}{2}(u_- + k)).$$

When $k = u_\pm$ the condition means that

$$0 \leq \tfrac{1}{2}(u_+ + u_-) - s \leq 0,$$

that is, we have the Rankine-Hugoniot condition. For k outside $[u_-, u_+]$ we have

$$|u_+ - k|(u_+ + k) - |u_- - k|(u_- + k) = \text{sgn}(u_+ - k)(u_+^2 - u_-^2)$$
$$= 2s(u_+ - u_-)\text{sgn}(u_+ - k); \quad |u_+ - k| - |u_- - k| = (u_+ - u_-)\text{sgn}(u_+ - k)$$

so there is equality then. If $k = (u_+ + u_-)/2 = s$ then the two sides are $\tfrac{1}{2}|u_+ - u_-|$ times $s - (u_\pm + s)/2 = (s - u_\pm)/2$, so the condition becomes $u_+ \leq u_-$. When that is true we have the inequality also in the whole interval (u_+, u_-), since the two sides are quadratic and equal at the end points.

The contraction property in Theorem 2.4.3 implies that for the Hopf solution of (2.3.1) we have

$$\int |u(t, x+y) - u(t,x)| dx \leq \int |u_0(x+y) - u_0(x)| dx,$$

for the space of solutions is translation invariant. In particular, this implies that u is of bounded variation with respect to x if u_0 is:

Lemma 2.4.4. *If f is a function of bounded variation on $\mathbf{R}$, then*

(2.4.13) $$\int |f(x+y) - f(x)| dx \leq |y|V$$

where V is the total variation. Conversely, if $f \in L^1(\mathbf{R})$ and

$$(2.4.14) \qquad \varliminf_{y \to 0} \int |f(x+y) - f(x)| \, dx / |y| = V < \infty,$$

then f is of bounded variation with total variation V, and the limit exists in (2.4.14).

Proof. If f is of bounded variation and $y > 0$ then

$$\omega(y) = \int |f(x+y) - f(x)| \, dx \le \int dx \int_x^{x+y} |df(t)| = \int |df(t)| \int_{t-y}^t dx = |y|V,$$

which proves (2.4.13), and $\omega(y)$ is even in y. If $\varphi \in C_0^1$ then

$$\left| \int -\frac{d\varphi}{dx} f(x) \, dx \right| = \lim_{y \to 0} \left| \int -\frac{\varphi(x+y) - \varphi(x)}{y} f(x) \, dx \right|$$

$$= \lim_{y \to 0} \left| \int \varphi(x) \frac{f(x) - f(x-y)}{y} \, dx \right| \le V_0 \sup |\varphi|,$$

where $V_0 = \varliminf_{y \to 0} \omega(y)/|y|$. If $V_0 < \infty$ then f' is a measure in the sense of distribution theory, of total mass $\le V_0$, that is, f is of bounded variation with total variation $V \le V_0$. Since $V_0 \le \varlimsup_{y \to 0} \omega(y)/|y| \le V \le V_0$ by (2.4.13), the proof is complete.

Finally we shall discuss the asymptotic behavior of $u(t,x)$ as $t \to \infty$. In doing so we assume that

$$M = \int_{-\infty}^{\infty} u_0(y) \, dy$$

exists as a Riemann-Lebesgue integral. We shall show that

$$U_t(X) = t^{1/2} u(t, X t^{1/2})$$

has a limit as $t \to \infty$. Note that $U_t(X) = X - Y$ where Y is a minimum point of

$$\int_{-\infty}^{Y t^{1/2}} u_0(\eta) \, d\eta + (X - Y)^2/2.$$

If

$$N = \inf_y \int_{-\infty}^y u_0(\eta) \, d\eta,$$

then the integral approaches the function F of Y which is 0 for $Y < 0$, M for $Y > 0$ and N for $Y = 0$; more precisely, in any neighborhood of 0 the infimum converges to N. (Note that $N \le \min(0, M)$.) With $X_{\pm} = \max(\pm X, 0)$ we have

$$\inf_{Y<0} (X - Y)^2/2 = X_+^2/2, \quad \inf_{Y>0} (M + (X - Y)^2/2) = M + X_-^2/2.$$

When $X < -\sqrt{-2N}$ or $X > \sqrt{2(M - N)}$ the minimum of $F(Y) + (X - Y)^2/2$ is taken for $Y = X$ but otherwise it is taken when $Y = 0$, so $X - Y$ is 0 outside the interval $(-\sqrt{-2N}, \sqrt{2(M - N)})$ and equal to X in that interval. From this it is easy to conclude that

$$(2.4.15) \qquad t^{1/2} u(t, X t^{1/2}) \to X \quad \text{if } -\sqrt{-2N} < X < \sqrt{2(M - N)}),$$

and that the limit is 0 otherwise. Note that the function

$$(2.4.16) \qquad U(t, x) = \begin{cases} x/t, & \text{when } -\sqrt{-2Nt} < x < \sqrt{2(M - N)t}, \\ 0, & \text{otherwise}, \end{cases}$$

is a weak solution of Burgers' equation with only negative jumps. In fact, for the endpoints $x(t)$ we have

$$dx(t)/dt = x(t)/2t = (u(t, x(t) + 0) + u(t, x(t) - 0))/2,$$

so the Rankine-Hugoniot condition is fulfilled. The solution (2.4.16) of Burgers' equation is called an N-wave although it looks more like И, a cyrillic I. Note that it follows from (2.4.15) that M and N are preserved under the evolution defined by Burgers' equation.

The preceding result is not very illuminating when $M = N = 0$, so we shall examine the asymptotic behavior in greater detail then. Set

$$U_0(y) = \int_{-\infty}^{y} u_0(\eta)\, d\eta.$$

Then we have $U_0 \geq 0$ since $N = 0$ and $U_0(y) \to 0$ as $y \to \pm\infty$ since $M = 0$. Let us assume for the sake of simplicity that u_0 has compact support. Then U_0 has compact support too, and $u(t, x \pm 0) = (x - y_{\pm}(t, x))/t$ where $y_{\pm}(t, x)$ are the largest and the smallest minimum points of $U_0(y) + (x - y)^2/2t$. When $U_0(x) = 0$ the unique minimum point is $y = x$, so $u(t, x) = 0$ then. When $U_0(x) > 0$ on the other hand, we have for the minimum points

$$2tU_0(y) + |x - y|^2 \leq d(x)^2, \quad d(x) = \min_{U_0(z)=0} |x - z|.$$

Hence $|x - y| \leq d(x)$ and it follows that $y_{\pm}(t, x)$ converges to the point y with $U_0(y) = 0$ closest to x, if it is unique. Thus $tu(t, x) \to \pm d(x)$ with the sign such that $\pm(x - y) \geq 0$ for the closest point with $U_0(y) = 0$. There is a jump $-2d(x)$ at the center of each interval where $U_0 > 0$.

2.5. General strictly convex conservation laws. As observed by Lax [2] the formulas for $u(t, x)$ obtained in Section 2.4 from the viscous Burgers' equation can be modified to the Cauchy problem for a general conservation law

$$(2.5.1) \qquad \partial u/\partial t + \partial f(u)/\partial x = 0; \quad u(0, x) = u_0(x);$$

where $f \in C^2$ is strictly convex, that is, $f''(u) > 0$ (or strictly concave). One just has to change the definition of $F(t, x, y)$. To see how F should be chosen we first note that when $u_0 \in C^1$ the classical solution – when it exists – is given by

$$u(t, x) = u_0(y) \quad \text{when } x = y + tf'(u_0(y)).$$

Since f' is increasing it has an inverse. If g is a primitive function of the inverse, then the condition $x = y + tf'(u_0(y))$ can be written $\partial F(t, x, y)/\partial y = 0$, if

$$(2.5.2) \qquad F(t, x, y) = tg((x - y)/t) + \int_{0}^{y} u_0(\eta)\, d\eta.$$

This is quite analogous to the function F used to study Burgers' equation. It is well known that (up to a constant term) g is the conjugate function (Legendre transform) $\tilde{f}$ of f,

$$\tilde{f}(v) = \sup_{u}(vu - f(u)).$$

In fact, in the interior of the interval where $\tilde{f}$ is finite, we have $\tilde{f}(v) = vu - f(u)$ where $f'(u) = v$; hence by differentiation, $\tilde{f}'(v) = u$ so $\tilde{f}'$ and f' are inverse functions.

Put for arbitrary (t, x) with $t > 0$

$$\varphi_\mu(t, x) = (4\pi\mu t)^{-1/2} \int_{-\infty}^{\infty} e^{-F(t,x,y)/2\mu} \, dy,$$

$$u_\mu(t, x) = -2\mu\partial \log \varphi_\mu(t, x)/\partial x$$
$$= \int_{-\infty}^{\infty} \partial F(t, x, y)/\partial x \, e^{-F(t,x,y)/2\mu} \, dy \Big/ \int_{-\infty}^{\infty} e^{-F(t,x,y)/2\mu} \, dy.$$

Then we have

$$\partial u_\mu(t, x)/\partial t + \partial f_\mu(t, x)/\partial x = 0 \quad \text{if } f_\mu(t, x) = \mu/t + 2\mu\partial \log \varphi_\mu(t, x)/\partial t,$$

that is,

$$f_\mu(t, x) = -\int_{-\infty}^{\infty} \partial F(t, x, y)/\partial t \, e^{-F(t,x,y)/2\mu} \, dy \Big/ \int_{-\infty}^{\infty} e^{-F(t,x,y)/2\mu} \, dy.$$

With $z = (x - y)/t$ and $g'(z) = s$ we have

$$\partial F(t, x, y)/\partial x = g'(z) = s$$
$$-\partial F(t, x, y)/\partial t = g'(z)z - g(z) = \tilde{g}(s) = f(s)$$

and $z = f'(s)$. It follows that $u_\mu(t, x) \to g'((x - y(t, x))/t) = u(t, x)$ at every point where $F(t, x, \cdot)$ has a unique minimum point $y(t, x)$, and $f_\mu(t, x) \to f(u(t, x))$ then. Hence

$$\partial u/\partial t + \partial f(u)/\partial x = 0$$

in the weak sense. It is now an easy exercise to repeat the arguments of Hopf presented in Section 2.4 and show that Theorem 2.4.2 remains valid for the Cauchy problem (2.5.1). So does the subsequent discussion of the shock curves, with $(x - y)/t$ replaced everywhere by $g'((x - y)/t)$. Leaving this for the reader we shall now discuss another existence proof using difference approximations.

In the half plane $\overline{\mathbf{R}}_+ \times \mathbf{R}$ we consider a grid

$$\{(kh, nl); \, k, n \in \mathbf{Z}, \, k \geq 0\}$$

where h, l are small positive numbers. We approximate the differential equation (2.5.1) by the difference equation

$$(2.5.3) \qquad (u_n^{k+1} - (u_{n+1}^k + u_{n-1}^k)/2)/h + (f(u_{n+1}^k) - f(u_{n-1}^k))/2l = 0$$

for the values of u at the grid points. This was first proposed by Lax [3], and the properties of the aproximation were established by Olejnik [1]. (In fact, she also allowed equations depending on t and x and with a constant term. However, following Smoller [1] we just present the arguments in a simple but typical case here.) Note that the first term in (2.5.3) is not just a time difference quotient but contains in addition a term

$$(2u_n^k - u_{n+1}^k - u_{n-1}^k)/2h.$$

For a smooth function u we have

$$(2u(t, x) - u(t, x + l) - u(t, x - l))/2 = -u''_{xx}(t, x)l^2/2 + O(l^3)$$

by Taylor's formula. This suggests that the solution of (2.5.3) approximates the solution of the equation

$$\partial u/\partial t + \partial f(u)/\partial x = (l^2/2h)\partial^2 u/\partial x^2$$

at the grid points. We shall let l and h tend to 0 with the same speed so $l^2/2h \to 0$, but the modified time difference quotient still has a stabilizing effect.

As initial conditions for the solution of (2.5.3) we choose

$$(2.5.4) \qquad u_n^0 = \int_{(n-1)l}^{(n+1)l} u_0(x)\, dx/2l.$$

If $M = \sup |u_0|$ it is then clear that $|u_n^0| \leq M$. Let

$$(2.5.5) \qquad A = \max_{|u| \leq M} |f'(u)|$$

be the largest possible speed of the characteristics for classical solutions.

Lemma 2.5.1. If $|u_0| \leq M$ it follows that $|u_n^k| \leq M$ for all k, n provided that

$$(2.5.6) \qquad Ah \leq l \quad \text{(the Courant-Friedrichs-Lewy condition)}.$$

The condition is a natural one: it states that the mesh widths are chosen so that the values at $((k+1)h, nl)$ depend on the values at time kh in an interval containing the points which could influence a classical solution.

Proof. Assume that the bound has been proved up to a certain value k; it is true when $k = 0$. By the mean value theorem

$$f(u_{n+1}^k) - f(u_{n-1}^k) = (u_{n+1}^k - u_{n-1}^k)\mu$$

where $|\mu| \leq A$, so we can write (2.5.3) in the form

$$2lu_n^{k+1} = u_{n+1}^k(l - \mu h) + u_{n-1}^k(l + \mu h).$$

By (2.5.6) the parentheses on the right are positive, so u_n^{k+1} is a weighted mean of $u_{n\pm1}^k$, hence bounded by M in absolute value.

Next we establish an analogue of the admissibility condition (2.4.8). In doing so we assume that

$$(2.5.7) \qquad \kappa = \min_{|u| \leq M} f''(u)/2 > 0,$$

that is, that f is strictly convex in $[-M, M]$.

Lemma 2.5.2. *If* $c = \min(\kappa/2, A/4M)$ *and* (2.5.6) *holds, then*

$$(2.5.8) \qquad (u_{n+1}^k - u_{n-1}^k)/l \le 1/ckh.$$

Note that kh is the time coordinate at the grid point, so this is precisely an analogue of (2.4.8).

Proof. Writing $z_n^k = u_{n+1}^k - u_{n-1}^k$ we have

$$2z_n^{k+1} = z_{n+1}^k + z_{n-1}^k + \frac{h}{l}(2f(u_n^k) - f(u_{n+2}^k) - f(u_{n-2}^k)).$$

Now Taylor's formula gives if $u, v \in [-M, M]$

$$f(v) - f(u) \ge (v - u)f'(u) + \kappa(v - u)^2,$$

hence $f(u) - f(v) \le (u - v)f'(u) - \kappa(u - v)^2$, so it follows that

$$
\begin{aligned}
2lz_n^{k+1} &\le l(z_{n+1}^k + z_{n-1}^k) \\
&\quad + h(2u_n^k - u_{n+2}^k - u_{n-2}^k)f'(u_n^k) - \kappa h((u_{n+2}^k - u_n^k)^2 + (u_{n-2}^k - u_n^k)^2) \\
&= l(z_{n+1}^k + z_{n-1}^k) + h(z_{n-1}^k - z_{n+1}^k)f'(u_n^k) - \kappa h((z_{n+1}^k)^2 + (z_{n-1}^k)^2).
\end{aligned}
$$

Again because of (2.5.6), it follows that z_n^{k+1} is bounded by a weighted average of $\varphi(z_{n\pm1}^k)$, with weights $(1 \mp f'(u_n^k)h/l)/2 \ge 0$, if $\varphi(z) = z - cz^2$ and $c \le \kappa h/2l$. By Lemma 2.5.1 we have $z_n^k \le 2M$ for all k, n, and

$$\varphi'(z) = 1 - 2cz \ge 0 \quad \text{for } z \le 2M \text{ if } 4cM \le 1.$$

Both conditions are fulfilled if $c = \min(\kappa h/2l, 1/4M)$. Writing

$$M_k = \max(0, \max_n z_n^k)$$

we conclude that

$$M_{k+1} \le \varphi(M_k) = M_k - cM_k^2.$$

This can be compared with the solution of the Cauchy problem

$$dM(t)/dt = -cM(t)^2, \quad M(0) = M_0,$$

that is, $M(t) = M_0/(1 + cM_0 t)$ (cf. Section 1.3). Since $M' = -cM^2$ is increasing, we have

$$M(k + 1) \ge M(k) - cM(k)^2 = \varphi(M(k)),$$

and we conclude inductively using the monotonicity of φ that

$$M_k \le M(k) = M_0/(1 + cM_0 k) \le 1/ck = \max(2l/\kappa kh, 4M/k).$$

Since $M \le Ml/Ah$, the estimate (2.5.8) is proved.

For a bounded function a bound on the positive variation implies a bound on the negative one. Thus we have

Lemma 2.5.3. *With assumptions and notation as in Lemma 2.5.2 we have*

$$(2.5.9) \qquad \sum_{|nl|<X} |u_{n+1}^k - u_{n-1}^k| \leq 4M + (4X + l)/(ckh), \quad X > 0.$$

Proof. By (2.5.8) we know that

$$v_n^k = u_n^k - nl/2ckh$$

decreases when n increases by 2, so $|u_{n+1}^k - u_{n-1}^k| \leq v_{n-1}^k - v_{n+1}^k + l/ckh$ and

$$\sum_{-N}^{N} |u_{n+1}^k - u_{n-1}^k| \leq \sum_{-N}^{N} (v_{n-1}^k - v_{n+1}^k) + (2N+1)l/ckh.$$

In the sum on the right all terms except at most 4 cancel, so we have the bound

$$4(M + Nl/2ckh) + (2N+1)l/ckh = 4M + (4N+1)l/ckh.$$

When $N < X/l$ the estimate (2.5.9) follows.

For $kh = t > 0$ Lemma 2.5.3 gives good control of the difference approximation in the x direction. Just as for classical solutions of the differential equation in (2.5.1) an estimate for $\int |\partial u/\partial x|\, dx$ gives an estimate of $\int |\partial u/\partial t|\, dx$, we get such a bound for the differences in the t direction.

Lemma 2.5.4. *If $p > k$ and $p - k$ is even, then*

$$(2.5.10) \qquad \sum_{|n|\leq X/l} |u_n^p - u_n^k| l \leq (4M + (4X + 5l)/ckh)(p-k)h(l/h).$$

Proof. It suffices to prove this when $p = k+2$. From the proof of Lemma 2.5.1 we know that u_n^{k+2} is a weighted average of $u_{n\pm1}^{k+1}$. Repeating this observation we see that u_n^{k+2} is a weighted average of u_{n-2}^k, u_n^k, u_{n+2}^k, hence that

$$|u_n^{k+2} - u_n^k| \leq |u_{n-2}^k - u_n^k| + |u_n^k - u_{n+2}^k|.$$

Hence

$$\sum_{|n|\leq X/l} |u_n^{k+2} - u_n^k| \leq 2 \sum_{|n|\leq(X+l)/l} |u_{n+1}^k - u_{n-1}^k|$$
$$\leq 2(4M + (4X + 5l)/ckh).$$

This completes the proof.

We have displayed the constant in such detail to show that the time has come to introduce also an upper bound on the ratio l/h in addition to (2.5.6). From now on we therefore define $l = Ah$. The right-hand side of (2.5.10) is then

$$A(4M + (4X + 5l)/(ckh))(ph - kh).$$

Before letting $h \to 0$ we extend the values u_n^k at the grid points to a function in the upper half plane defined by

$$(2.5.11) \quad u_h(t,x) = u_n^k \quad \text{when } kh \leq t < (k+1)h, \ (n-1)l \leq x < (n+1)l, \ n-k \text{ even.}$$

Theorem 2.5.5. *Let $u_0 \in L^\infty$, $|u_0| \leq M$, and define u_h by (2.5.11), (2.5.3), (2.5.4) with $l = Ah$ and A defined by (2.5.5). Assume that $f \in C^2$ is strictly convex in $[-M, M]$. Then $|u_h| \leq M$, and*

$$\lim_{h \to 0} u_h(t, x) = u(t, x)$$

exists except in a set which is countable for fixed t. The limit $u(t, x)$ satisfies (2.4.8) with a constant factor on the right, and

$$(2.5.12) \quad -\iint_{t>0} (u \partial\varphi/\partial t + f(u)\partial\varphi/\partial x)\, dt\, dx = \int u_0(x)\varphi(0, x)\, dx, \quad \varphi \in C_0^\infty(\mathbf{R}^2).$$

Proof. Lemma 2.5.3 gives a bound for the total variation of $u_h(t, x)$ when $|x| \leq X$ which is uniform in h for t in a compact set in the open positive real axis. In view of Lemma 2.4.4 this implies a uniform Lipschitz condition for

$$y \mapsto u_h(t, x + y) \in L^1(-X, X),$$

as a function of x, and by Lemma 2.5.4 we have uniform Lipschitz continuity in t also. It follows that every sequence of values of $h \to 0$ has a subsequence h_ν such that $u(t, \cdot) = \lim_{\nu \to \infty} u_{h_\nu}(t, \cdot)$ exists in L^1_{loc} for all $t > 0$. By Lemma 2.5.2 and Lemma 2.5.6 below applied to $Cl[x/l] - u_{h_\nu}(t, x)$ for C large enough, we can assume that $u(t, \cdot)$ satisfies (2.4.8) and that $u_{h_\nu}(t, x) \to u(t, x)$ except for countably many x when t is fixed. If we prove (2.5.12) it follows from Olejnik's uniqueness theorem (Theorem 2.2.1) that the limit of subsequences is unique, hence that the limit exists as claimed when $h \to 0$. (Note that $f(u) - f(v) = a(u - v)$ where $a = \int_0^1 f'(\lambda u + (1 - \lambda)v)\, d\lambda$ and f' is increasing. If u and v satisfy (2.4.8) with some constant it follows that a satisfies (2.4.8) with some other constant).

Let $\varphi \in C_0^\infty(\mathbf{R}^2)$ and set $\varphi_n^k = \varphi(kh, nl)$. Multiplication of (2.5.3) by $2lh\varphi_n^k$ and summation for odd $k - n$ and $k \geq 0$ gives

$$\sum_{k>0} 2lu_n^k(\varphi_n^{k-1} - (\varphi_{n+1}^k + \varphi_{n-1}^k)/2) - 2l\sum_n u_n^0(\varphi_{n+1}^0 + \varphi_{n-1}^0)/2$$

$$+ \sum_{k\geq 0} hf(u_n^k)(\varphi_{n-1}^k - \varphi_{n+1}^k) = 0.$$

In these sums $n - k$ is even. Since, with n even in the sums,

$$\sum_n 2lu_n^0(\varphi_{n+1}^0 + \varphi_{n-1}^0)/2 = \sum_n \int_{(n-1)l}^{(n+1)l} u(x)(\varphi(0, (n+1)l) + \varphi(0, (n-1)l))\, dx/2$$

differs from $\int u(x)\varphi(0, x)\, dx$ by at most $O(l)$, this sum converges to the boundary term in (2.5.12). It is equally clear that the other sums converge to the double integrals in (2.5.12). This completes the proof when we have verified the following lemma.

Lemma 2.5.6. *Let u_ν be a sequence of increasing functions in the compact interval $I \subset \mathbf{R}$ such that $|u_\nu| \leq M$ and $u_\nu \to u$ in $L^1(I)$. Then u can be changed on a null set so that u is increasing, $|u| \leq M$, and then we have $u_\nu(x) \to u(x)$ for all $x \in I$ where u is continuous, hence except at countably many points.*

Proof. If $x, y \in I$ and $x < y$ then

$$u_\nu(x) \leq \int_x^y u_\nu(z)\, dz/(y - x) \leq u_\nu(y).$$

Hence

$$\varliminf_{\nu \to \infty} u_\nu(x) \le \int_x^y u(z)\,dz/(y-x) \le \varliminf_{\nu \to \infty} u_\nu(y).$$

At any Lebesgue point x it follows that

$$\varlimsup_{\nu \to \infty} u_\nu(x) \le u(x) \le \varliminf_{\nu \to \infty} u_\nu(x),$$

so $\lim u_\nu(x)$ exists almost everywhere and is an increasing function. Thus $v(x) = \lim u_\nu(x)$ is equal to $u(x)$ almost everywhere and is clearly increasing, so it has only countably many discontinuities. Every point of continuity is of course a Lebesgue point, which completes the proof.

The Rankine-Hugoniot condition can be given an extended formulation.

Theorem 2.5.7. *Let u be a locally bounded weak solution of*

$$\partial u/\partial t + \partial f(u)/\partial x = 0$$

for $a < t < b$, and assume that u is of bounded variation when t is fixed. Let $x(t)$, $a < t < b$, be a Lipschitz continuous real valued function. Then

$$(2.5.13) \qquad f(u(t, x(t)+0)) - f(u(t, x(t)-0)) = x'(t)(u(t, x(t)+0) - u(t, x(t)-0))$$

for almost all $t \in (a, b)$.

Proof. We have

$$(2.5.14) \qquad \iint (u\partial\varphi/\partial t + f(u)\partial\varphi/\partial x)\,dt\,dx = 0$$

for all Lipschitz continuous φ with compact support in $(a, b) \times \mathbf{R}$. This is true by hypothesis if $\varphi \in C_0^\infty$. If φ is just Lipschitz continuous, we obtain (2.5.14) by applying (2.5.14) to standard regularizations φ_ε and letting $\varepsilon \to 0$, for the first derivatives of φ_ε converge boundedly to those of φ almost everywhere. In fact, if

$$\varphi_\varepsilon(t, x) = \iint \varphi(s, y)\chi((t-s)/\varepsilon, (x-y)/\varepsilon)\,ds\,dy/\varepsilon^2,$$

where $0 \le \chi \in C_0^\infty$, $\iint \chi(s, y)\,ds\,dy = 1$, and if $\chi_1(t, x) = \partial\chi(t, x)/\partial t$, then

$$\partial\varphi_\varepsilon(t, x)/\partial t = \iint \varphi(s, y)\chi_1((t-s)/\varepsilon, (x-y)/\varepsilon)\,ds\,dy/\varepsilon^3$$

$$= \iint (\varphi(t-\varepsilon s, x-\varepsilon y) - \varphi(t, x))\varepsilon^{-1}\chi_1(s, y)\,ds\,dy.$$

Hence $|\partial\varphi_\varepsilon(t, x)/\partial t| \le C \iint (|s| + |y|)|\chi_1(s, y)|\,ds\,dy$ if C is a Lipschitz constant for φ, and $\partial\varphi_\varepsilon(t, x)/\partial t \to \partial\varphi(t, x)/\partial t$ as $\varepsilon \to 0$ if φ is differentiable at (t, x), which is true almost everywhere by Rademacher's theorem. The derivative with respect to x is of course handled in the same way.

Now choose

$$\varphi(t, x) = \psi((x - x(t))/\varepsilon)\varrho(t),$$

where $\psi \in C_0^\infty(\mathbf{R})$, $\varrho \in C_0^\infty((a, b))$ and $\psi(0) = 1$. Then we obtain

$$\int \varrho(t)\,dt \int (f(u) - x'(t)u)\psi'((x - x(t))/\varepsilon)\,dx/\varepsilon + \iint u\psi((x - x(t))/\varepsilon)\varrho'(t)\,dt\,dx = 0.$$

The last integral is $O(\varepsilon)$. Since

$$\int_{x>0} \psi'(x)\, dx = -1, \quad \int_{x<0} \psi'(x)\, dx = 1,$$

and $f(u) - x'(t)u$ has right and left limits as a function of x for every fixed t, such that $x'(t)$ exists, we obtain as $\varepsilon \to 0$

$$\int \big(f(u(t, x(t) - 0)) - f(u(t, x(t) + 0)) - x'(t)(u(t, x(t) - 0) - u(t, x(t) + 0)) \big) \varrho(t)\, dt = 0.$$

Since ϱ is arbitrary, this proves the theorem.

Corollary 2.5.8. *Let the hypotheses of Theorem 2.5.7 be fulfilled and assume that $dx(t)/dt = f'(u(t, x))$ in the sense of Theorem 1.4.1. Then we have for almost all t*

$$x'(t) = \big(f(u(t, x(t) + 0)) - f(u(t, x(t) - 0)) \big) / (u(t, x(t) + 0) - u(t, x(t) - 0)),$$

if the denominator is not 0 and otherwise

$$x'(t) = f'(u(t, x(t) \pm 0)).$$

From (2.4.8) it follows as before that there is a unique forward integral curve through every point. By analyzing carefully the maximal and the minimal backward solutions one can recover the qualitative properties established for Burgers' equation in Section 2.4. We refer to Dafermos [1] for a very careful discussion of these matters.

CHAPTER III

SCALAR FIRST ORDER EQUATIONS WITH SEVERAL VARIABLES

3.1. Introduction. In this chapter we shall discuss a single conservation law

$$(3.1.1) \qquad \partial u/\partial t + \sum_1^n \partial f_i(u)/\partial x_i = 0; \quad u(0, x) = u_0(x);$$

where u is a function in $\mathbf{R}_+ \times \mathbf{R}^n$. Guided by the discussion of Burgers' equation in Chapter II we shall start by solving the corresponding diffusion equation

$$(3.1.2) \qquad \partial u/\partial t + \sum_1^n \partial f_i(u)/\partial x_i = \mu \Delta u; \quad u(0, x) = u_0(x);$$

for $\mu > 0$. In Section 3.2 we supply the minimum background required to show that (3.1.2) has a solution and prove compactness properties which guarantee the existence of convergent subsequences as $\mu \to 0$. We then show in Section 3.3, essentially as for Burgers' equation in Section 2.4, that every limit as $\mu \to 0$ must satisfy an entropy condition. The uniqueness of weak solutions of (3.1.1) satisfying the entropy condition is also proved in Section 3.4 essentially as in Section 2.4, so we obtain existence and uniqueness of such a solution for $u_0 \in L^\infty$.

3.2. Parabolic equations. We shall now discuss rather briefly the solution of the Cauchy problem

$$(3.2.1) \qquad \partial u/\partial t + \sum_1^n \psi_i(t, x, u)\partial u/\partial x_i = \mu \Delta u; \quad u(0, x) = u_0(x);$$

where $u_0 \in L^\infty(\mathbf{R}^n)$, $\Delta = \sum_1^n \partial^2/\partial x_i^2$ is the Laplacian in $\mathbf{R}^n$, and $\mu > 0$. At first we assume that $u_0 \in C_0^\infty(\mathbf{R}^n)$ and that

$$(3.2.2) \qquad |\psi_i| \leq C, \ |\partial_t \psi_i| \leq C, \ |\partial_x \psi_i| \leq C, \ |\partial_u \psi_i| \leq C \quad \text{for } t \geq 0.$$

Let $E(t, x)$ be the fundamental solution of the heat equation obtained when all ψ_i vanish,

$$E(t, x) = E_\mu(t, x) = (4\pi\mu t)^{-n/2} e^{-|x|^2/4\mu t}, \quad t > 0.$$

For reasons of homogeneity

$$\int |\partial_x^\alpha E(t, x)| \, dx = C_\alpha (\mu t)^{-|\alpha|/2}, \quad t > 0,$$

where $C_0 = 1$.

In order to solve (3.2.1) by successive approximation, we define u^0, u^1, ... so that for $\nu \geq 0$

(3.2.3) $\qquad \partial u^\nu / \partial t + \sum_1^n \psi_i(t, x, u^{\nu-1}) \partial u^{\nu-1} / \partial x_i = \mu \Delta u^\nu; \quad u^\nu(0, x) = u_0(x).$

Here u^{-1} should be read as 0. Since

$$u^\nu(t, x) = -\iint_0^t E(t - s, x - y) \sum_1^n \psi_i(s, y, u^{\nu-1}(s, y)) \partial u^{\nu-1}(s, y) / \partial y_i \, ds dy$$

$$+ \int E(t, x - y) u_0(y) \, dy$$

it follows inductively that all u_ν are in C^∞ for $t \geq 0$ and are rapidly decreasing as $x \to \infty$. Set

$$\Psi_i(t, x, u) = \int_0^u \psi_i(t, x, v) \, dv.$$

Then, with $E_i(t, x) = \partial E(t, x) / \partial x_i$ and $\Psi_i'(t, x, u) = \partial \Psi_i(t, x, u) / \partial x_i$,

$$u^\nu(t, x) = -\iint_0^t \sum_1^n E_i(t - s, x - y) \Psi_i(s, y, u^{\nu-1}(s, y)) \, ds \, dy$$

$$+ \iint_0^t E(t - s, x - y) \sum_1^n \Psi_i'(s, y, u^{\nu-1}(s, y)) \, ds \, dy + u^0(t, x).$$

Set $u^{\nu+1} - u^\nu = v^{\nu+1}$. For $\nu \geq 0$ we have

$$v^{\nu+1}(t, x) = -\iint_0^t \sum_1^n E_i(t - s, x - y)(\Psi_i(s, y, u^\nu(s, y)) - \Psi_i(s, y, u^{\nu-1}(s, y))) ds dy$$

$$+ \iint_0^t E(t - s, x - y) \sum_1^n \left(\Psi_i'(s, y, u^\nu(s, y)) - \Psi_i'(s, y, u^{\nu-1}(s, y)) \right) ds \, dy,$$

hence

$$\|v^{\nu+1}(t, \cdot)\|_\infty \leq \int_0^t C'(\mu^{-1/2}(t - s)^{-1/2} + 1) \|v^\nu(s, \cdot)\|_\infty \, ds.$$

We have $|v^0| = |u^0| \leq M = \|u_0\|_\infty$, and it follows inductively that for $t \leq T$

(3.2.4) $\qquad |v^\nu(t, x)| \leq M(C'\sqrt{\pi})^\nu (\sqrt{t/\mu} + t)^\nu / \Gamma((\nu + 2)/2).$

In fact, this is true when $\nu = 0$, and since

$$t^{-1/2} \int_0^t s^{\nu/2} \, ds / \Gamma((\nu + 2)/2) \leq \int_0^t (t - s)^{-1/2} s^{\nu/2} \, ds / \Gamma((\nu + 2)/2)$$

$$= t^{(\nu+1)/2} B(1/2, (\nu + 2)/2) / \Gamma((\nu + 2)/2) = t^{(\nu+1)/2} \Gamma(1/2) / \Gamma((\nu + 3)/2),$$

where B is the beta function and $\Gamma(1/2) = \sqrt{\pi}$, the estimate (3.2.4) with ν replaced by $\nu+1$ follows from (3.2.4) which implies $|v^\nu(s, x)| \leq M(C'\sqrt{\pi})^\nu (\mu^{-1/2} + t^{1/2})^\nu s^{\nu/2} / \Gamma((\nu + 2)/2)$ when $0 \leq s \leq t$. Hence it follows that $\sum_0^\infty v^\nu(t, x)$ is uniformly convergent for $t \leq T$, that is, $u(t, x) = \lim_{\nu \to \infty} u^\nu(t, x)$ exists with uniform convergence in any strip.

From the equations

$$\partial_j u^\nu(t, x) = -\iint_0^t E_j(t - s, x - y) \sum_1^n \psi_i(s, y, u^{\nu-1}(s, y)) \partial u^{\nu-1}(s, y) / \partial y_i \, ds \, dy$$

$$+ \partial_j u^0(t, x)$$

we obtain if

$$M_\nu(t) = \sup_x \max_{0 \le s \le t} \max_j |\partial u^\nu(s, x)/\partial x_j|$$

that

$$M_\nu(t) \le C\mu^{-1/2} \int_0^t (t - s)^{-1/2} M_{\nu-1}(s) \, ds + M_0(t).$$

Choose B so that $M_0(t) \le B/2$. Then it follows inductively that $M_\nu(t) \le Be^{Kt/\mu}$ if K is so large that

$$C \int_0^\infty s^{-1/2} e^{-Ks} \, ds < \tfrac{1}{2}.$$

Thus we have a uniform bound for $M_\nu(t)$ when t/μ is bounded. Using the recursion formulas we can also obtain a uniform estimate for the Hölder continuity of $\partial_j u^\nu(t, x)$ with respect to x, for

$$\int |\partial_j(E(t - s, x - y) - E(t - s, x + \Delta x - y))| \, dy \le C\mu^{-(1+\theta)/2}(|\Delta x|/\sqrt{t - s})^\theta/\sqrt{t - s},$$

if $0 \le \theta \le 1$, and the integral of the right-hand side with respect to s from 0 to t is $\le 2C|\Delta x|^\theta t^{(1-\theta)/2}\mu^{-(1+\theta)/2}/(1 - \theta)$ if $\theta < 1$. With $E_{jk}(t, x) = \partial^2 E(t, x)/\partial x_j \partial x_k$ we have

$$\left|\partial_{x_k} \iint_0^t E_j(t - s, x - y) f(s, y) \, ds \, dy\right| = \left|\iint_0^t E_{jk}(t - s, x - y)(f(s, y) - f(s, x)) \, ds \, dy\right|$$

$$\le \sup_{0<s<t, x\neq y} |f(s, y) - f(s, x)|/|x - y|^\theta \iint_0^t |E_{jk}(s, y)||y|^\theta \, ds \, dy$$

$$\le C_\theta \mu^{\theta/2-1} t^{\theta/2} \sup_{0<s<t, x\neq y} |f(s, y) - f(s, x)|/|x - y|^\theta,$$

which for fixed μ gives a uniform bound also for the derivatives $\partial_x^\alpha u^\nu$ with $|\alpha| = 2$, hence a bound for $\partial_t u^\nu$. We can now differentiate the recursion formula for $\partial_j u^\nu(t, x)$ with respect to x so that the derivative falls on $\psi_i \partial_i u^{\nu-1}$. This proves that $\partial_x^\alpha u^\nu$ also is uniformly Hölder continuous in x of every order $\theta < 1$ when $|\alpha| \le 2$. Together with the bounds for $\partial_t u^\nu$ this implies Hölder continuity in t of order $\theta/(\theta + 2)$, for if $\chi \in C_0^\infty(\mathbf{R}^n)$ and $\int \chi \, dx = 1$, then

$$\left|\partial_x^\alpha u^\nu(t, x) - \int \partial_y^\alpha u^\nu(t, y)\chi((x - y)/\delta)\delta^{-n} \, dy\right| \le C\delta^\theta,$$

$$\left|\int (u^\nu(t, y) - u^\nu(s, y))(\partial^\alpha \chi)((x - y)/\delta)\delta^{-n-|\alpha|} \, dy\right| \le C|s - t|\delta^{-|\alpha|},$$

when μ is fixed, $|\alpha| \le 2$, and s, t are bounded. This gives $|\partial_x^\alpha u^\nu(t, x) - \partial_x^\alpha u^\nu(s, x)| \le C'|s - t|^{\theta/(\theta+|\alpha|)}$ if we take $\delta = |s - t|^{1/(\theta+|\alpha|)}$. Now it follows from (3.2.3) that also $\partial u^\nu/\partial t$ is uniformly Hölder continuous in (t, x), and we conclude that the limit u is a classical solution of (3.2.1).

If (3.2.2) is strengthened to bounds for $\partial_{t,x,u}^{\alpha}\psi_i$ when $|\alpha| \leq k$, $k > 1$, we obtain in the same way bounds for $\partial_x^{\alpha} u^{\nu}$ when $|\alpha| \leq k+1$ by differentiating the recursion formula for $\partial_j u^{\nu}$ so that the derivatives under the integral sign fall on $\psi_i \partial_i u^{\nu-1}$. Estimates for time derivatives are then obtained from (3.2.3).

The *maximum principle* applied to the equation in (3.2.1) shows that the supremum of u in the strip $\{(t, x); 0 \leq t \leq T\}$ is taken when $t = 0$. (For the proof just note that if

$$(t, x) \mapsto u(t, x) - \varepsilon(2\mu t + |\varepsilon x|^2/2n),$$

where $\varepsilon > 0$, has a maximum at (t, x) and $0 < t \leq T$, then $\varepsilon^3 |x|^2$ has a bound independent of ε since u is bounded, and

$$\partial u/\partial x = \varepsilon^3 x/n, \quad \partial u/\partial t - 2\varepsilon\mu \geq 0, \quad \Delta u - \varepsilon^3 \leq 0,$$

which implies

$$\partial u/\partial t + \sum_1^n \psi_i(t, x, u)\partial u/\partial x_i - \mu\Delta u \geq 2\varepsilon\mu - C'\varepsilon^{3/2} - \mu\varepsilon^3 > 0$$

if ε is small. This is a contradiction so the maximum is attained when $t = 0$, hence

$$u(t, x) - \varepsilon(2\mu t + |\varepsilon x|^2/2n) \leq \sup u_0, \quad 0 \leq t \leq T,$$

if $\varepsilon > 0$ is small, so $u \leq \sup u_0$ when $0 \leq t \leq T$. In the same way, by changing the sign of ε, one finds that $u \geq \inf u_0$.) Thus $|u| \leq M$ if $|u_0| \leq M$, which also shows that the hypotheses on ψ_i only have to be made when $|u| \leq M$.

Using the maximum principle we can also prove a decay estimate for u outside the influence domain of the initial data for the equation with $\mu = 0$:

Lemma 3.2.1. *Let u satisfy (3.2.1) and assume that $|u_0| \leq e^{\Phi/\mu}$ where Φ is bounded above and has Lipschitz constant 1. Assume that*

$$\left(\sum_1^n |\psi_i(t, x, u)|^2\right)^{1/2} \leq A.$$

Then it follows with a constant C depending only on n that

$$|u(t, x)| \leq e^{Ct} e^{((A+1)t + \Phi(x) + 2)/\mu}.$$

Proof. Let $\Psi = 1 + \Phi * \chi$ where $0 \leq \chi \in C_0^{\infty}(\mathbf{R}^n)$, $\int \chi \, dx = 1$, and $\chi = 0$ outside the unit ball. Then $\Phi \leq \Psi \leq \Phi + 2$ and $|u_0| \leq e^{\Psi/\mu}$. The Lipschitz constant 1 is inherited by Ψ, and $|\Delta\Psi| < C$ where $C > \int |\chi'(x)| \, dx$ only depends on n. With

$$\mathcal{L} = \partial/\partial t + \sum_1^n \psi_i(t, x, u)\partial/\partial x_i - \mu\Delta$$

we obtain

$$\mathcal{L}(e^{((A+1+C\mu)t + \Psi(x))/\mu} \pm u(t, x)) > 0$$

since

$$(A + 1 + C\mu) - A|\partial\Psi/\partial x| - |\partial\Psi/\partial x|^2 - \mu\Delta\Psi > 0.$$

If Φ is bounded then a negative infimum of $e^{((A+1+C\mu)t+\Psi(x))/\mu} \pm u(t,x)$ when $0 \leq t \leq T$ would have to be attained. As in the proof of the maximum principle we conclude that it can only be attained when $t = 0$ which is against the hypothesis. Hence

$$\pm u(t,x) \leq e^{Ct}e^{((A+1)t+\Psi(x))/\mu},$$

which proves the lemma when Φ is bounded. It follows in general if we replace Φ by $\max(\Phi, a)$ and let $a \to -\infty$.

If $|u_0| \leq M$ one can choose Φ equal to $\mu \log M$ minus the distance to the support of u_0. In particular, if u_0 vanishes outside the ball with radius R and center at 0, we obtain

$$|u(t,x)| \leq \min(M, Me^{Ct}e^{((A+1)t+R+2-|x|)/\mu}),$$

which shows a uniform exponential decay to 0 outside the influence domain of the initial data as $\mu \to 0$. The following lemma gives a dual estimate:

Lemma 3.2.2. *If w is a solution of the Cauchy problem*

$$\partial w/\partial t + \sum_1^n \partial(a_i(t,x)w)/\partial x_i = \mu\Delta w; \quad w(0,x) = w_0(x);$$

where $w_0 \in C_0^\infty$, w decreases rapidly as $x \to \infty$, $a \in C^1$, $1 + |a| \leq K$, $|a'|$ is bounded, and $\mu > 0$, then

$$(3.2.5) \qquad\qquad \int |w(T,x)|\, dx \leq \int |w_0(x)|\, dx,$$

and when μ is small then

$$(3.2.6) \qquad\qquad \int_{|x|<R} |w(T,x)|\, dx \leq \int |w_0(x)|e^{-(|x|-R-2-KT)_+/\mu}\, dx.$$

Proof. Let g be the solution of the adjoint equation

$$\partial g/\partial t + \sum_1^n a_i\partial g/\partial x_i + \mu\Delta g = 0$$

with Cauchy data $g(T,\cdot) = \gamma \in C_0^\infty(\{x; |x| < R\})$. With $T - t$ as a new variable this is of the form discussed in Lemma 3.2.1 and the remarks after the proof, so if $|\gamma| \leq 1$ then

$$|g(t,x)| \leq e^{C(T-t)}e^{-(|x|-R-2-K(T-t))_+/\mu}, \quad |g(t,x)| \leq 1.$$

Since the equations satisfied by w and g give

$$\int w(T,x)\gamma(x)\, dx - \int w_0(x)g(0,x)\, dx = \iint_0^T (\partial w/\partial t g + w\partial g/\partial t)\, dt\, dx = 0,$$

it follows that $|\int w(T,x)\gamma(x)\, dx| \leq e^{CT}\int |w_0(x)|e^{-(|x|-R-2-KT)_+/\mu}\, dx$ which proves the estimate (3.2.6), and that $|\int w(T,x)\gamma(x)\, dx| \leq \int |w_0(x)|\, dx$, which proves (3.2.5).

When $\mu \to 0$ we see that the L^1 norm on a compact set at time T is essentially bounded by the L^1 norm of the initial data in the influence domain.

3.3. The conservation law with viscosity. We shall now study the Cauchy problem

$$(3.3.1) \qquad \partial u/\partial t + \sum_1^n \partial f_i(u)/\partial x_i = \mu \Delta u; \quad u(0, x) = u_0(x);$$

where $u_0 \in C_0^3$, for the sake of simplicity. We assume throughout that $f_i \in C^2$. Let $|u_0| \le M$. Modifying f_i at first when $|u| > M$ so that the second order derivatives are bounded on $\mathbf{R}$, we have proved that (3.3.1) has a solution u when $t \ge 0$ such that $|u| \le M$, $\partial_x^\alpha u$ and $\partial_t u$ are Hölder continuous in x for $|\alpha| \le 2$. Let v be a solution with these properties of

$$(3.3.2) \qquad \partial v/\partial t + \sum_1^n \partial f_i(v)/\partial x_i = \mu \Delta v; \quad v(0, x) = v_0(x)$$

with similar hypotheses on v_0. Then $w = u - v$ satisfies the equation

$$(3.3.3) \qquad \partial w/\partial t + \sum_1^n \partial(a_i w)/\partial x_i = \mu \Delta w; \quad w(0, x) = u_0(x) - v_0(x)$$

where

$$a_i = (f_i(u) - f_i(v))/(u - v) \in C^1,$$

and $|a_i| \le A$ if $|f_i'| \le A$. Hence we can apply Lemma 3.2.2 and obtain from (3.2.5)

$$(3.3.4) \qquad \int |u(t, x) - v(t, x)| \, dx \le \int |u_0(x) - v_0(x)| \, dx,$$

so we have established the same contraction properties as for Burgers' equation. Summing up:

Theorem 3.3.1. *For every $\mu > 0$ the Cauchy problem (3.3.1) with $u_0 \in C_0^3$ has a classical solution for $t \ge 0$ with the same bound as u_0. The derivatives $\partial_x^\alpha u$ and $\partial_t u$ are bounded for bounded t if $|\alpha| \le 2$. The solution is unique, and we have the L^1 contraction property (3.3.4).*

3.4. The entropy solution of the conservation law. Still with $u_0 \in C_0^3$ we let u_μ denote the solution of (3.3.1) obtained in Theorem 3.3.1. Before letting $\mu \to 0$ we must establish some compactness properties which allow us to pass to the limit. To do so we introduce the L^1 modulus of continuity

$$\omega(h) = \sup_{|y| < h} \int |u_0(x + y) - u_0(x)| \, dx.$$

Lemma 3.4.1. *For every $\mu > 0$ and $t > 0$ we have*

$$\int |u_\mu(t, x + y) - u_\mu(t, x)| \, dx \le \omega(|y|), \quad \int |u_\mu(t, x)| dx \le \int |u_0(x)| \, dx.$$

Proof. This is the contraction property (3.3.4) applied to $u(t, x) = u_\mu(t, x)$ and $v(t, x) = u_\mu(t, x + y)$ or $v(t, x) = 0$.

We shall use the differential equation to estimate the continuity with respect to t. Set

$$w(x) = u_\mu(t+h, x) - u_\mu(t, x),$$

where $h > 0$. First we note that if $\varphi \in C^2(\mathbf{R}^n)$ and the derivatives are bounded at infinity, and if $f_i(0) = 0$, as we can assume without restriction of generality, then

$$\int w(x)\varphi(x)\, dx = \iint_t^{t+h} \partial u_\mu(s, x)/\partial s\, \varphi(x)\, dx\, ds$$

$$= \mu \iint_t^{t+h} u_\mu(s, x)\Delta\varphi\, dx\, ds + \sum_{i=1}^n \iint_t^{t+h} f_i(u_\mu(s, x))\partial\varphi/\partial x_i\, dx\, ds.$$

Hence

$$\left| \int w(x)\varphi(x)\, dx \right| \le h(\mu \sup |\Delta\varphi| + C' \sup |\varphi'|) \int |u_0(x)|\, dx.$$

if $|u_0| \le M$ and $|f(u)| \le C'|u|$ when $|u| \le M$. To estimate $\int |w(x)|\, dx$ we shall choose φ as a regularization of $w^s = \operatorname{sgn} w$,

$$\varphi(x) = \int w^s(x - \varepsilon y)\chi(y)\, dy,$$

where $0 \le \chi \in C_0^\infty$, $\int \chi\, dy = 1$, and $|y| \le 1$ in supp χ. Then $|\varphi'| \le C/\varepsilon$, $|\Delta\varphi| \le C/\varepsilon^2$, where C only depends on the dimension, and

$$\left| \int w(x)\varphi(x)\, dx - \int |w(x)|\, dx \right| = \left| \iint (w(x) - w(x - \varepsilon y))w^s(x - \varepsilon y)\chi(y)\, dy\, dx \right|.$$

When $|y| \le 1$ we have

$$\int |w(x) - w(x - \varepsilon y)|\, dx \le 2\omega(\varepsilon),$$

so it follows that

$$\left| \int w(x)\varphi(x)\, dx - \int |w(x)|\, dx \right| \le 2\omega(\varepsilon),$$

$$\int |w(x)|\, dx \le 2\omega(\varepsilon) + C(\mu/\varepsilon^2 + C'/\varepsilon)h \int |u_0(x)|\, dx.$$

We have now proved

Lemma 3.4.2. *For every $\mu > 0$ and $t, h > 0$, we have*

$$\int |u_\mu(t+h, x) - u_\mu(t, x)|\, dx \le \min_{\varepsilon > 0} \left(2\omega(\varepsilon) + C(\mu/\varepsilon^2 + C'/\varepsilon)h \int |u_0(x)|\, dx \right).$$

The right-hand side tends to 0 as $h \to 0$.

As in Section 2.5 we can now find a sequence $\mu_j \to 0$ such that $u_{\mu_j}(t, \cdot)$ converges in L^1_{loc} for every t, locally uniformly in t. The limit $u(t, \cdot)$ converges to u_0 in L^1_{loc} as $t \to 0$, and it satisfies (3.1.1) weakly since

$$u_{\mu_j} \to u, \quad f_i(u_{\mu_j}) \to f_i(u) \quad \text{in } L^1_{\text{loc}}.$$

The derivation of the entropy condition in Section 2.4 can be applied with no essential change. In fact, if we multiply the equation (3.3.1) for u_μ by $\Phi'(u_\mu)$, we obtain

$$\partial\Phi(u_\mu)/\partial t + \sum_i \partial Y_i(u_\mu)/\partial x_i = \mu\Delta\Phi(u_\mu) - \mu\Phi''(u_\mu)|\partial u_\mu/\partial x|^2,$$

where

$$Y_i'(s) = \Phi'(s)f_i'(s).$$

As in Section 2.4 we conclude when Φ is convex that

(3.4.1)
$$\partial\Phi(u)/\partial t + \sum_1^n \partial Y_i(u)/\partial x_i \leq 0.$$

(If u is a C^1 solution of (3.1.1), then this is equal to 0.) Although the derivation assumes that $\Phi \in C^2$ a passage to the limit shows that we can apply (3.4.1) to

$$\Phi(s) = |s - k|, \quad Y_i(s) = \int_k^s f_i'(\sigma)\,\mathrm{sgn}(\sigma - k)\,d\sigma = (f_i(s) - f_i(k))\,\mathrm{sgn}(s - k),$$

where $k \in \mathbf{R}$. With

$$f_i(s, k) = (f_i(s) - f_i(k))/(s - k)$$

defined as $f_i'(k)$ for $s = k$, we obtain by repeating the proof of (2.4.11), that if u and v are two locally L^∞ functions satisfying (3.4.1), then

(3.4.2)
$$\iint |u(s, y) - v(s, y)|(\partial\psi(s, y)/\partial s + \sum_1^n f_i(u(s, y), v(s, y))\partial\psi(s, y)/\partial y_i)\,ds\,dy \geq 0$$

if $0 \leq \psi \in C_0^\infty(\mathbf{R}_+ \times \mathbf{R}^n)$. Let $|u| \leq M$, $|v| \leq M$, and let

(3.4.3)
$$\sum_1^n f_i(s, t)^2 \leq A^2 \quad \text{when } |s| \leq M, |t| \leq M.$$

Then it follows as before that

(3.4.4)
$$\int_{|x - x_0| \leq R - At} |u(t, x) - v(t, x)|\,dx$$

is a decreasing function of t for arbitrary x_0 and R. In particular, this proves that all convergent subsequences of u_μ as $\mu \to 0$ have the same limit.

Theorem 3.4.3. *If $u_0 \in C_0^3$ then the Cauchy problem (3.3.1) has a unique classical solution u_μ which decreases rapidly as $x \to \infty$. When $\mu \to 0$ it converges in L_{loc}^1 for every fixed $t \geq 0$, locally uniformly in t, to a weak solution of the Cauchy problem (3.1.1), taking the boundary values in the topology of L_{loc}^1. It satisfies the entropy condition (3.4.1) for every convex Φ. For every $u_0 \in L^\infty$ there is a unique solution of the Cauchy problem which is a continuous function of t with values in L_{loc}^1 and satisfies the entropy condition (3.4.1). For such solutions (3.4.4) is always a decreasing function of t.*

Proof. Only the last statement remains to be proved. If $u_0 \in L^\infty$, $|u_0| \leq M$, then regularization and a cutoff far away gives a sequence $u_0^\nu \in C_0^\infty(\mathbf{R}^n)$ with $|u_0^\nu| \leq M$, converging to u_0 in L_{loc}^1. Let u^ν be the corresponding solution of the Cauchy problem (3.1.1). If we apply (3.4.4) to u^ν and $u^{\nu'}$ we see that $u^\nu(t, \cdot)$ is a Cauchy sequence in L_{loc}^1

for every t. The limit satisfies the entropy inequality which implies the uniqueness and extends the fact that (3.4.4) is decreasing.

We shall now examine the meaning of the entropy condition at a simple discontinuity on a C^1 surface S. Let $\nu = (\nu_0, \nu_1, \ldots, \nu_n)$ be a normal to S, and let $u_\pm$ be the boundary value of u on the side to which $\pm \nu$ is pointing. Then the entropy condition (3.4.1) with $\Phi(s) = |s - k|$ means, if we express (3.4.1) explicitly as in (2.4.10)' and apply Gauss' formula on each side of S, that

$$(3.4.5) \qquad |u_+ - k|(\nu_0 + \sum_1^n f_i(u_+, k)\nu_i) \le |u_- - k|(\nu_0 + \sum_1^n f_i(u_-, k)\nu_i)$$

for arbitrary $k \in \mathbf{R}$. If we write

$$|u_\pm - k| f_i(u_\pm, k) = (f_i(u_\pm) - f_i(k)) \operatorname{sgn}(u_\pm - k)$$

we see that for k outside $[u_-, u_+]$ (3.4.5) reduces to

$$\pm((u_+ - u_-)\nu_0 + \sum_1^n (f_i(u_+) - f_i(u_-))\nu_i) \le 0$$

which is equivalent to the Rankine-Hugoniot condition

$$(3.4.6) \qquad [u]\nu_0 + \sum_1^n [f_i(u)]\nu_i = 0$$

with the usual notation $[\cdot]$ for the jump. If $u_- < k < u_+$ then (3.4.5) becomes

$$(u_+ + u_- - 2k)\nu_0 \le \sum_1^n (2f_i(k) - f_i(u_+) - f_i(u_-))\nu_i.$$

On the left we have a linear function of k which by the Rankine-Hugoniot condition is equal to the right-hand side for $k = u_\pm$. Hence the condition means that the function

$$(3.4.7) \qquad [u_-, u_+] \ni k \mapsto \sum_1^n f_i(k)\nu_i$$

is at least equal to the linear interpolation between the end values. If $u_+ < k < u_-$ it must lie below this line. When $n = 1$ and f_1 is strictly convex we therefore recover the condition that the jumps must be ≤ 0.

Theorem 3.4.4. *Assume that u satisfies the entropy condition (3.4.1) when Φ is convex, and that u is in C^1 on each side of a C^1 surface S with normal $\nu = (\nu_0, \nu_1, \ldots, \nu_n)$ where the boundary values are $u_\pm$ in the direction $\pm \nu$. Then*

   (i) *the Rankine-Hugoniot condition (3.4.6) holds;*
   (ii) *the function*

$$[u_-, u_+] \ni k \mapsto \operatorname{sgn}(u_+ - u_-) \sum_1^n f_i(k)\nu_i$$

lies above the linear interpolation between its values at $u_\pm$.

The "entropy condition" in (ii) was introduced by Olejnik [2] when $n = 1$ and was given by Vol'pert [1] and Kružkov [1] in general. In Theorem 3.4.4 we have assumed the entropy condition (3.4.1) for *all* convex Φ. However, if $\sum_1^n f_i(u)\nu_i$ is known to be strictly convex or concave it suffices to assume (3.4.1) for *one* strictly convex Φ in addition to the Rankine-Hugoniot condition. In fact, then we obtain instead of (3.4.5)

$$\nu_0[\Phi(u)] + \Big[\sum_1^n Y_i(u)\nu_i \Big] \leq 0,$$

that is, by the Rankine-Hugoniot condition,

$$[\Phi(u)]\Big[\sum_1^n f_i(u)\nu_i \Big]/[u] \geq \Big[\sum_1^n Y_i(u)\nu_i \Big].$$

Choose the orientation of ν so that $u_- < u_+$ and let $g(u) = \sum_1^n f_i'(u)\nu_i$, which is then strictly monotonic while $h(u) = \Phi'(u)$ is strictly increasing. Then the preceding condition means that

$$\int_{u_-}^{u_+} h(u)\, du \int_{u_-}^{u_+} g(u)\, du \geq (u_+ - u_-) \int_{u_-}^{u_+} h(u)g(u)\, du,$$

that is,

$$\int_{u_-}^{u_+} \int_{u_-}^{u_+} (h(u) - h(v))(g(u) - g(v))\, du\, dv \leq 0.$$

Thus g is decreasing so $\sum_1^n f_i(u)\nu_i$ is concave in $[u_-, u_+]$.

Even when $n = 1$ the solution obtained when f is not convex may have much more complicated singularities than those of Burgers' equation discussed in Section 2.4. One reason is that *contact discontinuities* occur where the shock speed is equal to the characteristic speed at one side or the other of the shock. This does not give the smoothing effect which we observed in the strictly convex case. For a detailed study of the case where f'' has just one change of sign we refer to Dafermos [2].

CHAPTER IV

FIRST ORDER SYSTEMS OF CONSERVATION LAWS
WITH ONE SPACE VARIABLE

4.1. Introduction. We shall begin in Sections 4.2 and 4.3 by a discussion of classical solutions of the Cauchy problem for a first order quasilinear system, not necessarily of conservation form. The purpose is to determine the lifespan when the data are small. This is done by showing that the equations decouple approximately into scalar equations. In Section 4.4 we then discuss the solution of the Riemann problem, for which the Cauchy data are constant on each half axis. This prepares for the proof in Section 4.5 by Glimm's method that the Cauchy problem has a global weak solution for arbitrary data of small total variation. It satisfies entropy conditions when suitable entropy pairs can be constructed, which is not always the case as we shall see in Section 4.6. However, the uniqueness of the solutions obtained with Glimm's method has been proved only recently and with methods which do not rely on entropy conditions but on refinements of the estimates in Sections 4.2 and 4.5. They are beyond the scope of this chapter.

4.2. Generalities on first order systems. We shall now discuss the Cauchy problem for the hyperbolic system

$$(4.2.1) \qquad \partial u/\partial t + a(u)\partial u/\partial x = 0,$$

where $u = (u_1, u_2, \ldots, u_N)$ and $a(u) = (a_{jk}(u))_{j,k=1}^N$ is an $N \times N$ matrix, with C^∞ entries for the sake of simplicity. We assume that $a(0)$ has real distinct eigenvalues. Note that if a is independent of u then (4.2.1) is satisfied by

$$u(t, x) = \sum_1^N \phi_j(x - \lambda_j t) r_j$$

if r_j are the eigenvectors of a, with eigenvalues λ_j. We have $u(0, x) = u_0(x)$ if

$$u_0(x) = \sum_1^N \phi_j(x) r_j$$

that is, if ϕ_j is the component of u_0 along r_j. In this linear case we therefore have an extremely elementary solution of the Cauchy problem.

In the nonlinear case it follows from our hypothesis that for all u in a neighborhood of 0 the matrix $a(u)$ has real eigenvalues

$$\lambda_1(u) < \lambda_2(u) < \cdots < \lambda_N(u)$$

and corresponding eigenvectors $r_1(u)$, $r_2(u)$, $\ldots$, $r_N(u)$ depending smoothly on u. The equation (4.2.1) makes perfect sense if u takes its values in a manifold M of dimension N and a is a section of $\text{Hom}(TM, TM)$, that is, $a(u)$ is a linear map in the tangent space $T_u M$ which depends smoothly on u. This proves the invariance of the eigenvector directions and is useful to keep in mind later on when we change the dependent variables u. Recall that this is only legitimate when we consider C^1 solutions (see the beginning of Section 2.4). However, for the time being we shall assume that u takes its values in $\mathbf{R}^N$.

The terms in the decomposition of the solutions in the linear case given above have an analogue in the nonlinear case:

Definition 4.2.1. A C^1 solution of (4.2.1) is called *simple* if the range belongs to a C^1 curve.

This definition and much of the following discussion comes from John [1], but we shall also follow Schatzman [1] in part. The definition means that if $s \mapsto U(s)$ is a parametrization of the curve, then $u(t, x) = U(s(t, x))$ where $s \in C^1$. The equation (4.2.1) can now be written

$$\partial s/\partial t U' + \partial s/\partial x \, a(U)U' = 0.$$

Thus U' is an eigenvector of $a(U)$, so for some j we have $a(U)U' = \lambda_j(U)U'$ and

$$(4.2.2) \qquad \partial s/\partial t + \lambda_j(U(s))\partial s/\partial x = 0,$$

which is a scalar equation of the form discussed in Chapter II. Conversely, let $s \mapsto U(s)$ be an integral curve of the eigenvector field $r_j(u)$, thus $U'(s) = r_j(U(s))$, and let s satisfy (4.2.2). Then $u(t, x) = U(s(t, x))$ satisfies (4.2.1), and we shall say that u is *j-simple*. We can find a j-simple solution in a neighborhood of any point with u prescribed there.

Let u be a C^2 solution of (4.2.1) taking values in $\mathbf{R}^N$, and let $w = \partial u/\partial x$, thus $\partial u/\partial t = -a(u)w$. If f is any C^1 function in $\mathbf{R}^N$ then the derivatives

$$\partial f(u)/\partial x = f'(u)\partial u/\partial x = f'(u)w, \quad \partial f(u)/\partial t = -f'(u)a(u)w$$

are linear functions of w. Thus $\partial a(u)/\partial x$ is a linear function of w, so differentiation of (4.2.1) with respect to x gives

$$(4.2.3) \qquad \partial w/\partial t + a(u)\partial w/\partial x = \gamma(u, w)$$

where γ is a quadratic form in w.

(4.2.3) is more useful if one splits w into its components along the eigenvectors, so we write

$$(4.2.4) \qquad \partial u(t, x)/\partial x = w(t, x) = \sum_i w_i(t, x)r_i(u(t, x)).$$

Since $ar_i = \lambda_i r_i$ and the derivatives of $r_i(u)$ are linear in w, we obtain from (4.2.3)

$$(4.2.3)' \qquad L_i w_i = \sum_{j,k} \gamma_{ijk}(u)w_j w_k, \quad i = 1, \ldots, N,$$

where

$$(4.2.5) \qquad L_i = \partial/\partial t + \lambda_i(u(t, x))\partial/\partial x$$

denotes differentiation with respect to t along the ith characteristic for the solution u, and γ_{ijk} are smooth functions of u.

When $N = 1$ we have $w_i(dx - \lambda_i(u)dt) = du_i$. To derive an analogue for general N we calculate the differential

$$d(w_i(dx - \lambda_i(u)dt)) = (\partial w_i/\partial t + \lambda_i(u)\partial w_i/\partial x + w_i\partial \lambda_i(u)/\partial x)dt \wedge dx$$

$$(4.2.6) \qquad = (\sum_{j,k} \gamma_{ijk}(u)w_j w_k + w_i\langle \partial\lambda_i(u)/\partial u, \sum_k w_k r_k(u)\rangle)dt \wedge dx$$

$$= \sum_{j,k} \Gamma_{ijk}(u)w_j w_k dt \wedge dx,$$

where the last equality defines Γ_{ijk}; $\Gamma_{ijk} = \Gamma_{ikj}$. In particular, (4.2.6) holds for any j simple solution u. Then we have $w_i = 0$ for $i \neq j$, and $w_j = \partial s/\partial x$, $-\lambda_j(u)w_j = \partial s/\partial t$ by (4.2.2), so $w_j(dx - \lambda_j(u)dt) = ds$ is exact, with the notation above. Hence

$$(4.2.7) \qquad\qquad \Gamma_{ijj} = 0 \quad \text{for } i, j = 1, \ldots, N.$$

By (4.2.6) this means that

$$(4.2.8) \qquad\qquad \gamma_{ijj} = -\delta_{ij}\langle \partial\lambda_i(u)/\partial u, r_i(u)\rangle,$$

where the Kronecker symbol δ_{ij} is the unit matrix. When using (4.2.6) to estimate $|w_i|$ it is useful to note that $|w_i|$ is Lipschitz continuous, hence differentiable almost everywhere with differential sgn $w_i\, dw_i$. Thus we can write (4.2.6) in the form

$$(4.2.6)' \qquad d(|w_i|(dx - \lambda_i(u)dt)) = \text{sgn } w_i \sum_{j,k} \Gamma_{ijk}(u)w_j w_k dt \wedge dx.$$

Stokes' formula is applicable to Lipschitz continuous forms.

John [1] proved $(4.2.3)'$, (4.2.8) by explicit computations. Here we have instead proved (4.2.6), (4.2.7) with arguments intended to show why there must exist such formulas. However, as pointed out by Hans Lindblad, the result can also be obtained rather quickly as follows. Let $l_1, \ldots, l_n$ be the differential forms on M biorthogonal to the eigenvectors $r_1, \ldots, r_n$. (These covectors are eigenvectors of the transpose of a.) The pullback of l_i by the map $(t, x) \mapsto u(t, x)$ is equal to

$$\langle l_i, wdx - a(u)wdt\rangle = w_i(dx - \lambda_i(u)dt),$$

for $du = wdx - a(u)wdt$. We write $l_i = \sum_j l_{ij}du_j$ and denote by e_k the kth basis vector in $\mathbf{R}^N$. Then

$$dl_i = \sum_{j<k}(\partial l_{ij}/\partial u_k - \partial l_{ik}/\partial u_j)du_k \wedge du_j$$

and the pullback of du_k is

$$\langle e_k, wdx - a(u)wdt\rangle = \sum_\nu r_{\nu k}w_\nu(dx - \lambda_\nu(u)dt), \quad r_{\nu k} = \langle e_k, r_\nu\rangle.$$

Since the differential of the pullback of l_i is the pullback of dl_i, we obtain

$$d(w_i(dx - \lambda_i(u)dt)) = \sum_{j<k}(\partial l_{ij}/\partial u_k - \partial l_{ik}/\partial u_j)\sum_{\nu,\mu}(\lambda_\mu - \lambda_\nu)r_{\nu k}r_{\mu j}w_\nu w_\mu\, dt \wedge dx,$$

which means that (4.2.6) is valid with

$$(4.2.7)' \qquad \Gamma_{i\nu\mu} = (\lambda_\mu - \lambda_\nu)\sum_{j<k}(\partial l_{ij}/\partial u_k - \partial l_{ik}/\partial u_j)r_{\nu k}r_{\mu j}.$$

This implies (4.2.7). For later reference we note that if $\Gamma_{i\nu\mu} = 0$ for all i, ν, μ at a point u_0 then the sum on the right vanishes for all ν, μ for this follows by antisymmetry if $\nu = \mu$. If we multiply it by $l_{\nu a}l_{\mu b}$ and sum over ν, μ it follows that $dl_i = 0$ at u_0 for every i. Now one can always replace r_i and l_i by r_i/c_i and $l_i c_i$ for arbitrary smooth $c_i \neq 0$. If this makes all $\Gamma_{ijk} = 0$ then $c_i dl_i = l_i \wedge dc_i$. Thus we see that we can make all Γ_{ijk} vanish at u_0 by appropriate choice of c_i if and only if $l_i \wedge dl_i = 0$ at u_0 for every i.

As a first application of $(4.2.6)'$ we prove a uniqueness theorem:

Theorem 4.2.2. *If $u \in C^2$ satisfies (4.2.1) for $0 \leq t < T$ and $u(0, x) = 0$ for $\alpha \leq x \leq \beta$, then $u(t, x) = 0$ when $0 \leq t < T$ and $\alpha + \lambda_N(0)t \leq x \leq \beta + \lambda_1(0)t$.*

Proof. Let D_s be the set of all (t, x) with $0 \leq t \leq s$ to the right of the integral curve of L_N through $(0, \alpha)$ and to the left of that of L_1 through $(0, \beta)$. This is well defined for small s. Set

$$Y(s) = \sum_i \iint_{D_s} |w_i(t, x)| \, dt \, dx.$$

Then $Y(0) = 0$ and if $\tau_s = \{x; (s, x) \in \partial D_s\}$ we have

$$
\begin{aligned}
(4.2.9) \qquad 0 \leq Y'(s) &= \sum_i \int_{\tau_s} |w_i(s, x)| dx \\
&\leq \sum_i \int_{\partial D_s} |w_i(t, x)| (dx - \lambda_i(u)dt) \leq CY(s).
\end{aligned}
$$

Here we have used first that $dx - \lambda_i(u)dt \geq 0$ on the characteristic boundaries of D_s and that the Cauchy data are 0, and then that $(4.2.6)'$ holds. Any constant C will do for small s. But this proves that $Y(s) = 0$ for small s, hence that u is constant there. In particular the characteristic boundaries are defined by $x = \alpha + \lambda_N(0)t$ and $x = \beta + \lambda_1(0)t$. By the standard "continuous induction argument" the theorem follows.

Note in particular, that for a solution with Cauchy data of compact support the support is always bounded when t is bounded. Next to the boundary of the support the solution is always simple:

Theorem 4.2.3. *Let u be a C^2 solution of (4.2.1) when $0 \leq t < T$, and assume that $\partial u / \partial x$ is 0 or at least is proportional to $r_N(u)$ in $\{(0, x); \alpha \leq x \leq \beta\}$. Then it follows that u is N-simple in the set D bounded by the lines $t = 0$, $t = T$, and the orbits of L_{N-1} through $(0, \alpha)$ and of L_1 through $(0, \beta)$ provided that $a(u(t, x))$ has distinct eigenvalues for all $(t, x) \in D$.*

Proof. Let D_s be the part of D where $t \leq s$. We modify the definition of $Y(s)$ in the proof of Theorem 4.2.3 by summing only for $i < N$. Then we still have (4.2.9) for suitable C and small s, because $dx - \lambda_i(u)dt \geq 0$ on the characteristic boundaries for $i \neq N$, and the sum in $(4.2.6)'$ contains no term involving w_N^2. Hence we obtain as before that $Y(s) = 0$, which proves the theorem.

In the scalar case we saw that the total variation with respect to x of a solution is a decreasing function of t. For a solution u of the system (4.2.1) we shall therefore examine a similar quantity,

$$(4.2.10) \qquad \mathcal{L}(t) = \sum_i \int |w_i(t, x)| \, dx.$$

For the sake of simplicity we assume that $u(0, x)$ has compact support, hence that $u(t, x)$ vanishes for large $|x|$ (Theorem 4.2.2). By $(4.2.6)'$

$$\mathcal{L}(t_2) - \mathcal{L}(t_1) = \sum_{i,j,k} \iint_{t_1}^{t_2} \operatorname{sgn} w_i \, \Gamma_{ijk}(u) w_j w_k dt dx,$$

which means that for almost all t

$$(4.2.11) \qquad d\mathcal{L}(t)/dt = \sum_{i,j,k} \int \operatorname{sgn} w_i \, \Gamma_{ijk}(u) w_j w_k \, dx.$$

There is no reason why this should be ≤ 0 in general. In fact, for fixed t the functions $w_i(x) = w_i(t, x)$ are arbitrary functions in C_0^∞ such that the differential equation

$$v'(x) = \sum_i w_i(x) r_i(v(x))$$

has a solution vanishing for large $|x|$ (see Theorem 4.2.5 below). Replacing w_i by εW_i where $W_i \in C_0^\infty(\mathbf{R})$ and v by εV, we obtain the equation

$$V' = \sum_i W_i r_i(\varepsilon V).$$

Choose $\chi \in C_0^\infty$ with $\int \chi \, dx = 1$, and let $V(x, a, \varepsilon)$, $a \in \mathbf{R}^N$, be the solution of

$$\partial V/\partial x = \sum_i (W_i - a_i \chi) r_i(\varepsilon V)$$

which is 0 for $x \ll 0$. This is well defined for small ε, and

$$V(x, a, 0) = \sum_i \left(\int_{-\infty}^\infty W_i \, dx - a_i \right) r_i(0)$$

when $x \gg 0$. Hence, by the implicit function theorem, there is a solution V of compact support if $a = a(\varepsilon)$ where a is in C^1 and $a_i(0) = \int W_i \, dx$. If $\partial \mathcal{L}/\partial t$ is always ≤ 0, it follows that

$$\sum_{i,j,k} \int \operatorname{sgn} W_i \, \Gamma_{ijk}(0) W_j W_k \, dx \leq 0$$

for all $W_i \in C_0^\infty$ with $\int W_i \, dx = 0$. The last condition is superfluous, for we can always achieve it by adding to each W_i a suitable translation with a change of sign so that the added terms contribute nothing, because they have disjoint supports and $\Gamma_{ijj} = 0$. Letting each W_j approximate a constant on the same interval we conclude that

$$\sum_{i,j,k} \operatorname{sgn} b_i \Gamma_{ijk}(0) b_j b_k \leq 0$$

for all $b_j \in \mathbf{R} \backslash 0$. Looking at even terms of degree 1 with respect to b_i we obtain $\Gamma_{iik}(0) = 0$, and looking at odd terms of degree 0 with respect to b_i we obtain $\Gamma_{ijk}(0) = 0$ if $j \neq i, k \neq i$. Altogether we obtain that $\Gamma_{ijk}(0) = 0$ at 0 for all i, j, k. We have shown above that this can be achieved by a change of normalization of the eigenvectors if and only if $l_i \wedge dl_i = 0$ at 0 for all i, where l_i are the biorthogonal one forms. When $N = 2$ this is no restriction, but when $N \geq 3$ the condition is not fulfilled for example by the one form $du_1 + u_2 du_3$ since the differential is $du_2 \wedge du_3$.

An idea of Glimm [1] adapted by Schatzman [1] to the situation at hand allows one to compensate for the fact that $\mathcal{L}$ may not decrease by considering also the function

(4.2.12) $$Q(t) = \sum_{i>j} \iint_{x<y} |w_i(t, x) w_j(t, y)| dx dy,$$

called "potential for future interaction". Note that $Q(t) \leq \mathcal{L}(t)^2/2$. The reason for the term is that waves with $i > j$ travel faster so those at x have a chance of catching up with those at y if $x < y$. In Sections 4.4 and 4.5 we shall return to this quantity in the original

situation of Glimm, for weak solutions with jump discontinuities. Here we note that by
$(4.2.6)'$

$$\int_{x<y} |w_i(t_2, x)|\, dx - \int_{x<y} |w_i(t_1, x)|\, dx + \int_{t_1}^{t_2} \lambda_i(u(t, y))|w_i(t, y)|\, dt$$
$$= \int_{-\infty}^{y} \int_{t_1}^{t_2} \text{sgn } w_i(t, x) \sum_{\nu, \mu} \Gamma_{i\nu\mu}(u(t, x))w_\nu(t, x)w_\mu(t, x)\, dt\, dx.$$

Using also the analogue for the integral when $x > y$, we obtain for almost all t

$$Q'(t) = \sum_{i>j} \int (\lambda_j(u(t, x)) - \lambda_i(u(t, x))|w_i(t, x)w_j(t, x)|\, dx$$
$$+ \sum_{i>j} \iint_{x<y} \Big(\text{sgn } w_i(t, x) \sum_{\nu, \mu} \Gamma_{i\nu\mu}(u(t, x))w_\nu(t, x)w_\mu(t, x)|w_j(t, y)|$$
$$+ |w_i(t, x)| \text{sgn } w_j(t, y) \sum_{\nu, \mu} \Gamma_{j\nu\mu}(u(t, y))w_\nu(t, y)w_\mu(t, y)\Big)\, dx\, dy.$$

Assume that

$$\lambda_i(u) - \lambda_j(u) \geq c > 0 \quad \text{if } i > j \text{ and } |u| \leq \delta,$$

and that $u(t, x)$ is a C^2 solution of (4.2.1) for $0 \leq t < T$ vanishing for large x. Set

$$(4.2.13) \qquad\qquad \Lambda(t) = \sum_{i>j} \int |w_i(t, x)w_j(t, x)|\, dx.$$

Since $\Gamma_{i\nu\nu} = 0$ it follows for some constant C that if $|u(t, \cdot)| \leq \delta$ then

$$(4.2.14) \qquad\qquad Q'(t) \leq -c\Lambda(t) + C\Lambda(t)\mathcal{L}(t), \quad \mathcal{L}'(t) \leq C\Lambda(t).$$

Lemma 4.2.4. *If $\mathcal{L}(0) < c/3C$ and $\mathcal{L}(0)/\delta$ is small enough, then*

$$\mathcal{L}(t) \leq \mathcal{L}(0) + 3C\mathcal{L}(0)^2/2c \leq 2\mathcal{L}(0), \quad \sup |u(t, \cdot)| \leq 2C'\mathcal{L}(0) < \delta, \quad \text{if } 0 \leq t < T,$$
$$(4.2.15)$$
$$c \int_0^T \Lambda(t)\, dt \leq 3\mathcal{L}(0)^2/2.$$

If $\gamma : t \mapsto \gamma(t)$, $0 \leq t \leq T$, is an orbit of L_j, then

$$(4.2.16) \qquad\qquad c \int_\gamma |w_i(t, x)|\, dt \leq 4\mathcal{L}(0), \quad i \neq j.$$

Proof. Using (4.2.14) we obtain for $0 \leq t < t_1$ if $|u(t, \cdot)| \leq \delta$ then

$$(3CQ + c\mathcal{L})' \leq (3C^2\mathcal{L} - 2cC)\Lambda.$$

Since u vanishes for large x, we have $|u(t, x)| \leq C'\mathcal{L}(t)$. If $\mathcal{L}(t) < \delta/C'$ and $\mathcal{L}(t) \leq 2c/3C$ for $0 \leq t < t_1$, it follows for $0 \leq t < t_1$ that

$$3CQ(t) + c\mathcal{L}(t) \leq 3CQ(0) + c\mathcal{L}(0) \leq 3C\mathcal{L}(0)^2/2 + c\mathcal{L}(0) \leq 2c\mathcal{L}(0),$$

hence $\mathcal{L}(t) \le 2\mathcal{L}(0)$. Hence it follows by "continuous induction" that the first part of (4.2.15) is valid for $0 \le t < T$ if $\mathcal{L}(0) < \min(c/3C, \delta/2C')$. Now

$$Q'(t) \le \Lambda(t)(2\mathcal{L}(0)C - c) \le -\Lambda(t)c/3,$$

which proves that

$$c \int_0^T \Lambda(t)\,dt \le 3Q(0) \le 3\mathcal{L}(0)^2/2.$$

Let D be the part of the strip with $0 < t < T$ to the left of γ. By (4.2.6)'

$$\int_{\partial D} |w_i|(dx - \lambda_i(u)dt) = \int_D \operatorname{sgn} w_i \sum \Gamma_{ijk}(u)w_j w_k \, dt \, dx.$$

On γ we have

$$|dx - \lambda_i(u)dt| = |(\lambda_j(u) - \lambda_i(u))dt| \ge c|dt|, \quad i \ne j,$$

so we obtain using (4.2.15)

$$c \int_\gamma |w_i(t,x)|\,dt \le \mathcal{L}(0) + \mathcal{L}(T) + C \int_0^T \Lambda(t)\,dt \le 3\mathcal{L}(0) + 3C\mathcal{L}(0)^2/2c \le 4\mathcal{L}(0)$$

which completes the proof.

Lemma 4.2.4 will be essential when we examine the lifespan of classical solutions in Section 4.3. We shall end this section with another preparation for that, a proof of a local existence theorem for classical solutions of the Cauchy problem with small data u_0. First assume that a classical solution is already known for $0 \le t < T$, say. If a small bound for u is known and

$$M_1(t) = \sup_{j,x} |w_j(t,x)|$$

it follows from (4.2.3)', integrated along the orbits of L_j, that

$$M_1(t) \le M_1(0) + C_1 \int_0^t M_1(s)^2 \, ds.$$

Thus $M_1(t) < Y(t)$ if $Y(0) = M_1(0)$ and $Y'(t) > C_1 Y(t)^2$, for

$$M_1(t) - Y(t) < C_1 \int_0^t (M_1(s)^2 - Y(s)^2)\,ds,$$

so we may conclude by "continuous induction" that $M_1(t) < Y(t)$ in $[0,T)$. Hence

$$M_1(t) \le M_1(0)/(1 - C_1 t M_1(0)), \quad 0 \le t < 1/C_1 M_1(0).$$

When $2C_1 M_1(0)t < 1$, we obtain $M_1(t) \le 2M_1(0)$, $\int_0^t M_1(s)\,ds \le 1/C_1$.

To estimate u itself we write

$$u = \sum_j u_j r_j(u)$$

and note that

$$a(u)\partial u/\partial x = \sum_j \partial u_j/\partial x\,\lambda_j(u)r_j(u) + \sum_j u_j a(u)\partial r_j(u)/\partial u\,w,$$

$$\partial u/\partial t = \sum_j \partial u_j/\partial t\,r_j(u) - \sum_j u_j \langle \partial r_j(u)/\partial u,\, aw\rangle.$$

Hence the equation (4.2.1) gives if we take the component along r_j that

$$|L_j u_j| \leq C|u||w|.$$

Set

$$M_0(t) = \sup_{j,x} |u_j(t, x)|.$$

Then

$$M_0(t) \leq C_0 \int_0^t M_0(s) M_1(s)\, ds + M_0(0) = I(t),$$

which implies that $I'(t) \leq C_0 M_1(t) I(t)$, $I(0) = M_0(0)$, hence

$$M_0(t) \leq M_0(0) \exp \left(\int_0^t C_0 M_1(s)\, ds \right) \leq M_0(0) e^{C_0/C_1},$$

when $2C_1 M_1(0)t < 1$. For small $M_0(0)$ this shows that u remains small, which was tacitly assumed in the estimates above.

Before estimating derivatives of higher order we recall some simple calculus inequalities. First we prove that if $v \in C^2(\mathbf{R})$ and the derivatives are bounded, then

(4.2.17) $$\sup |v'|^2 \leq 2 \sup |v| \sup |v''|.$$

By a change of scales of v and the independent variable x it suffices to prove that $|v'(0)|^2 \leq 2$ if $|v| \leq 1$, $|v''| \leq 1$ and $v(0) \geq 0$. By Taylor's formula we have for all $x \in \mathbf{R}$

$$1 \geq v(x) \geq v'(0)x - x^2/2,$$

which gives $|v'(0)|^2 \leq 2$ when $x = v'(0)$, as claimed. If $v \in C^k(\mathbf{R})$ and the derivatives are bounded, we obtain if (4.2.17) is applied to the derivatives of v that the sequence $2^{j^2/2} \sup |v^{(j)}|$ is logarithmically convex for $0 \leq j \leq k$, so we have

(4.2.17)' $$\sup |v^{(j)}| \leq 2^{j(k-j)/2} \sup |v^{(k)}|^{j/k} \sup |v|^{(k-j)/k}, \quad 0 < j < k.$$

If $F \in C^k(\mathbf{R})$ and $F(0) = 0$, it follows that

$$|d^k F(v)/dt^k| \leq C \sup |v^{(k)}|$$

where C only depends on F and an upper bound for v. In fact, $d^k F(v)/dt^k$ is a sum with bounded coefficients of products of derivatives of v with orders adding up to k, so each term has such a bound by (4.2.17)'. In Section 6.4 we shall discuss estimates of the preceding kind in greater generality, using L^p norms for a general p.

We are now ready to prove that also for $k > 1$

(4.2.18) $$\sup \sum_{|\alpha|=k} |\partial^\alpha u(t, x)| \leq C_k \sup |u_0^{(k)}|, \quad \text{if } 2C_1 M_1(0)t < 1.$$

Using the equation (4.2.1) and the interpolation inequalities above it is easy to see that it suffices to prove this with only derivatives with respect to x in the left-hand side. If we apply a partial differentiation ∂_x^k to (4.2.1) we obtain

$$(\partial_t + a(u)\partial_x)\partial_x^k u = f_k$$

where f_k only contains derivatives of u of order $\leq k$, the sum of the order of derivatives of u in the factors is $k+1$ in every term, and each term contains the product of at least two components of x derivatives of u. Set

$$\partial_x^k u = \sum_i w_i^k r_i,$$

and introduce

$$M_k(t) = \sup_{i,x} |w_i^k(t,x)|,$$

which is equivalent to $\sup |\partial_x^k u(t,x)|$. Since

$$\sum_i (L_i w_i^k) r_i = f_k - \sum_i w_i^k (\partial_t + a(u)\partial_x) r_i,$$

we have $L_i w_i^k = g_{ik}$ where g_{ik} has the properties stated above for f_k. Applying the interpolation inequalities (4.2.17)' to $\partial u/\partial x$, we can estimate every term in g_{ik} by a constant times $M_1^a M_k^b$ where $a+b \geq 2$ and $a + kb = k+1$ since homogeneity in u and ∂_x is preserved during this process. Hence $a + kb + 1 \leq k + a + b$, so $b \leq 1$. Now an application of (4.2.17)' to u allows us to estimate M_1^k by $M_0^{k-1} M_k$. Writing $M_1^a M_k^b = M_1 M_1^{k(1-b)} M_k^b$ and recalling that we have a bound for M_0 when $2C_1 M_1(0)t < 1$, we conclude that g_{ik} can then be estimated by a constant times $M_1 M_k$. Now an integration along the orbits of L_i gives

$$M_k(t) \leq C_k \int_0^t M_1(s) M_k(s)\, ds + M_k(0).$$

As in the case of M_0 we conclude from this inequality that

$$M_k(t) \leq M_k(0) e^{C_k/C_1},$$

which completes the proof of (4.2.18).

The preceding method for estimating a hypothetical solution easily gives an existence proof for given Cauchy data $u_0 \in C^\infty$ with all derivatives bounded. Indeed, put $u^0(t,x) = u_0(x)$ and define successively u^ν for $\nu \geq 1$ by solving the linear Cauchy problem

$$\partial u^\nu/\partial t + a(u^{\nu-1})\partial u^\nu/\partial x = 0, \quad u^\nu(0,x) = u_0(x).$$

Since the derivatives of $a(u)$ only contributed to the right-hand side of our estimates above we conclude inductively that the estimates of M_1 and of M_0 given for a hypothetical solution remain valid for all u^ν. The estimates (4.2.18) also hold uniformly. What remains is just to prove that u^ν converges in the maximum norm. Set $v = u^\nu - u^{\nu-1}$ and $V = u^{\nu+1} - u^\nu$. Then $v = V = 0$ when $t = 0$, and from the equations

$$\partial u^{\nu+1}/\partial t + a(u^\nu)\partial u^{\nu+1}/\partial x = 0, \quad \partial u^\nu/\partial t + a(u^{\nu-1})\partial u^\nu/\partial x = 0$$

it follows that

$$\partial V/\partial t + a(u^\nu)\partial V/\partial x = (a(u^{\nu-1}) - a(u^\nu))\partial u^\nu/\partial x.$$

The right-hand side can be estimated by a constant times v. As in the estimate of M_0 above we obtain that if $|v(t,x)| \leq m(t)$ when $2C_1 M_1(0)t < 1$, then

$$M(t) = \sup_x |V(t,x)|$$

satisfies in the same interval

$$M(t) \leq \int_0^t (C_0 M(s) + C m(s)) M_1(s) \, ds.$$

Thus M is bounded by the solution of the Cauchy problem

$$Y'(t) = (C_0 Y(t) + C m(t)) M_1(t), \quad Y(0) = 0.$$

Since $Y(t)$ is bounded by a constant times $\int_0^t m(s) \, ds$, it follows that

$$M(t) \leq C' \int_0^t m(s) \, ds.$$

Inductively we conclude that

$$|u^{\nu+1} - u^\nu| \leq C''^\nu t^{\nu-1}/(\nu - 1)!$$

which proves the uniform convergence of the sequence u^ν, hence of all its derivatives. Thus we have proved that the Cauchy problem for (4.2.1) has a solution with given Cauchy data u_0. The uniqueness follows from the preceding argument by an obvious modification. Extending the *a priori* estimates at the beginning of the discussion to an estimate of the modulus of continuity, we find that to get $u \in C^k$ it suffices to have $u_0 \in C^k$, with a uniform estimate of the derivatives of order $\leq k$ and of the modulus of continuity of the kth order derivative. Thus we have proved

Theorem 4.2.5. *For any $u_0 \in C^k$, $k \geq 1$, with bounded derivatives of order $\leq k$ and those of order ≤ 1 sufficiently small, the equation (4.2.1) has a C^k solution with $u(0, x) = u_0(x)$ for $0 \leq t < T$ provided that*

$$T \sup |u_0'| \leq c,$$

where c is a constant depending only on a; we have $\sup |u| \leq C \sup |u_0|$.

In the preceding result we may allow u_0 and u to depend on a parameter ε (or several parameters). The solution is in C^k as a function of (t, x, ε) if the data are. In fact, to obtain *a priori* estimates we just differentiate (4.2.1) with respect to ε which gives

$$\partial u_\varepsilon' / \partial t + a(u) \partial u_\varepsilon' / \partial x + a'(u) u_\varepsilon' w = 0.$$

Writing

$$u_\varepsilon' = \sum_j \psi_j r_j(u)$$

we obtain equations of the form

$$L_j \psi_j + \sum_k c_{jk} \psi_k = 0$$

where c_{jk} denotes a smooth function of u and w. Thus we obtain estimates of u_ε' in terms of C^1 bounds for the data, and by repeated differentiation we obtain C^k estimates. It is clear that these a priori estimates can also be used in the existence proofs above to show that the solution constructed becomes a C^k function of (x, ε).

4.3. The lifespan of classical solutions. In Section 2.3 we observed that the lifespan T of the classical solution of the Cauchy problem with one unknown

$$\partial u/\partial t + au\partial u/\partial x = 0, \quad u(0,x) = u_0(x),$$

with constant a is given by $1/T = \sup -au_0'$ if $u_0 \in C_0^1(\mathbf{R})$. The proof shows easily that more generally the lifespan T_ε of the classical solution of the Cauchy problem

$$\partial u/\partial t + a(u)\partial u/\partial x = 0, \quad u(0,x) = \varepsilon u_0(x),$$

is given asymptotically by

$$(4.3.1) \qquad\qquad \lim_{\varepsilon \to 0} 1/(\varepsilon T_\varepsilon) = \sup(-a'(0)u_0').$$

If we pose the Cauchy problem with small data in the form

$$(4.3.2) \qquad\qquad \partial u/\partial t + a(u)\partial u/\partial x = 0, \quad u(0,x) = u_0(\varepsilon, x),$$

where $u_0 \in C_0^2(\mathbf{R}^2)$ and $u_0(0, \cdot) = 0$, then

$$(4.3.1)' \qquad\qquad \lim_{\varepsilon \to 0} 1/(\varepsilon T_\varepsilon) = \sup(-a'(0)\partial^2 u(0, \cdot)/\partial\varepsilon\partial x).$$

For the solution u_ε we have

$$(4.3.3) \qquad\qquad \lim_{\varepsilon \to 0} u_\varepsilon(t/\varepsilon, x + ta(0)/\varepsilon)/\varepsilon = U(t, x)$$

where U is a solution of Burgers' equation

$$\partial U/\partial t + a'(0)U\,\partial U/\partial x = 0, \quad U(0,x) = \partial u_0(0,x)/\partial\varepsilon.$$

This greater generality is convenient when one wants to change variables as we shall do when we now pass to considering systems (4.3.2) where a is a $N \times N$ matrix such that $a(0)$ has real simple eigenvalues.

When $N = 2$ the situation becomes very simple if we choose coordinates in the manifold M where u takes its values so that the orbits of the eigenvector fields r_1, r_2 become the coordinate directions. (See the beginning of Section 4.2. To be quite correct one should note that r_j only defines a line field since only the direction is well defined. Locally there is no difficulty in integrating it, however.) Then a takes diagonal form, so (4.3.2) reduces to

$$(4.3.2)' \qquad \partial u_j/\partial t + \lambda_j(u)\partial u_j/\partial x = 0, \quad j = 1,2; \quad u(0,x) = u_0(\varepsilon, x).$$

By hypothesis $\lambda_1(0) < \lambda_2(0)$. From Theorem 4.2.5 we know that there is a unique solution for $t < c/\varepsilon$, such that $|u(t,x)| \leq C'\varepsilon$. For sufficiently small ε it follows that

$$\lambda_1(u) \leq c_1 < c_2 \leq \lambda_2(u), \quad t < c/\varepsilon, \quad \text{if } c_j = ((3-j)\lambda_1(0) + j\lambda_2(0))/3, \ j = 1,2.$$

If supp $u_0 \subset \mathbf{R} \times [a,b]$ we have $u_1(t,x) = 0$ for $x > b + c_1 t$ and $u_2(t,x) = 0$ for $x < a + c_2 t$. When $t \geq t_0 = (b-a)/(c_2 - c_1)$ we conclude that u_1 and u_2 have disjoint supports. They will therefore remain small as long as they exist, and the lifespan T_ε can be obtained from $(4.3.1)'$ if we replace T_ε by $T_\varepsilon - t_0$ and u_0 by $u_j(t_0, x)$ which vanishes when $\varepsilon = 0$ and has the derivative $\partial u_{0j}(0, x - \lambda_j(0)t_0)/\partial\varepsilon$ with respect to ε when $\varepsilon = 0$. Hence the limit of $1/(\varepsilon T_\varepsilon)$ is

$$\max_{j,y}(-\partial\lambda_j(0)/\partial u_j\,\partial^2 u_{0j}(0,y)/\partial\varepsilon\partial y).$$

We can interpret $\partial u_0(0, \cdot)/\partial\varepsilon$ as a map from $\mathbf{R}$ to the tangent space at 0 of the manifold M. This vector space is the direct sum of the one dimensional eigenspaces. The following theorem therefore puts the result just obtained in an invariant form, so that it is valid in the original coordinates when $N = 2$:

Theorem 4.3.1. *The largest T_ε such that the Cauchy problem (4.3.2), where we assume $u_0 \in C_0^\infty(\mathbf{R}^2, \mathbf{R}^N)$ and $u_0(0, \cdot) = 0$, has a C^2 solution for $0 \le t < T_\varepsilon$, is given asymptotically by*

$$(4.3.4) \qquad \lim_{\varepsilon \to 0} 1/(\varepsilon T_\varepsilon) = \max_{j, y} -\frac{d}{dy} \langle \lambda_j'(0), \pi_j \partial u_0(0, y)/\partial \varepsilon \rangle,$$

where π_j is the projection in the tangent space of M at 0 on the jth eigenspace which annihilates the others.

The full proof of this theorem will occupy the rest of the section. The difficulty when $N > 2$ is that it is usually impossible to choose coordinates so that all the line fields r_j obtain the directions of the coordinate axes. Nevertheless we shall show, using Lemma 4.2.4, that for small data the N equations will decouple approximately. By Lemma 4.2.4 we know already that $|u(t, x)| \le C\varepsilon$ when $0 \le t < T_\varepsilon$, so for small ε the orbits of the vector fields L_i defined by (4.2.5) pass from 0 to T_ε in a uniformly transversal way. Let $u_0(\varepsilon, x) = 0$ for $x < a$ and for $x > b$, and denote by R_i the strip covered by the orbits of L_i starting in $[a, b]$ when $t = 0$.

Lemma 4.3.2. *Let M be a fixed constant ≥ 1. There is a constant A independent of M such that for the solution of (4.3.2) we have for $\varepsilon < \varepsilon_M$*

$$(4.3.5) \qquad |w_i(t, x)| < A\varepsilon^2 \quad \text{if } 0 \le t < \min(T_\varepsilon, M/\varepsilon), \ (t, x) \notin R_i.$$

Proof. Assuming that this estimate is already known for $0 \le t < t_1$ where $t_1 < T_\varepsilon$, we shall prove that it remains valid for $t = t_1$. To do so we use the equation (4.2.3)' for an orbit Γ of L_i with $0 \le t < t_1$ starting outside $[a, b]$, so that the initial value of w_i is 0. The strips R_j are disjoint for $t > t_0$, say, and all w_k are $O(\varepsilon)$ when $t < t_0$. The integral

$$\int_\Gamma \gamma_{ijk}(u) w_j w_k \, dt$$

can be estimated by $C(A\varepsilon^2)^2 M/\varepsilon$ if $j = k = i$. For all other terms we have $j \ne k$ by (4.2.8), and the estimate $C(\varepsilon^2 + (A\varepsilon^2)^2 M/\varepsilon)$ is then obviously valid except for the integrals over $\{(t, x) \in \Gamma \cap (R_j \cup R_k), t > t_0\}$. In these intervals we can use (4.2.16) after estimating one factor by $A\varepsilon^2$. The conclusion is that

$$|w_i(t, x)| \le C(\varepsilon^2 + A^2 M \varepsilon^3 + A\varepsilon^3), \quad 0 \le t < t_1,$$

where C is independent of A. If we fix $A > C$ then the right-hand side is $< A\varepsilon^2$ for small ε, and (4.3.5) follows by "continuous induction".

Proof of Theorem 4.3.1. We shall first estimate T_ε from above by applying Lemma 1.3.2 to an orbit Γ of L_i inside R_i. We can write (4.2.3)' in the form

$$(4.3.6) \qquad L_i w_i = \gamma_{iii}(u) w_i^2 + a_1 w_i + a_2; \quad a_1 = 2\sum_{j \ne i} \gamma_{iij} w_j, \quad a_2 = \sum_{j, k \ne i} \gamma_{ijk} w_j w_k.$$

By the estimates just given we have if $t < T = \min(T_\varepsilon, M/\varepsilon)$

$$\int_0^t |a_2(t)| \, dt < C\varepsilon^2, \quad \int_0^t |a_1(t)| \, dt < CM\varepsilon,$$

where C is independent of M. Hence

$$K = \int_0^t |a_2(s)|\, ds \exp\left(\int_0^t |a_1(s)|\, ds\right) < 2C\varepsilon^2,$$

if ε is small. The integrals are of course taken along Γ. If $\gamma_{iii}(0) > 0$ and $w_i(0,x) > 2C\varepsilon^2$ at the initial point $(0,x)$ of Γ, it follows by Lemma 1.3.2 that

$$\int_0^T \gamma_{iii}(u)dt < e^{CM\varepsilon}(w_i(0,x) - 2C\varepsilon^2)^{-1}.$$

Here $w_i(0,x)$ is the component of $\partial u_0(\varepsilon,x)/\partial x$ along $r_i(u_0(\varepsilon,x))$, hence equal to $\varepsilon U_i(x) + O(\varepsilon^2)$ where

$$\sum_i U_i(x)r_i(0) = \partial^2 u(0,x)/\partial x \partial \varepsilon;$$

we have used that $u = O(\varepsilon)$. If we multiply by ε it follows that

$$\varlimsup_{\varepsilon \to 0} \min(\varepsilon T_\varepsilon, M)\gamma_{iii}(0)U_i(x) \le 1.$$

By (4.2.8) we have $\gamma_{iii}(0) = -\langle \partial \lambda_i(0)/\partial u, r_i(0)\rangle$, and since M is arbitrary we obtain

$$(4.3.7) \qquad \varlimsup_{\varepsilon \to 0} \varepsilon T_\varepsilon(-\langle \partial \lambda_i(0)/\partial u, U_i(x)r_i(0)\rangle) \le 1.$$

Changing the sign of r_i we have the same conclusion if U_i and $\gamma_{iii}(0)$ are both negative. Since (4.3.7) is trivial when the signs are opposite, this completes the proof of the upper bound for T_ε implied by (4.3.4).

It remains to prove that if

$$(4.3.8) \qquad \max_{i,x} -\langle \partial \lambda_i(0)/\partial u, U_i(x)r_i(0)\rangle = B_0 < B_1,$$

then $\varepsilon T_\varepsilon > 1/B_1$ for small ε. To do so we shall continue our estimates above to derive a uniform bound for w_i when $t < T \le (B_1\varepsilon)^{-1}$ and ε is small, assuming that $T \le T_\varepsilon$. Once we have such a bound we can deduce as in Section 4.2 that $u \in C^\infty$ for $t \le T$, hence that $T_\varepsilon > T$ and so $T_\varepsilon > (B_1\varepsilon)^{-1}$. To estimate w_i in R_i we use the differential equation (4.3.6) for w_i on an orbit Γ of L_i as in the first part of the proof, but now we appeal to Lemma 1.3.3. Assume that the initial value w_{i0} of w_i on Γ is ≥ 0 for example. By (4.3.8) we have $\gamma_{iii}(u)w_{i0} < \varepsilon(B_1 + B_0)/2$ on the orbit if ε is small. Thus, if $a_0 = \gamma_{iii}$,

$$w_{i0}\int_0^T a_0^+\, dt \le \varepsilon T(B_1 + B_0)/2 \le (B_1 + B_0)/2B_1 < 1.$$

As before $K < 2C\varepsilon^2$, and

$$\int_0^T |a_1(t)|\, dt = O(\varepsilon), \qquad \int_0^T |a_0(t)|\, dt = O(1/\varepsilon),$$

so (1.3.6) and (1.3.7) hold with $x_0 = w_{i0}$ when ε is small. The right-hand side of (1.3.8) is at least $((B_1 - B_0)/2B_1 - O(\varepsilon))/(w_{i0} + K)$, and that of (1.3.9) is at least $(1 - O(\varepsilon))/K$. Hence we conclude that w_i is bounded by ε times a constant independent of T. By the remarks above the proof is now completed by establishing bounds for the higher order derivatives of u in a completely straightforward manner. We leave this for the reader.

The results of this section are taken from Hörmander [1] but go back to John [1].

4.4. The Riemann problem. Keeping the hypotheses made at the beginning of Section 4.2 we shall now discuss the solution of the *Riemann problem* for (4.2.1) where the Cauchy data are constant on each half axis,

$$(4.4.1) \qquad\qquad u(0, x) = u_\pm \quad \text{when } \pm x > 0,$$

and $u_\pm$ are small. Since it will now be necessary to consider discontinuous solutions we shall assume that the equation is *given* in conservation form

$$(4.4.2) \qquad\qquad \partial u/\partial t + \partial f(u)/\partial x = 0,$$

where $f \in C^3$ in a neighborhood of 0, $f'(u) = a(u)$. In general it is of course not possible to write the equation (4.2.1) in the form (4.4.2), so this is an added hypothesis on the system.

Recall that for Burgers' equation the solution was a continuous *rarefaction wave* if $u_+ > u_-$. More generally, for a scalar equation (4.2.1) we have the solution

$$u(t, x) = \begin{cases} u_-, & \text{for } x < a(u_-)t \\ u_+, & \text{for } x > a(u_+)t \\ a^{-1}(x/t), & \text{for } a(u_-)t < x < a(u_+)t, \end{cases}$$

if $u_+ > u_-$ and a is strictly increasing with inverse a^{-1}. For systems one may also find j-simple rarefaction waves. Let $s \mapsto U(s)$ be the orbit of the eigenvector field r_j starting at u_-,

$$(4.4.3) \qquad\qquad dU(s)/ds = r_j(U(s)), \quad U(0) = u_-.$$

Recall that $u(t, x) = U(s(t, x))$ then satisfies (4.2.1) if (4.2.2) is valid. Assume now that the system is *genuinely nonlinear* in the strong sense that

$$\langle \lambda_j'(0), r_j(0) \rangle \neq 0,$$

that is, $\gamma_{jjj}(0) \neq 0$ with the notation in (4.2.3)'. Changing the sign of r_j if necessary we may then assume that the sign is positive, so that

$$(4.4.4) \qquad\qquad \langle \lambda_j'(u), r_j(u) \rangle > 0$$

in a neighborhood of 0. Then the equation (4.2.2) has a rarefaction wave solution which gives a Lipschitz continuous solution of the Riemann problem (4.4.1), (4.4.2) for $t > 0$ provided that $u_+ = U(s)$ for some $s > 0$ and that $u_\pm$ are small enough.

In the scalar case the Riemann problem had the solution

$$(4.4.5) \qquad\qquad u(t, x) = u_\pm \quad \text{for } \pm (x - ct) > 0,$$

if the speed c is chosen so that

$$(4.4.6) \qquad\qquad c(u_+ - u_-) = f(u_+) - f(u_-).$$

This solution was rejected in favor of the rarefaction wave when $u_+ > u_-$ and $a = f'$ is increasing, that is, f is convex. Let us now look for solutions of the form (4.4.5) also in the case of systems. The Rankine-Hugoniot condition for u to be a weak solution of (4.4.2) is still the equation (4.4.6), but now it cannot be satisfied by just a suitable choice of the speed c; it implies $N - 1$ conditions on u_+ for given u_-. To study (4.4.6) we note that by Taylor's formula

$$(4.4.7) \qquad f(u) - f(v) = A(u, v)(u - v); \quad A(u, v) = \int_0^1 f'(v + t(u - v))\, dt.$$

Since $A(u, u) = f'(u) = a(u)$ has only simple real eigenvalues for $u = 0$, it follows that $A(u, v)$ has only simple real eigenvalues $\lambda_1(u, v) < \lambda_2(u, v) < \cdots < \lambda_N(u, v)$ which are C^2 functions of (u, v) in a neighborhood of $(0, 0)$. We shall denote corresponding eigenvectors by $r_j(u, v)$. To fix a normalisation for them we observe that since

$$A(u, v) = f'((u + v)/2) + O(|u - v|^2),$$

we have

$$A(u, v) r_j((u + v)/2) = (\lambda_j r_j)((u + v)/2) + O(|u - v|^2).$$

If we *define* $r_j(u, v)$ as the projection of $r_j((u + v)/2)$ on the jth eigenvector of $A(u, v)$ along the space spanned by the others, then we still have $r_j \in C^2$, and

$$(4.4.8) \qquad r_j(u, v) - r_j((u + v)/2) = O(|u - v|^2),$$

or equivalently

$$(4.4.8)' \qquad r_j(u, v) - (r_j(u) + r_j(v))/2 = O(|u - v|^2);$$

for the eigenvalues we have

$$(4.4.9) \qquad \lambda_j(u, v) = \lambda_j((u + v)/2) + O(|u - v|^2) = (\lambda_j(u) + \lambda_j(v))/2 + O(|u - v|^2).$$

If $u_+ \neq u_-$ and both are small then the condition (4.4.6) means that, for some $\sigma \in \mathbf{R}$ and some j,

$$(4.4.6)' \qquad c = \lambda_j(u_+, u_-), \quad u_+ - u_- = \sigma r_j(u_+, u_-).$$

Writing u instead of u_+ we have N equations

$$(4.4.10) \qquad u - u_- - \sigma r_j(u, u_-) = 0$$

for the $N + 1$ unknowns (u, σ). Since the Jacobian matrix with respect to u is the identity when $\sigma = 0$, it follows from the implicit function theorem that u is a C^2 function of σ with $du/d\sigma = r_j(u_-)$ when $\sigma = 0$. Thus a solution of the form (4.4.5) exists if and only if u_+ lies on a certain C^2 curve through u_-, which is tangent to the orbit of r_j. We shall now examine how close the contact is. Differentiation of the equations (4.4.10) gives

$$(1 - \sigma \partial r_j(u, u_-)/\partial u)du/d\sigma = r_j(u, u_-),$$

hence $du/d\sigma = r_j(u)$ when $\sigma = 0$, as already observed, and another differentiation gives when $\sigma = 0$

$$d^2u/d\sigma^2 = 2\partial r_j(u, u_-)/\partial u\, du/d\sigma = \partial r_j(u)/\partial u\, du/d\sigma,$$

by (4.4.8), so the curve defined by (4.4.10) is tangent of higher order to the orbit of r_j. If w is a j-*Riemann invariant*, that is, if w is constant along the orbits of r_j, and if w is equal to 0 at u_- then $w = O(\sigma^3)$ on the curve (4.4.10).

For small $\sigma \neq 0$ it follows from (4.4.4), (4.4.9) and (4.4.6)' that

$$(4.4.11) \qquad\qquad \lambda_j(u_-) < c < \lambda_j(u_+) \quad \text{if } \sigma > 0$$
$$(4.4.12) \qquad\qquad \lambda_j(u_+) < c < \lambda_j(u_-) \quad \text{if } \sigma < 0$$

If (4.4.11) holds then the j characteristics point away from the shock at both sides, which was rejected in the scalar case in favor of a rarefaction wave. We shall therefore require $\sigma < 0$ so that (4.4.12) holds. Since $u_\pm$ are small, this is equivalent to the *Lax shock condition*

$$\lambda_j(u_+) < c < \lambda_{j+1}(u_+), \quad \lambda_{j-1}(u_-) < c < \lambda_j(u_-).$$

For larger shocks these may be stronger but can be motivated heuristically by demanding that only $N - 1$ characteristics go out from the shock curve so that the shock speed and the outgoing waves can be determined from the incoming waves. (See Lax [2, p. 555] or Smoller [1, p. 261].)

Definition 4.4.1. A solution (4.4.5) of the Riemann problem is called an *admissible j-shock* if (4.4.12) holds, or equivalently, the number σ in (4.4.6)' is < 0 and (4.4.4) is valid.

Now define

$$\Phi_j(\varepsilon)u_- = u_+,$$

if $\varepsilon \geq 0$ and u_+ is the value for $s = \varepsilon$ of the solution of (4.4.3), or $\varepsilon < 0$ and u_+ is the solution of (4.4.6)' with $\sigma = \varepsilon$. Thus the Riemann problem with $u_+ = \Phi_j(\varepsilon)u_-$ is solved by a j simple rarefaction wave if $\varepsilon \geq 0$ and by an admissible j-shock (4.4.5) if $\varepsilon < 0$. The function $(\varepsilon, u_-) \mapsto \Phi_j(\varepsilon)u_-$ is in C^2. In fact, it is clearly a C^2 function for $\varepsilon \neq 0$, with continuous boundary values for the derivatives as $\varepsilon \to \pm 0$, and we have seen that the limits of the derivatives of order ≤ 2 of the derivatives with respect to ε agree.

If

$$(4.4.13) \qquad\qquad u_+ = \Phi_N(\varepsilon_N) \cdots \Phi_1(\varepsilon_1)u_-$$

and all ε_j are small, then we obtain a solution of the Riemann problem consisting in order of increasing j from left to right by a j rarefaction wave (or an admissible j-shock) of strength ε_j connecting the constant states

$$\Phi_{j-1}(\varepsilon_{j-1}) \cdots \Phi_1(\varepsilon_1)u_- \quad \text{and} \quad \Phi_j(\varepsilon_j) \cdots \Phi_1(\varepsilon_1)u_-, \quad j = 1, \ldots, N.$$

These do not interfere at all with each other since they move from the origin with quite different speeds, close to $\lambda_1(0), \ldots, \lambda_N(0)$ respectively. Since

$$\Phi_j(\varepsilon)v = v + \varepsilon r_j(v) + O(\varepsilon^2)$$

it follows from (4.4.13) that

$$\partial u_+ / \partial \varepsilon_j = r_j(u_-) \quad \text{when } \varepsilon = 0,$$

and since the vectors r_j are linearly independent the implicit function theorem shows that for small u_+, u_- the equation (4.4.13) determines $\varepsilon = (\varepsilon_1, \ldots, \varepsilon_N)$ uniquely as a C^2 function of (u_+, u_-). Thus we have

Theorem 4.4.2. *If u_-, u_+ are sufficiently small, then the Riemann problem* (4.4.1), (4.4.2) *has a unique solution consisting from left to right for increasing j of a small j simple rarefaction wave or a small j shock, $j = 1, \ldots, N$, admissible in the sense that* (4.4.12) *holds, or equivalently, that $\sigma < 0$ in* (4.4.6)'. *(We assume* (4.4.4).)

The maps $\Phi_j(\varepsilon_j)$ in (4.4.13) do not commute, except for a fixed value of j and ε_j nonnegative. However, for small ε and δ we have a uniquely determined $\gamma(\varepsilon, \delta, u_-)$, which is a C^2 function of $(\varepsilon, \delta, u_-)$, such that

$$(4.4.14) \qquad \Phi_N(\delta_N) \ldots \Phi_1(\delta_1)\Phi_N(\varepsilon_N) \ldots \Phi_1(\varepsilon_1)u_- = \Phi_N(\gamma_N) \ldots \Phi_1(\gamma_1)u_-.$$

From the differentiability it follows that $\gamma = \varepsilon + \delta + O(|\varepsilon|^2 + |\delta|^2)$, but it will be essential when estimating the interaction of waves in Section 4.5 to have a better control of the error term. To that end we note that $\gamma = \varepsilon + \delta$ if for some k we have

$$(4.4.15) \qquad \delta_j = 0 \quad \text{for } j < k; \ \varepsilon_j = 0 \quad \text{for } j > k; \ \varepsilon_k > 0 \text{ and } \delta_k > 0, \text{ or } \varepsilon_k \delta_k = 0.$$

This condition is equivalent to the vanishing of

$$(4.4.16) \qquad \Delta(\varepsilon, \delta) = \sum_{i > j} |\varepsilon_i \delta_j| + \sum_{\varepsilon_i < 0 \text{ or } \delta_i < 0} |\varepsilon_i \delta_i|.$$

In fact, suppose that $\Delta(\varepsilon, \delta) = 0$ and that $\varepsilon \neq 0$. If k is the largest j with $\varepsilon_j \neq 0$, then $\varepsilon_j = 0$ for $j > k$, and $\delta_j = 0$ for $j < k$ since the first sum is 0. If $\delta_k \neq 0$ then $\varepsilon_k > 0$ and $\delta_k > 0$ since the second sum is 0. That (4.4.15) implies $\Delta(\varepsilon, \delta) = 0$ is obvious. To exploit this conclusion we need the following slightly extended and simplified version of a lemma from Schatzman [1].

Lemma 4.4.3. *Let $I \subset \{1, \ldots, m\} \times \{1, \ldots, n\}$ and let $f \in C^2(\overline{\mathbf{R}}_+^{n+m})$ be a function with $f(x, y) = 0$ when $\Delta(x, y) = \sum_{(i,j) \in I} x_i y_j = 0$; $x = (x_1, \ldots, x_m)$, $y = (y_1, \ldots, y_n)$. Then we have*

$$(4.4.17) \qquad |f(x, y)| \leq \tfrac{1}{2}(3^m - 1)\Delta(x, y) \sup_{x, y, i, j} |\partial^2 f(x, y)/\partial x_i \partial y_j|; \quad x \in \overline{\mathbf{R}}_+^m, y \in \overline{\mathbf{R}}_+^n.$$

Proof. Assume that $\Delta \not\equiv 0$ and that the supremum in the right-hand side is equal to 1. If $m = 1$ then $\Delta(x, y) = x_1 \sum_1^k y_j$ if the coordinates y_j are suitably labelled; we may assume that $k = n$, regarding the other coordinates as parameters. Then $f(x, y) = 0$ if $x_1 = 0$ or $y = 0$. By Taylor's formula

$$f(x, y) = \int_0^1 \int_0^1 \sum_{j=1}^k x_1 y_j \, \partial_{x_1} \partial_{y_j} f(tx_1, sy) \, dt \, ds,$$

which proves that $|f(x, y)| \leq \Delta(x, y)$. Assume now that $m > 1$ and that the lemma has already been proved for smaller values of m. Let $(1, j) \in I$ for $1 \leq j \leq k$ but $(1, l) \notin I$ for $l > k$. Set $y' = (y_1, \ldots, y_k)$, $x'' = (x_2, \ldots, x_m)$, $y'' = (y_{k+1}, \ldots, y_n)$. Then

$$f(x_1, x'', y', y'') - f(0, x'', y', y'') - f(x_1, x'', 0, y'') + f(0, x'', 0, y'')$$

vanishes when $x_1 = 0$ or $y' = 0$, so we can estimate by $\Delta(x_1, 0, y', 0)$ as in the first part of the proof. Furthermore, $f(0, x'', y', y'')$, $f(x_1, x'', 0, y'')$ and $f(0, x'', 0, y'')$ all vanish when $\Delta(0, x'', y) = 0$, for

$$\Delta(x_1, x'', 0, y'') = \Delta(0, x'', 0, y'') \leq \Delta(0, x'', y).$$

By inductive hypothesis we conclude that they can be estimated by $\Delta(0, x'', y)$ times $(3^{m-1} - 1)/2$, and since $1 + 3(3^{m-1} - 1)/2 = (3^m - 1)/2$, the induction is complete.

If we apply the lemma to the function $\gamma(\varepsilon, \delta, x_-) - \varepsilon - \delta$ in each quadrant, after a suitable cutoff, considering x_- as a parameter, we obtain Glimm's interaction estimate:

Theorem 4.4.4. *For the function* $\gamma(\varepsilon, \delta, x_-)$ *we have in a neighborhood of* $(0,0)$

$$(4.4.18) \qquad\qquad |\gamma(\varepsilon, \delta, x_-) - \varepsilon - \delta| \leq C\Delta(\varepsilon, \delta),$$

where Δ *is defined by* (4.4.16).

4.5. Glimm's existence theorem. In this section we shall prove that the Cauchy problem

$$(4.5.1) \qquad\qquad \partial u/\partial t + \partial f(u)/\partial x = 0; \quad u(0,x) = u_0(x);$$

has a weak solution for $t \geq 0$ if $f'(0)$ has real distinct eigenvalues, the equation is genuinely nonlinear in the sense (4.4.4), and

$$\sup|u| + \int_{-\infty}^{\infty} |du|$$

is small enough. The proof due to Glimm [1] is based on successive solution of Riemann problems alternating with a procedure for making the solution piecewise constant. First we choose small mesh widths l and h in the x and t directions respectively, as in Section 2.5, which satisfy the Courant-Friedrichs-Lewy condition

$$(4.5.2) \qquad\qquad Ah \leq l$$

where A is chosen so that $|f'| \leq A$ in a neighborhood U of the origin where the results of Section 4.4 are applicable. The approximate solutions $u^{h,l}$ constructed will always take their values in U. We introduce now the mesh points

$$\{(kh, nl); k, n \in \mathbf{Z}, \, k - n \text{ even }\}$$

and define piecewise constant initial values by averaging

$$(4.5.3) \qquad u^{h,l}(0,x) = \int_{nl}^{(n+2)l} u_0(y)\, dy/2l, \quad nl < x < (n+2)l, \, n \text{ even}.$$

From Section 4.4 we know how to solve (4.5.1) with these initial data when $t \leq h$, for by (4.5.2) the solutions of the Riemann problems corresponding to the jumps at $(0, nl)$, n even, will not interact during this time interval. Thus we have defined $u^{h,l}(t,x)$ for $0 \leq t < h$. Next we define $u^{h,l}(h,x)$ when $nl < x < (n+2)l$, n odd, by sampling the values of $u^{h,l}(h-0,x)$ at an appropriate point in the interval. The value is thus determined by the data at $(0, (n+1)l \pm 0)$. More precisely, with $\theta_1 \in (-1,1)$ to be chosen later, we set

$$u^{h,l}(h,x) = u^{h,l}(h-0, (n+1+\theta_1)l), \quad \text{if } nl < x < (n+2)l \text{ and } n \text{ is odd}.$$

We can continue in the same way, solving Riemann problems for $kh \leq t < (k+1)h$ and taking

$$u^{h,l}((k+1)h, x) = u^{h,l}((k+1)h - 0, (n+1+\theta_{k+1})l)$$
$$\text{if } nl < x < (n+2)l \text{ and } n - k \text{ is odd}.$$

Our next goal is to prove that the maximum and the total variation of $u^{h,l}$ remain uniformly small if they are small initially. The method of proof is similar to that of Lemma 4.2.4 and was in fact introduced first by Glimm [1] in the present context.

The step function $u^{h,l}(kh, x)$ is fully described by a sequence $U(k, n)$,

$$u^{h,l}(kh, x) = U(k, n) \quad \text{if } (k + 2n)l < x < (k + 2n + 2)l.$$

(We suppress the parameters h, l now in order to shorten notation.) With the notation

$$\Phi(\gamma) = \Phi_N(\gamma_N) \cdots \Phi_1(\gamma_1)$$

the solution of the Riemann problem from $t = kh$ to $t = (k + 1)h$ consists in writing

$$U(k, n) = \Phi(\gamma(n - 1))U(k, n - 1)$$

and get from $\gamma(n - 1)$ the waves emanating from $(kh, (k + 2n)l)$ as described in Section 4.4. We define $U(k + 1, n - 1)$ to be the value of the resulting wave at the sampling point. (If it is a shock we have to choose a specific side of it.) This means that

$$U(k + 1, n - 1) = \Phi(\varepsilon(n - 1))U(k, n - 1),$$

where $\gamma(n) = \varepsilon(n) + \delta(n)$ is a decomposition satisfying (4.4.15), thus

(4.5.4) $\gamma(n) = \varepsilon(n) + \delta(n); \quad \varepsilon(n)_j \delta(n)_j \geq 0 \quad \text{for } j = 1, \ldots, N.$

(Of course $\gamma, \varepsilon, \delta$ depend on k too.) We get

$$U(k + 1, n) = \Phi(\varepsilon(n))U(k, n) = \Phi(\varepsilon(n))\Phi(\gamma(n - 1))U(k, n - 1)$$
$$= \Phi(\varepsilon(n))\Phi(\delta(n - 1))\Phi(\varepsilon(n - 1))U(k, n - 1) = \Phi(\varepsilon(n))\Phi(\delta(n - 1))U(k + 1, n - 1).$$

For the analogue Γ of γ at the level $k + 1$,

$$U(k + 1, n) = \Phi(\Gamma(n - 1))U(k + 1, n - 1),$$

this means that

$$\Phi(\Gamma(n)) = \Phi(\varepsilon(n + 1))\Phi(\delta(n)).$$

Thus it follows from (4.4.18) that

(4.5.5) $|\varrho(n)| \leq C\Delta(\delta(n), \varepsilon(n + 1)), \quad \text{if } \varrho(n) = \Gamma(n) - \varepsilon(n + 1) - \delta(n).$

The total variation of $u^{k,l}(kh, \cdot)$ is equivalent to

(4.5.6) $$\mathcal{L}(k) = \sum_n |\gamma(n)| = \sum_n |\varepsilon(n)| + \sum_n |\delta(n)|,$$

where the norms are l^1 norms and the second equality follows from (4.5.4). The notation has been chosen to emphasize the analogy with (4.2.10). By (4.5.5)

(4.5.7) $$\sum_n |\varrho(n)| \leq C\Lambda(k); \quad \Lambda(k) = \sum_n \Delta(\delta(n), \varepsilon(n + 1)).$$

Hence

(4.5.8) $$\mathcal{L}(k + 1) \leq \mathcal{L}(k) + C\Lambda(k).$$

As in Section 4.2 we shall also consider another quantity $Q(k)$, "the potential for future interaction", which will decrease in a way which compensates for an occasional increase of $\mathcal{L}(k)$. Thus we set

$$Q(k) = \sum_{n<m} \Delta(\gamma(n), \gamma(m)).$$

Since Q is linear in each argument in every "octant", it follows from (4.5.4) that

$$(4.5.9) \quad Q(k) = \sum_{n<m} \left(\Delta(\varepsilon(n), \varepsilon(m)) + \Delta(\delta(n), \delta(m)) + \Delta(\varepsilon(n), \delta(m)) + \Delta(\delta(n), \varepsilon(m)) \right).$$

To estimate

$$Q(k+1) = \sum_{n<m} \Delta(\Gamma(n), \Gamma(m))$$

we shall use that $\Delta(\varepsilon, \delta)$ is convex and positively homogeneous in each variable when the other is held fixed. Thus

$$\Delta(\Gamma(n), \Gamma(m)) \leq \Delta(\varepsilon(n+1), \Gamma(m)) + \Delta(\delta(n), \Gamma(m)) + \Delta(\varrho(n), \Gamma(m))$$
$$\leq \Delta(\varepsilon(n+1), \varepsilon(m+1)) + \Delta(\varepsilon(n+1), \delta(m)) + \Delta(\delta(n), \varepsilon(m+1))$$
$$+ \Delta(\delta(n), \delta(m)) + \Delta(\varepsilon(n+1), \varrho(m)) + \Delta(\delta(n), \varrho(m)) + \Delta(\varrho(n), \Gamma(m)).$$

When $n < m$ the first four terms occur in the right-hand side of (4.5.9) apart from the vanishing term $\Delta(\varepsilon(n+1), \delta(n+1))$. On the other hand, the terms $\Delta(\delta(n), \varepsilon(n+1))$ in (4.5.9) are missing. Hence it follows from (4.5.7), (4.5.8) that

$$(4.5.10) \qquad Q(k+1) + \Lambda(k) \leq Q(k) + C\Lambda(k)\mathcal{L}(k) + C\Lambda(k)(\mathcal{L}(k) + C\Lambda(k)).$$

If we start with data u_0 which are constant for large $|x|$, then all sums are finite and no convergence questions occur. We have therefore reached a difference analogue of the situation studied in Lemma 4.2.4.

Lemma 4.5.1. *Assume that $Q(k)$, $\mathcal{L}(k)$ and $\Lambda(k)$ are positive sequences such that $\Lambda(k) \leq Q(k) \leq \mathcal{L}(k)^2/2$ and (4.5.8), (4.5.10) hold. If $\mathcal{L}(0) < (5C)^{-1}$, it follows that*

$$(4.5.11) \qquad \mathcal{L}(k) \leq \mathcal{L}(0) + 3C\mathcal{L}(0)^2/2 \leq 2\mathcal{L}(0),$$

$$(4.5.12) \qquad \sum_0^\infty \Lambda(k) \leq 3\mathcal{L}(0)^2/2.$$

Proof. By (4.5.8) and (4.5.10) we have

$$\mathcal{L}(k+1) + 3CQ(k+1) \leq \mathcal{L}(k) + 3CQ(k) \quad \text{if } 1 + 3C(2\mathcal{L}(k) + C\Lambda(k)) \leq 3.$$

This is true if $3C(2\mathcal{L}(k) + C\Lambda(k)) \leq 2$. Now an inequality

$$\mathcal{L}(k) + 3CQ(k) \leq \mathcal{L}(0) + 3CQ(0) \leq \mathcal{L}(0) + 3C\mathcal{L}(0)^2/2$$

implies

$$3C(2\mathcal{L}(k) + C\Lambda(k)) \leq 6C(\mathcal{L}(k) + 3CQ(k)) \leq 6C(\mathcal{L}(0) + 3C\mathcal{L}(0)^2/2).$$

If the right-hand side is ≤ 2, hence if $\mathcal{L}(0) \leq 1/5C$, we conclude inductively that for all k

$$\mathcal{L}(k) + 3C\mathcal{Q}(k) \leq \mathcal{L}(0) + 3C\mathcal{Q}(0) \leq \mathcal{L}(0) + 3C\mathcal{L}(0)^2/2 \leq 2\mathcal{L}(0).$$

Since

$$\mathcal{Q}(k+1) - \mathcal{Q}(k) \leq \Lambda(k)(2C\mathcal{L}(k) + C^2\Lambda(k) - 1) \leq -\Lambda(k)/3,$$

it follows that

$$\sum_0^\infty \Lambda(k) \leq 3\mathcal{Q}(0) \leq 3\mathcal{L}(0)^2/2,$$

which completes the proof.

In the proof of these estimates we have tacitly assumed that all the values of $U(k,n)$ belong to a certain neighborhood U of 0. Now, if u_0 is constant near $\pm\infty$, then all $U(k,n)$ take the same values as u_0 for sufficiently large $|n|$, so

$$|U(k,n)| \leq \max(|u_0(-\infty)|, |u_0(+\infty)|) + C\mathcal{L}(k).$$

If $\mathcal{L}(0)$ and $|u_0(\pm\infty)|$ are sufficiently small the inductive proof above also yields that $U(k,n)$ stays in U. Since any $U(k,n)$ just depends on a finite number of $U(0,\nu)$, the estimates proved remain valid if u_0 is not constant at $\pm\infty$, so we have proved

Theorem 4.5.2. *Let U be a given neighborhood of 0. If $\sup|u_0|$ and the total variation of u_0 are sufficiently small, then Glimm's scheme with small h, l satisfying (4.5.2) defines a function $u^{h,l}(t,x)$ for $t \geq 0$ with values in U, such that the total variation for every fixed t is bounded by a constant times that of u_0.*

By Lemma 2.4.4 the bound for the total variation implies a uniform Lipschitz condition for the L^1 modulus of continuity of $u^{h,l}(t, \cdot)$. This is also true for time translations:

Lemma 4.5.3. *There is a constant C such that for $s, t \geq 0$*

$$(4.5.13) \qquad \int_{-\infty}^\infty |u^{h,l}(t,x) - u^{h,l}(s,x)|\, dx \leq C(|s-t| + l\lceil(|s-t|/h)) \int_{-\infty}^\infty |du_0|,$$

where $\lceil$ denotes the smallest larger integer.

Proof. If v is a solution of a Riemann problem with small initial data v_0, equal to $v_\pm$ when $\pm x > 0$ then, for $s, t \geq 0$,

$$\int |v(t,x) - v(s,x)|\, dx \leq C|t-s||v_+ - v_-|$$

for the derivative $\partial v(t,x)/\partial t$ in the distribution sense is a measure with total mass bounded by A times that of $\partial v/\partial x$, hence bounded by a constant times $|v_+ - v_-|$. Combining this with the total variation estimate in Theorem 4.5.2 we obtain (4.5.13) without the term $l\lceil(|s-t|/h)$ if no integer multiple of h lies between s and t. It remains to establish (4.5.13) when $t = kh$ and $s = kh - 0$. Since

$$|u^{h,l}(kh,x) - u^{h,l}(kh-0,x)| \leq \int_{x-2l}^{x+2l} |du^{h,l}(kh-0,x)|,$$

we obtain

$$\int |u^{h,l}(kh,x) - u^{h,l}(kh-0,x)|\, dx \leq 4l \int_{-\infty}^\infty |du^{h,l}(kh-0,x)|,$$

which completes the proof in view of Theorem 4.5.2.

We shall require from now on that there is equality in (4.5.2) and can then estimate $l\lceil(|s-t|/h)$ in (4.5.13) by $A|s-t|+l$. From Lemma 4.5.3 it follows then that for every such sequence of values of (h,l) converging to $(0,0)$, one can find a subsequence (h_ν, l_ν) such that $u^{h_\nu,l_\nu}(t,\cdot)$ converges in L^1_{loc} locally uniformly in t as $\nu \to \infty$. The next question is whether the limit $u(t,x)$ must satisfy (4.5.1) weakly.

So far we have introduced no conditions at all on the shifts θ_k; in fact, we could have allowed them to depend on n also. However, they must be chosen carefully if limits of the functions $u^{h,l}$ shall be solutions of (4.5.1). To see this we consider the example of a Riemann problem for a scalar equation where we know that a true solution is $u = u_-$ for $x < ct$ and $u = u_+$ for $x > ct$. After the first step the discontinuity has shifted the distance l to the right if $\theta_j l < ch$ and to the left in the opposite case. After k steps we have a shift to the right by the distance

$$kl - 2lN(k,\kappa)$$

where $N(k,\kappa)$ is the number of $\theta_1,\ldots,\theta_k$ between $\kappa = ch/l$ and 1. (Note that $|\kappa| < 1$ by (4.5.2).) The deviation from the displacement ckh of the true solution is

$$kl(1 - 2N(k,\kappa)/k - \kappa),$$

so we want $N(k,\kappa)/k$ to be close to $(1-\kappa)/2$, that is, the shifts should be equidistributed in $(-1,1)$.

After this motivating discussion we return to the functions $u^{h,l}$ in Theorem 4.5.2. If $\phi \in C_0^\infty(\mathbf{R}^2)$ we have

$$-\iint_{t>0} (u^{h,l}\partial\phi/\partial t + f(u^{h,l})\partial\phi/\partial x)\, dt\, dx$$

$$= \sum_{k=1}^\infty \int (u^{h,l}(kh,x) - u^{h,l}(kh-0,x))\phi(kh,x)\, dx + \int u^{h,l}(0,x)\phi(0,x)\, dx$$

since $u^{h,l}$ is a weak solution in any strip $kh < t < (k+1)h$. The last integral converges to $\int u_0(x)\phi(x)\, dx$ as $(h,l) \to (0,0)$. To prove that limits of $u^{h,l}$ are weak solutions of (4.5.1) we must verify that the sum of the terms

$$(4.5.14) \qquad E_k(l,h,\theta) = \int (u^{h,l}(kh,x) - u^{h,l}(kh-0,x))\phi(kh,x)\, dx$$

converges to 0. Here we have emphasized the dependence on $\theta = (\theta_1,\theta_2,\ldots)$, since the choice of θ will now be important, but we have suppressed the dependence on ϕ since ϕ is fixed in the following discussion.

From the proof of Lemma 4.5.3 we know that

$$(4.5.15) \qquad |E_k(l,h,\theta)| \le Cl.$$

Since the number of nonzero terms is $O(1/l)$ this gives a bound for $\sum E_k$ but not more. This was to be expected, for it remains to choose θ appropriately. Let

$$\Theta = \prod_1^\infty [-1,1], \quad d\theta = \prod_1^\infty (d\theta_j/2).$$

Θ is a compact space, and $\int \psi(\theta)\, d\theta$ is defined in the obvious way for continuous functions ψ on Θ depending only on a finite number of θ_j. These are dense in $C(\Theta)$, so $d\theta$ extends to a measure on Θ of total mass 1. Since $E_k(l, h, \theta)$ only depends on $\theta_1, \ldots, \theta_k$ the definition of the integrals in the formula

$$(4.5.16) \quad \int \left| \sum_k E_k(l, h, \theta) \right|^2 d\theta$$

$$= \sum_k \int |E_k(l, h, \theta)|^2 d\theta + 2 \sum_{j<k} \int \langle E_j(l, h, \theta), E_k(l, h, \theta) \rangle d\theta$$

is completely elementary. By (4.5.15) the first sum is $O(l)$, for the number of terms is $O(1/l)$. In the second sum the number of terms is just $O(1/l^2)$, so we have to study the terms carefully. Since E_j is independent of θ_k the integration with respect to θ_k can be made on E_k. Writing $v(x) = u^{h,l}(kh - 0, x)$ and $I_n = (nl, (n+2)l)$, we have
(4.5.17)

$$\int E_k(l, h, \theta) \, d\theta_k/2 = \sum_{n; n-k \text{ even}} \int \left(\int_{I_n} (v((n+1+\theta_k)l) - v(x)) \phi(kh, x) \, dx \right) d\theta_k/2$$

$$= \sum_{n; n-k \text{ even}} \int_{I_n} \left(\int_{I_n} v(y) \, dy/2l - v(x) \right) \phi(kh, x) \, dx.$$

To estimate this we need an elementary lemma.

Lemma 4.5.4. *If v and ϕ are continuous functions in an interval $I \subset \mathbf{R}$ of length L, and if the oscillations are $\leq V$ and $\leq \Phi$ respectively, then*

$$\left| \int_I \left(\int_I v(y) \, dy/L - v(x) \right) \phi(x) \, dx \right| \leq LV\Phi.$$

Proof. Neither side changes if a constant is added to v or to ϕ. We may therefore assume that both have mean value 0 over I. But then we have $|v| \leq V$ and $|\phi| \leq \Phi$, which makes the assertion obvious.

Using Lemma 4.5.4 we can estimate the sum in (4.5.17) by Cl^2, for we have a bound for the total variation of v and for $|\partial \phi / \partial x|$. Using (4.5.15) with k replaced by j we conclude that the terms in the second sum in (4.5.16) are $O(l^3)$, hence the right-hand side is $O(l)$. Thus we have proved that

$$\int \left| \sum_k E_k(l, h, \theta) \right|^2 d\theta \to 0 \quad \text{when } l, h \to 0.$$

Hence every sequence of $l_\nu = Ah_\nu \to 0$ has a subsequence for which $\sum E_k(l, h, \theta)$ converges to 0 for almost all $\theta \in \Theta$. If $u(t, x)$ is the limit of some subsequence of $u^{h,Ah}(t, x)$ with $h \to 0$, it follows that

$$-\iint_{t>0} (u\partial\phi/\partial t + f(u)\partial\phi/\partial x) dt dx = \int u_0(x)\phi(0, x) \, dx$$

for almost all $\theta \in \Theta$, when $\phi \in C_0^\infty(\mathbf{R}^2)$ is fixed. For almost all θ the equality must therefore hold for all ϕ in a countable dense subset of C_0^1, which proves that u is a weak solution of (4.5.1). Hence we have proved

Theorem 4.5.5. *For almost all $\theta \in \Theta$ every limit of the functions $u^{h,Ah}(t,x)$ constructed using Glimm's method with u_0 small and of small total variation is a weak solution of the Cauchy problem (4.5.1).*

Remark. The proof above shows that almost all $\theta \in \Theta$ are equidistributed in $(-1,1)$.

In Section 3.4 we found a class of solutions to a scalar conservation law characterized by a family of entropy conditions. For systems it may not always be possible to find entropies. By analogy with the scalar case we need to find smooth *entropy pairs* $\Phi(u)$ and $Y(u)$, that is, real valued functions such that for classical solutions of (4.5.1) we have an additional conservation law

$$\partial \Phi(u)/\partial t + \partial Y(u)/\partial x = 0.$$

If $w = \partial u/\partial x$ then

$$\partial u/\partial t = -f'(u)w, \quad \partial \Phi(u)/\partial t = -\langle \Phi'(u), f'(u)w \rangle, \quad \partial Y(u)/\partial x = \langle Y'(u), w \rangle$$

so this means that

$$(4.5.18) \qquad\qquad \langle Y'(u), w \rangle = \langle \Phi'(u), f'(u)w \rangle.$$

This is a system of N differential equations which may have no solutions beyond the obvious ones. The existence of such entropy pairs will be discussed in Section 4.6. Here we shall instead examine if inequalities of the form

$$(4.5.19) \qquad\qquad \partial \Phi(u)/\partial t + \partial Y(u)/\partial x \leq 0$$

can be expected for the solutions given by Theorem 4.5.5. Let us first look at the case of a simple discontinuity on a C^1 curve where $u \in C^1$ on each side. Assume that it is of the jth family. If the shock speed is c then condition (4.5.19) means that

$$(4.5.20) \qquad\qquad Y(u_+) - Y(u_-) - c(\Phi(u_+) - \Phi(u_-)) \leq 0$$

if u_+, u_- are the values to the right and left of the shock. We assume that the shock strength is small and know then that u_+ must lie on the curve parametrized by σ defined by (4.4.10). We write $u_+ = u(\sigma)$ and $c = c(\sigma)$. The left-hand side of (4.5.20) is then a function $J(\sigma)$ with

$$J'(\sigma) = \langle Y'(u(\sigma)), u'(\sigma) \rangle - c(\sigma)\langle \Phi'(u(\sigma)), u'(\sigma) \rangle - c'(\sigma)(\Phi(u(\sigma)) - \Phi(u_-)).$$

Using (4.5.18) and the fact that differentiation of the Rankine-Hugoniot condition

$$c(\sigma)(u(\sigma) - u_-) = f(u(\sigma)) - f(u_-)$$

gives that

$$c(\sigma)u'(\sigma) + c'(\sigma)(u(\sigma) - u_-) = f'(u(\sigma))u'(\sigma),$$

we obtain by Taylor's formula since $\langle Y'(u(\sigma)), u'(\sigma) \rangle = \langle \Phi'(u(\sigma)), f'(u(\sigma))u'(\sigma) \rangle$

$$J'(\sigma) = -c'(\sigma)\big(\Phi(u(\sigma)) - \Phi(u_-) - c'(\sigma)\langle \Phi'(u(\sigma)), u_- - u(\sigma)\rangle\big)$$
$$= c'(0)\Phi''(u(\sigma))(u_- - u(\sigma))(u_- - u(\sigma))/2 + O(\sigma^3).$$

Here $c'(0) = \langle \lambda_j'(u_-), r_j(u_-)\rangle/2 > 0$ by (4.4.4), (4.4.6)' and (4.4.9). If Φ is strictly convex in the direction $r_j(0)$, it follows when u is small enough that $J(\sigma)$ has the sign of σ, hence $J(\sigma) < 0$ precisely when $\sigma < 0$. This is the admissibility condition in (4.4.11), (4.4.12). Thus we have proved

Theorem 4.5.6. *If u is a small weak solution of the system* (4.5.1) *which is in C^1 on each side of a C^1 curve, then the admissibility condition* (4.4.11), (4.4.12) *is equivalent to* (4.5.19) *if $\Phi(u)$, $Y(u)$ is an entropy pair such that Φ is strictly convex in the eigenvector directions $r_j(0)$.*

In particular this is true for the solution of the Riemann problem constructed in Section 4.4. We shall now show that Glimm's solutions inherit the entropy inequality.

Theorem 4.5.7. *For almost all $\theta \in \Theta$ the solutions of* (4.5.1) *constructed by Glimm's method satisfy* (4.5.19) *if Φ, Y is an entropy pair such that Φ is strictly convex in the eigenvalue directions $r_j(0)$.*

Proof. Take the test function ϕ in the proof of Theorem 4.5.5 nonnegative and with support in the open half plane where $t > 0$. Since (4.5.19) holds for solutions of the Riemann problem, we then obtain

$$- \iint (\Phi(u^{h,l})\partial\phi/\partial t + Y(u^{h,l})\partial\phi/\partial x)\, dt\, dx$$

$$\leq \sum_{k=1}^{\infty} \int (v^{h,l}(kh, x) - v^{h,l}(kh - 0, x))\phi(kh, x)\, dx,$$

where $v^{h,l} = \Phi(u^{h,l})$. The uniform bound for the total variation of $u^{h,l}$ as a function of x implies one for the functions $v^{h,l}$ since Φ is Lipschitz continuous. Hence we obtain as before for almost all θ that the sum converges to 0 as $l = Ah \to 0$, for all ϕ in a dense subset of the nonnegative functions in C_0^1. This completes the proof.

Even for systems having entropy pairs with the required convexity properties it is not known if (4.5.18) characterizes the solutions of (4.5.1) uniquely. The question may no longer be so important for the uniqueness of the Glimm solution has now been proved in general with other methods. (See the survey paper Bressan [1] and the references there.)

In this section we have to some extent followed DiPerna [1]; the discussion of entropies here and in Section 4.6 has been taken from Lax [4]. A deterministic version of the construction has been given by Liu [1].

4.6. Entropy pairs. With the notation used in Section 4.5 two real valued functions $\Phi(u)$ and $Y(u)$ defined in a neighborhood of 0 in $\mathbf{R}^N$ are called an entropy pair with respect to the system (4.5.1) if (4.5.18) holds, or explicitly,

$$(4.6.1) \qquad \partial Y/\partial u_j = \sum_{k=1}^{N} \partial f_k/\partial u_j\, \partial\Phi/\partial u_k; \quad j = 1, \ldots, N.$$

These are N equations for two unknown functions so one can expect that in general there are no entropy pairs for $N > 2$ except the trivial ones where $\Phi(u) = \sum_1^N a_k u_k$ and $Y(u) = \sum_1^N a_k f_k(u)$ for some constants $a_1, \ldots, a_N$. Let us therefore assume at first that $N = 2$. The characteristic equations of the linear system (4.6.1) are

$$(4.6.2) \qquad Y\eta_j = \sum_{k=1}^{2} \partial f_k/\partial u_j\, \Phi\eta_k, \quad j = 1, 2,$$

where (η_1, η_2) is a dual vector to u. For a nontrivial solution we must have $\Phi \neq 0$, and η must be an eigenvector of the transpose of the Jacobian matrix f' with eigenvalue Y/Φ. The annihilators of these two directions are the directions of the eigenvectors $r_1(u)$, $r_2(u)$.

Thus the system is hyperbolic with respect to all curves which are not tangent to the eigenvector directions.

We can eliminate Y from the system (4.6.1) for locally the only relation between $\partial Y/\partial u_1$ and $\partial Y/\partial u_2$ is that $\partial(\partial Y/\partial u_1)/\partial u_2 = \partial(\partial Y/\partial u_2)/\partial u_1$. This gives

$$(4.6.3) \qquad \sum_1^2 \partial f_k/\partial u_1 \partial^2\Phi/\partial u_k\partial u_2 - \sum_1^2 \partial f_k/\partial u_2 \partial^2\Phi/\partial u_k\partial u_1 = 0$$

which must also be hyperbolic. The characteristic equation

$$\sum_1^2 \partial f_k/\partial u_1 \eta_k\eta_2 = \sum_1^2 \partial f_k/\partial u_2 \eta_k\eta_1$$

is just what we get by eliminating Y and Φ from (4.6.2). By a linear change of the u coordinates we can assume that at the origin the coordinate axes are tangent to r_j, so that the equation takes the form $\partial^2\Phi/\partial u_1\partial u_2 = 0$ at the origin. Now solve the Cauchy problem for (4.6.3) with the same Cauchy data as $u_1^2 + u_2^2$ when $u_1 + u_2 = 0$. Since $\partial^2(u_1^2+u_2^2)/\partial u_1\partial u_2 = 0$ it follows that $\Phi(u) - u_1^2 - u_2^2$ vanishes of third order at the origin. Hence Φ is convex in a neighborhood so we have proved

Theorem 4.6.1. When $N = 2$ it is possible to find entropy pairs Φ, Y locally for the system (4.5.1) so that Φ is strictly convex.

When $N > 2$ elimination of Y from (4.6.1) leads to $\binom{N}{2}$ equations

$$(4.6.4) \qquad \sum_k \partial f_k/\partial u_j \partial^2\Phi/\partial u_k\partial u_i = \sum_k \partial f_k/\partial u_i \partial^2\Phi/\partial u_k\partial u_j, \quad 1 \le i < j \le N.$$

These mean that the Hessian of Φ is diagonalized at any point where the eigenvectors r_j are tangential to the coordinate axes. In general we obtain if we multiply by the coordinates $r_{a,i}$ and $r_{b,j}$ of r_a and r_b and sum that if $a \ne b$ then

$$\sum_{i,j} \partial^2\Phi/\partial u_i\partial u_j r_{a,i}r_{b,j} = 0.$$

If $c \ne a$ and $c \ne b$ and we let $\sum r_{c,k}\partial/\partial u_k$ operate here, we obtain at a point where r_ν is the unit vector along the νth coordinate axis for every ν

$$\partial^3\Phi/\partial u_a\partial u_b\partial u_c + \sum_i \partial^2\Phi/\partial u_i\partial u_b\partial r_{a,i}/\partial u_c + \sum_j \partial^2\Phi/\partial u_a\partial u_j \partial r_{b,j}/\partial u_c = 0.$$

The first term is unchanged if we interchange a and c, so we get one more condition on the second order derivatives if we just do so and subtract. In general they show that entropy pairs with Φ strictly convex do not exist.

However, as observed by Friedrichs and Lax [1] there are many systems for which entropy pairs do exist. For example, by (4.6.1) it is possible to take $\Phi(u) = (u_1^2 + \cdots + u_N^2)/2$ if the equations

$$\partial Y/\partial u_j = \sum_k u_k\partial f_k/\partial u_j$$

can be solved, that is, if the equations

$$\partial(Y - \sum_k u_k f_k)/\partial u_j = -f_j, \quad j = 1, \ldots, N,$$

can be solved. This is possible locally if and only if

$$\partial f_j/\partial u_k = \partial f_k/\partial u_j, \quad j, k = 1, \ldots, N,$$

that is, if (4.5.1) is a *symmetric* hyperbolic system.

Equations which occur in physics often have naturally defined entropy pairs with a physical interpretation.

CHAPTER V

COMPENSATED COMPACTNESS

5.1. Introduction. In the existence proofs in Chapters III and IV via diffusion equations with small viscosity or Glimm's method it was essential that we could establish compactness properties in L^1_{loc} for the approximations and not only a priori L^∞ bounds. There are a (very) few other situations where L^∞ bounds are available but not more, so it is of interest to examine how far one can get then using weak L^∞ convergence. The difficulty is that even if $u_\nu \to u$ weakly in L^∞ it does not follow that $f(u_\nu) \to f(u)$ weakly in L^∞. Such facts are explained in a striking way by the so-called Young measures which we shall discuss at some length in Section 5.2. (The presentation follows Tartar [1] to a large extent.) In Section 5.3 we prove some results of Murat and Tartar on the consequence of differential equations satisfied by u_ν. We return in Sections 5.4 and 5.5 to the discussion of conservation laws from this point of view, following DiPerna [2, 3]. As an application we give in Section 5.6 an existence theorem for a system from elasticity theory without assuming the data to be small.

5.2. Weak convergence in L^∞. Let $\Omega \subset \mathbf{R}^n$ be an open set and let $u_\nu \in L^\infty(\Omega, K)$ be the set of bounded measurable functions with values in a set $K \subset \mathbf{R}^N$. If K is bounded it is always possible to select a weakly convergent subsequence, so assume that u_ν is weakly convergent. However, that does not imply that $f(u_\nu)$ is weakly convergent if f is a continuous function on K, nor that the limit is $f(\lim u_\nu)$ if it exists. Let us consider a basic example.

Lemma 5.2.1. *Let v_j, $j = 1, \ldots, J$, be measurable functions in $\mathbf{R}^n$, periodic with period 1 in each variable and taking only the values 0 and 1, and assume that $\sum_1^J v_j(x) \equiv 1$. Let $y_j \in L^\infty(\mathbf{R}^n, K)$, $j = 1, \ldots, J$, and set $u_\nu = \sum_1^J v_j(\nu\cdot)y_j \in L^\infty(\mathbf{R}^n, K)$. If f is a continuous real valued function on K then, as $\nu \to \infty$,*

$$f(u_\nu) = \sum_1^J v_j(\nu\cdot)f(y_j) \to \sum_1^J \lambda_j f(y_j), \quad \lambda_j = \int_T v_j(x)\,dx,$$

where the convergence is weak in $L^\infty(\mathbf{R}^n, \mathbf{R})$ and $T = \mathbf{R}^n/\mathbf{Z}^n$. One can choose v_j such that $\lambda_1, \ldots, \lambda_J$ are arbitrary numbers ≥ 0 with $\sum_1^J \lambda_j = 1$.

Proof. Assume at first that $y_1, \ldots, y_J$ are constant. If $\varphi \in C_0(\mathbf{R}^n)$, which is a dense subset of $L^1(\mathbf{R}^n)$, then

$$\int f(u_\nu(x))\varphi(x)\,dx = \sum_1^J f(y_j) \int_T \sum_{g \in \mathbf{Z}^n} \nu^{-n}\varphi((x+g)/\nu)v_j(x)\,dx$$

$$\to \sum_1^J f(y_j) \int_{\mathbf{R}^n} \varphi(z)\,dz \int_T v_j(x)\,dx,$$

for the Riemann sums converge uniformly in T to the integral of φ. Since $f(u_\nu)$ is uniformly bounded, this proves the asserted weak convergence when y_j are constant. If $y_1, \ldots, y_J$

take only finitely many values we can write any $\varphi \in L^1(\mathbf{R}^n)$ as a sum $\varphi = \sum_1^m \varphi_\mu$ where $\varphi_\mu \in L^1(\mathbf{R}^n)$ and $y_1, \ldots, y_J$ are constant when $\varphi_\mu \neq 0$. By the special case already proved

$$\int f(u_\nu)\varphi\,dx = \sum_{\mu=1}^m \int f(u_\nu)\varphi_\mu\,dx \to \sum_{\mu=1}^m \int \sum_{j=1}^J \lambda_j f(y_j)\varphi_\mu\,dx = \int \sum_{j=1}^J \lambda_j f(y_j)\varphi\,dx,$$

so the lemma is valid for such functions $y_1, \ldots, y_J$. Since they are dense in $L^\infty(\mathbf{R}^n, K)$, the statement follows in general.

Theorem 5.2.2. *If $K \subset \mathbf{R}^N$ then the weak closure of $L^\infty(\Omega, K)$ in $L^\infty(\Omega, \mathbf{R}^N)$ is equal to $L^\infty(\Omega, \widehat{K})$ where $\widehat{K}$ is the closed convex hull of K.*

Proof. First we prove that $L^\infty(\Omega, \widehat{K})$ is weakly closed. To do so we note that if L is a linear form on $\mathbf{R}^N$ and C a constant such that $C \leq L(y)$ when $y \in K$, and if $L^\infty(\Omega, \widehat{K}) \ni u_\nu \rightharpoonup u$, that is, $u_\nu \to u$ weakly, then

$$\int \varphi(x)L(u(x))\,dx = \lim_{\nu\to\infty} \int \varphi(x)L(u_\nu(x))\,dx \geq C \int \varphi(x)\,dx, \quad 0 \leq \varphi \in L^1(\Omega),$$

so $L(u(x)) \geq C$ almost everywhere. This is true for a countable set of pairs L_j, C_j such that $\widehat{K} = \cap_j \{y \in \mathbf{R}^n; L_j(y) \geq C_j\}$, which proves that $u(x) \in \widehat{K}$ almost everywhere.

Now let $u \in L^\infty(\Omega, \widehat{K})$. For $\varepsilon > 0$ we can decompose Ω into the disjoint union of measurable sets $\Omega_1, \ldots, \Omega_I$ such that for some Y_i in the convex hull of K we have $|u-Y_i| < \varepsilon$ in Ω_i for $i = 1, \ldots, I$. Hence it suffices to prove that for each i we can find $u_\nu \in L^\infty(\Omega_i, K)$ such that $u_\nu \rightharpoonup Y_i$ in $L^\infty(\Omega_i)$. This is possible by Lemma 5.2.1.

Theorem 5.2.3. *If F is a continuous real valued function on the convex set $K \subset \mathbf{R}^N$, then $u \mapsto \int_\Omega F(u)\,dx$, where $\Omega \subset \mathbf{R}^n$ is open and $m(\Omega) < \infty$, is lower semicontinuous on $L^\infty(\Omega, K)$ with the weak topology if and only if F is convex. Then $u_\nu \rightharpoonup u$ and $F(u_\nu) \rightharpoonup f$ implies $f \geq F(u)$ almost everywhere.*

Proof. We prove the last statement first, so assume that F is convex. If L is affinely linear with $L \leq F$ then $F(u_\nu) \geq L(u_\nu) \rightharpoonup L(u)$, hence $f \geq L(u)$ almost everywhere. Choosing countably many such linear L_j with $\sup L_j = F$ we conclude that $f \geq F(u)$ almost everywhere. Hence

$$\lim_{\nu\to\infty} \int_\Omega F(u_\nu)\,dx = \int_\Omega f\,dx \geq \int_\Omega F(u)\,dx$$

which proves the lower semicontinuity of the functional. Now *assume* instead that the functional is lower semi-continuous and apply it to the sequence constructed in Lemma 5.2.1. Then we obtain for $y_1, \ldots, y_J \in K$

$$\int_\Omega F\Big(\sum_1^J \lambda_j y_j\Big)\,dx \leq \lim_{\nu\to\infty} \int_\Omega F(u_\nu)\,dx = m(\Omega)\sum_1^J \lambda_j F(y_j),$$

which proves that F is convex.

Corollary 5.2.4. *The functional in Theorem 5.2.3 is continuous if and only if F is affinely linear.*

The preceding results are greatly clarified by the introduction of the *Young measure* corresponding to a function. (See Young [1].) Let $u \in L^\infty(\Omega, K)$ where $\Omega \subset \mathbf{R}^n$ is open and bounded and $K \subset \mathbf{R}^N$ is compact. We can associate with u the Young measure ν_u in $\Omega \times K$ defined by

$$(5.2.1) \qquad \int \varphi(x, y) \, d\nu_u(x, y) = \int \varphi(x, u(x)) \, dx, \quad \varphi \in C_0(\Omega \times K),$$

or equivalently

$$\int g(x)h(y) \, d\nu_u(x, y) = \int g(x)h(u(x)) \, dx, \quad h \in C(K), \ g \in C_0(\Omega).$$

If $\pi : \Omega \times K \to \Omega$ is the projection then

$$(5.2.2) \qquad \pi_* d\nu_u = dx,$$

where dx is the Lebesgue measure in Ω, for if $(\pi^* \psi)(x, y) = \psi(x)$ then

$$(5.2.2)' \qquad \int \pi^* \psi \, d\nu_u = \int \psi(x) \, dx, \quad \psi \in C_0(\Omega),$$

by (5.2.1). More generally, $\pi_* h(y) d\nu_u(x, y) = h(u(x)) \, dx$. From (5.2.2) it follows that if E is a null set in Ω, then $\pi^{-1}E$ is a null set with respect to $d\nu_u$. It is clear that $d\nu_u$ is supported by the closure of the graph of u. That the complement of the graph has measure 0 is therefore obvious if u is continuous, and it follows in general since if $u_j \in C(\Omega)$, $u_j \to u$ in $L^1(\Omega)$, and $|u_j| \leq C$, then

$$\int |u_j(x) - y| \, d\nu_u(x, y) = \int |u_j(x) - u(x)| \, dx \to 0;$$

if $u_j(x) \to u(x)$ almost everywhere with respect to dx, hence almost everywhere with respect to $d\nu_u$, we conclude by Fatou's lemma that $\int_{u(x) \neq y} d\nu_u(x, y) = 0$.

Theorem 5.2.5. *For any sequence $u_j \in L^\infty(\Omega, K)$ we can choose a subsequence $u_{j(k)}$ such that the Young measures $d\nu_{u_{j(k)}}$ converge weakly as measures to a positive measure $d\nu$ in $\Omega \times K$. Thus*

$$\int f(x, u_{j(k)}(x)) \, dx \to \int f(x, y) \, d\nu(x, y)$$

for every $f \in C_0(\Omega \times K)$, and $\pi_* d\nu = dx$. If $f(x, y) = g(x)h(y)$ with $g \in C_0(\Omega)$ and $h \in C(K)$, then

$$(5.2.3) \qquad \left| \int g(x)h(y) \, d\nu(x, y) \right| \leq \sup |h| \int |g(x)| \, dx$$

so

$$\langle g \otimes h, d\nu \rangle = \langle g, H_h \rangle$$

for some $H_h \in L^\infty(\Omega)$ with $\sup |H_h| \leq \sup |h|$; we have

$$h(u_{j(k)}) \rightharpoonup H_h \quad \text{as } k \to \infty.$$

Conversely, for every positive measure $d\nu$ in $\Omega \times \mathbf{R}^N$ with $\pi_* d\nu = dx$ and $\operatorname{supp} d\nu \subset \Omega \times K$ there exists a sequence $u_j \in L^\infty(\Omega, K)$ such that $d\nu_{u_j} \rightharpoonup d\nu$.

Proof. The total mass of $d\nu_{u_j}$ is uniformly bounded so the existence of a weakly convergent subsequence is clear, and (5.2.2)' follows for $d\nu$ since it is valid for all $d\nu_{u_j}$. The positivity of $d\nu$ gives

$$\left| \int g(x)h(y) \, d\nu(x, y) \right| \leq \sup |h| \int \pi^* |g(x)| \, d\nu(x, y) = \sup |h| \int |g(x)| \, dx,$$

which proves (5.2.3) and the following statement. Now let

$$M = \{d\nu_u; u \in L^\infty(\Omega, K)\},$$

and let $\overline{M}$ be the weak closure in the space of bounded measures on $\Omega \times K$. First we prove that $\overline{M}$ is convex. Let $u_1, \ldots, u_J \in L^\infty(\Omega, K)$ and $\lambda_1, \ldots, \lambda_J \geq 0, \sum_1^J \lambda_j = 1$. Choose v_j as in Lemma 5.2.1 and set

$$U_k(x) = \sum_1^J v_j(kx)u_j(x).$$

By Lemma 5.2.1 $U_k \to \sum_1^J \lambda_j u_j$ as $k \to \infty$. Since $h(U_k(x)) = \sum_1^J v_j(kx)h(u_j(x))$ we have $h(U_k) \to \sum_1^J \lambda_j h(u_j)$, so

$$\int g(x)h(y)\, d\nu_{U_k} = \int g(x)h(U_k(x))\, dx \to \sum_1^J \lambda_j \int g(x)h(u_j(x))\, dx$$

$$= \sum_1^J \lambda_j \int g(x)h(y)\, d\nu_{u_j},$$

which means that $d\nu_{U_k}$ converges weakly to $\sum_1^J \lambda_j d\nu_{u_j}$ and proves the convexity of $\overline{M}$.

What remains now is to determine the weakly closed convex hull of M, which is the intersection of the weakly closed half spaces containing M, by the Hahn-Banach theorem. Such a half space is defined as the set of measures $d\nu$ with

$$\int \varphi(x, y)\, d\nu(x, y) \leq C$$

where $\varphi \in C_0(\Omega \times K)$ and C is a constant. It contains M if

$$\int \varphi(x, u(x))\, dx \leq C \quad \text{for all } u \in L^\infty(\Omega, K).$$

This is equivalent to

$$\int \Phi(x)\, dx \leq C, \quad \text{where } \Phi(x) = \max_{y \in K} \varphi(x, y).$$

One way this is clear since $\varphi(x, y) \leq \Phi(x)$. In the other direction it follows since we can choose $u \in L^\infty(\Omega, K)$ taking only finitely many values so that $\varphi(x, u(x)) - \Phi(x)$ is as small as we please in Ω. If $d\nu$ is any positive measure in $\Omega \times K$ with $\pi_* d\nu = dx$, then

$$\int \varphi(x, y)\, d\nu(x, y) \leq \int \Phi(x)\, d\nu(x, y) = \int \Phi(x)\, dx \leq C$$

which completes the proof.

The structure of the measure $d\nu$ is very simple:

Proposition 5.2.6. *The following two properties of a positive measure $d\nu$ in $\Omega \times K$ are equivalent:*

(i) $\pi_* d\nu = dx$,

(ii) *For $x \in \Omega$ there exist probability measures ν_x on K depending measurably on x in the sense that $x \mapsto \int h(y)\, d\nu_x(y)$ for every $h \in C(K)$ is a measurable function of x such that*

$$\int h(x, y)\, d\nu(x, y) = \int dx \int h(x, y)\, d\nu_x(y), \quad h \in C_0(\Omega \times K).$$

Proof. (ii) ⇒ (i) for if h is independent of y then integration with respect to $d\nu_x$ just gives $h(x)$, since $d\nu_x$ is a probability measure. Assume now that (i) holds. If h is a continuous function on K, then $H_h \in L^\infty(\Omega)$ and $|H_h| \leq \sup |h|$ if H_h is defined by

$$\int g(x)h(y)\, d\nu(x,y) = \int g(x) H_h(x)\, dx.$$

We make H_h defined everywhere in Ω so that $|H_h| \leq \sup |h|$ and $H_h \geq 0$ if $h \geq 0$. Choose a countable dense subset $\mathcal{H}$ of $C(K)$ which is a vector space over the rationals containing the function $h \equiv 1$, and for each $h \in \mathcal{H}$ a function H_h with these properties and $H_1 \equiv 1$ by (i). For $x \notin E$ where E is a null set we have for $h \in \mathcal{H}$, $h_1, h_2 \in \mathcal{H}$ and rational t_1, t_2

$$H_h(x) \geq 0 \text{ if } h \geq 0, \quad H_{t_1 h_1 + t_2 h_2}(x) = t_1 H_{h_1}(x) + t_2 H_{h_2}(x), \quad H_1(x) = 1.$$

For $x \notin E$ this means that $\mathcal{H} \ni h \rightarrow H_h(x)$ defines a positive measure $d\nu_x$ of total mass 1, and the statements in (ii) follow at once.

If $d\nu_x$ is a Dirac measure for almost every x then it is located at

$$u(x) = \int y\, d\nu_x(y)$$

so $u(x)$ is measurable with values in K and

$$\int h(x,y)\, d\nu(x,y) = \int dx \int h(x,y)\, d\nu_x(y) = \int h(x, u(x))\, dx.$$

Thus $d\nu$ is the Young measure of some $u \in L^\infty(\Omega, K)$ if and only if $d\nu_x$ is a Dirac measure for almost every $x \in \Omega$.

Proposition 5.2.7. *If $u, u_k \in L^\infty(\Omega, K)$, $k = 1, 2, \ldots$, then $d\nu_{u_k} \rightarrow d\nu_u$ if and only if $u_k \rightarrow u$ in L^p for every $p \in [1, \infty)$.*

Proof. Assume at first that $u_k \rightarrow u$ in L^1. If $h \in C_0^1(\mathbf{R}^{n+N})$ then

$$\left| \int (h(x, u_k(x)) - h(x, u(x)))\, dx \right| \leq C \int |u_k(x) - u(x)|\, dx \rightarrow 0 \text{ as } k \rightarrow \infty.$$

Hence $d\nu_{u_k} \rightarrow d\nu_u$. On the other hand, assume that we know that $d\nu_{u_k} \rightarrow d\nu_u$ weakly as measures. Then

$$u_k \rightharpoonup u, \quad |u_k|^2 \rightharpoonup |u|^2,$$

which implies that

$$\int |u_k - u|^2\, dx = \int |u_k|^2\, dx + \int |u|^2\, dx - 2 \int (u_k, u)\, dx \rightarrow 0.$$

Hence $u_k \rightarrow u$ in L^2, which implies $u_k \rightarrow u$ in L^p for $2 < p < \infty$ by the uniform bounds and for $1 \leq p < 2$ by the Cauchy-Schwarz inequality.

Proposition 5.2.7 shows that one can obtain compactness for a sequence in a strong L^p topology if one is able to analyze the weak limit of the corresponding Young measures.

5.3. Weak convergence of solutions of linear differential equations. In this section we shall reexamine the results of Section 5.2 when $u \in L^\infty(\Omega, \mathbf{R}^N)$, $\Omega \subset \mathbf{R}^n$, satisfies a system of partial differential equations with constant coefficients

$$(5.3.1) \qquad \sum_1^n A_k \partial u / \partial x_k = 0$$

where A_k is a $M \times N$ matrix, so that we have M first order differential equations for the N unknowns. Weak limits of solutions are still solutions (in the sense of distribution theory). If (5.3.1) were elliptic then all solutions would be in C^∞ and weak convergence is equivalent to convergence in C^∞. It is therefore clear that the strength of the consequences of (5.3.1) must depend very much on the characteristics of (5.3.1).

For $\xi \in \mathbf{R}^n$ we denote the kernel of $\sum_1^n A_k \xi_k$ by $\Lambda(\xi)$. Thus $\xi \in \mathbf{R}^n \setminus 0$ is a characteristic if and only if $\Lambda(\xi) \neq \{0\}$. Set $\Lambda = \cup_{\xi \neq 0} \Lambda(\xi)$, and assume that Ω is open and bounded.

Proposition 5.3.1. *If $K \subset \mathbf{R}^N$ and the set of all $u \in L^\infty(\Omega, K)$ satisfying (5.3.1) is sequentially weakly closed, then K is closed and $K \cap (c + \Lambda(\xi))$ is convex for all $c \in \mathbf{R}^N$ and $\xi \in \mathbf{R}^n \setminus 0$. If F is a continuous function on K such that in addition*

$$L^\infty(\Omega, K) \ni u \mapsto \int F(u) \, dx$$

is weakly sequentially lower semicontinuous when u satisfies (5.3.1), then F is convex in $K \cap (c + \Lambda(\xi))$. The same is true if for every sequence $u_\nu \in L^\infty(\Omega, K)$ with $u_\nu \rightharpoonup u$ and $F(u_\nu) \rightharpoonup f$ we have $f \geq F(u)$ almost everywhere.

Proof. That K must be closed follows if we just consider the constant solutions of (5.3.1). Let $c_j \in \Lambda(\xi)$, $c + c_j \in K$, $j = 1, \ldots, J$. Then

$$u(x) = \sum_1^J c_j v_j(\langle x, \xi \rangle)$$

is a solution of (5.3.1) for arbitrary $v_j \in L^\infty(\mathbf{R})$, since $c_j \in \Lambda(\xi)$. Choose v_j as in Lemma 5.2.1. Then $\sum_1^J v_j = 1$ and

$$u_\nu(x) = \sum_1^J (c + c_j) v_j(\nu\langle x, \xi \rangle) = c + \sum_1^J c_j v_j(\nu\langle x, \xi \rangle)$$

is a solution of (5.3.1) which belongs to $L^\infty(\Omega, K)$. When $\nu \to \infty$ it converges weakly to

$$u = \sum_1^J \lambda_j(c + c_j)$$

so this constant must belong to K which proves the convexity. We have

$$\int_\Omega F(u_\nu) \, dx = \int_\Omega \sum_1^J F(c + c_j) v_j(\nu\langle x, \xi \rangle) \, dx \to \sum_1^J \lambda_j F(c + c_j) m(\Omega),$$

while

$$\int_\Omega F(u) \, dx = m(\Omega) F\left(\sum_1^J \lambda_j(c + c_j)\right).$$

This proves that F must be convex in $K \cap (c + \Lambda(\xi))$ if the functional is sequentially lower semicontinuous. The last statement follows at the same time.

When F is a quadratic form the convexity condition reduces to $F(\lambda) \geq 0$ for $\lambda \in \Lambda$. We shall then prove that it is also sufficient (Tartar [1, p. 170]). In fact, we state a more general result involving standard Sobolev spaces $H_{(s)}$ of functions with s derivatives in L^2; more precisely, $v \in H_{(s)}(\mathbf{R}^n)$ if $v \in \mathcal{S}'(\mathbf{R}^n)$ and $(1 + |\xi|^2)^{s/2}\hat{v}(\xi) \in L^2(\mathbf{R}^n)$. In particular, $H_{(-1)}(\mathbf{R}^n)$ is the set of distributions u in $\mathbf{R}^n$ which can be written in the form $u = u_0 + \sum_1^n \partial_j u_j$ where $u_0, \ldots, u_n \in L^2(\mathbf{R}^n)$. (See e.g. Hörmander [4, Section 7.9] or Section 8.2 below for more details.) If instead one requires that $u_0, \ldots, u_n \in L^p(\mathbf{R}^n)$, this space of distributions is denoted by $W^{-1,p}(\mathbf{R}^n)$; it will occur in Lemma 5.3.4 below. If $\Omega \subset \mathbf{R}^n$ is an open set then $H^{loc}_{(-1)}(\Omega)$ is the set of distributions u in Ω such that $\varphi u \in H_{(-1)}(\mathbf{R}^n)$ for every $\varphi \in C_0^\infty(\Omega)$. The restriction to Ω of any distribution in $H_{(-1)}(\mathbf{R}^n)$ is in $H^{loc}_{(-1)}(\Omega)$, but not all distributions in $H^{loc}_{(-1)}(\Omega)$ are such restrictions since they have unrestricted growth at the boundary. Similarly one defines $W^{-1,p}_{loc}(\Omega)$ when $p \in (1, \infty)$.

Theorem 5.3.2. *Let u_ν be a sequence in $L^2_{loc}(\Omega, \mathbf{R}^N)$ such that $u_\nu \rightharpoonup u$ in $L^2_{loc}(\Omega, \mathbf{R}^N)$ and $\sum_{k=1}^n A_k \partial u_\nu / \partial x_k$ is precompact in $H^{loc}_{(-1)}(\Omega, \mathbf{R}^M)$. If Q is a quadratic form in $\mathbf{R}^N$ such that $Q(\lambda) \geq 0$ for $\lambda \in \Lambda$, it follows that*

$$(5.3.2) \qquad dq \geq Q(u)$$

for every weak measure limit dq of the sequence $Q(u_\nu)$.

Proof. (5.3.2) is equivalent to

$$\int \varphi^2 \, dq \geq \int \varphi^2 Q(u) \, dx = \int Q(\varphi u) \, dx$$

for all $\varphi \in C_0^\infty(\Omega, \mathbf{R})$. Since $\varphi u_\nu \rightharpoonup \varphi u$, and

$$\sum_{k=1}^n A_k \partial (\varphi u_\nu) / \partial x_k = \varphi \sum_{k=1}^n A_k \partial u_\nu / \partial x_k + \sum_{k=1}^n A_k u_\nu \partial \varphi / \partial x_k$$

is precompact in $H_{(-1)}$, it is enough to prove that

$$\int dq \geq \int Q(u) \, dx$$

if supp u_ν belongs to a fixed compact subset of Ω. Writing $u_\nu = u + v_\nu$ we have $v_\nu \rightharpoonup 0$ in L^2, and

$$Q(u_\nu) = Q(u) + Q(v_\nu) + 2\tilde{Q}(u, v_\nu)$$

if $\tilde{Q}$ is the polarized form of Q. Since $\int \tilde{Q}(u, v_\nu) \, dx \to 0$ it is therefore sufficient to prove that

$$(5.3.3) \qquad \lim_{\nu \to \infty} \int Q(v_\nu) \, dx \geq 0.$$

To prove (5.3.3) we observe that the hypothesis on Q means that for every $\varepsilon > 0$ there is a constant C_ε such that

$$(5.3.4) \qquad Q(\lambda) + \varepsilon |\lambda|^2 + C_\varepsilon (|A(\xi)\lambda|^2 + |\lambda|^2)/(1 + |\xi|^2) \geq 0, \quad \xi \in \mathbf{R}^n, \lambda \in \mathbf{R}^N.$$

In fact, this is obvious when $|\xi| \leq 1$ and also for λ in a conic neighborhood of Λ since $Q(\lambda) + \varepsilon |\lambda|^2 > 0$ there. Outside such a neighborhood we have a positive lower bound for $|A(\xi)\lambda|$ if $|\xi| = 1$, $|\lambda| = 1$, which proves (5.3.4) for reasons of homogeneity when $|\xi| \geq 1$ since $1 + |\xi|^2 \leq 2|\xi|^2$ then. If we now take $v \in L^2(\mathbf{R}^n, \mathbf{R}^N)$, apply (5.3.4) with $\lambda = \operatorname{Re} \hat{v}(\xi)$ and $\lambda = \operatorname{Im} \hat{v}(\xi)$ where $\hat{v}$ is the Fourier transform of v, integrate and apply Parseval's formula, it follows that

$$(5.3.5) \qquad \int (Q(v) + \varepsilon |v|^2)\, dx + C_\varepsilon \left(\left\| \sum_1^n A_k \partial v / \partial x_k \right\|_{(-1)}^2 + \|v\|_{(-1)}^2 \right) \geq 0.$$

We apply this to the sequence v_ν in the first part of the proof, noting that $v_\nu \rightharpoonup 0$ and that $\sum A_k \partial v / \partial x_k \to 0$ strongly in $H_{(-1)}$. Then it follows that

$$\varliminf_{\nu \to \infty} \int (Q(v_\nu) + \varepsilon |v_\nu|^2)\, dx \geq 0.$$

Since v_ν is bounded in L^2 and ε is any positive number, the inequality (5.3.3) is proved and the proof is complete.

Corollary 5.3.3. *Let u_ν and v_ν be p and q forms respectively in Ω, such that $u_\nu \rightharpoonup u$ and $v_\nu \rightharpoonup v$ in L^2_{loc} and $du_\nu \to du$, $dv_\nu \to dv$ strongly in $H^{\mathrm{loc}}_{(-1)}$. Then it follows that $u_\nu \wedge v_\nu \rightharpoonup u \wedge v$ in the weak topology of measures.*

Proof. (u_ν, v_ν) takes its values in the vector space $\wedge^p \mathbf{R}^n \oplus \wedge^q \mathbf{R}^n$, and

$$\wedge^p \mathbf{R}^n \oplus \wedge^q \mathbf{R}^n \ni (U, V) \mapsto U \wedge V \in \wedge^{p+q} \mathbf{R}^n$$

is a vector valued quadratic form. For the differential operator $(u, v) \mapsto (du, dv)$ the set $\Lambda(\xi)$ becomes the set of all (U, V) such that $U \wedge \xi = 0$ and $V \wedge \xi = 0$, that is, $U = U_1 \wedge \xi$ and $V = V_1 \wedge \xi$ for some forms U_1 and V_1 of degree $p - 1$ and $q - 1$. Hence $U \wedge V = 0$, that is, all components of $Q(U, V) = \pm U \wedge V$ satisfy the condition in Theorem 5.3.2, so $u_\nu \wedge v_\nu \rightharpoonup u \wedge v$ weakly as measures.

The special case of forms of degree 1 and $n - 1$ is known as the "div curl lemma of Murat and Tartar"; also the general corollary is well known.

We shall end this section with a lemma of Murat (see Tartar [1]) which will be used to verify the hypotheses of Theorem 5.3.2 in a later application.

Lemma 5.3.4. *Let Ω be open in $\mathbf{R}^n$, let E_1 be a compact subset of $H^{\mathrm{loc}}_{(-1)}(\Omega)$, E_2 a bounded set in the space of measures in Ω, and E_3 a bounded set in the Sobolev space $W^{-1,p}_{\mathrm{loc}}(\Omega)$ for some $p > 2$. Then it follows that $(E_1 + E_2) \cap E_3$ is precompact in $H^{\mathrm{loc}}_{(-1)}(\Omega)$.*

Proof. It suffices to prove the statement for elements with support in a fixed compact set $K \subset \Omega$. Let $g_j \in E_j$, $j = 1, 2, 3$, and $g_3 = g_1 + g_2$. Set

$$h_j = (1 + |D|^2)^{-1/2} g_j, \text{ that is, } \hat{h}_j(\xi) = \hat{g}_j(\xi) / (1 + |\xi|^2)^{1/2}$$

where $\hat{\ }$ denotes Fourier transforms. Thus $h_j = F * g_j$ where $\widehat{F}(\xi) = (1 + |\xi|^2)^{-\frac{1}{2}}$, which implies that F is in $C^\infty(\mathbf{R}^n \setminus \{0\})$ and rapidly decreasing at infinity since the Fourier transform $(i\partial_\xi)^\alpha \widehat{F}(\xi)$ of $x^\alpha F$ is in L^1 when $|\alpha| + 1 > n$. At the origin we have $F(x) = O(|x|^{1-n})$ if $n > 1$, $F(x) = O(\log |x|)$ if $n = 1$, for the Fourier transform of $\varepsilon^{n-1} F(\varepsilon x)$ is $(\varepsilon^2 + |\xi|^2)^{-\frac{1}{2}}$ which implies that $\varepsilon^{n-1} F(\varepsilon x)$ can be estimated by $C \int_{|\xi| < 1} (\varepsilon^2 + |\xi|^2)^{-\frac{1}{2}}\, d\xi$ when $|x| = 1$.

By assumption h_1 belongs to a compact subset of L^2, and we have a uniform exponential decrease outside K, so h_1 belongs to a compact subset of L^q for every $q \in [1,2]$. Moreover, h_2 belongs to a compact subset of L^q if $1 \le q < n/(n-1)$ since $F \in L^q$ then. By hypothesis h_3 is bounded in L^p for some $p > 2$ and since $h_3 = h_1 + h_2$ we have proved compactness in L^q if $1 \le q \le 2$ and $(n-1)q < n$. Now a sequence h_3^ν which converges in L^q for some $q \in [1,2)$ and is bounded in L^p for some $p > 2$ must converge in L^2 by Hölder's inequality (the logarithmic convexity in r of the L^r norm). But convergence of h_3^ν in L^2 is equivalent to convergence of g_3^ν in $H_{(-1)}$ which completes the proof.

Remark. An exposition of the results in this section from the point of view of harmonic analysis and refinements of them can be found in Coifman *et al.* [1].

5.4. A scalar conservation law with one space variable. Following Tartar [1] we shall now show how the results obtained in Sections 5.2 and 5.3 can be applied to the study of the Cauchy problem

$$(5.4.1) \qquad \partial u/\partial t + \partial f(u)/\partial x = 0, \quad u(0,x) = u_0(x),$$

where $u_0 \in L^\infty(\mathbf{R})$ is small at infinity. First we note that weak L^∞ limits of weak solutions of the equation need not be solutions. In fact, choose any solution of the Rankine-Hugoniot relation

$$f(u_2) - f(u_1) = s(u_2 - u_1).$$

Any function which is equal to u_2 resp. u_1 on the two sides of a line with speed s is then a weak solution. Let $0 < \theta < 1$ and denote by v_1, v_2 the characteristic functions of $(0,\theta)$, $(\theta,1)$, continued with period 1. Set for positive integers ν

$$u^\nu(t,x) = \begin{cases} u_1, & \text{for } x < st, \\ \sum_1^2 u_j v_j(\nu(x-st)), & \text{for } 0 \le x - st \le 1, \\ u_2, & \text{for } x > st + 1, \end{cases}$$

which is a weak solution of the equation in (5.4.1). By Lemma 5.2.1 we have $u^\nu \rightharpoonup u$ as $\nu \to \infty$, where

$$u(t,x) = \begin{cases} u_1, & \text{for } x < st, \\ \theta u_1 + (1-\theta)u_2, & \text{for } 0 \le x - st \le 1, \\ u_2, & \text{for } x > st. \end{cases}$$

The Rankine-Hugoniot condition is not fulfilled unless

$$f(\theta u_1 + (1-\theta)u_2) - f(u_2) = s(\theta u_1 + (1-\theta)u_2 - u_2),$$

that is,

$$f(u_2 + \theta(u_1 - u_2)) = f(u_2) + s\theta(u_1 - u_2), \quad 0 < \theta < 1,$$

which means that f is linear in $[u_1, u_2]$.

Returning to (5.4.1) we introduce a viscosity term and consider the solution of the diffusion equation

$$(5.4.2) \qquad \partial u^\mu/\partial t + \partial f(u^\mu)/\partial x = \mu \partial^2 u^\mu/\partial x^2; \quad u^\mu(0,x) = u_0(x).$$

We recall that there is a solution which is small at infinity, and that $|u^\mu| \le \sup |u_0| = M$. Let Φ be a C^2 strictly convex function and introduce Y as before so that

$$(5.4.3) \qquad Y'(u) = f'(u)\Phi'(u).$$

Then (5.4.2) gives

$$(5.4.4) \qquad \partial\Phi(u^\mu)/\partial t + \partial Y(u^\mu)/\partial x = \frac{\partial}{\partial x}(\mu\Phi'(u^\mu)\partial u^\mu/\partial x) - \mu\Phi''(u^\mu)(\partial u^\mu/\partial x)^2,$$

or after integration for $0 \le t \le T$

$$\mu \iint_{0<t<T} \Phi''(u^\mu)(\partial u^\mu/\partial x)^2 \, dt \, dx \le \int \Phi(u_0(x)) \, dx$$

if $\Phi \ge 0$. Since $\Phi''(u)$ has a positive lower bound for $|u| \le M$ we conclude that

$$\iint_{t>0} |\partial u^\mu/\partial x|^2 \, dt dx = O(1/\mu),$$

hence that

$$\iint_{t>0} |\mu\Phi'(u^\mu)\partial u^\mu/\partial x|^2 \, dt dx = O(\mu).$$

The right-hand side of (5.4.4) is therefore the sum of one term which tends to 0 in $H^{loc}_{(-1)}$ as $\mu \to 0$ and one which is a negative measure with uniformly bounded total mass if $\Phi(0) = 0$. On the other hand, the left-hand side is bounded in $W^{-1,p}_{loc}$ for any p since $|u^\mu| \le M$. The same is true for the equation (5.4.2). If we choose a sequence $\mu_j \to 0$ such that the Young measure of u^{μ_j} converges weakly when $t > 0$ to a measure $d\nu$, then Lemma 5.3.4 shows that the sequence

$$v^j = (u^{\mu_j}, f(u^{\mu_j}), \Phi(u^{\mu_j}), Y(u^{\mu_j})) = (v_1^j, v_2^j, v_3^j, v_4^j)$$

satisfies the hypotheses of Theorem 5.3.2 for the operator defined by

$$A_0\partial v/\partial t + A_1\partial v/\partial x = (\partial v_1/\partial t + \partial v_2/\partial x, \partial v_3/\partial t + \partial v_4/\partial x).$$

The characteristic equations for this system are

$$v_1\tau + v_2\xi = 0, \quad v_3\tau + v_4\xi = 0,$$

which implies $v_1v_4 - v_2v_3 = 0$ when $(\tau, \xi) \ne (0,0)$. Hence the quadratic forms $Q(v) = v_1v_4 - v_2v_3$ and $-Q(v)$ satisfy the hypotheses of Theorem 5.3.2, which proves that

$$u^{\mu_j}Y(u^{\mu_j}) - f(u^{\mu_j})\Phi(u^{\mu_j}) \longrightarrow uY(u) - f(u)\Phi(u)$$

in the weak measure topology if u is the weak limit of u^{μ_j}. Since

$$\int g(t,x)h(\lambda)d\nu(t,x,\lambda) = \lim_{j \to \infty} \int g(t,x)h(u^{\mu_j}(t,x)) \, dt \, dx,$$

if $g \in C_0(\mathbf{R}^2)$ and $h \in C(\mathbf{R})$, we obtain with $h(\lambda) = \lambda Y(\lambda) - f(\lambda)\Phi(\lambda)$

$$\int g(t,x)h(\lambda) \, d\nu(t,x,\lambda) = \int g(t,x)(u(t,x)Y(u(t,x)) - f(u(t,x))\Phi(u(t,x))) \, dt \, dx.$$

If $\nu_{t,x}$ is the probability measure defined by ν at (t,x) it follows that for almost all (t,x),

$$(5.4.5) \qquad \begin{aligned} &\langle \lambda Y(\lambda) - f(\lambda)\Phi(\lambda), d\nu_{t,x}(\lambda)\rangle \\ &= \langle \lambda, d\nu_{t,x}(\lambda)\rangle\langle Y(\lambda), d\nu_{t,x}(\lambda)\rangle - \langle f(\lambda), d\nu_{t,x}(\lambda)\rangle\langle \Phi(\lambda), d\nu_{t,x}(\lambda)\rangle, \end{aligned}$$

for $u(t,x) = \langle \lambda, d\nu_{t,x}(\lambda)\rangle, \ldots, \Phi(u(t,x)) = \langle \Phi(\lambda), d\nu_{t,x}(\lambda)\rangle$.

Lemma 5.4.1. *A compactly supported probability measure $d\nu_{t,x}$ satisfies (5.4.5) for all positive C^∞ strictly convex functions Φ when Y is defined by (5.4.3) if and only if f' is constant in the smallest interval containing supp $d\nu$.*

Proof. Let us first note that adding a constant to Y or to Φ changes both sides of (5.4.5) in the same way, and (5.4.5) is valid for $\Phi(u) = u$, $Y(u) = f(u)$. The positivity of Φ is therefore not essential. Nor is the convexity, for every Φ is the difference between two strictly convex functions. Dropping the positivity and convexity conditions now we also drop the subscript t, x in order to shorten notation, and we write

$$\bar{u} = \langle \lambda, d\nu \rangle, \quad \bar{f} = \langle f(\lambda), d\nu \rangle$$

for the expected values of λ and of $f(\lambda)$, which are almost everywhere equal to the weak limits of u^{μ_j} and of $f(u^{\mu_j})$. Then

$$(\lambda - \bar{u})d\nu = G', \quad (f(\lambda) - \bar{f})d\nu = H',$$

where G and H have bounded variation and supports contained in the smallest interval $[\alpha, \beta]$ containing supp $d\nu$. We can write (5.4.5) in the form

$$\langle G', Y \rangle = \langle H', \Phi \rangle$$

for the right-hand side of (5.4.5) is $\bar{u}\langle Y, d\nu \rangle - \bar{f}\langle \Phi, d\nu \rangle$, so an integration by parts gives

$$\langle H, \Phi' \rangle = \langle G, Y' \rangle = \langle G, f'\Phi' \rangle.$$

Since Φ' is arbitrary this means that $f'G = H$. Recalling that

(5.4.6) $$(f(\lambda) - \bar{f})G' = (\lambda - \bar{u})H'$$

we conclude that

(5.4.7) $$(f(\lambda) - \bar{f})G = (\lambda - \bar{u})H$$

for the derivatives agree and both sides have compact support. From the equations (5.4.6) and (5.4.7) it follows that

$$G'H - GH' = 0 \quad \text{when } \lambda \neq \bar{u}.$$

Since $G' = (\lambda - \bar{u})d\nu$ and $d\nu \geq 0$ we have $G < 0$ in (α, β), hence $f' = H/G$ is a constant in $(\alpha, \bar{u})$ and in $(\bar{u}, \beta)$, so f' is a constant s in (α, β) for reasons of continuity. This implies $f(\lambda) - \bar{f} = s(\lambda - \bar{u})$ in $[\alpha, \beta]$. (These statements are trivial if $\alpha = \beta$.) The proof is complete.

In the preceding proof we also saw that $\bar{f} = f(\bar{u})$. This means that

$$f(u^{\mu_j}) \rightharpoonup f(u),$$

so u is a weak solution of the equation (5.4.1). Since

$$\iint_{t>0} (u^\mu \partial\psi/\partial t + f(u^\mu)\partial\psi/\partial x) \, dt \, dx + \int u_0(x)\psi(0, x) \, dx$$

$$= \mu \iint_{t>0} \partial u^\mu/\partial x \partial\psi/\partial x \, dt \, dx \to 0, \quad \mu \to 0, \text{ if } \psi \in C_0^1(\mathbf{R}^2),$$

we conclude using the uniform bound for u^μ that u is a weak solution of the Cauchy problem (5.4.1). Moreover,

$$f'(u^{\mu_j}) \to f'(u)$$

in L^p_{loc} for $1 \leq p < \infty$ by Proposition 5.2.7 and Lemma 5.4.1. If there is no interval where f is affine, then $u^{\mu_j} \to u$ too. In that case we also conclude that u satisfies the entropy condition

$$\partial \Phi(u)/\partial t + \partial Y(u)/\partial x \leq 0$$

for every convex Φ.

Altogether we see that the results obtained are weaker than those of Chapter III. Besides the intrinsic interest of the methods, the justification for the discussion is the potential for obtaining additional results for systems where only fairly weak *a priori* estimates are available. This will be the topic of the next two sections.

An alternative to the proof of Lemma 5.4.1 can be given if one notes that the proof of (5.4.5) shows that

(5.4.5)′
$$\begin{aligned} \langle \Phi_1(\lambda), d\nu \rangle \langle Y_2(\lambda), d\nu \rangle &- \langle Y_1(\lambda), d\nu \rangle \langle \Phi_2(\lambda), d\nu \rangle \\ &= \langle \Phi_1(\lambda) Y_2(\lambda) - Y_1(\lambda) \Phi_2(\lambda), d\nu \rangle \end{aligned}$$

for all entropy pairs (Φ_j, Y_j), that is, solutions of $Y_j' = f'\Phi_j'$. (5.4.5) is the special case where one of them is the trivial pair $(u, f(u))$. Assume that supp $d\nu$ is not a point, and let $[\alpha, \beta]$ be the smallest interval containing supp $d\nu$. Choose $\gamma \in (\alpha, \beta)$ and Φ_1, Φ_2 vanishing to the right and to the left of γ respectively, and set

$$Y_j(\lambda) = \int_\gamma^\lambda f'(t) \Phi_j'(t)\, dt.$$

Then supp Y_j is located in the same way, so the right-hand side of (5.4.5)′ is equal to 0. If Φ_2 is chosen so that $\langle \Phi_2, d\nu \rangle \neq 0$ and $c = \langle Y_2, d\nu \rangle / \langle \Phi_2, d\nu \rangle$, it follows that

$$\langle Y_1 - c\Phi_1, d\nu \rangle = 0.$$

The derivative of $Y_1 - c\Phi_1$ is $(f' - c)\Phi_1'$. If $f' - c$ does not vanish identically in (α, γ) we can choose Φ_1 so that $(f' - c)\Phi_1'$ is ≤ 0 and not identically 0. But this implies $Y_1 - c\Phi_1 \geq 0$ with strict inequality at α, so $\langle Y_1 - c\Phi_1, d\nu \rangle > 0$, which is a contradiction. Hence $f' = c$ in (α, γ), and since γ was any point in (α, β) it follows that $f' = c$ in $[\alpha, \beta]$. Thus the preceding argument leads to the same conclusion as in Lemma 5.4.1, and it is closer to the arguments we shall use in the following section for systems of two equations.

5.5. Probability measures associated with a system of two equations. In this section we shall discuss the analogue of Lemma 5.4.1 for a hyperbolic system of two conservation laws

(5.5.1) $$\partial u/\partial t + \partial f(u)/\partial x = 0,$$

where $u = (u_1, u_2)$ and $f = (f_1, f_2)$. As pointed out in Section 4.6 there are numerous entropy pairs (Φ, Y) at least locally then. To simplify the discussion we introduce new coordinates (v_1, v_2) (Riemann coordinates) instead of (u_1, u_2) so that v_1 (resp. v_2) is constant along the orbits of the eigenvector field r_2 (resp. r_1). This is always possible locally and can be done globally for many systems. As observed at the beginning of Section 4.3 the equation (5.5.1) for classical solutions then becomes

(5.5.2) $$\partial v_j/\partial t + \lambda_j(v)\partial v_j/\partial x = 0, \quad j = 1, 2,$$

and (Φ, Y), considered as function of v, is an entropy pair if these equations imply

$$\partial\Phi/\partial t + \partial Y/\partial x = 0,$$

that is, if

(5.5.3) $$\partial Y/\partial v_j = \lambda_j(v)\partial\Phi/\partial v_j, \quad j = 1, 2.$$

We can always assume that λ_j are defined in $\mathbf{R}^2$ and that $\lambda_1 \neq \lambda_2$ everywhere.

Eliminating Y from the linear system (5.5.3) we get

$$\partial(\lambda_1\partial\Phi/\partial v_1)/\partial v_2 = \partial(\lambda_2\partial\Phi/\partial v_2)/\partial v_1,$$

that is,

$$(\lambda_1 - \lambda_2)\partial^2\Phi/\partial v_1\partial v_2 = \partial\lambda_2/\partial v_1\partial\Phi/\partial v_2 - \partial\lambda_1/\partial v_2\partial\Phi/\partial v_1.$$

The characteristics are the parallels of the coordinate axes. Our problem is to determine all probability measures $d\nu$ with compact support such that

(5.5.4) $$\langle\Phi_1, d\nu\rangle\langle Y_2, d\nu\rangle - \langle\Phi_2, d\nu\rangle\langle Y_1, d\nu\rangle = \langle\Phi_1 Y_2 - \Phi_2 Y_1, d\nu\rangle,$$

for arbitrary entropy pairs (Φ_j, Y_j).

In the analogous problem for one dimension we know from Lemma 5.4.1 that the constant coefficient case is exceptional: the measure is arbitrary then. Let us therefore look first at the case where λ_j are constant. Then the equations (5.5.3) mean that

$$Y - \lambda_1\Phi = h(v_2)(\lambda_1 - \lambda_2); \quad Y - \lambda_2\Phi = g(v_1)(\lambda_1 - \lambda_2),$$

or equivalently,

$$\Phi(v) = g(v_1) - h(v_2); \quad Y = \lambda_1 g(v_1) - \lambda_2 h(v_2).$$

If $(\widetilde{\Phi}, \widetilde{Y})$ is another entropy pair, represented by $\tilde{g}$ and $\tilde{h}$, then the determinant $\Phi\widetilde{Y} - \widetilde{\Phi}Y$ is equal to $(\lambda_1 - \lambda_2)(g\tilde{h} - \tilde{g}h)$. Let $d\nu_j$ be the projection of $d\nu$ on the v_j axis. Then (5.5.4) becomes

$$((g, d\nu_1) - \langle h, d\nu_2\rangle)(\lambda_1\langle\tilde{g}, d\nu_1\rangle - \lambda_2\langle\tilde{h}, d\nu_2\rangle) - (\langle\tilde{g}, d\nu_1\rangle - \langle\tilde{h}, d\nu_2\rangle)(\lambda_1\langle g, d\nu_1\rangle$$
$$-\lambda_2\langle h, d\nu_2\rangle) = (\lambda_1 - \lambda_2)\langle g(v_1)\tilde{h}(v_2) - \tilde{g}(v_1)h(v_2), d\nu\rangle$$

which simplifies to

$$\langle g, d\nu_1\rangle\langle\tilde{h}, d\nu_2\rangle - \langle\tilde{g}, d\nu_1\rangle\langle h, d\nu_2\rangle = \langle g(v_1)\tilde{h}(v_2) - \tilde{g}(v_1)h(v_2), d\nu\rangle.$$

Thus $d\nu$ is the product of the measures $d\nu_j$, and any such product of probability measures satisfies (5.5.4).

Following DiPerna [2] we shall now study (5.5.4) for general variable λ_j by applying (5.5.4) to solutions of (5.5.3) obtained from the asymptotic expansions of geometrical optics. (See Lax [4].) Thus we shall first discuss formal asymptotic solutions of the form

$$\Phi^\omega(v) \sim e^{\omega v_1}\sum_0^\infty \Phi_j(v)\omega^{-j}, \quad Y^\omega(v) \sim e^{\omega v_1}\sum_0^\infty Y_j(v)\omega^{-j},$$

where $\omega \to \pm\infty$. This is a formal solution of (5.5.3) if

$$Y_0 = \lambda_1 \Phi_0; \quad Y_j - \lambda_1 \Phi_j + \partial Y_{j-1}/\partial v_1 - \lambda_1 \partial \Phi_{j-1}/\partial v_1 = 0, \; j > 0,$$
$$\partial Y_j/\partial v_2 = \lambda_2 \partial \Phi_j/\partial v_2, \quad j \geq 0.$$

In particular $Y_0 = \lambda_1 \Phi_0$ and

$$\partial(\lambda_1 \Phi_0)/\partial v_2 = \lambda_2 \partial \Phi_0/\partial v_2,$$

that is,

$$(\lambda_1 - \lambda_2)\partial \Phi_0/\partial v_2 + \Phi_0 \partial \lambda_1/\partial v_2 = 0$$

(the transport equation). Similarly we have for $j > 0$

$$(\lambda_1 - \lambda_2)\partial \Phi_j/\partial v_2 + \Phi_j \partial \lambda_1/\partial v_2 = \frac{\partial}{\partial v_2}(\partial Y_{j-1}/\partial v_1 - \lambda_1 \partial \Phi_{j-1}/\partial v_1).$$

Together with the equations $Y_j - \lambda_1 \Phi_j + \partial Y_{j-1}/\partial v_1 - \lambda_1 \partial \Phi_{j-1}/\partial v_1 = 0$ the equations can be solved recursively with given initial values for all Φ_j for some constant value of v_2.

Choose Φ^ω with the asymptotic expansion just determined. It will not be an exact solution of (5.5.3) but rather a solution of a corresponding inhomogeneous equation with inhomogeneity $O(e^{\omega v_1} \omega^{-N})$ for every N, and similarly for the derivatives. Now we can use any of the standard existence theorems for second order linear hyperbolic differential equations to solve this inhomogeneous system of differential equations with Cauchy data 0 on a line $v_1 + v_2 = \text{const}$ to the left of supp $d\nu$. Since the solution only depends on the inhomogeneity for smaller values of v_1 it follows if $\omega > 0$ that the solution has a similar bound. Thus there exists an exact solution of (5.5.3) with the calculated asymptotic expansion in a neighborhood of supp $d\nu$. (We argue similarly when $\omega \to -\infty$, taking Cauchy data zero on a line to the right of supp $d\nu$.) From now on Φ^ω, Y^ω will denote this solution.

Let us now consider a triple of entropy pairs (Φ_j, Y_j), $j = 1, 2, 3$. (Note that the meaning of subscripts is changed in this discussion.) Obviously the determinant

$$\begin{vmatrix} \langle \Phi_1, d\nu \rangle & \langle Y_1, d\nu \rangle & \langle \Phi_1, d\nu \rangle \\ \langle \Phi_2, d\nu \rangle & \langle Y_2, d\nu \rangle & \langle \Phi_2, d\nu \rangle \\ \langle \Phi_3, d\nu \rangle & \langle Y_3, d\nu \rangle & \langle \Phi_3, d\nu \rangle \end{vmatrix}$$

is equal to 0 since two columns are equal. If we expand by the last column and use the condition (5.5.4), it follows that

(5.5.5)
$$\langle \Phi_1 Y_2 - \Phi_2 Y_1, d\nu \rangle \langle \Phi_3, d\nu \rangle - \langle \Phi_1 Y_3 - \Phi_3 Y_1, d\nu \rangle \langle \Phi_2, d\nu \rangle$$
$$+ \langle \Phi_2 Y_3 - \Phi_3 Y_2, d\nu \rangle \langle \Phi_1, d\nu \rangle = 0.$$

Choose $\Phi_2 = \Phi^\omega$, $Y_2 = Y^\omega$, $\Phi_3 = \Phi^{-\omega}$, $Y_3 = Y^{-\omega}$ where ω is large positive, and make the leading coefficient Φ_0 positive which just requires a choice of positive initial data for an ordinary differential equation. If v_1^+ and v_1^- are the maximum and the minimum of v_1 in supp $d\nu$, then it follows if $v_1^- < v_1^+$ that $\langle \Phi^\omega, d\nu \rangle \langle \Phi^{-\omega}, d\nu \rangle$ tends to ∞ exponentially as $\omega \to +\infty$. If we divide by this product the last term in (5.5.5) converges to 0 as $\omega \to \infty$. The measures $\Phi^{\pm\omega} d\nu / \langle \Phi^{\pm\omega}, d\nu \rangle$ get more and more concentrated at $v_1 = v_1^\pm$. For a suitable sequence $\omega_j \to +\infty$ they will converge to probability measures $d\mu_\pm$ with support there. Noting that $Y^{\pm\omega} d\nu / \langle \Phi^{\pm\omega}, d\nu \rangle$ converges to $\lambda_1 d\mu_\pm$ for the same sequence, we conclude that

(5.5.6)
$$\langle \lambda_1 \Phi - Y, d\mu_+ \rangle = \langle \lambda_1 \Phi - Y, d\mu_- \rangle$$

for an arbitrary entropy pair (Φ, Y). (We have dropped the subscript 1 here.)

Now DiPerna [2] observes that one can of course apply (5.5.6) also to the solutions (Φ^ω, Y^ω). Note that, with Φ_j, Y_j again denoting the coefficients in the asymptotic expansion,

$$\lambda_1 \Phi^\omega - Y^\omega = e^{\omega v_1}((\lambda_1 \Phi_1 - Y_1)\omega^{-1} + O(\omega^{-2}));$$
$$\lambda_1 \Phi_1 - Y_1 = \partial Y_0 / \partial v_1 - \lambda_1 \partial \Phi_0 / \partial v_1 = \Phi_0 \partial \lambda_1 / \partial v_1.$$

If the system of conservation laws is *genuinely nonlinear* then $\partial \lambda_1 / \partial v_1$ has a constant sign, so the term involving $d\mu_+$ in (5.5.6) dominates completely when $\omega \to +\infty$. This is a contradiction showing that in fact $v_1^+ = v_1^-$. Similarly we obtain that v_2 is constant in supp $d\nu$. Hence we have

Theorem 5.5.1. *For a genuinely nonlinear hyperbolic system of two conservation laws the only probability measures satisfying (5.5.4) for arbitrary entropy pairs are the Dirac measures.*

If only the first characteristic family is genuinely nonlinear, the preceding argument still shows that v_1 is constant in supp $d\nu$. Let us choose coordinates so that $v_1 = 0$ there. Now all functions (Φ_0, Y_0) of v_2 with

$$\lambda_2(0, v_2)\Phi_0'(v_2) = Y_0'(v_2)$$

are restrictions to $v_1 = 0$ of entropy pairs. In fact, we can find formal solutions of (5.5.3) of the form

$$\Phi(v) \sim \sum_0^\infty \Phi_j(v_2)v_1^j, \quad Y(v) \sim \sum_0^\infty Y_j(v_2)v_1^j \quad \text{as } v_1 \to 0,$$

for this means solving recursively equations

$$\Phi_j' \lambda_2(0, \cdot) - Y_j' = R_j, \quad j \geq 0; \qquad \Phi_j \lambda_1(0, \cdot) - Y_j = S_j, \quad j > 0,$$

where R_j and S_j are determined by $\Phi_0, Y_0, \ldots, \Phi_{j-1}, Y_{j-1}$ and $R_0 = 0$. Eliminating Y_j we get an ordinary differential equation for Φ_j which can be solved. Now choose (Φ, Y) in C^∞ with these Taylor expansions. Then (5.5.3) is fulfilled with an error vanishing of infinite order when $v_1 = 0$. We can write it as a sum of one term with support where $v_1 \geq 0$ and one with support where $v_1 \leq 0$. The corresponding inhomogeneous equations (5.5.3) have C^∞ solutions with supports in the same half planes. Subtracting them from (Φ, Y) we obtain a solution of (5.5.3) with the given restriction (Φ_0, Y_0) to $v_1 = 0$.

Now we are in exactly the situation discussed in Section 5.4 and conclude that $\lambda_2(0, v_2)$ must be constant in the smallest interval containing supp $d\nu$. Hence we have

Theorem 5.5.2. *If one of the characteristic families of the system (5.5.1) of two conservation laws is genuinely nonlinear, then every probability measure satisfying (5.5.4) is supported by an arc on an orbit of the other eigenvector field where the corresponding eigenvalue is constant.*

DiPerna [2] has observed that all probability measures satisfying (5.5.4) can be determined by the preceding arguments if the system is genuinely nonlinear except on a curve Γ where the Riemann invariants v_1 and v_2 are strictly monotonic. Indeed, the proof of Theorem 5.5.1 shows that if $v_1^- < v_1^+$ then Γ must intersect each of the sides $v_1 = v_1^\pm$ of the smallest rectangle with sides parallel to the coordinate axes containing supp $d\nu$. Similarly Γ must intersect the other two sides $v_2 = v_2^\pm$. Unless $d\nu$ is a Dirac measure it follows that we have a nondegenerate rectangle with two opposite corners γ_1 and γ_2 on Γ. The measures $d\mu_\pm$ in the proof of Theorem 5.5.1 must be Dirac measures at these points, so $\lambda_1 \Phi - Y$ takes the same value at γ_1 as at γ_2 for every entropy pair. In the same way we see that this is true for $\lambda_2 \Phi - Y$, so $\Phi(\gamma_1) = \Phi(\gamma_2)$ and $Y(\gamma_1) = Y(\gamma_2)$ for all entropy pairs. However, this does not even hold for all solutions constructed above by means of the geometrical optics expansion, so we have proved:

Theorem 5.5.3. *Assume that the system* (5.5.1) *is genuinely nonlinear except on a curve where the Riemann invariants are strictly monotonic. Then it follows that every probability measure satisfying* (5.5.4) *is a Dirac measure.*

5.6. Existence of weak solutions for a system of two equations. To prove the existence of solutions to a system (5.5.1) of two equations we shall as in Chapter III study the Cauchy problem for the corresponding diffusion equation

$$(5.6.1) \qquad \partial u/\partial t + \partial f(u)/\partial x = \mu \partial^2 u/\partial x^2; \quad u(0,x) = u_0(x);$$

where $\mu > 0$ and u, f take values in $\mathbf{R}^2$. This requires an extension of the maximum principle (see Weinberger [1], Chueh, Conley and Smoller [1]). The maximum principle for a scalar equation says that any (semi-infinite) interval containing the range of the initial data must also contain the range of the solution. We want to have an analogue for (5.6.1) with the interval replaced by a suitable set $\Sigma \subset \mathbf{R}^2$. Let us assume that

$$\Sigma = \{u \in \mathbf{R}^2; G_1(u) \leq 0, \ldots, G_r(u) \leq 0\},$$

where $G_j \in C^2(\mathbf{R}^2)$, $G_j' \neq 0$ in $\{u \in \partial\Sigma, G_j(u) = 0\}$, and Σ is compact. Suppose that u is a solution of (5.6.1) with u_0 and u rapidly decreasing as $x \to \infty$ and $u_0(x) \in \Sigma$ for all x. Let T be the infimum of all $t \geq 0$ such that $u(t,x) \in \partial\Sigma$ for some x. Then $u_T(y) = u(T,y) \in \Sigma$ for all y, so $G_j(u_T(x)) = 0$ implies

$$\langle G_j'(u_T(x)), u_T'(x)\rangle = 0, \ G_j''(u_T(x); u_T'(x), u_T'(x)) + \langle G_j'(u_T(x)), \partial^2 u_T(x)/\partial x^2\rangle \leq 0,$$

where $u_T'(x) = \partial u_T(x)/\partial x$. Hence we have then

$$\partial G_j(u(t,x))/\partial t|_{t=T} = \langle G_j'(u_T(x)), \mu \partial^2 u_T(x)/\partial x^2 - f'(u_T(x))u_T'(x)\rangle$$
$$\leq -\mu G_j''(u_T(x); u_T'(x), u_T'(x)) - \langle {}^t f'(u_T) G_j'(u_T(x)), u_T'(x)\rangle.$$

The second term on the right vanishes if $G_j'(u_T(x))$ is an eigenvector of ${}^t f'(u_T(x))$, hence if $\{u; G_j(u) = G_j(u_T(x))\}$ is an integral curve of r_1 or r_2 in a neighborhood of u_T. The first term on the right is nonpositive if the curvature is nonnegative when the normal is oriented towards the interior of Σ. Thus we see that $\partial G_j(u(t,x))/\partial t \leq 0$ at (T,x) if Σ is convex and $\partial\Sigma$ consists of a finite number of orbits of the eigenvector fields. This does not quite suffice to conclude that the entire range of u is contained in Σ. However, assuming that 0 is an interior point of Σ we obtain for a solution u of the modified Cauchy problem

$$(5.6.2) \qquad \partial u/\partial t + \partial f(u)/\partial x = \mu \partial^2 u/\partial x^2 - \varepsilon u; \quad u(0,x) = u_0(x)$$

with $\varepsilon > 0$ that

$$\partial G_j(u)/\partial t \leq -\langle G_j'(u_0), \varepsilon u_0\rangle < 0 \quad \text{if } u \in \partial\Sigma, \ G_j(u) = 0,$$

when $t = T$. Hence $u(t,x)$ is in the interior of Σ for small positive t, and the infimum T of all $t > 0$ such that $u(t,x) \in \partial\Sigma$ must be equal to $+\infty$. Now it follows by simple modifications of the proofs in Section 3.2 that (5.6.2) has global solutions which are rapidly decreasing as $x \to \infty$, and for fixed μ we have uniform bounds in C^α for some $\alpha > 2$. From the solutions u_ε of (5.6.2) we can therefore find a subsequence converging to a limit $u \in C^\alpha$ as $\varepsilon \to 0$, and its range belongs to Σ. We sum up the result:

Proposition 5.6.1. *If $u_0 \in C_0^3$ and the range is contained in a convex compact neighborhood Σ of the origin in $\mathbf{R}^2$ bounded by finitely many orbits of the eigenvector fields r_1, r_2, then (5.6.1) has a solution for $t \geq 0$ with range contained in Σ.*

Following DiPerna [2] we shall now discuss the system

$$(5.6.3) \qquad \partial v/\partial t - \partial u/\partial x = 0, \quad \partial u/\partial t - \partial \sigma(v)/\partial x = 0$$

as an example. The equation for the eigenvectors is

$$\sigma'(v)v' = \lambda u', \, u' = \lambda v'$$

which means that $\lambda^2 = \sigma'(v)$ and that (v', u') is proportional to $r = (1, \lambda)$. We assume that $\sigma' > 0$ which implies strict hyperbolicity. Differentiation gives

$$2\lambda\langle\lambda', r\rangle = \sigma''(v),$$

so we have genuine nonlinearity when $\sigma'' \neq 0$. Assume now that σ'' has just one change of sign. By a linear change of the variable v we can attain that $v\sigma''(v) > 0$ for $v \neq 0$. The orbit of r is defined by $du/dv = \lambda(v)$, so

$$d^2u/dv^2 = \lambda'(v) = \sigma''(v)/2\lambda(v).$$

The second derivative has the sign of $v\lambda(v)$. A set Σ satisfying the conditions in Proposition 5.6.1 is therefore obtained if we take Σ bounded by integral curves with $\lambda < 0$ in the first and third quadrant and their reflections in the v axis in the other quadrants. Such convex sets can absorb any set, for $du = \pm\sqrt{\sigma'(v)}\,dv$ on the boundary curves and $\sqrt{\sigma'(v)}$ is not integrable on any half axis since $\sigma' \geq \sigma'(0)$. We shall be able to apply Theorem 5.5.3 since the Riemann invariants $u \pm \int \sqrt{\sigma'(v)}\,dv$ are strictly increasing on the u axis and the system is genuinely nonlinear elsewhere.

For any given $u_0 \in L^\infty$ we can find $u_0^\mu \in C_0^3$ converging weakly to u_0 as $t \to 0$ and with range contained in a set Σ of the type just constructed containing the range of u_0. By Proposition 5.6.1 we know that (5.6.1) with u_0 replaced by u_0^μ has a solution u^μ for $t \geq 0$ with range contained in Σ. We can now find a sequence $\mu_j \to 0$ such that u^{μ_j} converges weakly to some $u \in L^\infty(\mathbf{R}_+ \times \mathbf{R})$. By Theorem 5.3.2, Lemma 5.3.5 and Theorem 5.5.3 the Young measure of u is the weak limit of the Young measure of u^{μ_j}, so $u^{\mu_j} \to u$ in L_{loc}^p for every $p < \infty$, by Proposition 5.2.7. Hence u is a weak solution of (5.6.3) with Cauchy data u_0. Beyond the results of Chapter IV we have thus found that for the system (5.6.3) there is a weak solution of the Cauchy problem with arbitrary bounded initial data. This is a rather meager result of all the work in this chapter, but we emphasize again that the main point has been to explain the functional analytic ideas involved which may have a much wider application.

CHAPTER VI

NONLINEAR PERTURBATIONS OF THE WAVE EQUATION

6.1. Introduction. In this chapter we shall discuss the solution of a nonlinear Cauchy problem in $\mathbf{R}^{1+n}$,

$$(6.1.1) \qquad F(u, u', u'') = 0; \quad u(0, \cdot) = u_0, \ \partial_t u(0, \cdot) = u_1;$$

where u_j are small and of compact support or at least decrease fast at infinity. The variables will usually be denoted by $t \in \mathbf{R}$ and $x \in \mathbf{R}^n$, but sometimes we write x_0 instead of t to obtain symmetrical notation, and $\vec{x} = (x_1, \ldots, x_n)$ to avoid confusion. We shall assume that $u = 0$ is a solution of the equation $F(u, u', u'') = 0$ and that the linearization there is the wave equation. Thus

$$F(u, u', u'') = \square u + f(u, u', u'')$$

where

$$(6.1.2) \qquad \square = \partial_t^2 - \Delta$$

is the wave operator, also called the d'Alembertian, and f vanishes of second order at 0. Additional conditions will be imposed on f in low dimensions.

As an orientation we shall first rephrase a simple special case of Theorem 4.3.1 as a result on the Cauchy problem (6.1.1) when $n = 1$. Thus consider the Cauchy problem

$$(6.1.3) \qquad \sum_{j,k=0}^{1} g_{jk}(u') \partial_j \partial_k u = 0; \quad u(0, \cdot) = \varepsilon u_0, \ \partial_t u(0, \cdot) = \varepsilon u_1;$$

where $\sum_{j,k=0}^{1} g_{jk}(0) \partial_j \partial_k = \partial_0^2 - \partial_1^2$. With $U_1 = \partial_t u$, $U_2 = \partial_x u$, this is equivalent to a system

$$\partial_t U + a(U) \partial_x U = 0; \quad U_1 = \varepsilon u_1, \ U_2 = \varepsilon u_0' \text{ if } t = 0.$$

Here

$$a(U) = \begin{pmatrix} (g_{01} + g_{10})/g_{00} & g_{11}/g_{00} \\ -1 & 0 \end{pmatrix}.$$

The eigenvalues are given by

$$\lambda^2 g_{00} - \lambda(g_{01} + g_{10}) + g_{11} = 0.$$

When $U = 0$ this reduces to $\lambda^2 = 1$, and the corresponding eigenvectors of $a(0)$ are $(1, -\lambda)$. The projections of (u_1, u_0') on these directions are $f_\lambda(1, -\lambda)$ where $2f_1 = u_1 - u_0'$ and $2f_{-1} = u_0' + u_1$. The differentials of the eigenvalues at 0 are given by

$$2\lambda \, d\lambda + \lambda^2 dg_{00} - \lambda d(g_{01} + g_{10}) + dg_{11} = 0,$$

that is,

$$2\langle d\lambda, (1, -\lambda)\rangle = \sum_{j,k,l=0}^{1} g_{jkl}(-\lambda)^{j+k+l+1}, \quad g_{jkl} = \partial g_{jk}(0)/\partial u'_l.$$

If T_ε is the lifespan of the classical solution of (6.1.3), with $u_j \in C_0^\infty(\mathbf{R})$, then Theorem 4.3.1 states that

$$(6.1.4) \qquad \lim_{\varepsilon \to 0} 1/(\varepsilon T_\varepsilon) = \max_{\lambda=\pm 1} -\tfrac{1}{2} \sum_{j,k,l=0}^{1} g_{jkl}(-\lambda)^{j+k+l+1} f'_\lambda(y).$$

To interpret this result we observe that the solution of the unperturbed wave equation with Cauchy data (u_0, u_1) can be written in the form $F_1(x-t) + F_{-1}(-x-t)$ where F_1 is the wave moving to the right and F_{-1} is the wave moving to the left. To determine $F_{\pm 1}$ we have the equations

$$F_1(x) + F_{-1}(-x) = u_0(x), \quad F'_1(x) + F'_{-1}(-x) = -u_1(x),$$

and after differentiation of the first formula we find that $F'_\lambda(\lambda x) = -f_\lambda(x)$ for $\lambda = \pm 1$. Hence we can rewrite (6.1.4) in the form

$$(6.1.5) \qquad \lim_{\varepsilon \to 0} 1/(\varepsilon T_\varepsilon) = \max \tfrac{1}{2} \sum_{j,k,l=0}^{1} g_{jkl} \hat{\lambda}_j \hat{\lambda}_k \hat{\lambda}_l F''_\lambda(y)$$

where $\hat{\lambda} = (-1, \lambda)$ and the maximum is taken with respect to $y \in \mathbf{R}$ and $\lambda = \pm 1$.

Our aim will be to prove results similar to (6.1.5) when $n = 2$ or $n = 3$, with T_ε replaced by $\sqrt{T_\varepsilon}$ and $\log T_\varepsilon$ respectively while there will be global existence theorems for $n > 3$. The reason for the improved behavior when the dimension n gets large is that solutions of the linear Cauchy problem in $\mathbf{R}^{1+n}$ with smooth data of compact support can be estimated by $(1+t)^{(1-n)/2}$, and that for the solution of the ordinary differential equation $du(t)/dt = \varepsilon(1+t)^{(1-n)/2} u(t)$ we have

$$
\begin{aligned}
u(t) &= u(0) \exp \varepsilon t, && \text{if } n = 1; \\
u(t) &= u(0) \exp\left(2\varepsilon(\sqrt{1+t} - 1)\right), && \text{if } n = 2; \\
u(t) &= u(0)(1+t)^\varepsilon, && \text{if } n = 3; \\
u(t) &\leq |u(0)| e^{2\varepsilon}, && \text{if } n > 3.
\end{aligned}
$$

In view of the crucial role of the linear wave equation we shall start with a thorough study of its solutions in Section 6.2. This can be based on Fourier transforms or the completely explicit fundamental solution. The Friedlander radiation field which describes the asymptotic behavior of the solutions is studied in Section 6.2, first using the fundamental solution of $\square$, and then using a conformal compactification. However, for equations with variable coefficients or nonlinear equations we must use the energy integral method to derive estimates. It is presented in Section 6.3 where we also outline a proof of existence and uniqueness of solutions to the Cauchy problem for linear hyperbolic equations with variable coefficients. In Section 6.4 we then discuss briefly some "interpolation inequalities" and Sobolev inequalities which are indispensible in nonlinear problems, and we show how they work in a simple case by proving local existence theorems for the nonlinear Cauchy problem (6.1.1). Global existence theorems are proved in Section 6.5 using an idea of Klainerman [3] which exploits the infinitesimal generators of the Lorentz group, that is, vector fields commuting with $\Box$. In dimensions 2 and 3 the results only give at first the order of magnitude of a finite lower bound for the lifespan in terms of the size of the initial data, but they are then improved to asymptotic lower bounds for the lifespan similar to (6.1.5). It is plausible that these really give the asymptotic behavior as $\varepsilon \to 0$ of the lifespan but this is only known from work of F. John [2] in some special cases closely related to (6.1.5). (See also the remarks at the end of Section 6.5.) In Section 6.6 the results of Christodoulou [1] and Klainerman [4] on existence of solutions for all $t \geq 0$ when $n = 3$ and the so called null condition is fulfilled are proved with the methods of Klainerman. Section 6.7 gives the proof of Christodoulou based on a conformal compactification, which explains the role of the null condition better.

6.2. The linear wave equation. In this section we shall discuss the behavior at infinity of the solution of the wave equation in $\mathbf{R}^{1+n}$, $n \geq 1$,

$$(6.2.1) \qquad\qquad \Box u = (\partial_t^2 - \Delta)u = 0,$$

with Cauchy data

$$(6.2.2) \qquad\qquad u = f, \ \partial_t u = g \quad \text{when } t = 0,$$

where $f, g \in C_0^\infty(\mathbf{R}^n)$. At first we assume that $f = 0$, which implies that in the sense of distribution theory

$$u(t, x) = \int E(t, x - y)g(y)\, dy, \quad t > 0,$$

where E is the fundamental solution (see e.g. Hörmander [4, Section 6.2])

$$E = \tfrac{1}{2}\pi^{\frac{1-n}{2}}\chi_+^{\frac{1-n}{2}}(t^2 - |x|^2).$$

Here $x = (x_1, \ldots, x_n)$ and

$$\chi_+^a(s) = s^a/\Gamma(a+1), \quad s > 0, \quad \chi_+^a(s) = 0, \quad s \leq 0, \quad \text{if } \mathrm{Re}\, a > -1,$$
$$d\chi_+^a/ds = \chi_+^{a-1} \quad \text{for all } a \in \mathbf{C}.$$

Thus $\chi_+^{-k} = \delta_0^{(k-1)}$, $k = 1, 2, \ldots$, is supported by the origin.

Set $x = r\omega$ where $r = |x|$ and $\omega \in S^{n-1}$. Then $r \leq t + M$ in supp u if $|y| \leq M$ in supp g, for $|x| \leq |x - y| + |y| \leq t + M$ in the support of the "integrand". When n is odd we also have $r \geq t - M$, for $|x| \geq |x - y| - |y| \geq t - M$ (Huygens' principle). When n is even we note instead that $u(t, x) = O(t^{1-n})$ if $|x| < t/2$ and $t \to \infty$, for E is homogeneous of degree $1 - n$. Differentiation gives a faster decrease, $\partial^\alpha u(t, x) = O(t^{1-n-|\alpha|})$ when $|x| < t/2$. The main contributions to u must therefore always occur when $r - t$ is small compared to t, so we set $r = t + \varrho$ where $-r \leq \varrho \leq M$, if $2r \geq t \geq r - M$. Then

$$t^2 - |x - y|^2 = (r - \varrho)^2 - |r\omega - y|^2 = 2r(\langle\omega, y\rangle - \varrho) + \varrho^2 - |y|^2,$$

and we obtain by the homogeneity of E

$$(2\pi r)^{\frac{n-1}{2}}u(t, x) = \tfrac{1}{2}\int \chi_+^{\frac{1-n}{2}}(\langle\omega, y\rangle - \varrho + (\varrho^2 - |y|^2)/2r)g(y)\, dy$$
$$= \tfrac{1}{2}\int \chi_+^{\frac{1-n}{2}}(s - \varrho + \varrho^2/2r)G(\omega, r^{-1}, s)\, ds$$
$$= \tfrac{1}{2}\int \chi_+^{\frac{1-n}{2}}(s + (t^2 - r^2)/2r)G(\omega, r^{-1}, s)\, ds.$$

Here

$$G(\omega, z, s) = \int \delta(s - \langle \omega, y \rangle + |y|^2 z/2)g(y)\, dy$$

is a C^∞ function in $S^{n-1} \times [0, 1/2M] \times \mathbf{R}$ with $|s| \leq 5M/4$ in the support and $G(\omega, 0, s) = R(\omega, s; g)$, where

$$(6.2.3) \qquad R(\omega, s; g) = \int \delta(s - \langle \omega, y \rangle)g(y)\, dy = \int_{\langle \omega, y \rangle = s} g(y)\, dS(y)$$

is the *Radon transform* of g, which has support in $S^{n-1} \times [-M, M]$. Hence

$$(6.2.4) \qquad r^{\frac{n-1}{2}} u(t, x) = F(\omega, r^{-1}, \varrho)$$

where the convolution

$$(6.2.5) \qquad F(\omega, z, \varrho) = \tfrac{1}{2}(2\pi)^{\frac{1-n}{2}} \int \chi_+^{\frac{1-n}{2}}(s - \varrho + \varrho^2 z/2)G(\omega, z, s)ds$$

is a C^∞ function in $S^{n-1} \times [0, 1/2M] \times \mathbf{R}$ with $\varrho \leq M$ in the support. This result is due to Friedlander [1, 2] who only assumed that u satisfies the wave equation for large $|x|$. The restriction F_0 of F to $z = 0$ is the *Friedlander radiation field*,

$$F_0(\omega, \varrho) = \tfrac{1}{2}(2\pi)^{\frac{1-n}{2}} \int \chi_+^{\frac{1-n}{2}}(s - \varrho)R(\omega, s; g)\, ds.$$

From the homogeneity of $\chi_+^{\frac{1-n}{2}}$ it follows at once that $F_0(\omega, \varrho)$ is a polyhomogeneous symbol in ϱ of degree $(1 - n)/2$ (cf. Hörmander [4, Def. 18.1.5]). We have, still with $\varrho = r - t$ and $x = r\omega$,

$$(6.2.6) \qquad |u(t, x) - r^{\frac{1-n}{2}} F_0(\omega, \varrho)| \leq C((1 + \varrho)/r)(r(1 + \varrho))^{\frac{1-n}{2}}, \quad \text{if } r > t/2 > 1.$$

This follows at once from the differentiability of F when $\varrho \geq -2M$, say. When n is even we must also use that for $\varrho < -2M$ we have

$$|\partial F(\omega, z, \varrho)/\partial z| \leq C(1 + |\varrho|)^{\frac{3-n}{2}},$$

since differentiation of (6.2.5) shows that $\partial F/\partial z$ can be estimated by a constant times $\varrho^2(1 + |\varrho|)^{-\frac{1+n}{2}} + (1 + |\varrho|)^{\frac{1-n}{2}}$.

So far we have only studied the solution of (6.2.1), (6.2.2) when $f = 0$. However, an approximating radiation field always exists. In fact, choosing $\psi \in C_0^\infty(\mathbf{R})$ equal to 1 in $[1, \infty)$ and 0 in $(-\infty, 0)$ we obtain when $t > 1$

$$u = E * K, \quad K = \square(\psi(t)u) = 2\psi'(t)\partial u/\partial t + \psi''(t)u.$$

Thus $K \in C_0^\infty(\mathbf{R}^{1+n})$ if $f, g \in C_0^\infty(\mathbf{R}^n)$, and in the sense of distribution theory

$$u(t, x) = \int_0^1 ds \int E(t - s, x - y)K(s, y)dy$$

is a superposition of solutions of the form discussed already. If (6.2.6) is valid for a solution u of (6.2.1) then (6.2.6) remains valid if u is replaced by $u(\cdot - s, \cdot)$ and F_0 is replaced by $F_0(\cdot, \cdot + s)$. In view of the uniformity in g of the estimate (6.2.6) in the case where we have proved it, we conclude that there is always some F_0, obviously uniquely defined, such that (6.2.6) holds and F_0 is a symbol in ϱ of order $(1-n)/2$, which is smooth in ω. The radiation field depends continuously on the initial data f, g. If F_0 is the radiation field of u we have just observed that the radiation field of $(u(\cdot + s, \cdot) - u)/s$ is $(F_0(\cdot, \cdot - s) - F_0)/s$, so we conclude that the radiation field of $\partial u/\partial t$ is $-\partial F_0/\partial \varrho$. If u satisfies (6.2.1), (6.2.2) with $f = 0$, then $\partial u/\partial t$ satisfies (6.2.1) and (6.2.2) with f replaced by g and g replaced by 0. It follows that the radiation field is given in general by

$$(6.2.7) \qquad F_0(\omega, \cdot) = \tfrac{1}{2}(2\pi)^{\frac{1-n}{2}} \chi_-^{\frac{1-n}{2}} * (R(\omega, \cdot; g) - R(\omega, \cdot; f)'), \qquad \chi_-^{\frac{1-n}{2}} = \chi_+^{\frac{1-n}{2}}(-\cdot),$$

with convolution and differentiation taken in the variable ϱ indicated by a dot.

By (6.2.6) we have for bounded ϱ, small h and large r

$$u(t, x+h) = r^{\frac{1-n}{2}}(F_0(\omega, \varrho) + \partial F_0(\omega, \varrho)/\partial \varrho \langle \varrho', h \rangle + O(|h|^2 + 1/r)),$$

so the radiation field of $u(t, x+h)$ is $F_0(\omega, \varrho) + \partial F_0(\omega, \varrho)/\partial \varrho \langle \omega, h \rangle + O(|h|^2)$. In view of the continuous dependence just pointed out it follows that the radiation field of $\partial u/\partial x_j$ is $\omega_j \partial F_0/\partial \varrho$; the other terms obtained by formal differentiation of (6.2.6) are absorbed by the error term. However, to give a precise description of the behavior of u at infinity we must also apply other differential operators which exploit the invariance of the wave operator under the Lorentz group and homotheties. These are the vector fields

$$(6.2.8) \qquad Z_{jk} = \lambda_j x_j \partial/\partial x_k - \lambda_k x_k \partial/\partial x_j, \qquad j, k = 0, \ldots, n,$$

where $\lambda = (1, -1, \ldots, -1)$, which all commute with $\square$, and the radial vector field

$$(6.2.9) \qquad Z_0 = \sum_0^n x_j \partial/\partial x_j,$$

for which $[\square, Z_0] = \square Z_0 - Z_0 \square = 2\square$. For an arbitrary product Z^I of such vector fields it is clear that $\square Z^I u = 0$ if $\square u = 0$. We have

$$Z_{jk}(|x|-t) = 0, \quad j, k \neq 0; \quad Z_{0k}(|x|-t) = (t-|x|)x_k/|x|, \quad k \neq 0; \quad Z_0(|x|-t) = |x| - t.$$

Thus

$$Z^I(|x|^{\frac{1-n}{2}} F_0(x/|x|, |x|-t)) = |x|^{\frac{1-n}{2}} F_I(x/|x|, |x|-t),$$

where $F_I(\omega, \varrho)$ is also a polyhomogeneous symbol of order $(1-n)/2$ in ϱ. In fact, when $Z = Z_{jk}$ or Z_0 operates on a homogeneous function it gives another one of the same degree, and when it operates on $\varrho = |x| - t$ we have just seen that Z acts as the operator $\varrho\partial/\partial\varrho$ followed by multiplication with a homogeneous function of x of degree 0. If $s \mapsto T(s)$ is the one parameter group of linear transformations generated by Z, then we deduce as above from (6.2.6) that the radiation field of $u \circ T(s)$ is $r^{\frac{n-1}{2}}(1 + sZ)r^{\frac{1-n}{2}} F_0(\omega, \varrho) + O(s^2)$, so the radiation field of Zu is $r^{\frac{n-1}{2}} Z r^{\frac{1-n}{2}} F_0(\omega, \varrho)$. Repeating the argument we find that (6.2.6) implies the following result:

Theorem 6.2.1. *If u is a solution of the Cauchy problem (6.2.1), (6.2.2) with $f, g \in C_0^\infty(\mathbf{R}^n)$, then*

$$(6.2.10) \qquad |\partial^\alpha Z^I u| \leq C_{\alpha, I}(1 + |t| + |t^2 - |x|^2|)^{\frac{1-n}{2}}$$

for arbitrary α and I. More precisely, if F_0 is the radiation field of u, then

$$(6.2.11) \qquad |\partial^\alpha Z^I(u(t, x) - r^{\frac{1-n}{2}} F_0(\omega, \varrho))| \leq C(1 + \varrho)^{\frac{3-n}{2}} t^{-\frac{(1+n)}{2}}, \quad \text{if } r > t/2 > 1.$$

When $n = 1$ then (6.2.6) simplifies to

$$u(t, x) = F(\operatorname{sgn} x, |x| - t)$$

for large t when $x \neq 0$, so $F(\omega, \cdot)$ is the function F_ω of the introduction for $\omega = \pm 1$. When $n = 2, 3$ we shall obtain results similar to (6.1.5) for the lifespan of the solution of a nonlinear Cauchy problem where F_λ is replaced by the radiation field. The following theorem will be important in the interpretation of the result:

Theorem 6.2.2. *If $f, g \in C_0^\infty(\mathbf{R}^n)$ then the radiation field F_0 of the solution u of (6.2.1), (6.2.2) is not identically 0 unless f and g are identically 0.*

Proof. If F_0 is identically 0 it follows from the theorem of supports and (6.2.7) that

$$R(\omega, \varrho; g) - dR(\omega, \varrho; f)/d\varrho \equiv 0.$$

Since $R(\omega, \varrho; \cdot) = R(-\omega, -\varrho; \cdot)$, this is equivalent to

$$R(-\omega, -\varrho; g) - dR(-\omega, -\varrho; f)/d\varrho \equiv 0.$$

If we carry out the differentiation and replace $(-\omega, -\varrho)$ by (ω, ϱ) afterwards, it follows that

$$R(\omega, \varrho; g) + dR(\omega, \varrho; f)/d\varrho \equiv 0,$$

so $R(\omega, \varrho; g) \equiv 0$ and $dR(\omega, \varrho; f)/d\varrho \equiv 0$, which implies $R(\omega, \varrho; f) \equiv 0$. Hence f and g vanish identically.

Remark. The proof shows that the projection of the support of $R(\omega, \varrho; g) - R(\omega, \varrho; f)'$ on S^{n-1} cannot omit two antipodal points. In fact, if $R(\omega, \varrho; g)$ vanishes for all ω in an open set, then it vanishes identically since the Fourier transform of g will vanish in the open cone which it generates.

When $n = 3$ and f, g are functions of $r = |x|$ only, hence C^∞ functions of r^2, then the solution of (6.2.1), (6.2.2) is a function of t and r, and the equation (6.2.1) can be written

$$(ru)''_{tt} - (ru)''_{rr} = 0.$$

Hence

$$u(t, x) = r^{-1} F_0(r - t), \quad t > M, \ r = |x|,$$

so the left-hand side of (6.2.6) vanishes for large t. It is easy to compute F_0, for

$$R(\omega, \varrho; g) = \int_{|\varrho|}^\infty g(t) d\pi(t^2 - \varrho^2) = 2\pi \int_\varrho^\infty tg(t) \, dt,$$

and similarly for $R(\omega, \varrho; f)$. It follows that the radiation field, which only depends on ϱ, is given by

(6.2.12) $$dF_0(\varrho)/d\varrho = (d(\varrho f(\varrho))/d\varrho - \varrho g(\varrho))/2.$$

The arguments in this section have all been based on the properties of the fundamental solution of $\square$, essentially as in Friedlander [1]. We shall now discuss another approach from Friedlander [3] which exploits the conformal map from the Minkowski space M^{1+n}, that is, $\mathbf{R}^{1+n}$ with the standard Lorentz metric, to the Einstein universe $\mathbf{R} \times S^n$, defined in Section A.4 of the appendix. The scalar curvature of the Einstein universe is minus that of S^n with the standard metric, so it is $-n(n-1)$ by (A.2.10). The conformal d'Alembertian (see Section A.3) is therefore

$$\partial_T^2 - \widetilde{\Delta} + n(n-1)(n-1)/4n = \widetilde{\square} + (n-1)^2/4$$

where $\widetilde{\Delta}$ is the Laplace operator on S^n and $\widetilde{\square}$ is the d'Alembertian on $\mathbf{R} \times S^n$. If u is a solution of $\square u = 0$ in M^{1+n}, and with the notation in Section A.4

(6.2.13) $$\tilde{u}(T, X) = (\cos T + X_0)^{(1-n)/2} u(\Psi(T, X)), \quad \text{when } \cos T + X_0 > 0, \ 0 \le T < \pi,$$

it follows that $(\tilde{\Box} + (n-1)^2/4)\tilde{u} = 0$ and conversely. If the Cauchy data of u are in $\mathcal{S}(\mathbf{R}^n)$, then those of $\tilde{u}$ are in $C^\infty(S^n)$, and they vanish of infinite order at the pole corresponding to infinity in $\mathbf{R}^n$. We shall prove in Section 6.3 that this linear Cauchy problem has a solution $\tilde{u} \in C^\infty(\mathbf{R} \times S^n)$, and below we shall also outline a proof using an explicit fundamental solution of $\tilde{\Box} + (n-1)^2/4$. Since

$$\cos T + X_0 = \Omega = 2\big((1 + (|x| - t)^2)(1 + (|x| + t)^2)\big)^{-\frac{1}{2}},$$

it follows at once that

$$|u(t,x)| \le C\big((1 + (|x| - t)^2)(1 + (|x| + t)^2)\big)^{\frac{1-n}{2}},$$

which is the estimate (6.2.10) (without differentiations). To conclude (6.2.11) we must determine the limit of $\Psi^{-1}(t, (t + \varrho)\omega)$, $\omega \in S^{n-1}$, as $t \to +\infty$. The corresponding coordinates T, α are defined by

$$\sin T = \Omega t, \;\; \sin \alpha = \Omega(t + \varrho), \;\; \cos \alpha = \tfrac{1}{2}\Omega(1 - \varrho(t + r)),$$

and since $\Omega \sim t^{-1}(1 + \varrho^2)^{-\frac{1}{2}}$, it follows that

$$\sin \alpha \to (1 + \varrho^2)^{-\frac{1}{2}}, \quad \cos \alpha \to -\varrho(1 + \varrho^2)^{-\frac{1}{2}}, \quad \sin T \to (1 + \varrho^2)^{-\frac{1}{2}}.$$

We have $T + \alpha \to \pi$ since $T > 0$ and $\Psi(T, X)$ remains finite when $|T| + \alpha$ has a bound $< \pi$. Hence

$$(6.2.14) \quad t^{\frac{n-1}{2}}(1 + \varrho^2)^{\frac{n-1}{4}} u(t, (t + \varrho)\omega) \to \tilde{u}(\pi - \alpha, (\cos\alpha, \sin\alpha\,\omega)); \quad 0 < \alpha < \pi, \cot\alpha = -\varrho,$$

which means that

$$(6.2.15) \qquad F_0(\omega, \varrho)(1 + \varrho^2)^{\frac{n-1}{4}} = \tilde{u}(\pi - \alpha, (\cos\alpha, \sin\alpha\,\omega)).$$

For bounded ϱ the difference between (T, α) and the limit is $O(1/t)$, so the difference between the two sides in (6.2.14) is actually $O(1/t)$, which gives (6.2.11) when no derivatives are present. What remains is to examine what application of the differential operators Z and ∂ mean in the Einstein model; the full result (6.2.11) will then be a consequence of the fact that $\tilde{u} \in C^\infty$, with vanishing of infinite order at the infinitely distant point $T = 0$, $\alpha = \pi$. We leave this for the interested reader but will return to the conformal d'Alembertian in Section 6.7.

It would not have been necessary to use the results of Section 6.3 here, for we can write down the fundamental solution for the conformal d'Alembertian in $\mathbf{R} \times S^n$ explicitly. To do so we first take the pole at $T = 0$ and the point in S^n corresponding to $\alpha = 0$. Recall that the fundamental solution of $\Box$ in M^{1+n} is for $t > 0$

$$E = \tfrac{1}{2}\pi^{\frac{1-n}{2}} \chi_+^{\frac{1-n}{2}}(t^2 - |x|^2).$$

With the notation in (A.4.2) we set $t = \Omega^{-1}\sin T$, $x = \Omega^{-1}X$, $\Omega = \cos T + \cos \alpha$, and note that

$$t^2 - |x|^2 = (\sin^2 T - \sin^2 \alpha)/\Omega^2 = (\cos\alpha - \cos T)/\Omega.$$

Hence $E(\Psi(T, X)) = \tfrac{1}{2}\pi^{(1-n)/2}\Omega^{(n-1)/2}\chi_+^{(1-n)/2}(\cos\alpha - \cos T)$. Now

$$(\tilde{\Box} + (n-1)^2/4)(\Omega^{\frac{1-n}{2}} E(\Psi(T, X))) = \Omega^{-\frac{n+3}{2}}\Psi^* \Box E = 2^{\frac{n-1}{2}}\delta_0$$

where δ_0 should be replaced by $\delta_0/\sqrt{g}$ if the coordinates in the Einstein universe are not chosen geodesic. (Recall that δ_0 is a distribution density.) Hence

$$(6.2.16) \qquad (\tilde{\square} + (n-1)^2/4)\tilde{E} = \delta_0, \quad \text{if } \tilde{E} = \tfrac{1}{2}(2\pi)^{\frac{1-n}{2}}\chi_+^{\frac{1-n}{2}}(\cos\alpha - \cos T),$$

provided that $\alpha + |T| < \pi$ and we define $\tilde{E}(T, \cdot) = 0$ when $T \leq 0$. (6.2.16) remains true for $T < \pi$. If n is odd then $\chi_+^{(1-n)/2}$ is even or odd, so replacing T by $\pi - T$ and α by $\pi - \alpha$ shows that we have a solution of the homogeneous conformal d'Alembertian when $\pi < \alpha + T$ and $T < \pi$, and the solution has the same distribution limits on the characteristic surface $\alpha + T = \pi$ from both sides. If n is even, then we obtain by this substitution the distribution $\tfrac{1}{2}(2\pi)^{(1-n)/2}\chi_-^{(1-n)/2}(\cos\alpha - \cos T)$, which satisfies the homogeneous conformal d'Alembertian for $\alpha + |T| < \pi$ since $\square\chi_-^{(1-n)/2}(t^2 - |x|^2) = 0$ in $\mathbf{R}^{1+n}$ for even n. (See Hörmander [4, Theorem 6.2.1].) Again we conclude that (6.2.16) is a fundamental solution for $T < \pi$.

If the pole on S^n defined by $\alpha = 0$ is replaced by another point, we just have to replace α in (6.2.16) by the geodesic distance s to that point along S^n. We can therefore write down the fundamental solution for $T < \pi$ with pole at an arbitrary point in S^n. This suffices to conclude that the Cauchy problem for the conformal d'Alembertian with C^∞ initial data when $T = 0$ has a solution in C^∞ for $T < \pi$, and iteration of this conclusion proves that there is a C^∞ solution for all $T \geq 0$.

It is in fact easy to obtain a global fundamental solution. Assume first that n *is odd*. Then the fundamental solution arrives as $T \to \pi - 0$ as $(-1)^{(3-n)/2}$ times the backward fundamental solution at the antipode, and it must be continued as $(-1)^{(1-n)/2}$ times the forward fundamental solution with pole at the antipode and time π. At time 2π it arrives as minus the backward fundamental solution at the original point, so the fundamental solution then repeats with period 2π in T. Assume now that n *is even*. Then the fundamental solution continues beyond the antipodal point at time π with support outside the characteristic conoid there, and it arrives at time 2π as the backward fundamental solution at the original point in S^n. It is then continued with a change of sign to the next interval $2\pi < T < 4\pi$ and so on, with a period of 4π. The fundamental solution is for every n a continuous function of T with values in $\mathcal{D}'(S^n)$ for all T, and when $T \neq 0$ it is infinitely differentiable with values in $\mathcal{D}'$.

6.3. The energy integral method. The basic energy estimate for the d'Alembertian $\square$ is obtained from the identity

$$(6.3.1) \qquad 2\partial_0 u \square u = \partial_0 |u'|^2 - 2\sum_1^n \partial_j(\partial_0 u \partial_j u),$$

where

$$|u'|^2 = \sum_0^n |\partial_j u|^2, \quad \partial_j = \partial/\partial x_j.$$

If $u \in C^2$ and u vanishes for large $\vec{x} = (x_1, \ldots, x_n)$, then integration with respect to $\vec{x}$ gives

$$\partial_0 \|u'(x_0, \cdot)\|^2 = 2\int \partial_0 u \square u \, d\vec{x} \leq 2\|u'(x_0, \cdot)\| \|\square u(x_0, \cdot)\|,$$

where the norms are L^2 norms with respect to $\vec{x}$. Thus

$$\partial_0 \|u'(x_0, \cdot)\| \leq \|\square u(x_0, \cdot)\|,$$

which gives after integration

$$(6.3.2) \qquad \|u'(x_0, \cdot)\| \leq \|u'(0, \cdot)\| + \int_0^{x_0} \|\square u(t, \cdot)\| dt.$$

In particular it follows that $u \equiv 0$ if $\square u = 0$ and the Cauchy data vanish, which also follows at once by taking Fourier transforms in $\tilde{x}$.

The strength of the energy integral method is its stability under perturbations of the equation whereas methods based on the Fourier transformation break down at once. As a first step towards more general energy identities we consider a hyperbolic operator $\sum_{j,k=0}^n g^{jk} \partial_j \partial_k$ with a constant symmetric matrix (g^{jk}). If K^i are constants then (6.3.1) generalizes to

$$(6.3.3) \qquad 2 \sum_{i=0}^n K^i \partial_i u \sum_{j,k=0}^n g^{jk} \partial_j \partial_k u = \sum_{i,j=0}^n \partial_j (T_i^j(u) K^i),$$

$$(6.3.4) \qquad T_i^j(u) = 2 \sum_{k=0}^n g^{jk} \partial_k u \partial_i u - \delta_i^j \sum_{k,l=0}^n g^{kl} \partial_k u \partial_l u.$$

The definition of T_i^j means that if ν is a covector, then

$$(6.3.4)' \qquad \sum_{i,j=0}^n T_i^j(u) K^i \nu_j = 2 \langle K, u' \rangle \langle gu', \nu \rangle - \langle K, \nu \rangle \langle gu', u' \rangle.$$

If A denotes the bilinear form defined by the inverse of g and $N = g\nu$, $U = gu'$ are the vectors corresponding to the covectors ν, u', then the *energy* on a surface with conormal ν, that is, the integrand obtained there by integrating out the right-hand side of (6.3.3), becomes

$$(6.3.5) \qquad \sum_{i,j=0}^n T_i^j(u) K^i \nu_j = 2 A(U, N) A(U, K) - A(U, U) A(N, K)$$

$$= A(2 U A(U, K) - K A(U, U), N).$$

Lemma 6.3.1. *(6.3.5) is positive definite in U if A has Lorentz signature and*

$$A(K, K) > 0, \quad A(N, N) > 0, \quad A(N, K) > 0,$$

that is, if K and N are in the same open Lorentz half cone.

Proof. Set $V = 2U A(K, U) - K A(U, U)$. Then

$$A(V, V) = A(K, K) A(U, U)^2 \geq 0, \quad A(V, K) = A(K, U)^2 + A(K, U)^2 - A(K, K) A(U, U),$$

where the difference of the last two terms is positive if $U \notin \mathbf{R}K$, by the reversed Cauchy-Schwarz inequality which follows from the Lorentz signature. Hence $A(V, K) > 0$ and $A(V, V) \geq 0$, which implies $A(V, N) > 0$.

One can also prove the lemma by an easy computation. If K is normalized with $A(K, K) = 1$, we can choose coordinates so that $A(x) = x_0^2 - x_1^2 - \cdots - x_n^2$, $K = (1, 0, \ldots, 0)$, and obtain

$$V = (U_0^2 + r^2, 2U_0 r \omega) \quad \text{if } (U_1, \ldots, U_n) = r\omega, \ |\omega| = 1.$$

We have

$$(U_0^2 + r^2, 2U_0 r) = (U_0^2 + r^2)(1, \sin 2\theta),$$

if θ is the polar angle in a Euclidean (U_0, r) plane. If $N = (N_0, \vec{N})$ with $|\vec{N}| < N_0$ then

$$A(V, N) = (U_0^2 + r^2) N_0 - 2 U_0 r \langle \omega, \vec{N} \rangle \geq (N_0 - |\vec{N}|)(U_0^2 + r^2).$$

If g and K are allowed to depend on x, we get additional terms in (6.3.3) compensating those where ∂_j acts on the coefficients of T_i^j. The following simple but useful proposition shows that these can often be taken care of:

Proposition 6.3.2. *Let $u \in C^2$ satisfy a differential equation*

$$\Box u + \sum_{j,k=0}^{n} \gamma^{jk}(x)\partial_j\partial_k u = f, \quad 0 \le x_0 < T,$$

and assume that $u = 0$ for large $\vec{x}$. If

$$|\gamma| = \sum_{j,k=0}^{n} |\gamma^{jk}| \le \tfrac{1}{2}, \quad 0 \le x_0 < T,$$

it follows for $0 \le x_0 < T$ that

$$(6.3.6) \qquad \|u'(x_0,\cdot)\| \le 2\Big(\|u'(0,\cdot)\| + \int_0^{x_0} \|f(t,\cdot)\|\,dt\Big) \exp\Big(\int_0^{x_0} 2|\gamma'(t)|\,dt\Big),$$

where the norms are L^2 norms with respect to $\vec{x}$ and

$$|\gamma'(t)| = \sum_{i,j,k=0}^{n} \sup |\partial_i\gamma^{jk}(t,\cdot)|.$$

Proof. We shall use (6.3.3) with the modification just indicated, with g^{jk} equal to γ^{jk} plus the coefficients of $\Box$, and $K = (1,0,\ldots,0)$. Then

$$T_0^0(x,u) = |u'|^2 + \gamma^{00}(\partial_0 u)^2 - \sum_{k,l=1}^{n} \gamma^{kl}\partial_k u \partial_l u \ge |u'|^2/2.$$

With

$$R(x,u) = 2\sum_{j,k=0}^{n} \partial_j\gamma^{jk}\partial_k u\partial_0 u - \sum_{k,l=0}^{n} \partial_0\gamma^{kl}\partial_k u\partial_l u$$

$$= \partial_0\gamma^{00}(\partial_0 u)^2 - \sum_{k,l=1}^{n} \partial_0\gamma^{kl}\partial_k u\partial_l u + 2\sum_{j=1}^{n}\sum_{k=0}^{n} \partial_j\gamma^{jk}\partial_k u\partial_0 u$$

we obtain

$$\partial_0 \int T_0^0(x,u)d\vec{x} \le 2\|f(x_0,\cdot)\|\|\partial_0 u(x_0,\cdot)\| + \int R(x,u)\,d\vec{x},$$

hence

$$\partial_0 E(x_0)^2 \le 2\sqrt{2}\|f(x_0,\cdot)\|E(x_0) + 4|\gamma'(x_0)|E(x_0)^2,$$

if $E(x_0)^2 = \int T_0^0(x,u)\,d\vec{x}$. Thus

$$\partial_0 E(x_0) \le \sqrt{2}\|f(x_0,\cdot)\| + 2|\gamma'(x_0)|E(x_0).$$

If we multiply by the integrating factor $\exp(-\int_0^{x_0} 2|\gamma'(t)|\,dt)$ and integrate, the estimate (6.3.6) follows.

The estimate (6.3.6) suffices to prove most of the existence theorems for nonlinear perturbations of $\Box$ in this chapter. However, it has the weakness that no estimate of u itself is obtained, and for the proof in Section 6.6 of some more refined results when $n = 3$

one also needs a more sophisticated estimate which we shall now discuss. (Alternative proofs in Section 6.7 do not require this estimate.)

Let us first assume that g^{jk} are constant and look for the variable vector fields K such that for some variable scalar L

$$
\begin{aligned}
(6.3.3)' \quad & 2\Big(\sum_i K^i \partial_i u + Lu\Big) \sum_{j,k} g^{jk} \partial_j \partial_k u = \sum_j \partial_j \Big(\sum_i T^j_i(u) K^i + 2Lu \sum_k g^{jk} \partial_k u\Big) \\
& -2 \sum_{i,j,k} \partial_j K^i g^{jk} \partial_k u \partial_i u + \Big(\sum_i \partial_i K^i - 2L\Big) \sum_{j,k} g^{jk} \partial_j u \partial_k u - 2 \sum_{j,k} u g^{jk} \partial_j L \partial_k u
\end{aligned}
$$

is an exact divergence. With the notation $K' = (\partial_j K^i)$ we see that cancellation requires that

$$ K'g + g\,{}^t K' = \Big(\sum_i \partial_i K^i - 2L\Big) g, $$

which means that K is a conformal vector field with respect to the metric defined by the dual quadratic form A. When $n \geq 2$ there are not many:

Proposition 6.3.3. *If $n \geq 2$ then all smooth vector fields K such that*

$$ (6.3.7) \qquad\qquad K'g + g\,{}^t K' = 2Fg $$

are of the form

$$ (6.3.8) \qquad\qquad K = 2A(x,\theta)x - A(x,x)\theta + K_0 $$

where θ is a constant vector, $F(x) = 2A(x,\theta) + c$ with a constant c, and K_0 is the sum of $c \sum_0^n x_j \partial_j$, a constant vector field, and a linear combination of the vector fields

$$ (6.3.9) \qquad (\partial_j A(x))\partial_k - (\partial_k A(x))\partial_j; \quad j,k = 0,\dots,n. $$

Proof. We may assume that g is diagonal and write g^i instead of g^{ii}. Then (6.3.7) can be written

$$ \partial_j K^i g^j + g^i \partial_i K^j = 0, \quad i \neq j; \quad \partial_i K^i = F. $$

If i,j,k are different indices (recall that we have assumed that the dimension $n+1$ is at least 3), we obtain

$$ \partial_k \partial_j K^i / g^i = -\partial_k \partial_i K^j / g^j = -\partial_i \partial_k K^j / g^j = \partial_i \partial_j K^k / g^k = -\partial_j \partial_k K^i / g^i. $$

Hence $\partial_k \partial_j K^i = 0$ if i,j,k are different indices, so $\partial_k \partial_j F = 0$ when $j \neq k$, and

$$ g^j \partial_j^2 F = g^j \partial_j^2 \partial_i K^i = -\partial_j \partial_i^2 K^j g^i = -g^i \partial_i^2 F = g^k \partial_k^2 F = -g^j \partial_j^2 F. $$

Thus F is affine linear. When K is defined by (6.3.8) with $K_0 = 0$, then

$$ \partial_j K^i g^j = 2\theta^j x_i - 2x_j \theta^i + 2A(x,\theta)\delta_{ij} g^j $$

since $A(x,\theta) = \sum x_j \theta^j / g^j$, so (6.3.7) holds with $F = 2A(x,\theta)$. (The coordinates of x should have been denoted by x^j for consistency but that would conflict with our notation elsewhere.) This is an arbitrary linear form. If $K_0 = \sum_0^n x_j \partial_j$ then (6.3.7) is valid with $F = 1$. In view of the linearity of (6.3.7) it just remains to study the case where $F = 0$, that is,

$$ \partial_j K^i g^j + \partial_i g^i K^j = 0, \quad i,j = 0,\dots,n. $$

Thus $\partial_j(K^i/g^i)$ is a skew symmetric matrix Z_{ij}, and it is constant since

$$\partial_k\partial_j K^i/g^i = -\partial_k\partial_i K^j/g^j = -\partial_i\partial_k K^j/g^j,$$

which must vanish since the sign changes after three circular permutations of i, k, j. It follows that up to a constant vector field

$$\sum_{i=0}^n K^i\partial_i = \sum_{i,j=0}^n Z_{ij}x_j g^i\partial_i = \tfrac{1}{2}\sum_{i,j=0}^n Z_{ij}(x_j/g^j\,\partial_i - x_i/g^i\,\partial_j)g^i g^j$$

which is of the form (6.3.9).

Remark. That the vector fields in (6.3.8) appear is no surprise since they are obtained from constant vector fields by an inversion.

Now we choose $L = \sum_0^n \partial_i K^i/2 - F = (n-1)(A(\cdot,\theta) + c/2)$. Then the last term in (6.3.3)' simplifies to

$$-(n-1)\sum_0^n \theta^k\partial_k u^2,$$

so (6.3.3)' can be written

$$(6.3.3)'' \quad 2\Big(\sum_{i=0}^n K^i\partial_i u + (n-1)(A(x,\theta) + c/2)u\Big)\sum_{j,k=0}^n g^{jk}\partial_j\partial_k u$$

$$= \sum_{j=0}^n \partial_j\Big(\sum_{i=0}^n T_i^j(u)K^i + (n-1)((2A(x,\theta)u + cu)\sum_{k=0}^n g^{jk}\partial_k u - \theta^j u^2)\Big).$$

The vector field K_0 in (6.3.8) does not contribute much more than translations of the origin. In the following discussion we therefore take $K_0 = 0$. (Later on we shall let $K_0 = (1,0,\ldots,0)$ to take advantage also of the energy estimate (6.3.2).) The vector field K is then in the span of the vector fields Z_{jk} and Z_0 in (6.2.8) and (6.2.9) (with (6.2.8) interpreted as (6.3.9) divided by 2 for general A). In fact, an easy calculation gives

$$(6.3.10) \qquad \langle K,\partial\rangle = A(x,\theta)Z_0 + \sum_{j,k=0}^n \theta^j x_k Z_{jk}.$$

Note that the coefficients are linear in x. It is also remarkable and important that K agrees with the vector field V in the proof of Lemma 6.3.1 if the vector K there is replaced by θ. Thus the present vector field K belongs to the closed light cone if θ is chosen in its interior; $K(x)$ is isotropic (or 0) if and only if x is isotropic (or 0). Taking for A the standard Lorentz form and $\nu = \theta = (1,0,\ldots,0)$ we shall now compute the leading term

$$(6.3.11) \qquad e = \sum_{i,j=0}^n T_i^j K^i \nu_j = \sum_{i=0}^n T_i^0 K^i$$

which will appear when we integrate (6.3.3)'' with respect to $\bar{x}$ to derive an energy estimate. We know already that e is nonnegative. As already observed after the proof of Lemma 6.3.1 we have

$$K = (x_0^2 + \cdots + x_n^2, 2x_0 x_1, \ldots, 2x_0 x_n).$$

If we write $x = (x_0, \vec{x})$, $gu' = U = (U_0, \vec{U})$, it follows that (see (6.3.5))

$$
\begin{aligned}
e = 2U_0 A(U, K) - K_0 A(U, U) &= 2U_0(U_0(x_0^2 + |\vec{x}|^2) - 2x_0\langle \vec{U}, \vec{x}\rangle) - (x_0^2 + |\vec{x}|^2)(U_0^2 - |\vec{U}|^2) \\
&= U_0^2(x_0^2 + |\vec{x}|^2) - 4x_0\langle \vec{U}, \vec{x}\rangle U_0 + x_0^2|\vec{U}|^2 + |\vec{x}|^2|\vec{U}|^2 \\
&= A(x, U)^2 + U_0^2|\vec{x}|^2 - 2x_0 U_0\langle \vec{x}, \vec{U}\rangle + x_0^2|\vec{U}|^2 + |\vec{x}|^2|\vec{U}|^2 - \langle \vec{x}, \vec{U}\rangle^2 \\
&= A(x, U)^2 + \tfrac{1}{2}\sum_{j,k=0}^{n}(x_j U_k - x_k U_j)^2.
\end{aligned}
$$

If we return to $u' = AU$, we obtain

$$
e = |\langle x, \partial u\rangle|^2 + |x \wedge A^{-1}\partial u|^2.
$$

Since $Ax \wedge \partial u$ has the components $(\partial_k A \, \partial_j u - \partial_j A \, \partial_k u)/2$, we obtain

Lemma 6.3.4. *If K is defined by (6.3.8) with $K_0 = 0$ and $\theta = (1, 0, \ldots, 0)$, and if g is the standard Lorentz form, then the energy form e defined by (6.3.11) can be written*

$$
e = |Z_0 u|^2 + \sum_{j<k}|Z_{jk}u|^2
$$

where Z_{jk} and Z_0 are defined by (6.2.8) and (6.2.9). The norm of $\partial u \mapsto \langle K, \partial u\rangle$ with respect to the quadratic form e in ∂u is $\leq |x|$.

Proof. The first statement was just proved, and it implies the second in view of (6.3.10).

Keeping the same assumptions on A and K we shall now prove the positivity of the complete energy expression which comes from (6.3.3)″:

Lemma 6.3.5. *If $n > 2$ and $u \in C_0^\infty$ we have*

(6.3.12)
$$
1/41 \leq \frac{(\int (e(u) + (n-1)(2x_0 u\partial_0 u - u^2))\,d\vec{x}}{\|Z_0 u\|^2 + \sum_{j<k}\|Z_{jk}u\|^2 + \|(n-1)u\|^2} \leq 2.
$$

Proof. Writing $2x_0 u\partial_0 u = 2u Z_0 u - \sum_1^n x_j \partial_j u^2$, we obtain

$$
2\int x_0 u\partial_0 u \, d\vec{x} = 2\int u Z_0 u \, d\vec{x} + n\|u\|^2.
$$

Thus the numerator $E(u)$ in (6.3.12) can be written

(6.3.13)
$$
E(u) = \|Z_0 u + (n-1)u\|^2 + \sum_{j<k}\|Z_{jk}u\|^2,
$$

which proves the upper bound in (6.3.12), also when $n = 2$. To obtain the lower bound it suffices to establish a bound for $\|u\|^2$. With polar coordinates $\vec{x} = r\omega$ we can write the integrand in E explicitly as follows by using the second expression for e given above

$$
\begin{aligned}
&(\partial_t u)^2(t^2 + r^2) + 4tr\partial_t u\partial_r u + (t^2 + r^2)|\partial_{\vec{x}} u|^2 + 2(n-1)tu\partial_t u - (n-1)u^2 \\
&\geq -(2r\partial_r u + (n-1)u)^2 t^2/(t^2 + r^2) + (t^2 + r^2)(\partial u/\partial r)^2 - (n-1)u^2,
\end{aligned}
$$

with equality in the radial case for an appropriate choice of $\partial_t u$. (Note that we have $U = (\partial_t u, -\partial_{\bar{x}} u)$.) Writing $r^{\frac{n-1}{2}} u = v$ to remove the factor r^{n-1} in the volume element, we obtain

$$E \geq \int d\omega \int \left(-4r^2 t^2 |\partial_r v|^2/(t^2 + r^2) + (t^2 + r^2)(\partial v/\partial r + (1-n)v/2r)^2 - (n-1)v^2 \right) dr$$

$$= \int d\omega \int \left((r^2 - t^2)^2 (r^2 + t^2)^{-1}(\partial v/\partial r)^2 + (n-1)(n-3)4^{-1}(1 + t^2/r^2)v^2 \right) dr,$$

after expansion and integration by parts. We know that this is still positive when $n = 2$ and $v = O(r)$ at 0, so the integral of the first term is at least equal to $\int (1 + t^2/r^2)v^2 \, dr/4$, which proves that $E \geq (n-2)^2 \|u\|^2/4$ when $n > 2$. Since the inequalities

$$\|a + b\|^2 + c^2 \leq E, \quad \|b\|^2 \leq \kappa^2 E, \quad \kappa = 2(n-1)/(n-2) \leq 4$$

imply that $\|a\|^2 = \|a + b - b\|^2 \leq (1 + \kappa)\|a + b\|^2 + (1 + \kappa^{-1})\|b\|^2$, hence

$$\|a\|^2 + \|b\|^2 + c^2 \leq (1 + \kappa)\|a + b\|^2 + (2 + \kappa^{-1})\|b\|^2 + c^2 \leq (1 + \kappa + 2\kappa^2 + \kappa)E \leq 41E,$$

the lower bound in (6.3.12) follows when $a = Z_0 u$, $b = (n-1)u$, $c^2 = \sum_{j<k} \|Z_{jk} u\|^2$.

Remark. It is easy to show that no positive lower bound exists in (6.3.12) when $n = 2$.

When u is a solution of $\Box u = 0$ with Cauchy data in C_0^∞ it follows from (6.3.3)″ that $E(u)$ is independent of x_0. In view of Lemma 6.3.5 we conclude that

$$\|Z_0 u(x_0, \cdot)\|^2 + \sum \|Z_{jk} u(x_0, \cdot)\|^2 + \|u(x_0, \cdot)\|^2$$

can be estimated for all x_0 by a constant times the value for $x_0 = 0$. More generally, if $\Box u = f$ we can use the identity (6.3.3)″ just as in the proof of (6.3.2) if we observe that by (6.3.10)

$$|\langle K, \partial u \rangle + (n-1)x_0 u|^2 = |x_0(Z_0 u + (n-1)u) + \sum_1^n x_k Z_{0k} u|^2$$

$$\leq |x|^2 (|Z_0 u + (n-1)u|^2 + \sum_1^n |Z_{0k} u|^2).$$

In view of (6.3.13) it follows from (6.3.3)″ that

$$dE(x_0; u)/dx_0 \leq 2\|F(x_0, \cdot)\| E(x_0; u)^{\frac{1}{2}}$$

where $F(x) = |x| f(x)$. Hence

(6.3.14) $$E(x_0; u)^{\frac{1}{2}} \leq E(0; u)^{\frac{1}{2}} + \int_0^{x_0} \|F(t, \cdot)\| dt.$$

Combined with (6.3.12) this gives improved control of the behavior of u at infinity when we have additional information on the decay of f.

(6.3.14) can be extended to operators with variable coefficients just as Proposition 6.3.2 extended (6.3.2). However, the hypotheses needed are far more complicated now so we shall postpone the discussion until we are ready for an application which justifies them. Instead we end this section with a brief sketch of how Proposition 6.3.2 leads to existence and uniqueness theorems for the Cauchy problem for linear second order hyperbolic equations. Local results follow from global ones, so we consider an equation

$$(6.3.15) \quad Lu = \sum_{j,k=0}^{n} g^{jk}(x)\partial_j \partial_k u(x) + \sum_{j=0}^{n} b^j(x)\partial_j u(x) + c(x)u(x) = f(x), \quad 0 \le x_0 \le T,$$

such that all derivatives of the coefficients are bounded in $[0,T] \times \mathbf{R}^n$ and $\sum |g^{jk}(x) - \lambda^{jk}| \le \frac{1}{2}$, where $\sum \lambda^{jk}\partial_j \partial_k = \Box$. If $M_1(x_0) = \|u(x_0, \cdot)\| + \|u'(x_0, \cdot)\|$, then it follows from (6.3.6) if $u \in C^\infty$ for $0 \le x_0 \le T$ and vanishes for large $\tilde{x}$ that for such x_0

$$(6.3.16) \qquad M_1(x_0) \le C(M_1(0) + \int_0^{x_0} (\|f(t,\cdot)\| + M_1(t))dt),$$

since

$$\|u(x_0, \cdot)\| \le \|u(0, \cdot)\| + \int_0^{x_0} \|\partial_t u(t, \cdot)\| dt.$$

Now we can apply *"Gronwall's lemma"*:

Lemma 6.3.6. *If φ, k and E are nonnegative, E is increasing and*

$$\varphi(t) \le E(t) + \int_0^t \varphi(\tau)k(\tau)\,d\tau, \quad 0 \le t \le T,$$

then

$$\varphi(t) \le E(t)\exp\left(\int_0^t k(\tau)\,d\tau\right), \quad 0 \le t \le T.$$

Proof. It is enough to prove this when $t = T$, and E may be replaced by $E(T)$ then, so we may assume that E is constant. Writing

$$F(t) = E + \int_0^t \varphi(\tau)k(\tau)\,d\tau$$

we have

$$F'(t) = \varphi(t)k(t) \le F(t)k(t),$$

since $\varphi \le F$, hence

$$F(t)\exp\left(-\int_0^t k(\tau)\,d\tau\right) \le F(0) = E,$$

which completes the proof.

(6.3.16) implies in view of Lemma 6.3.6 that with another constant C_0 we have

$$(6.3.17) \qquad \sum_{|\alpha| \le 1} \|\partial^\alpha u(x_0, \cdot)\| \le C_0\Big(\sum_{|\alpha| \le 1} \|\partial^\alpha u(0, \cdot)\| + \int_0^{x_0} \|f(t, \cdot)\|dt\Big).$$

We claim that, more generally, for any integer $s \ge 0$ there is a constant C_s such that

$$(6.3.18) \qquad \sum_{|\alpha| \le s+1} \|\partial^\alpha u(x_0, \cdot)\| \le C_s\Big(\sum_{|\alpha| \le s+1} \|\partial^\alpha u(0, \cdot)\| + \int_0^{x_0} \sum_{|\alpha| \le s} \|\partial^\alpha f(t, \cdot)\|dt\Big).$$

(We assume that $0 \leq x_0 \leq T$, and all constants may depend on T.) In the proof we may assume that $s > 0$ and note that

$$L\partial^\alpha u = \partial^\alpha f + [L, \partial^\alpha]u,$$

where the commutator $[L, \partial^\alpha]$ is of order $\leq s+1$ when $|\alpha| \leq s$. If we apply (6.3.17) to all such equations and write

$$M_s(f; x_0) = \sum_{|\alpha| \leq s} \|\partial^\alpha f(x_0, \cdot)\|,$$

and similarly for u, it follows that

$$M_{s+1}(u; x_0) \leq C_s'(M_{s+1}(u; 0) + \int_0^{x_0} (M_s(f; t) + M_{s+1}(u; t)) \, dt).$$

By Gronwall's lemma this gives (6.3.18). Note that if x_0 has a fixed positive lower bound, then we can estimate $\sum_{|\alpha| < s} \|\partial^\alpha f(0, \cdot)\|$ by the integral in (6.3.18). Using the equation $Lu = f$ we can therefore restrict $\sum_{|\alpha| \leq s+1} \|\partial^\alpha u(0, \cdot)\|$ to terms of order ≤ 1 with respect to x_0.

We may assume without restriction that the coefficient of ∂_0^2 in L is identically 1, for by our hypotheses both this coefficient and its reciprocal as well as their derivatives have uniform bounds. Then the commutator $[L, \partial^\alpha]$ is of order ≤ 1 with respect to x_0 if ∂^α has no such derivative. Hence we obtain by repeating the proof of (6.3.18)

$$(6.3.18)' \qquad \sum_{|\alpha| \leq 1} \|\partial^\alpha u(x_0, \cdot)\|_{(s)} \leq C_s(\sum_{|\alpha| \leq 1} \|\partial^\alpha u(0, \cdot)\|_{(s)} + \int_0^{x_0} \|f(t, \cdot)\|_{(s)} dt),$$

where

$$\|v\|_{(s)} = ((2\pi)^{-n} \int |\hat{v}(\xi)|^2 (1 + |\xi|^2)^s \, d\xi)^{\frac{1}{2}}, v \in C_0^\infty(\mathbf{R}^n)$$

is equivalent to $\sum_{|\alpha| \leq s} \|\partial_x^\alpha v\|$ when s is a positive integer (and $\|\cdot\|$ is the L^2 norm). As in Section 5.3 we denote by $H_{(s)}(\mathbf{R}^n)$ the Hilbert space of all temperate distributions v in $\mathbf{R}^n$ for which the Fourier transform $\hat{v}$ is a function such that $\|v\|_{(s)} < \infty$. The estimate (6.3.18)' can be extended to all real s, but for the sake of simplicity we shall only do so when s is a negative integer. Then we define $U(x_0, \cdot) \in \mathcal{S}(\mathbf{R}^n)$ by

$$U(x_0, \cdot) = (1 - \Delta)^s u(x_0, \cdot), \quad \text{that is, } u(x_0, \cdot) = (1 - \Delta)^{-s} U(x_0, \cdot),$$

where Δ is the Laplace operator in $\mathbf{R}^n$. The estimate (6.3.18)' remains valid for such functions, with s replaced by $-s$, which is a positive integer. Set

$$M(x_0) = \sum_{|\alpha| \leq 1} \|\partial^\alpha U(x_0, \cdot)\|_{(-s)} = \sum_{|\alpha| \leq 1} \|\partial^\alpha u(x_0, \cdot)\|_{(s)}.$$

Then

$$M(x_0) \leq C_s(M(0) + \int_0^{x_0} \|LU(t, \cdot)\|_{(-s)} dt).$$

We have

$$f = Lu = L((1 - \Delta)^{-s} U) = (1 - \Delta)^{-s} LU + RU,$$

where R is a differential operator of order $1 - 2s$ and order ≤ 1 with respect to x_0. If we write $R = \sum \partial^\beta c_{\alpha\beta\gamma} \partial^\alpha \partial^\gamma$ with $|\beta| \leq -s$, $|\gamma| \leq -s$, $|\alpha| \leq 1$, and $\beta_0 = \gamma_0 = 0$, we see that

$$\|RU(t, \cdot)\|_{(s)} \leq \sum \|c_{\alpha\beta\gamma} \partial^\alpha \partial^\gamma U(t, \cdot)\| \leq CM(t),$$

so we obtain
$$\|LU(t,\cdot)\|_{(-s)} \le \|f(t,\cdot)\|_{(s)} + CM(t),$$

and $(6.3.18)'$ follows as before.

The importance of the estimate $(6.3.18)'$ when $s < 0$ is primarily that it allows one to prove existence theorems in the spaces $H_{(s)}$ with $s > 0$. To show how that is done we first observe that if s is a positive integer we can apply $(6.3.18)'$ with s replaced by $-s-1$, L replaced by the adjoint L^* and x_0 replaced by $T - x_0$. This gives

$$\|\varphi(x_0,\cdot)\|_{(-s)} \le \sum_{|\alpha|\le 1} \|\partial^\alpha \varphi(x_0,\cdot)\|_{(-s-1)} \le C \int_{x_0}^T \|L^*\varphi(t,\cdot)\|_{(-s-1)}\,dt, \quad 0 \le x_0 \le T,$$

if $\varphi \in C_0^\infty(\mathbf{R}^{n+1})$ and $\varphi = 0$ for $x_0 > T$. If $f \in L^1([0,T]; H_{(s)}(\mathbf{R}^n))$, then

$$|(f,\varphi)| = \left| \int_0^T (f(t,\cdot), \varphi(t,\cdot))\,dt \right| \le C \int_0^T \|L^*\varphi(t,\cdot)\|_{(-s-1)}\,dt,$$

so by the Hahn-Banach theorem we can find $u \in L^\infty([-\infty,T]; H_{(s+1)}(\mathbf{R}^n))$ such that

$$(f,\varphi) = (u, L^*\varphi), \quad \text{if } \varphi(x_0,T) = 0 \text{ for } x_0 \ge T,$$

and $u = 0$ when $x_0 < 0$. Thus $Lu = f$ for $0 < x_0 \le T$, in the sense of distribution theory. Set $\partial_0 u = v$. If we single out the terms containing some factor ∂_0, then the differential equation gives

$$\partial_0 v + \sum_1^n a_j(x)\partial_j v + a_0 v \in L^\infty([-\infty,T]; H_{(s-1)}).$$

If we change variables so that the characteristics $dx_j/dx_0 = a_j(x)$ become straight lines, integrate the equation and return to the original variables, then we see that $\partial_0 u = v \in L^\infty([-\infty,T]; H_{(s-1)})$, hence using the equation we find that $\partial_0^k u \in L^\infty([-\infty,T]; H_{(s-k)})$ if $0 \le k \le s$ and f is say smooth and vanishes outside a compact set when $0 \le x_0 \le T$. For $s \ge 2$ we conclude that the Cauchy data of u when $x_0 = 0$ must vanish, and since $(6.3.17)$ extends by continuity to all u with the derivatives of order ≤ 2 in L^2, the solution obtained does not depend on s, so it is in C^∞. Now it follows by continuity in view of $(6.3.18)$ that for every f such that $\partial^\alpha f \in L^1([0,T]; L^2)$ for $|\alpha| \le s$ there is a solution of the equation $Lu = f$ with $\partial^\alpha u \in L^\infty([0,T]; L^2)$ when $|\alpha| \le s+1$ and Cauchy data 0. A solution with arbitrary Cauchy data is of course obtained if one chooses any function u_0 with the given Cauchy data and introduces $u - u_0$ as unknown instead of u.

The existence and uniqueness theorems we have now proved imply local existence and uniqueness theorems. In fact, if L is just given in a neighborhood Ω of the origin, with smooth coefficients, we can choose $\chi \in C_0^\infty(\Omega)$ with $0 \le \chi \le 1$ so that $\chi = 1$ in another neighborhood of the origin. Then $\chi(x)L(x,\partial) + (1 - \chi(x))L(0,\partial)$ will satisfy the global hypotheses made above if the support of χ is small enough, and the global existence theorems proved above imply local existence theorems for L. If we have a solution of $Lu = 0$ in a neighborhood of 0 with vanishing Cauchy data when $x_0 = 0$, then we just introduce $x_0 + x_1^2 + \cdots + x_n^2$ as a new variable instead of x_0 to guarantee that $x_0 > 0$ in the support except at 0. If we then extend L to a global operator as just indicated we obtain a solution of the extended homogeneous equation when $0 \le x_0 \le T$ if T is small enough, and the Cauchy data vanish when $x_0 = 0$. This gives local uniqueness of solutions of the hyperbolic Cauchy problem. In what follows we shall take such results in the case of linear equations for granted; they can be proved in many different ways. (See e.g. Hörmander [4, Chapter XXIII].)

6.4. Interpolation and Sobolev inequalities. Already for the elementary existence theorems proved at the end of Section 4.2 we needed inequalities like (4.2.17)' to estimate the norm of composite functions. In the case of several space variables we must use L^2 estimates, obtained from the energy integral method, rather than L^∞ estimates, so we need the following analogous and more general interpolation inequalities. They are special cases of those of Gagliardo [1] and Nirenberg [1].

Theorem 6.4.1. Let $1 \leq r \leq \infty$, $1 \leq q \leq \infty$, and let m be an integer ≥ 2. If $u \in L^q(\mathbf{R}^n)$ and $\partial^\alpha u \in L^r(\mathbf{R}^n)$ when $|\alpha| = m$, then $\partial^\alpha u \in L^{p(\alpha)}(\mathbf{R}^n)$ for $|\alpha| \leq m$, if

$$(6.4.1) \qquad m/p(\alpha) = (m - |\alpha|)/q + |\alpha|/r;$$

moreover, if $\| \cdot \|_s$ denotes the L^s norm, then

$$(6.4.2) \qquad \sup_{|\alpha|=j} \|\partial^\alpha u\|_{p(\alpha)} \leq 4^{|\alpha|(m-|\alpha|)} \Big(\sup_{|\alpha|=m} \|\partial^\alpha u\|_r \Big)^{|\alpha|/m} \|u\|_q^{(m-|\alpha|)/m}.$$

The general result will follow from the special case $m = 2$, $n = 1$. We begin with a simple lemma.

Lemma 6.4.2. Let I be a finite interval on $\mathbf{R}$ of length $|I|$, and let $u \in L^q(I)$, $u'' \in L^r(I)$ for some $q, r \in [1, \infty]$. Then $u' \in L^p(I)$, $1 \leq p \leq \infty$, and

$$(6.4.3) \qquad \|u'\|_p |I|^{1-1/p} \leq \|u''\|_r |I|^{2-1/r} + 4\|u\|_q |I|^{-1/q}.$$

Proof. Since the inequality is invariant under affine changes of variables, we may assume that $I = (-\frac{1}{2}, \frac{1}{2})$. It is then obvious that the strongest case of (6.4.3) is obtained when $r = q = 1$ and $p = \infty$. Assume that

$$\max_{x \in I} u'(x) = \int_I |u''|\, dx + M$$

for some $M > 0$. Then $u'(x) \geq M$ in I, and it follows that for some $x_0 \in \mathbf{R}$

$$|u(x)| \geq M|x - x_0|, \quad x \in I,$$

where $x_0 \in I$ if u has a zero in I, and $x_0 < -\frac{1}{2}$ $(x_0 > \frac{1}{2})$ if $u > 0$ $(u < 0)$ in I. Now

$$2\int_I |x - x_0|\, dx = \int_I (|x - x_0| + |x + x_0|)\, dx \geq 2\int_I |x|\, dx = 1/2,$$

hence $\int_I |u|\, dx \geq M/4$, which proves the lemma. (The estimate (6.4.3) is easily seen to be optimal when $r = q = 1$ and $p = \infty$, but the estimate derived in Lemma 6.4.3 does not contain the best possible constants.)

Lemma 6.4.3. Let $u \in L^q(\mathbf{R}_+)$, $u'' \in L^r(\mathbf{R}_+)$ where $q, r \in [1, \infty]$. If $2/p = 1/r + 1/q$, it follows that $u' \in L^p(\mathbf{R}_+)$ and that

$$(6.4.4) \qquad \|u'\|_p \leq 4(\|u''\|_r \|u\|_q)^{\frac{1}{2}}.$$

Proof. In an interval $I \subset \mathbf{R}_+$ where the two terms in the right-hand side of (6.4.3) are equal, we can write (6.4.3) in the form

$$(6.4.4)' \qquad \|u'\|_{p,I} \leq 4(\|u''\|_{r,I} \|u\|_{q,I})^{\frac{1}{2}},$$

where the norms are taken in I, for $2 - 1/r - 1/q = 2 - 2/p$ by hypothesis. If E is a disjoint union of intervals I_j for which (6.4.4)' holds, then

$$\|u'\|_{p,E}^p \le 4^p \sum \|u''\|_{r,I_j}^{p/2} \|u\|_{q,I_j}^{p/2} \le 4^p \Big(\sum \|u''\|_{r,I_j}^r\Big)^{p/2r} \Big(\sum \|u\|_{q,I_j}^q\Big)^{p/2q}$$

by Hölder's inequality, for $p/2r + p/2q = 1$. Hence (6.4.4)' is valid with I replaced by E. (This is obvious when $p = q = r = \infty$.) If we consider intervals I with given left end point a, then the first term in (6.4.3) is the largest one if $|I|$ is large (unless $u' = 0$ in (a, ∞)), for otherwise

$$\|u'\|_{p,I} \le 8\|u\|_{q,I}|I|^{1/p - 1/q - 1} \to 0 \text{ as } |I| \to \infty,$$

since $1/p - 1 - 1/q = (1/r - 2 - 1/q)/2 \le -\frac{1}{2}$. Whenever the first term is the largest one then

$$\|u'\|_{p,I}|I|^{-1/p} \le 2\|u''\|_{r,I}|I|^{1-1/r},$$

$$4\|u\|_{q,I}|I|^{-1/q} \le \|u''\|_{r,I}|I|^{2-1/r},$$

which implies $u(a) = u'(a) = 0$ if I can be arbitrarily small. If 0 is not a critical value of u this can never happen, so we get a sequence of intervals $I_j = [a_j, a_{j+1})$ for which (6.4.4)' holds because the two terms in the right-hand side of (6.4.3) are equal. $a_j \to \infty$ since u and u' would vanish at an accumulation point by the argument just given. This proves (6.4.4) when 0 is not a critical value of u. Now the set of critical values of $e^x u(x)$ is of measure 0, and if ε is not a critical value then 0 is not a critical value for $u(x) - \varepsilon e^{-x}$. Hence (6.4.4) holds with $u(x)$ replaced by $u(x) - \varepsilon e^{-x}$ for a sequence of values of ε converging to 0, which proves (6.4.4) in general.

Proof of Theorem 6.4.1. Assuming at first that $n = 1$ we set

$$M_j = \|u^{(j)}\|_{p(j)}.$$

If all M_j with $0 < j < m$ are finite, then it follows from (6.4.4) that

$$M_j \le 4(M_{j+1} M_{j-1})^{\frac{1}{2}},$$

hence $A_j^2 \le A_{j+1} A_{j-1}$ if $A_j = 4^{j^2} M_j$. This logarithmic convexity implies that $A_j \le A_m^{j/m} A_0^{(m-j)/m}$ and gives (6.4.2). If we do not already know that $u^{(j)} \in L^{p(j)}$, we can at least by standard regularization reduce the proof to functions with $u^{(j)} \in L^\infty$ for all j. Choose $\chi \in C_0^\infty$ with $0 \le \chi \le 1$ and $\chi = 1$ in a neighborhood of 0 and set $u_\varepsilon(x) = \chi(\varepsilon x)u(x)$. Then

$$\|u_\varepsilon\|_q \le \|u\|_q, \quad \|u_\varepsilon^{(m)}\|_r \le \|u^{(m)}\|_r + O(\varepsilon^{1-1/r}) = O(1) \text{ as } \varepsilon \to 0.$$

Hence we conclude from the first part of the proof that $u^{(j)} \in L^{p(j)}$, and that (6.4.2) holds.

When the number n of variables is larger than 1, we conclude from Hölder's inequality as in the proof of Lemma 6.4.3 that (6.4.2) holds if all derivatives are taken with respect to the same variable x_k. By repeated use of that result we find that $\partial^\alpha u \in L^{p(\alpha)}$ for $|\alpha| \le m$. If we now introduce

$$M_j = \sup_{|\alpha|=j} \|\partial^\alpha u\|_{p(\alpha)}$$

and repeat the arguments used above in the one dimensional case, the estimate (6.4.2) follows in general.

Corollary 6.4.4. *If* $u, v \in L^\infty(\mathbf{R}^n)$ *and* $\partial^\alpha u, \partial^\alpha v \in L^r(\mathbf{R}^n)$ *when* $|\alpha| = m$, *then* $\partial^\alpha(uv) \in L^r(\mathbf{R}^n)$ *when* $|\alpha| = m$, *and*

$$(6.4.5) \qquad \sum_{|\alpha|=m} \|\partial^\alpha(uv)\|_r \leq C_m \Big(\sum_{|\alpha|=m} \|\partial^\alpha u\|_r \|v\|_\infty + \|u\|_\infty \sum_{|\alpha|=m} \|\partial^\alpha v\|_r \Big).$$

Proof. $\partial^\alpha(uv)$ consists of 2^m terms of the form $\partial^\beta u \partial^\gamma v$ with $|\beta| + |\gamma| = m$. With the notation in Theorem 6.4.1, $q = \infty$, we have $1/p(\beta) + 1/p(\gamma) = 1/r$, so the L^r norm of each term is at most $16^{|\beta||\gamma|} B^{|\beta|/m} C^{|\gamma|/m}$, where B and C are the two expressions on the right-hand side of (6.4.5). Hence we obtain (6.4.5) with $C_m = 2^m 2^{m^2} \binom{m+n-1}{m}$.

Remark. A more general version of the preceding estimates is sometimes useful: If $v_1, \ldots, v_j \in L^\infty(\mathbf{R}^n)$ and $\partial^\alpha v_1, \ldots, \partial^\alpha v_j \in L^r(\mathbf{R}^n)$, then

$$(6.4.5)' \qquad \|\partial^{\alpha_1} v_1 \cdots \partial^{\alpha_j} v_j\|_r \leq 2^{jm^2/2} \max_{1 \leq i \leq j} \prod_{k \neq i} \|v_k\|_\infty \sup_{|\alpha|=m} \|\partial^\alpha v_i\|_r, \quad \text{if } \sum_1^j |\alpha_i| = m.$$

Again by Theorem 6.4.1 we have

$$\|\partial^{\alpha_i} v_i\|_{r/\lambda_i} \leq 2^{m^2/2} A_i^{\lambda_i} B_i^{1-\lambda_i}, \quad A_i = \sup_{|\alpha|=m} \|\partial^\alpha v_i\|_r, \ B_i = \|v_i\|_\infty,$$

where $\lambda_i = |\alpha_i|/m$, thus $0 \leq \lambda_i$ and $\sum \lambda_i = 1$. Hölder's inequality gives the bound $2^{jm^2/2} \prod A_i^{\lambda_i} B_i^{1-\lambda_i}$ for the left-hand side of (6.4.5)'. This convex function of λ in a simplex takes its maximum at a vertex, that is, when one $\lambda_i = 1$ and the others are 0, which proves (6.4.5)'.

Corollary 6.4.5. *Let* $u \in L^\infty(\mathbf{R}^n, \mathbf{R}^N)$, *let* $F \in C^m(\mathbf{R}^N)$, *and assume that* $\partial^\alpha u \in L^r(\mathbf{R}^n)$ *when* $|\alpha| = m$. *Then* $\partial^\alpha F(u) \in L^r(\mathbf{R}^n)$ *when* $|\alpha| = m$, *and*

$$(6.4.6) \qquad \sup_{|\alpha|=m} \|\partial^\alpha F(u)\|_r \leq C_m \sup_{1 \leq |\gamma| \leq m} |F^{(\gamma)}(u)| \|u\|_\infty^{|\gamma|-1} \sup_{|\alpha|=m} \|\partial^\alpha u\|_r,$$

if $m > 0$, *while for* $m = 0$

$$\|F(u) - F(0)\|_r \leq M \|u\|_r,$$

if M is a Lipschitz constant for F in the range of u.

Proof. It suffices to prove the estimate when u is smooth. We assume that $m > 0$, for the statement is obvious when $m = 0$. It is clear that $\partial^\alpha F(u)$ is a linear combination of terms of the form

$$F^{(\gamma)}(u) \partial^{\alpha_1} u_{i_1} \ldots \partial^{\alpha_j} u_{i_j},$$

where $|\alpha_1| + \cdots + |\alpha_j| = |\alpha|$, $|\alpha_k| > 0$ for $k = 1, \ldots, j$, $|\gamma| = j$. By (6.4.5)' the L^r norm of such a term can be estimated by

$$C^{|\gamma|} \sup |F^{(\gamma)}(u)| \|u\|_\infty^{|\gamma|-1} \sup_{|\alpha|=m} \|\partial^\alpha u\|_r,$$

which proves (6.4.6).

Remark. Note the somewhat surprising fact that the estimate is linear in the norms of the derivatives of u of order m once a bound for u is known. We shall actually need a more general version of (6.4.6) where $F \in C^m(\mathbf{R}^{n+m})$ also depends on x,

$$(6.4.6)' \quad \sup_{|\alpha|=m} \|\partial^\alpha(F(x,u(x)) - F(x,0))\|_r$$

$$\leq C'_m \sup_{\substack{1 \leq |\gamma| \leq |\alpha| \\ |\alpha|+|\beta|=m}} \sup_{|v| \leq \|u\|_\infty} |\partial^\beta_x \partial^\gamma_v F(x,v)| \|u\|_\infty^{|\gamma|-1} \|\partial^\alpha u\|_r$$

$$+ C''_m \sup_{|\alpha|=m} \sup_{|v| \leq \|u\|_\infty} |\partial^\alpha_x \partial_v F(x,v)| \|u\|_r.$$

The terms in $\partial^\alpha(F(x,u(x)) - F(x,0))$ where no derivative falls on u are estimated by the last expression, and for the others the estimates in the proof of (6.4.6) can be applied with m replaced by $m - |\beta|$ if ∂^β acts directly on the first argument of F.

We shall also need *Sobolev's lemma*. The following version (see Aubin [1]) is very precise and of great geometrical interest.

Proposition 6.4.6. *If $u \in \mathcal{E}'(\mathbf{R}^n)$ and $\partial_j u \in L^1(\mathbf{R}^n)$, $j = 1, \ldots, n$, then it follows that $u \in L^{n/(n-1)}(\mathbf{R}^n)$ and that*

$$(6.4.7) \quad \left(\int |u|^{n/(n-1)} dx \right)^{(n-1)/n} \leq C_n \int \left(\sum_1^n |\partial_j u|^2 \right)^{\frac{1}{2}} dx.$$

Here $C_n = (\omega_n/n)^{(n-1)/n}/\omega_n$ where ω_n is the area of the unit sphere S^{n-1}.

Proof. When $n = 1$ this is just the obvious statement that $|u| \leq \|u'\|_1/2$, so we assume that $n > 1$. By a regularisation we can reduce the proof to the case where $u \in C_0^\infty$. Then $|u|$ is Lipschitz continuous and $|u'| = \||u|'|$ almost everywhere. Hence it is no restriction to assume that $u \geq 0$. This will imply (6.4.7) even if $\partial_j u$ are only measures. In that generality the geometrical meaning is more clear: if u is the characteristic function of a set E with C^1 boundary, then u' is the unit normal to ∂E multiplied by the surface measure $d\sigma$, so (6.4.7) means that

$$\left(m(E)/(\omega_n/n) \right)^{(n-1)/n} \leq \sigma(\partial E)/\omega_n,$$

which is the isoperimetric inequality stating that among all sets with given volume the ball has the smallest boundary area. To prove (6.4.7) in general for $0 \leq u \in C_0^\infty$ we denote by χ_t the characteristic function of $\{x; u(x) > t\}$. Then $u = \int \chi_t dt$, so Minkowski's inequality gives

$$\|u\|_{n/(n-1)} \leq \int \|\chi_t\|_{n/(n-1)} \, dt.$$

When t is not a critical value of u, then χ_t is the characteristic function of a set with smooth boundary and, as just observed, the isoperimetric inequality gives

$$\|\chi_t\|_{n/(n-1)} \leq C_n \int |\chi_t'| \, dx,$$

where the right-hand side should be understood as the total mass of a measure. Now $\chi_t = H(u - t)$ where H is the Heaviside function, so $\chi_t' = u'\delta(u - t)$. Thus

$$\int |\chi_t'| \, dx = \int |u'|\delta(u - t) \, dx.$$

Since the set of critical values of u is closed and of measure 0, we obtain (6.4.7) now by integrating over the complement, for $\int \delta(u - t) \, dt = 1$.

To complete the proof we digress to give a beautiful proof of the Brunn-Minkowski inequality, following Federer [1]:

Proposition 6.4.7. *If A and B are compact subsets of $\mathbf{R}^n$ and $A+B$ denotes the sum $\{x + y; x \in A, y \in B\}$, then*

$$(6.4.8) \qquad m(A + B)^{1/n} \geq m(A)^{1/n} + m(B)^{1/n}.$$

Proof. It suffices to prove this when A and B are unions of finitely many disjoint intervals, that is, sets of the form $\{x \in \mathbf{R}^n; \alpha_j \leq x_j \leq \beta_j \text{ for } j = 1, \dots, n\}$. This can be done by induction over $a + b$ if a and b are the number of intervals which constitute A and B. In fact, if A and B are both intervals, with side lengths a_i, b_i, $i = 1, \dots, n$, then

$$m(A)^{1/n} + m(B)^{1/n} = \prod_1^n a_j^{1/n} + \prod_1^n b_j^{1/n} \leq (\sum_1^n a_j + \sum_1^n b_j)/n = 1 = \prod_1^n (a_j + b_j)^{1/n}$$

by the inequality between geometric and arithmetic means, if $a_j + b_j = 1$ for every j. By homogeneity reasons this gives (6.4.8) in general if A and B are both intervals. Now assume that $a > 1$. Then there is a plane $x_j = $ constant separating two of the intervals defining A. In view of the translation invariance it is no restriction to assume that it is the plane $x_j = 0$, and that

$$m(A_+)/m(A_-) = m(B_+)/m(B_-),$$

if $A_\pm$ and $B_\pm$ are the intersections of A and B with the half spaces defined by $x_j \geq 0$ and $x_j \leq 0$ respectively. These are constituted by at most $a - 1$ and at most b intervals, so by the inductive hypothesis

$$
\begin{aligned}
m(A + B) &\geq m(A_+ + B_+) + m(A_- + B_-) \\
&\geq (m(A_+)^{1/n} + m(B_+)^{1/n})^n + (m(A_-)^{1/n} + m(B_-)^{1/n})^n \\
&= (m(A)^{1/n} + m(B)^{1/n})^n
\end{aligned}
$$

which completes the proof.

If we take for A a set with C^2 boundary and for B the unit ball, then

$$m(A + rB) = m(A) + r\sigma(\partial A) + O(r^2),$$

hence

$$m(A + rB)^{1/n} - m(A)^{1/n} = r\sigma(\partial A)m(A)^{(1-n)/n}/n + O(r^2), \quad r > 0,$$

and when $r \to 0$ it follows from (6.4.8) that $\sigma(\partial A) \geq nm(A)^{(n-1)/n}m(B)^{1/n}$, which is the isoperimetric inequality.

Returning to Sobolev's inequality we apply (6.4.7) to a power $|u|^{r+1}$ where $r \geq 0$. (We assume that u is smooth with compact support in the calculation.) This gives

$$\left(\int |u|^{(r+1)n/(n-1)} dx \right)^{(n-1)/n} \leq C_n(1 + r) \int |u'||u|^r \, dx.$$

Set $q = (r + 1)n/(n - 1)$ and apply Hölder's inequality with the exponents p and q/r, where

$$1/p + r/q = 1, \text{ that is, } 1/p - 1/q = 1/n.$$

Note that the condition $r \geq 0$ means that $p \geq 1$. Then we obtain

$$\|u\|_q^{r+1} \leq C_n(1 + r)\|u'\|_p\|u\|_q^r,$$

that is,

$$(6.4.9) \qquad \|u\|_q \leq C_n q(1 - 1/n)\|u'\|_p, \text{ if } 1/p = 1/q + 1/n, \ 1 \leq p < q < \infty.$$

Iteration of this inequality gives the *Sobolev estimate*

$$(6.4.10) \qquad \|u\|_q \leq C_{n,m,q} \sum_{|\alpha|=m} \|\partial^\alpha u\|_p, \text{ if } 1/p = 1/q + m/n, \ 1 \leq p < q < \infty.$$

Hence $u \in \mathcal{E}'$ and $\partial^\alpha u \in L^p$ for $|\alpha| = m$ implies $u \in L^q$ then. When $p > n$ the estimate (6.4.9) is no longer applicable but there is a Hölder estimate instead:

Proposition 6.4.8. *If $u' \in L^p(\mathbf{R}^n)$ and $p > n$, then u is a continuous function and*

$$(6.4.11) \qquad |u(x) - u(y)| \leq C_{n,p}|x - y|^{1-\frac{n}{p}}\|u'\|_p.$$

Proof. By a regularization we reduce the proof to the case where $u \in C^\infty$. In view of the homogeneity under scale changes we may also assume that $|x - y| = 1$ in the proof. First we shall prove that

$$(6.4.12) \qquad |u * \varphi(x) - u(x) \int \varphi(y)dy| \leq C_{p,n}\|\varphi\|_\infty\|u'\|_p,$$

if $p > n$ and φ vanishes outside the unit ball. It is sufficient to do so when $x = 0$. With polar coordinates r, ω we have to estimate

$$I = \iint (u(r\omega) - u(0))\varphi(-r\omega)r^{n-1}\, dr\, d\omega.$$

Let $\partial\Phi(r,\omega)/\partial r = r^{n-1}\varphi(-r\omega)$ and $\Phi(r,\omega) = 0$ when $r > 1$. Then $|\Phi| \leq \|\varphi\|_\infty$, and

$$I = -\iint \partial u(r\omega)/\partial r\, \Phi(r,\omega)\, dr\, d\omega.$$

Now

$$\|u'\|_p \geq \left(\int_0^1 \int |\partial u/\partial r|^p r^{n-1}\, dr\, d\omega\right)^{1/p},$$

so Hölder's inequality gives

$$|I| \leq \|u'\|_p\|\varphi\|_\infty\left(\int_0^1 \int r^{(1-n)q/p}\, dr\, d\omega\right)^{1/q}$$

where $1/q + 1/p = 1$, hence $1 + (1-n)q/p = q - nq/p = q(p-n)/p > 0$ if $p > n$. Thus the integral converges then and (6.4.12) is proved. Since

$$|(u * \varphi)'| = |u' * \varphi| \leq \|\varphi\|_q\|u'\|_p, \quad \text{hence } |u * \varphi(x) - u * \varphi(y)| \leq |x - y|\|\varphi\|_q\|u'\|_p,$$

it follows from (6.4.12) if φ is fixed with integral equal to 1 that for some other constant

$$|u(x) - u(y)| \leq C_{p,n}\|u'\|_p, \quad \text{if } |x - y| = 1,$$

and this completes the proof.

Corollary 6.4.9. *If m is a positive integer and $n/m < p \leq \infty$, then $\partial^\alpha u \in L^p(\mathbf{R}^n)$ for $|\alpha| \leq m$ implies that u is a continuous function and that*

$$(6.4.13) \qquad \sup |u| \leq C_{n,p,m}\sum_{|\alpha|\leq m}\|\partial^\alpha u\|_p.$$

Proof. It suffices to prove the estimate in the unit ball. Choose $\chi \in C_0^\infty(\mathbf{R}^n)$ equal to 1 in the unit ball and set $v = \chi u$. Then $v = u$ in the unit ball and $\partial^\alpha v \in L^p$ for $|\alpha| \leq m$, with the norms estimated by the right-hand side of (6.4.13). If $p > n$ it follows from Proposition 6.4.8 that v is continuous and that (6.4.13) is valid for v, which proves the corollary for $m = 1$. If $m/n > 1/p$ but $p \leq n$, we can choose $\tilde{p} < p$ so that $m/n > 1/\tilde{p}$. Then $\partial^\alpha v \in L^{\tilde{p}}$ for $|\alpha| \leq m$, and if $1/q = 1/\tilde{p} - 1/n$ it follows from (6.4.9) that $\partial^\alpha v \in L^q$

for $|\alpha| \le m - 1$, with the norms estimated by the right-hand side of (6.4.13), and we have $1/q < (m - 1)/n$. Hence the corollary follows by induction with respect to m.

We shall often refer to (6.4.9), (6.4.10) and Corollary 6.4.9 as *Sobolev's lemma*. Note that (6.4.13) applied to $\chi(x)u(Rx)$ where $\chi \in C_0^\infty(\{x; |x| < 1\})$ and $\chi(0) = 1$ gives the frequently useful estimate

$$(6.4.13)' \quad R^{n/p}|u(0)| \le C'_{n,p,m} \sum_{|\alpha|\le m} R^{|\alpha|}\Big(\int_{|x|<R} |\partial^\alpha u(x)|^p \, dx\Big)^{1/p}, \quad \text{if } n/m < p \le \infty.$$

If Γ is an open cone we even have for the same exponents p

$$(6.4.13)'' \quad R^{n/p}|u(0)| \le C'_{n,p,m,\Gamma} \sum_{|\alpha|\le m} R^{|\alpha|}\Big(\int_{|x|<R, x\in\Gamma} |\partial^\alpha u(x)|^p \, dx\Big)^{1/p}.$$

It suffices to prove this when $|x| < R$ in supp u, for u can be replaced by $\chi(\cdot/R)u(\cdot)$. With polar coordinates $x = r\omega$, where $r > 0$ and $|\omega| = 1$, we have

$$u(0) = (-1)^m \int_0^R r^{m-1}\partial_r^m u(r\omega)dr/(m-1)!,$$

hence

$$u(0)\int_{\omega\in\Gamma} d\omega = (-1)^m \int_{\omega\in\Gamma}\int_0^R r^{m-1}\partial_r^m u(r\omega)\, dr\, d\omega/(m-1)!.$$

With $1/p + 1/p' = 1$ it follows from Hölder's inequality that

$$|u(0)| \le C\Big(\int_{\omega\in\Gamma}\int_0^R |\partial_r^m u(r\omega)|^p r^{n-1}\, dr\, d\omega\Big)^{1/p}\Big(\int_{\omega\in\Gamma}\int_0^R r^{(m-n/p)p'-1}\, dr\, d\omega\Big)^{1/p'},$$

which proves (6.4.13)''. (One can also deduce (6.4.13)'' from (6.4.13)'.) When $p = 1$ we can take $m = n$: if Q is the parallelepiped $\{x \in \mathbf{R}^n; 0 \le x_j \le a_j\}, j = 1,\ldots,n$, then

$$(6.4.13)''' \quad \prod_1^n a_j \sup_Q |u| \le \sum_{\alpha_1,\ldots,\alpha_n\in\{0,1\}} a^\alpha \int_Q |\partial^\alpha u(x)|\, dx.$$

This is an immediate consequence of the one dimensional case which follows by integrating the inequality $|u(x_1)| \ge \sup_{[0,a_1]} |u| - \int_0^{a_1} |u'(t)|\, dt$ from 0 to a_1.

Remark. From Proposition 6.4.8 one also obtains Hölder continuity of order $\gamma > 0$ if $\gamma \le m - n/p$ and γ is not an integer. We leave the verification as an exercise.

We are now prepared for the proof of *local* existence and uniqueness theorems for solutions of nonlinear hyperbolic equations. Uniqueness for solutions with sufficiently high regularity is easily reduced to the linear case:

Theorem 6.4.10. Let $u \in C^3$ be a solution of the differential equation

$$(6.4.14) \qquad\qquad F(x, u, u', u'') = 0$$

in a neighborhood of $0 \in \mathbf{R}^{1+n}$, which is hyperbolic at the origin with respect to the plane $x_0 = 0$, that is, assume that

$$\sum_{j,k=0}^n F_{jk}(0, u(0), u'(0), u''(0))\xi_j\xi_k$$

is hyperbolic in the direction of the ξ_0 axis if $F_{jk} = \partial F/\partial u_{jk}$, $u'' = (u_{jk})$. Let $F \in C^2$. Then any other solution $v \in C^3$ of (6.4.14) with $\partial^\alpha u = \partial^\alpha v$ when $x_0 = 0$ and $|\alpha| \leq 2$ is equal to u in a neighborhood of 0.

Proof. By Taylor's formula we can write

$$F(x, u, u', u'') - F(x, v, v', v'') = \sum_{|\alpha| \leq 2} a_\alpha(x, u, u', \ldots, v'') \partial^\alpha(u - v),$$

where $a_\alpha \in C^1$ as a function of $x, u, \ldots, v''$. Hence $u - v$ satisfies a linear differential equation with C^1 coefficients which by hypothesis is hyperbolic with respect to the plane $x_0 = 0$ at 0. Since the Cauchy data are 0 it follows from the results on linear equations proved in Section 6.3 that $u - v = 0$ in a neighborhood of 0. (See the remarks at the end of Section 6.3.)

Remark. The regularity condition used here is not minimal. For example, if the equation is quasilinear, that is, linear in the highest derivatives, the same proof works if $u, v \in C^2$.

To prove local existence theorems we shall argue essentially as in Section 6.3 for the linear case, using the interpolation inequalities above. Again we state a result which is global in the space variables but implies an existence theorem which is local in all the variables. For the sake of simplicity we just consider a quasilinear equation at first. Extensions and consequences will be discussed after the proof.

Theorem 6.4.11. *Consider the Cauchy problem*

$$(6.4.15) \qquad \Box u + \sum_{j,k=0}^{n} \gamma^{jk}(x, u, u') \partial_j \partial_k u = f(x, u, u'),$$

$$(6.4.16) \qquad u(0, \cdot) = u_0, \quad \partial_0 u(0, \cdot) = u_1, \quad \text{if } x_0 = 0,$$

where γ^{jk} and f are C^∞ functions, $\gamma^{00} = 0$, $f(x, 0, 0) = 0$, $\sum |\gamma^{jk}| < 1/2$. We assume that f and all derivatives of f or of γ^{jk} are bounded. If $u_0 \in H_{(s+1)}(\mathbf{R}^n)$ and $u_1 \in H_{(s)}(\mathbf{R}^n)$ for some integer $s > (n+2)/2$, then the Cauchy problem has for some $T > 0$ a solution

$$(6.4.17) \qquad u \in L^\infty([0, T]; H_{(s+1)}(\mathbf{R}^n)) \cap C^{0,1}([0, T]; H_{(s)}(\mathbf{R}^n)).$$

Here $C^{0,1}$ denotes the space of Lipschitz continuous functions so the second condition means that $\partial_0 u \in L^\infty([0, T]; H_{(s)})$. This implies that $u \in C^2([0, T] \times \mathbf{R}^n)$ and that $\partial^\alpha u$ is bounded when $|\alpha| \leq 2$. The supremum of all such T is equal to the supremum of all T such that the Cauchy problem has a C^2 solution with $\partial^\alpha u$ bounded for $0 \leq t \leq T$ and $|\alpha| \leq 2$.

Proof. We shall solve the Cauchy problem by successive approximation starting from the solution u^0 of $\Box u^0 = 0$ with Cauchy data (6.4.16). Since

$$\hat{u}^0(t, \xi) = \hat{u}_0(\xi) \cos(t|\xi|) + \hat{u}_1(\xi) \sin(t|\xi|)/|\xi|$$

and $(1 + |\xi|^2)^{\frac{1}{2}} |\sin(t|\xi|)| \leq (|t\xi|^2 + |\xi|^2)^{\frac{1}{2}} = |\xi|(1 + t^2)^{\frac{1}{2}}$, it is clear that (6.4.17) holds for u^0. The arguments will be simplified by assuming during the proof that $u_0, u_1 \in \mathcal{S}$, but only the norms corresponding to (6.4.17) will be used in our estimates.

By the results on linear equations in Section 6.3 we can define u^ν inductively by solving the Cauchy problem for

$$(6.4.18) \qquad \Box u^\nu + \sum_{j,k=0}^{n} \gamma^{jk}(x, J_1 u^{\nu-1}) \partial_j \partial_k u^\nu = f(x, J_1 u^{\nu-1}),$$

with the boundary condition (6.4.16). Here $J_1 u = (u, \partial u) = (u, \partial_0 u, \ldots, \partial_n u)$ denotes the 1-*jet* of u in all the variables. Apart from that we shall only work with smoothness in the space variables. All derivatives of every u^ν are continuous functions of t with values in $H_{(s)}$ for arbitrary s. We shall estimate them for $0 \le t \le T$ assuming that we already know uniform bounds for the derivatives of order ≤ 2,

$$(6.4.19) \qquad \sum_{|\alpha| \le 2} |\partial^\alpha u^\nu(t, \cdot)| \le M, \quad 0 \le t \le T.$$

We shall then prove that (6.4.19) follows inductively for small T and large M. After that we shall prove convergence of the sequence.

There is of course no difficulty in choosing M and T so that (6.4.19) holds when $\nu = 0$. To estimate

$$M_\nu(t) = \|u^\nu(t, \cdot)\|_{(s+1)} + \|\partial_t u^\nu(t, \cdot)\|_{(s)} = \|J_1 u^\nu(t, \cdot)\|_{(s)},$$

we shall apply the *a priori* estimate (6.3.6) to the equations obtained when ∂_x^α is applied to (6.4.18) for all α with $|\alpha| \le s$. This yields the equations

$$(6.4.20) \qquad \begin{aligned} (\Box + \sum_{j,k} \gamma^{jk}(x, J_1 u^{\nu-1}) \partial_j \partial_k) \partial_x^\alpha u^\nu &= \partial_x^\alpha f(x, J_1 u^{\nu-1}) \\ &- \sum_{j,k} [\partial_x^\alpha, \gamma^{jk}(x, J_1 u^{\nu-1})] \partial_j \partial_k u^\nu. \end{aligned}$$

The terms in the sum on the right are linear combinations with bounded coefficients of terms of the form

$$(\partial_x^{\alpha'} \partial_x \gamma^{jk}(x, J_1 u^{\nu-1}))(\partial_x^{\alpha''} \partial_x \partial u^\nu), \quad |\alpha'| + |\alpha''| = |\alpha| - 1 \le s - 1.$$

Here we have used that $\gamma^{00} = 0$. For $0 \le t \le T$ it follows from (6.4.19) with ν replaced by $\nu - 1$ and the remark after Corollary 6.4.5 that

$$\|\partial_x \gamma^{jk}(x, J_1 u^{\nu-1})\|_\infty \le C(M); \quad \|\partial_x^\alpha(\gamma^{jk}(x, J_1 u^{\nu-1}) - \gamma^{jk}(x, 0))\|_2 \le C(M) M_{\nu-1}, \quad |\alpha| \le s.$$

Corollary 6.4.4 applied to $\partial_x \gamma^{jk}(x, J_1 u^{\nu-1})$ and $\partial_x \partial u^\nu$, with $m = s - 1$, now shows that the L^2 norm of the sum in the right-hand side of (6.4.20) can be estimated by $C(M)(M_\nu(t) + M_{\nu-1}(t))$. By the remark after Corollary 6.4.5 we have

$$\|\partial_x^\alpha f(x, J_1 u^{\nu-1})\|_2 \le C(M) M_{\nu-1}, \quad |\alpha| \le s,$$

so an application of the energy estimate (6.3.6) yields, again by using (6.4.19),

$$(6.4.21) \qquad M_\nu(t) \le C e^{CMt}\left(M_\nu(0) + C(M) \int_0^t (M_\nu(\tau) + M_{\nu-1}(\tau)) d\tau\right), \quad 0 \le t \le T.$$

Here $M_\nu(0) = M_0(0)$ is independent of ν. By Gronwall's lemma we conclude that

$$M_\nu(t) \le C e^{CMt}\left(M_0(0) + C(M) \int_0^t M_{\nu-1}(\tau)\, d\tau\right) \exp(tCC(M)e^{CMt}).$$

Let $A > CM_0(0)$ and let $A > M_0(t)$, $0 \le t \le T$. We can then choose $T_{M,A}$ with $0 < T_{M,A} \le T$ so that $M_0(t) \le A$, $0 \le t \le T_{M,A}$, and

$$M_\nu(t) \le A, \quad 0 \le t \le T_{M,A},$$

if this is true with ν replaced by $\nu - 1$. Now Sobolev's lemma (Corollary 6.4.9) shows that

$$\sum_{|\alpha| \leq 2} |\partial^\alpha u^\nu(t, \cdot)| \leq C(A) \quad \text{if } M_\nu(t) \leq A,$$

for $s > n/2 + 1$ so we can estimate the maximum of $J_1 u^\nu$ and $\partial_x(J_1 u^\nu)$, and $\partial_0^2 u$ is then estimated using (6.4.15). With $M = C(A)$ and T replaced by $T_{M,A}$ the estimate (6.4.19) is therefore also proved inductively, so we have found M and $T > 0$ such that for all ν we have (6.4.19) and

$$(6.4.22) \qquad M_\nu(t) \leq A, \quad 0 \leq t \leq T.$$

We shall now prove that u^ν converges in

$$C([0, T]; H_{(1)}(\mathbf{R}^n)) \cap C^1([0, T]; H_{(0)}(\mathbf{R}^n))$$

to a limit u which must then automatically satisfy (6.4.15), (6.4.16), (6.4.17). To do so we subtract two successive equations (6.4.18) and obtain

$$(\Box + \sum_{j,k} \gamma^{jk}(x, J_1 u^\nu)\partial_j \partial_k)(u^{\nu+1} - u^\nu) = \sum_{j,k}(\gamma^{jk}(x, J_1 u^{\nu-1})$$
$$-\gamma^{jk}(x, J_1 u^\nu))\partial_j \partial_k u^\nu + f(x, J_1 u^\nu) - f(x, J_1 u^{\nu-1}).$$

Recall that we have uniform bounds for the derivatives of u^ν of order ≤ 2 for all ν. The L^2 norm of the right-hand side can therefore be estimated by a constant times

$$m_\nu(t) = \|u^\nu(t, \cdot) - u^{\nu-1}(t, \cdot)\|_{(1)} + \|\partial_t(u^\nu(t, \cdot) - u^{\nu-1}(t, \cdot))\|_{(0)} = \|J_1(u^\nu - u^{\nu-1})(t, \cdot)\|.$$

We have uniform bounds for the derivatives of γ^{jk}, so (6.3.17) yields

$$(6.4.23) \qquad m_{\nu+1}(t) \leq C \int_0^t m_\nu(\tau)\, d\tau, \quad 0 \leq t \leq T,$$

since the Cauchy data of $u^{\nu+1} - u^\nu$ vanish. Hence we obtain inductively

$$m_\nu(t) \leq (Ct)^\nu \sup m_0/\nu!.$$

which completes the existence proof when the Cauchy data are in $\mathcal{S}$. This condition is eliminated by a standard approximation argument.

Let T_s be the supremum of all T such that the Cauchy problem has a solution satisfying (6.4.17). By Sobolev's lemma (6.4.17) implies uniform Hölder continuity with respect to x of $J_1 u$ and $\partial_x J_1 u$ when $0 \leq t \leq T$, and the differential equation then shows that $\partial_t^2 u(t, \cdot)$ is also uniformly Hölder continuous. This implies that $\partial^\alpha u$ is uniformly continuous and uniformly bounded for $0 \leq t \leq T$ if $0 \leq |\alpha| \leq 2$. What remains is therefore to show that there is no uniform bound for $0 \leq t < T_s$. Assume the contrary, so that we have (6.4.19) for a fixed M and every $T < T_s$. Then the inequality (6.4.21) holds for $0 \leq t < T_s$ with $M_\nu(t) = M(t)$ independent of ν, for we can take the sequence u^ν constantly equal to the solution of the exact Cauchy problem. But then it follows from Gronwall's lemma that $M(t)$ is uniformly bounded for $0 \leq t < T_s$. As we have just seen it follows that u has a C^2 extension to $0 \leq t \leq T_s$, and we have $u(T_s, \cdot) \in H_{(s+1)}$, $\partial_t u(T_s, \cdot) \in H_{(s)}$. Hence the Cauchy problem has a solution with these data when $t = T_s$, satisfying (6.4.17) up to some time $T > T_s$, which contradicts the definition of T_s and completes the proof.

Remark 1. The proof of Theorem 6.4.11 remains valid with an obvious modification if we replace the hypothesis $f(x, 0, 0) = 0$ by the assumption that the support is compact. This observation is useful because it allows one to deduce a general local existence theorem for an equation

$$\sum_{j,k=0}^{n} g^{jk}(x, u, u')\partial_j \partial_k u = f(x, u, u')$$

which is hyperbolic with respect to the plane $x_0 = 0$ at 0, for the given Cauchy data. After a linear change of variables preserving the plane $x_0 = 0$ the equation is then of the form (6.4.15) with $\gamma^{jk} = 0$ at 0. Cutting γ^{jk} and f off in a suitable way in all variables x, u, u' we obtain an equation satisfying the hypotheses of Theorem 6.4.10 as weakened at the beginning of this remark.

Remark 2. Using the equation (6.4.15) we can get bounds in L^2 for all derivatives of u of order $\leq s + 1$ and not only for those which are of order ≤ 1 with respect to t. In fact, if $|\alpha| + k \leq s$ we can write $\partial_x^\alpha \partial_t^k \partial u$ as a linear combination of terms of the form

$$a(x, J_1 u)\partial_x^{\alpha_1} J_1 u \ldots \partial_x^{\alpha_j} J_1 u, \quad |\alpha_1| + \cdots + |\alpha_j| \leq |\alpha| + k.$$

This follows if we keep replacing $\partial_t^2 u$ by the expression given by the equation (6.4.15) until all terms are of order at most equal to 1 with respect to t. Using (6.4.5)' we can then improve (6.4.17) to

$$\partial^\alpha u \in L^\infty([0, T]; L^2), \quad |\alpha| \leq s + 1.$$

Remark 3. The proof of Theorem 6.4.11 shows that given T we have a solution of the Cauchy problem for $0 \leq t \leq T$ satisfying (6.4.17) provided that $\|u_0\|_{(s+1)}$ and $\|u_1\|_{(s)}$ are small enough, for some $s > (n + 2)/2$. The relations between the size of the Cauchy data and the lifespan will be the subject of Section 6.5. Here we observe that in view of the finite propagation speed for solutions of hyperbolic equations it follows easily that the unboundedness of second order derivatives proved in Theorem 6.4.11 when t approaches the lifespan T_s for a solution satisfying (6.4.17) does not occur at infinity. Thus the solution does not have a C^2 extension beyond the time T_s at some point. From the discussion of first order systems in Chapter II and the reduction of second order equations in one space variable to first order systems given in the introduction, we see that in general first order derivatives may remain bounded up to the time T_s.

Remark 4. Theorem 6.4.11 and the preceding remarks remain valid if u takes its values in a finite dimensional vector space $\mathbf{R}^N$, the coefficients γ^{jk} are diagonal $N \times N$ matrices and f takes its values in $\mathbf{R}^N$. In particular, we can apply this remark to find a solution of a fully non-linear equation

(6.4.15)' $\Box u = F(x, u, u', u'')$

with Cauchy data (6.4.16). For the sake of simplicity we assume again that u is just a real valued function, and by solving for $\partial_0^2 u$ we can attain that ∂_0^2 only occurs in $\Box u$. Differentiation of the equation (6.4.15)' with respect to x_j gives a quasilinear equation for $u_j = \partial_j u$, $j = 0, \ldots, n$, which together with (6.4.15)' can be considered as a quasilinear system of equations for $(u, \partial_0 u, \ldots, \partial_n u)$. (The second order derivatives occurring as arguments of F in (6.4.15)', for example, are just first order derivatives of our new unknowns u_j.) From the given Cauchy data for u and the differential equation we can calculate Cauchy data for all u_j. If we assume one derivative more for the initial data than in the quasilinear case we therefore get a solution to our new system, and a vector valued version of Theorem 6.4.10 shows that $u_j = \partial_j u$ for the solution obtained. Thus Theorem 6.4.11 remains valid in the fully non-linear case if we require $s > (n+4)/2$; the supremum of all T such that a solution satisfying (6.4.17) exists in $[0, T]$ is equal to the supremum of all T such that there is a C^3 solution with $\partial^\alpha u$ bounded for $0 \le t \le T$ and $|\alpha| \le 3$.

If one is content with existence in a time interval depending on bounds for a larger number of derivatives of u_0 and u_1, then a much more elementary argument can be used, as pointed out by Klainerman [3]. We shall use his idea in the proof of global existence theorems to avoid the more intricate interpolation inequalities which would otherwise be needed then since one cannot confine oneself to discussing regularity in the space variables for fixed time. As a preparation we shall now present the idea as an alternative proof of Theorem 6.4.11.

With a positive integer s to be chosen later we shall try to establish uniform bounds for

$$(6.4.24) \qquad M_\nu(t) = \sum_{|\alpha| \le s+1} \|\partial^\alpha u^\nu(t, \cdot)\|,$$

when $0 \le t \le T$. Note that we now consider *all* derivatives of order $\le s+1$. By Sobolev's inequality, applied for fixed t, it follows that

$$(6.4.25) \qquad |\partial^\alpha u^\nu(t, \cdot)| \le C M_\nu(t), \quad |\alpha| + \kappa \le s+1,$$

if κ is the smallest integer $> n/2$. This estimate is applicable to all α with $|\alpha| \le 2$ if $s \ge \kappa + 1$. We want to prove inductively that

$$(6.4.26) \qquad M_\nu(t) \le M, \quad 0 \le t \le T.$$

Assume that this is already known with ν replaced by $\nu - 1$, and consider (6.4.20) again, now with ∂_x^α replaced by any ∂^α with $|\alpha| \le s$. In the last sum there is no term where $J_1 u^\nu$ or $J_1 u^{\nu-1}$ is differentiated more than s times. If N is an integer with $2(N+1) > s+1$, that is, $2N \ge s$, then no term contains two factors where $J u^\nu$ or $J u^{\nu-1}$ is differentiated more than N times. Thus (6.4.25) can be used in all factors except one if $N + \kappa \le s$, so we conclude that the L^2 norm of the right-hand side of (6.4.20) can be estimated by $C(M)(M_\nu(t) + 1)$, and the energy estimates give

$$M_\nu(t) \le C e^{CMt}\left(M_\nu(0) + C(M) \int_0^t (M_\nu(\tau) + 1) d\tau\right).$$

When $\nu \ge s$ the derivatives of $\partial^\alpha u^\nu$, $|\alpha| \le s$, are equal to the derivatives of the formal solution of the Cauchy problem when $t = 0$, hence independent of ν, and we conclude using Gronwall's lemma that

$$M_\nu(t) \le C e^{CMt}(M_\nu(0) + C(M)t) \exp(tCC(M)e^{CMt}), \quad 0 \le t \le T.$$

If $M > C M_\nu(0)$ and T is small enough it follows that (6.4.26) is valid, and the proof is finished as before.

The conditions $N + \kappa \le s \le 2N$ imply $N \ge \kappa$ and $s \ge 2\kappa$, so to estimate u^ν we must in this approach assume bounds on about twice as many derivatives of the initial data as in Theorem 6.4.11 . However, we shall use it all the same in the following section because of the simplicity of the argument.

6.5. Global existence theorems for nonlinear wave equations. In this section we shall study the Cauchy problem in $\mathbf{R}^{1+n}$

$$(6.5.1) \qquad \sum_{j,k=0}^n g^{jk}(u')\partial_j \partial_k u = f(u'),$$

$$(6.5.2) \qquad u(0, x) = \varepsilon u_0(x), \quad \partial_0 u(0, x) = \varepsilon u_1(x).$$

As usual we denote the variables in $\mathbf{R}^{1+n}$ by (t, x) but occasionally we write x_0 instead of t for the sake of symmetry. We assume that $u = 0$ is a solution of (6.5.1) and that the linearisation at this solution is the wave operator, that is, *we assume that* $\sum g^{jk}(0)\partial_j \partial_k = \square$ *and that* f *vanishes of second order at* 0. We shall prove that there is a global solution for small ε if $u_j \in C_0^\infty$ and $n > 3$ and estimate the lifespan of the solution when $n \leq 3$. The condition that u_j has compact support can be relaxed, but it is essential that u_j decays at infinity. For instance, if u_0 and u_1 are independent of one variable we actually have a problem involving one space variable less.

By Z we shall denote any one of the vector fields (6.2.8), (6.2.9) or ∂_j, $j = 0, \ldots, n$, and Z^I will denote any product of $|I|$ such vector fields. We shall use the energy integral method to estimate $\|Z^I u(t, \cdot)\|$ and deduce maximum norm estimates from the following proposition (see Klainerman [3] and also Hörmander [2]).

Propositition 6.5.1. *There is a constant C such that*

$$(6.5.3) \qquad (1 + |t| + |x|)^{n-1}(1 + \||t| - |x|\|)|u(t, x)|^2 \leq C \sum_{|I| \leq (n+2)/2} \|Z^I u(t, \cdot)\|^2,$$

if $u \in \mathcal{S}$ in $(t - 1, t + 1) \times \mathbf{R}^n$, say.

Proof. If $|t| + |x| < 1$, this is an immediate consequence of the standard Sobolev lemma since we can take Z^I equal to ∂_x^α for any α with $|\alpha| \leq (n + 2)/2$. (The integer part of $(n + 2)/2$ is the smallest integer $> n/2$.) So assume that $R = |t| + |x| > 1$. The vector fields (6.2.8), (6.2.9) form a basis for all vector fields when $t^2 \neq |x|^2$, for

$$\left(\sum_k \lambda_k x_k^2\right)\partial/\partial x_j = \lambda_j x_j Z_0 - \sum_k x_k Z_{jk},$$

so we can then write

$$\partial_j = \sum_\nu a_{j\nu}(t, x)Z_\nu, \quad j = 0, \ldots, n,$$

where Z_ν is any labelling of the vector fields (6.2.8), (6.2.9), and $a_{j\nu}$ is in C^∞ and homogeneous of degree -1 outside the light cone $\Lambda = \{(t, x) \in \mathbf{R}^{1+n}, |t| = |x|\}$. Hence

$$\partial^\alpha = \sum_{1 \leq |I| \leq |\alpha|} a_{\alpha I} Z^I, \quad \alpha \neq 0,$$

where $a_{\alpha I}$ is homogeneous of degree $-|\alpha|$. Let Γ be a closed cone which does not intersect Λ except at 0, and choose $\gamma > 0$ such that

$$(t, x) \in \Gamma \implies (t, x + y) \notin \Lambda \text{ if } |y| \leq \gamma R = \gamma(|t| + |x|).$$

Thus $|a_{\alpha I}(t, x + y)| \leq CR^{-|\alpha|}$ when $(t, x) \in \Gamma$ and $|y| < \gamma R$, and we conclude that

$$\int_{|y| \leq \gamma R} \sum_{|\alpha| \leq (n+2)/2} |R^{|\alpha|}\partial^\alpha u(t, x + y)|^2 dy \leq C \sum_{|I| \leq (n+2)/2} \|Z^I u(t, \cdot)\|^2, \quad (t, x) \in \Gamma.$$

With $z = y/R$ as new variable we obtain (6.5.3) from (6.4.13)'.

Now assume that $t/2 < |x| < 3t/2$. It is then convenient to use polar coordinates

$$x = r\omega, \text{ where } r > 0, \omega \in S^{n-1},$$

or rather use $t = x_0$, $q = r - x_0$, and ω as new variables. Note that this makes the light cone a coordinate plane $q = 0$. The vector fields (6.2.8) with $j, k \neq 0$ annihilate t and q and preserve homogeneity, so they can be regarded as vector fields in S^{n-1}. Since the orthogonal covectors are spanned by dr and dt, they span the vector fields in the unit sphere. Let us denote this set of vector fields by Ω. The remaining vector fields (6.2.8), (6.2.9) are

$$Z_{0k} = x_0 \partial_k + x_k \partial_0, \quad 0 < k \leq n; \quad \text{and } Z_0 = \sum_0^n x_j \partial_j.$$

Since $\partial/\partial r$ becomes $\partial/\partial q$ and $\partial/\partial x_0$ becomes $\partial/\partial t - \partial/\partial q$ in the new coordinates, we obtain

$$\sum_1^n \omega_k Z_{0k} = t\partial/\partial q + (t + q)(\partial/\partial t - \partial/\partial q) = (t + q)\partial/\partial t - q\partial/\partial q.$$

The radial vector field Z_0 is $t\partial/\partial t + q\partial/\partial q$ with these coordinates. It follows that in the conic neighborhood of Λ where $t/2 < |x| < 3t/2$ we have

$$t\partial/\partial t = a_0 Z_0 + \sum_1^n a_k Z_{0k}, \quad q\partial/\partial q = b_0 Z_0 + \sum_1^n b_k Z_{0k}, \quad \partial/\partial q = \sum_1^n \omega_j \partial_j,$$

where a_ν, b_ν and of course ω_j are homogeneous of degree 0. Writing $u(t, x) = v(t, q, \omega)$ we obtain

(6.5.4)
$$t^{n-1} \sum_{\alpha+\beta+|\gamma| \leq N} \int_{|q| < t/2} |(q\partial/\partial q)^\alpha (\partial/\partial q)^\beta (\partial/\partial \omega)^\gamma v(t, q, \omega)|^2 \, dq \, d\omega$$
$$\leq C \sum_{|I| \leq N} \|Z^I u(t, \cdot)\|^2,$$

where $N = (n + 2)/2$, for the Lebesgue measure becomes $r^{n-1} dr d\omega = (t + q)^{n-1} dq d\omega$. Taking $\alpha = 0$ we conclude using Sobolev's lemma that $t^{n-1}|v(t, q, \omega)|^2$ can be estimated by the right-hand side if $|q| < t/4$. (We can use local coordinates on the unit sphere, and $t > 2/5$ since $|x| + t \geq 1$ and $|x| < 3t/2$.)

Let $\chi \in C_0^\infty((-1/2, 1/2))$, $\chi(0) = 1$, and set

$$V_Q(t, q, \omega) = \chi((q - Q)/Q)v(t, q, \omega)$$

for some Q with $1 < |Q| < t/4$. We have $V_Q(t, Q, \omega) = v(t, Q, \omega)$, and $|q - Q| < |Q|/2$, hence $|Q|/2 < |q| < 3|Q|/2 < t/2$ in the support. Hence the square of the L^2 norm of

$$(Q\partial/\partial q)^\alpha (\partial/\partial \omega)^\gamma V_Q(t, q, \omega) = \left(\frac{Q}{q} q\partial/\partial q\right)^\alpha (\partial/\partial \omega)^\gamma \chi((q - Q)/Q)v(t, q, \omega),$$

can be estimated by a constant times the sum in the left-hand side of (6.5.4) with $\beta = 0$ when $\alpha + |\gamma| \leq N$, for

$$(q\partial/\partial q)((q - Q)/Q) = q/Q, \quad (q\partial/\partial q)(Q/q) = -Q/q,$$

are uniformly bounded in the support. Taking q/Q as a new variable instead of q we obtain another factor Q from the integration element, and Sobolev's lemma gives

$$t^{n-1} Q|v(t, Q, \omega)|^2 \leq C \sum_{|I| \leq (n+2)/2} \|Z^I u(t, \cdot)\|^2.$$

This completes the proof.

Remark. If u is a solution of the homogeneous unperturbed wave equation with Cauchy data in C_0^∞, then we know from (6.3.2) that $\|Z^I u'(t, \cdot)\|$ is uniformly bounded for any I, for $\Box Z^I u = 0$, too. (See Section 6.2.) By Proposition 6.5.1 this implies

$$\sup(1 + |t|)^{\frac{n-1}{2}}(1 + ||x| - |t||)^{\frac{1}{2}}|u'(t, x)| < \infty.$$

The weak Huygen's principle gives $u(t, x) = 0$ if $|x| - |t| >$ constant, so it follows that

$$\sup(1 + |t|)^{\frac{n-1}{2}}(1 + ||x| - |t||)^{-\frac{1}{2}}|u(t, x)| < \infty.$$

From Section 6.2 we know that these estimates have the right order of magnitude near the boundary of the light cone, that is, when $|x| - |t|$ is bounded. However, the best bounds are

$$\sup(1 + |t|)^{\frac{n-1}{2}}(1 + ||x| - |t||)^{\frac{n-1}{2}}(|u(t, x)| + |u'(t, x)|) < \infty,$$

and they cannot be obtained from L^2 estimates of $Z^I u$. Fortunately this flaw will not affect the following existence theorems much since the critical estimates concern the immediate neighborhood of the boundary of the light cone.

Theorem 6.5.2. *The Cauchy problem (6.5.1), (6.5.2) with $u_j \in C_0^\infty(\mathbf{R}^n)$ has a C^∞ solution for $t \geq 0$ if $n \geq 4$ and ε is sufficiently small.*

Proof. We know from Theorem 6.4.11 that the set of all T such that a smooth solution exists for $0 \leq t \leq T$ is open. The theorem will be proved by establishing estimates of the solution which are independent of T and imply that this set is also closed.

As at the end of Section 6.4 we choose positive integers s and N with

$$(6.5.5) \qquad\qquad N + \kappa \leq s \leq 2N,$$

where κ is the smallest integer $> n/2$. We want to prove that there is a constant M such that for small ε

$$(6.5.6) \qquad M_s(t) = \sum_{|I| \leq s} \|Z^I u'(t, \cdot)\| \leq M\varepsilon, \quad \text{if } 0 \leq t \leq T.$$

This is true for small T if M is large enough. We shall prove that for sufficiently large M the estimate (6.5.6) implies the same bound with M replaced by $M/2$. By "continuous induction" we may then conclude that (6.5.6) holds.

By Proposition 6.5.1 it follows from (6.5.6) that

$$(6.5.7) \qquad (1 + t)^{\frac{n-1}{2}}|Z^I u'(t, x)| \leq CM\varepsilon, \quad \text{if } 0 \leq t \leq T, |I| + \kappa \leq s.$$

Since $N + \kappa \leq s$ we can use this estimate when $|I| \leq N$. To estimate $\|Z^I u'(t, \cdot)\|$ when $|I| \leq s$ we apply Z^I to (6.5.1) and obtain (cf. (6.4.20)), with the notation $\gamma^{jk}(u') = g^{jk}(u') - g^{jk}(0)$,

$$(6.5.8) \qquad \begin{aligned} \Big(\Box + \sum_{j,k}\gamma^{jk}(u')\partial_j\partial_k\Big)Z^I u &= Z^I f(u') + [\Box, Z^I]u \\ &- \sum_{j,k}[Z^I, \gamma^{jk}(u')]\partial_j\partial_k u - \sum_{j,k}\gamma^{jk}(u')[Z^I, \partial_j\partial_k]u. \end{aligned}$$

From (6.5.7) we obtain in particular

$$|\gamma^{jk}(u')| \le CM\varepsilon(1+t)^{\frac{1-n}{2}} \le CM\varepsilon,$$

so the hypotheses of the energy estimate (6.3.6) are fulfilled for small ε. Moreover, if $\gamma^{jk}(u')$ is considered as a function of t, x we have with the notation used there

(6.5.9) $$|\gamma'(t)| \le CM\varepsilon(1+|t|)^{\frac{1-n}{2}}.$$

This is an integrable function when $n > 3$, which motivates this condition in the theorem. The L^2 norm of the last sum in (6.5.8) can be estimated by $CM\varepsilon(1+t)^{\frac{1-n}{2}} M_s(t)$, for $[Z^I, \partial_j \partial_k]$ is a linear combination of operators $Z^J \partial_i$ with $|J| \le s$ since $[Z, \partial_i]$ is always either 0 or equal to $\pm\partial_k$ for some k. We can write

$$[Z^I, \gamma^{jk}(u')]\partial_j \partial_k u$$

as a sum of derivatives of γ^{jk} multiplied by components of

$$Z^{J_1} u' \cdots Z^{J_r} u' Z^K \partial_j \partial_k u$$

where $|J_1| + \cdots + |J_r| + |K| \le s$, $J_i \ne 0$ for every i, and $r \ne 0$. At most one of the positive integers $|J_1|, \ldots, |J_r|, |K| + 1$ can be larger than N since $2(N+1) > s+1$ by (6.5.5), so we can estimate all factors except one using (6.5.7). There are at least two factors. For small ε it follows that the L^2 norm of the first sum of commutators in (6.5.8) can also be estimated by $CM\varepsilon(1+t)^{\frac{1-n}{2}} M_s(t)$. Since f vanishes of second order at 0, we can by Taylor's formula write

$$f(u') = \sum_{j,k} f_{jk}(u')\partial_j u \partial_k u$$

with smooth f_{jk}, and a similar estimate is then obtained for this term. Finally

$$[\square, Z^I] = \sum_{|J|<|I|} c_{I,J} Z^J \square u$$

with constant $c_{I,J}$, for $[\square, Z]$ is equal to 0 or $2\square$ for each factor. We express $Z^J \square u$ by means of the equation (6.5.1),

$$Z^J \square u = Z^J f(u') - Z^J \sum_{j,k} \gamma^{jk}(u')\partial_j \partial_k u.$$

Since $|J| < s$ all terms obtained are of the form already discussed. Hence it follows from (6.3.6) when ε is small that

$$\|\partial Z^I u(t, \cdot)\| \le 3\|\partial Z^I u(0, \cdot)\| + CM\varepsilon \int_0^t (1+\tau)^{\frac{1-n}{2}} M_s(\tau)\, d\tau, \quad |I| \le s,$$

for the exponential factor in (6.3.6) will be close to 1 by (6.5.9). We can write $Z^I \partial_j$ as a linear combination of $\partial_k Z^J$ with $|J| \le |I|$ by an argument used a moment ago, so it follows that

(6.5.10) $$M_s(t) \le C\Big(M_s(0) + M\varepsilon \int_0^t (1+\tau)^{\frac{1-n}{2}} M_s(\tau)\, d\tau\Big).$$

In view of Gronwall's lemma it follows that

$$(6.5.11) \qquad M_s(t) \leq CM_s(0)\exp\left(\int_0^t CM\varepsilon(1+\tau)^{\frac{1-n}{2}}\,d\tau\right).$$

Choose M so that $3CM_s(0) \leq M\varepsilon$. Then we have attained that (6.5.6) holds for small T and for any T implies the same estimate with M replaced by $M/2$ if ε is small enough. Thus (6.5.6) follows by continuous induction for every T such that the Cauchy problem has a solution for $0 \leq t \leq T$, so this set is closed. Since it is open by Theorem 6.4.11 the global existence follows.

Remark. The proof works if we just prescribe small Cauchy data such that $Z^I u'(0, \cdot) \in L^2$ for $|I| \leq s$ for some $s \geq 2\kappa$, and $\|Z^I u'(t, \cdot)\|$ is then bounded when $|I| \leq s$. If the support is compact this follows at once from the preceding proof. Otherwise one can choose $\chi \in C_0^\infty(\mathbf{R}^n)$ equal to 1 in a neighborhood of 0 and note that the bounds for the solution with Cauchy data $u_j\chi(\delta\cdot)$ are independent of δ as $\delta \to 0$.

If $n \leq 3$ the proof of Theorem 6.5.2 does not break down completely. Firstly, the existence proof remains valid if the perturbation vanishes of sufficiently high order and $n > 1$. Assume that $\gamma^{jk}(u')$ and $f(u')$ vanish of order p and $p+1$ respectively at 0. Then we get at least p factors for which we can use the estimate (6.5.7), and it follows that there is global existence for small ε if $p(n-1) > 2$. When $n = 3$ it is therefore only the quadratic terms that can cause difficulties. Secondly, even if the perturbation is just of second order we get an estimate for the lifespan of the solution by just looking carefully at the exponential factors which appear in (6.3.6) and in the application of Gronwall's lemma. Assume that for some small $\delta > 0$ we have

$$(6.5.12) \qquad \varepsilon \int_0^T (1+\tau)^{\frac{1-n}{2}}\,d\tau \leq \delta.$$

Then (6.5.10) is modified by a factor $e^{C\delta}$ in the right-hand side, and (6.5.11) is replaced by

$$M_s(t) \leq Ce^{C\delta}M_s(0)\exp(CMe^{C\delta}\delta).$$

Choose M as before so that $3CM_s(0) \leq M\varepsilon$. When δ is so small that

$$e^{C\delta}\exp(CMe^{C\delta}\delta) < 3/2,$$

we conclude that (6.5.6) holds for small T and always implies the same estimate with M replaced by $M/2$, if T satisfies (6.5.12) and ε is small. Hence we have proved

Theorem 6.5.3. *The Cauchy problem* (6.5.1), (6.5.2) *with* $u_j \in C_0^\infty(\mathbf{R}^n)$ *has for small ε a solution for* $0 \leq t \leq T_\varepsilon$ *if*

$$\varepsilon \int_0^{T_\varepsilon} (1+t)^{\frac{1-n}{2}}\,dt = c,$$

where $c > 0$ *depends on* u_0, u_1. *Equivalently, with some other* c, *a solution exists for* $0 \leq t \leq T_\varepsilon$ *where for some* $c > 0$

$$T_\varepsilon = \begin{cases} e^{c/\varepsilon}, & \text{if } n = 3, \\ c/\varepsilon^2, & \text{if } n = 2, \\ c/\varepsilon, & \text{if } n = 1. \end{cases}$$

If the perturbation vanishes of order $p+1$ *as above and* $p(n-1) \leq 2$, *then we get existence for* $0 \leq t \leq T_\varepsilon$ *if*

$$\varepsilon^p \int_0^{T_\varepsilon} (1+t)^{p(1-n)/2}\,dt \leq c$$

for a sufficiently small constant c. When $n = 2$ this gives "almost global existence" if $p = 2$. However, the most interesting problem is to determine how quadratic perturbations influence the lifespan T_ε. When $n = 1$ we know this from (6.1.5). We shall now prove that for $n = 2$ or $n = 3$ there is at least a similar *lower* bound for T_ε. The main point is that one can construct an approximate solution with an error which is small compared to ε as long as it exists. This will make the exponential factors controlled by (6.5.12) in the preceding proof harmless even if δ is not small. For the sake of brevity we assume that $f = 0$. The extension to general f was prepared in Lemma 2.3.1 though, and it will be carried out at the end of the section.

Assuming at first that $n = 3$ and that $u_0, u_1 \in C_0^\infty(\mathbf{R}^3)$, we shall look for an approximate solution to (6.5.1), (6.5.2) of the form

$$u(t, r\omega) = \varepsilon r^{-1} U(\omega, \varepsilon \log t, r - t), \quad |\omega| = 1, r > 0.$$

This is motivated by the asymptotic formula $\varepsilon r^{-1} F_0(\omega, r - t)$ for the solution of the unperturbed wave equation given in (6.2.6) and the hint from Theorem 6.5.3 that nonlinear effects start to be important when $\varepsilon \log t$ attains a certain value essentially independent of ε. Since

$$\Box u = r^{-1}((\partial_t - \partial_r)(\partial_t + \partial_r) - r^{-2}\Delta_\omega)ru,$$

with Δ_ω denoting the Laplacian in S^2, the main term in $\Box u$ is obtained when $\partial_t + \partial_r$ acts on the argument $s = \varepsilon \log t$ and $\partial_t - \partial_r$ acts on $q = r - t$, which gives

$$-2\varepsilon^2 (tr)^{-1} U''_{sq}(\omega, s, q).$$

Writing as in the introduction

(6.5.13) $$g^{jk}(u') = g^{jk}(0) + \sum_{l=0}^{n} g^{jkl}\partial_l u + O(|u'|^2),$$

we find that the main nonlinear terms in the equation (6.5.1) are

$$\varepsilon^2 r^{-2} G(\omega) U'_q U''_{qq},$$

where

(6.5.14) $$G(\omega) = \sum_{j,k,l=0}^{n} g^{jkl}\hat{\omega}_j\hat{\omega}_k\hat{\omega}_l; \quad \hat{\omega} = (-1, \omega_1, \omega_2, \omega_3) = q'.$$

Thus it is natural to choose U so that

$$2U''_{sq}(\omega, s, q) = G(\omega)U'_q(\omega, s, q)U''_{qq}(\omega, s, q).$$

If U vanishes for large q this is equivalent to

(6.5.15) $$4\partial U(\omega, s, q)/\partial s = G(\omega)(\partial U(\omega, s, q)/\partial q)^2.$$

When t is large but $\varepsilon \log t$ is still small, the nonlinear effects should not yet be important, so it is natural to require the initial condition

(6.5.16) $$U(\omega, 0, q) = F_0(\omega, q),$$

where F_0 is the Friedlander radiation field in (6.2.6). This Cauchy problem is easy to solve:

Lemma 6.5.4. *The Cauchy problem* (6.5.15), (6.5.16) *has a unique C^∞ solution for $0 \leq s < A$ where*

$$(6.5.17) \qquad A = (\max_{\omega, \varrho} \tfrac{1}{2} G(\omega) \partial^2 F_0(\omega, \varrho)/\partial \varrho^2)^{-1},$$

but the second order derivatives are unbounded when $s \to A$ if $A < \infty$, which is true unless $G \equiv 0$ or $(u_0, u_1) \equiv 0$.

Proof. If we set $u = \partial U/\partial q$ the equation (6.5.15) implies that u satisfies Burgers' equation with parameters

$$(6.5.15)' \qquad 2\partial u(\omega, s, q)/\partial s = G(\omega) u(\omega, s, q) \partial u(\omega, s, q)/\partial q.$$

We have $\int u(\omega, s, q)\, dq = 0$ for all (ω, s) since this is true when $s = 0$, and $|q| \leq M$ when $(\omega, s, q) \in \operatorname{supp} u$ if $|y| \leq M$ when $y \in \operatorname{supp} u_0 \cup \operatorname{supp} u_1$. Thus $\partial U/\partial q = u$ for a unique U with such support, and U satisfies (6.5.15). The lemma is now a consequence of the discussion of the lifespan of solutions of Burgers' equation in Section 2.3; it applies with no change when the parameters ω are present. (The lemma could also have been proved using the Hamilton-Jacobi integration theory.) That $A < \infty$ except in the trivial cases listed follows from Theorem 6.2.2, for the second order derivative of a function of compact support which is not identically 0 takes both positive and negative values.

We have now found a good approximation when t is fairly large, but it does not have the correct Cauchy data so we piece it together with the solution εw_0 of the homogeneous wave equation with Cauchy data (6.5.2). To do so we choose $\chi \in C^\infty(\mathbf{R})$ decreasing, equal to 1 in $(-\infty, 1)$ and equal to 0 in $(2, \infty)$, and we set for $0 \leq \varepsilon t < e^{A/\varepsilon}$

$$(6.5.18) \qquad w_\varepsilon(t, x) = w(t, x) = \varepsilon \big(\chi(\varepsilon t) w_0(t, x) + (1 - \chi(\varepsilon t)) r^{-1} U(\omega, \varepsilon \log(\varepsilon t), r - t)\big).$$

Thus we shift when $1 \leq \varepsilon t \leq 2$ from the solution of the homogeneous wave equation to the approximation constructed for large t. The initial conditions (6.5.2) are of course satisfied by w, and we shall now estimate how well (6.5.1) is fulfilled (with $f = 0$).

Lemma 6.5.5. *With w defined by* (6.5.18) *and*

$$(6.5.19) \qquad R = \sum_{j,k=0}^{n} g^{jk}(w') \partial_j \partial_k w,$$

we have $R, w \in C^\infty$ when $t < e^{A/\varepsilon}$, where A is defined by (6.5.17). *If $(t, x) \in \operatorname{supp} w$ we have*

$$||x| - t| \leq \sup\{|y|; y \in \operatorname{supp} u_0 \cup \operatorname{supp} u_1\}.$$

For fixed $B \in (0, A)$ we have for small ε and all I if $\varepsilon \log t \leq B$

$$(6.5.20) \qquad |Z^I w(t, x)| \leq C_{I,B} \varepsilon (1 + t)^{-1},$$

$$(6.5.21) \qquad |Z^I R(t, x)| \leq C_{I,B} \varepsilon^2 (1 + t)^{-2} (1 + \varepsilon t)^{-1}.$$

Proof. If w is replaced by the solution εw_0 of the wave equation with initial data εu_0, εu_1, then (6.5.20) follows from (6.2.10). To prove (6.5.20) when $\varepsilon t \geq 1$ and $\varepsilon \log t \leq B$ we note that $Z^I \chi(\varepsilon t)$ is uniformly bounded for any I when $|x| < t + C$, that $Z \log(\varepsilon t)$ is homogeneous of degree ≤ 0, and that Zq is either homogeneous of degree 0 or else equal to $-\omega_j q$, if $Z = x_j \partial_t + t \partial_j$. This implies that Z^I applied to the second term in (6.5.18) is a sum with bounded coefficients of derivatives of U multiplied by functions homogeneous of degree ≤ -1 and powers of $q = r - t$, which is bounded in the support. This proves (6.5.20).

To prove (6.5.21) we distinguish three different cases:

i) When $\varepsilon t \leq 1$ we have $w(t, x) = \varepsilon w_0(t, x)$, and

$$R = \sum_{j,k}(g^{jk}(w') - g^{jk}(0))\partial_j\partial_k w$$

since $\square w_0 = 0$. Hence (6.5.21) follows from (6.5.20), for the factor $1 + \varepsilon t$ plays no role.

ii) Now consider the transition zone where $1 \leq \varepsilon t \leq 2$. In addition to the arguments in case i) we must then also estimate

$$\square w = \square(w - \varepsilon w_0)$$
$$= \varepsilon\big(((1 - \chi(\varepsilon t))\square - 2\varepsilon\chi'(\varepsilon t)\partial_t - \varepsilon^2\chi''(\varepsilon t))(r^{-1}U(\omega, \varepsilon\log(\varepsilon t), r - t) - w_0(t, x))\big).$$

In the term where χ is differentiated twice the desired bound $O(\varepsilon^4)$ is immediately clear. In the term where χ is differentiated once we use that

$$r^{-1}(U(\omega, \varepsilon\log(\varepsilon t), r - t) - F_0(\omega, r - t)), \quad r^{-1}F_0(\omega, r - t) - w_0(t, x)$$

are $O(\varepsilon r^{-1})$ and $O(r^{-2})$ respectively, and that such bounds still hold after multiplication by any Z^I. In the first case this follows from the proof of (6.5.20) since $0 \leq \log(\varepsilon t) \leq \log 2$; in the second case it follows from (6.2.11). What remains is to study

$$\varepsilon(1 - \chi(\varepsilon t))r^{-1}((\partial_t - \partial_r)(\partial_t + \partial_r) - r^{-2}\Delta_\omega)U(\omega, \varepsilon\log(\varepsilon t), r - t).$$

Here $\partial_t + \partial_r$ must act on $\varepsilon\log(\varepsilon t)$, producing a factor ε/t, which gives the desired bound if we recall the proof of (6.5.20) once more.

iii) Let $2/\varepsilon \leq t \leq e^{B/\varepsilon}$. Then

$$\square w = \varepsilon r^{-1}((\partial_t - \partial_r)(\partial_t + \partial_r) - r^{-2}\Delta_\omega)U(\omega, \varepsilon\log(\varepsilon t), r - t),$$

and as just observed $\partial_t + \partial_r$ must act on $\varepsilon\log(\varepsilon t)$, which yields a factor ε/t. Writing $s = \varepsilon\log(\varepsilon t)$ and $q = r - t$, we obtain

$$|\square w + 2\varepsilon^2 r^{-1}t^{-1}U''_{sq}(\omega, s, q)| \leq C\varepsilon t^{-3}.$$

With the notation in (6.5.14) we have

$$|\partial^\alpha w - \varepsilon r^{-1}\hat{\omega}^\alpha\partial_q^{|\alpha|}U(\omega, s, q)| \leq C\varepsilon r^{-2},$$

$$\left|\sum_{j,k=0}^n(g^{jk}(w') - g^{jk}(0))\partial_j\partial_k w - \varepsilon^2 r^{-2}G(\omega)U'_q U''_{qq}\right| \leq C\varepsilon^2 r^{-3}.$$

Recalling that $2U''_{sq} = G(\omega)U'_q U''_{qq}$ we conclude that $|R| \leq C\varepsilon t^{-3}$, for $1/t - 1/r = O(1/t^2)$. This proves (6.5.21) when $I = 0$, and using (6.5.20) we obtain (6.5.21) for arbitrary I too.

We shall write the solution u of (6.5.1) (with $f = 0$) and (6.5.2) in the form $u = v + w$ where w is the approximate solution studied in Lemma 6.5.5. Then the Cauchy problem is restated as

$$(6.5.1)' \quad \sum_{j,k=0}^n g^{jk}(v' + w')\partial_j\partial_k v + R + \sum_{j,k=0}^n (g^{jk}(v' + w') - g^{jk}(w'))\partial_j\partial_k w = 0$$

$$(6.5.2)' \quad\quad\quad\quad\quad\quad v = \partial_t v = 0 \quad \text{when } t = 0.$$

Here R is defined by (6.5.19).

Lemma 6.5.6. *Assume that* $(6.5.1)'$, $(6.5.2)'$ *has a* C^∞ *solution for* $0 \le t \le T$ *where* $\varepsilon \log T \le B < A$, *with* A *defined by* $(6.5.17)$. *If* $0 < \varepsilon < \delta_B$, *it follows that*

$$(6.5.22) \qquad \|Z^I v'(t, \cdot)\| \le C_{I,B} \varepsilon^2 \log(1/\varepsilon), \quad 0 \le t \le T,$$

where δ_B *and* $C_{I,B}$ *are independent of* T *and* ε.

Proof. The proof is parallel to that of Theorem 6.5.2. We shall estimate $Z^I v'$ by applying the standard energy estimate $(6.3.6)$ to the equation obtained when $(6.5.1)'$ is multiplied by Z^I. By Lemma 6.5.5 we have

$$(6.5.23) \qquad \|Z^I R(t, \cdot)\| \le C_{I,B} \varepsilon^2 (1+t)^{-1}(1+\varepsilon t)^{-1},$$

for the measure of the support is $O(1+t)^2$, and we shall use that

$$(6.5.24) \qquad \int_0^\infty (1+t)^{-1}(1+\varepsilon t)^{-1} dt = (\varepsilon - 1)^{-1} \log \varepsilon.$$

From $(6.5.1)'$ and $(6.5.2)'$ we also find that $Z^I v = O(\varepsilon^2)$ for every I when $t = 0$.

Choose s and N satisfying $(6.5.5)$ and assume that

$$(6.5.25) \qquad N_s(t) = \sum_{|I| \le s} \|Z^I v'(t, \cdot)\| \le \varepsilon, \quad \text{if } 0 \le t \le T,$$

where the equality is of course a definition. If we prove $(6.5.22)$ under this assumption then we see that for small ε we actually have $(6.5.25)$ with ε replaced by $\varepsilon/2$, so $(6.5.25)$ and $(6.5.22)$ will follow by "continuous induction". By Proposition 6.5.1 it follows from $(6.5.25)$ that

$$(6.5.26) \qquad |Z^I v'(t, \cdot)| \le C N_s(t)(1+t)^{-1} \le C\varepsilon(1+t)^{-1}, \quad 0 \le t \le T, \ |I| \le N.$$

Combining $(6.5.20)$ and $(6.5.25)$, $(6.5.26)$ we obtain for some constant C'

$$(6.5.27) \qquad \begin{aligned} \|Z^I u'(t, \cdot)\| &\le C'\varepsilon, & 0 \le t \le T, \ |I| \le s, \\ |Z^I u'(t, \cdot)| &\le C'\varepsilon(1+t)^{-1}, & 0 \le t \le T, \ |I| \le N. \end{aligned}$$

When $|I| \le s$ we are now ready to estimate the right-hand side of the equation for $Z^I v$ obtained from $(6.5.1)'$

$$\sum_{j,k=0}^{n} g^{jk}(u') \partial_j \partial_k Z^I v = \sum_0^4 R_j,$$

$$R_0 = -Z^I R, \quad R_1 = [\Box, Z^I] v, \quad R_2 = \sum [\gamma^{jk}(u'), Z^I] \partial_j \partial_k v,$$

$$R_3 = \sum \gamma^{jk}(u') [\partial_j \partial_k, Z^I] v, \quad R_4 = -Z^I \sum (g^{jk}(v' + w') - g^{jk}(w')) \partial_j \partial_k w.$$

As before the hardest term is R_2, which is a sum of derivatives of γ^{jk} multiplied by components of

$$Z^{J_1} u' \cdots Z^{J_r} u' Z^K \partial_j \partial_k v$$

where $|J_1| + \cdots + |J_r| + |K| \le s$, $J_i \ne 0$ for every i, and $r \ne 0$. Estimating the last factor using $(6.5.25)$ or $(6.5.26)$ and the others by means of $(6.5.27)$ we obtain

$$\|R_2(t, \cdot)\| \le C\varepsilon(1+t)^{-1} N_s(t).$$

For R_3 and R_4 the same estimate is obtained even more easily if we use Taylor's formula to write $\gamma^{jk}(u')$ and $g^{jk}(v'+w') - g^{jk}(w')$ as scalar products with u' and v' respectively. After writing

$$R_1 = \sum_{|J|<|I|} c_{I,J} Z^J \Box v$$

and substituting the expression for $\Box v$ given by $(6.5.1)'$, we have the same estimate for R_1 too, and R_0 was estimated in $(6.5.23)$, $(6.5.24)$. Now it follows from $(6.5.27)$ that with the notation in $(6.3.6)$ we have

$$\int_0^T |\gamma'(t)| \, dt \le C\varepsilon \int_0^T dt/(1+t) = C\varepsilon \log(1+T) \le 2CB, \quad \text{if } \varepsilon \log T \le B.$$

Recalling that the Cauchy data of $Z^I v$ are $O(\varepsilon^2)$, we obtain using $(6.3.6)$

$$\|\partial Z^I v(t, \cdot)\| \le C\big(\varepsilon^2 \log(1/\varepsilon) + \int_0^t \varepsilon N_s(\tau) \, d\tau/(1+\tau)\big).$$

This implies with another C, depending on B, that

$$N_s(t) \le C\big(\varepsilon^2 \log(1/\varepsilon) + \int_0^t \varepsilon N_s(\tau) \, d\tau/(1+\tau)\big),$$

hence by Gronwall's lemma

$$N_s(t) \le C\varepsilon^2 \log(1/\varepsilon) \exp\Big(\int_0^t C\varepsilon \, d\tau/(1+\tau)\Big) \le C\varepsilon^2 \log(1/\varepsilon) \exp(2CB).$$

This completes the proof of $(6.5.22)$.

From Lemma 6.5.6 and the local existence theorem it follows at once that for small ε the Cauchy problem $(6.5.1)'$, $(6.5.2)'$ has a solution satisfying $(6.5.22)$ when $t \le \exp(B/\varepsilon)$, if $B < A$. By Proposition 6.5.1

$$|u'(t,x) - w'(t,x)| = |v'(t,x)| \le C(1+t)^{-1}(1+||x|-|t||)^{-\frac{1}{2}}\varepsilon^2 \log(1/\varepsilon), \quad 0 \le t \le e^{B/\varepsilon}.$$

Since $v(t,x) = 0$ when $|x| - |t| > \text{constant}$ it follows that

$$\varepsilon^{-1}(1+t)|u(t,x) - w(t,x)| \le C(1+||x|-|t||)^{\frac{1}{2}}\varepsilon \log(1/\varepsilon), \quad 0 \le t \le e^{B/\varepsilon}.$$

When $t = e^{s/\varepsilon}$ and $x = (t+q)\omega$, $|\omega| = 1$, we have

$$tw(t,x) = \varepsilon U(\omega, s + \varepsilon \log \varepsilon, q)t/(t+q)$$

when $\varepsilon e^{s/\varepsilon} > 2$, so we have proved:

Theorem 6.5.7. *The Cauchy problem* $(6.5.1)$ *(with $f = 0$), $(6.5.2)$ with $u_j \in C_0^\infty(\mathbf{R}^3)$ has a C^∞ solution u_ε for $0 \le t < T_\varepsilon$ where*

$$(6.5.28) \qquad \lim_{\varepsilon \to 0} \varepsilon \log T_\varepsilon \ge A = (\max_{\omega, \varrho} \tfrac{1}{2} G(\omega) \partial^2 F_0(\omega, \varrho)/\partial \varrho^2)^{-1}.$$

Here $\omega \in S^2$ and G is defined in $(6.5.14)$ while εF_0 is the Friedlander radiation field of the solution of the Cauchy problem for the unperturbed equation. If U is the solution of $(6.5.15)$, $(6.5.16)$, then

$$(6.5.29) \qquad \varepsilon^{-1} e^{s/\varepsilon} u_\varepsilon(e^{s/\varepsilon}, (e^{s/\varepsilon} + \varrho)\omega) - U(\omega, s, \varrho) \to 0, \quad \text{as } \varepsilon \to 0,$$

locally uniformly in $S^2 \times (0, A) \times \mathbf{R}$; in fact, the difference is locally uniformly $O(\varepsilon \log \varepsilon)$.

When g^{jk} only depend on $\partial u/\partial t$ and the Cauchy data are rotationally symmetric, then John [2] has proved that for the lifespan T_ε of the solution one can replace $\underline{\lim}$ in (6.5.28) by lim and inequality by equality. The idea of the proof is that in polar coordinates one actually has a problem with just one space variable where one can use a modification of the proof of Theorem 4.3.1. For the details of the proof we refer to John [2] or Hörmander [1]. The latter paper also contains Theorem 6.5.7, which was proved independently by John [3] with a different argument. Theorem 6.5.2 was proved by Klainerman [1] when $n \geq 6$ with a far more complicated proof. The proof given here is essentially that of Klainerman [3] who proved the theorem for $n > 4$ and also proved Theorem 6.5.3 which had been established somewhat earlier for the critical dimension $n = 3$ by John and Klainerman [1]. In Hörmander [1] the analogue of Theorem 6.5.7 for $n = 2$ was also proved with essentially the same arguments. The only additional complication is that the Friedlander radiation field does not have compact support in the radial variable, but that is compensated by its symbol properties. Thus we have the following result also:

Theorem 6.5.8. *The Cauchy problem* (6.5.1) *(with $f = 0$), (6.5.2) with $u_j \in C_0^\infty(\mathbf{R}^2)$ has a C^∞ solution u_ε for $0 \leq t < T_\varepsilon$ where*

$$(6.5.30) \qquad \lim_{\varepsilon \to 0} \varepsilon \sqrt{T_\varepsilon} \geq A = (\max_{\omega, \varrho} G(\omega) \partial^2 F_0(\omega, \varrho)/\partial \varrho^2)^{-1}.$$

Here $\omega \in S^1$ and G is defined by the analogue of (6.5.14), *while εF_0 is the Friedlander radiation field of the solution of the Cauchy problem for the unperturbed wave equation. If U is the solution of* (6.5.15), (6.5.16) *then*

$$(6.5.31) \qquad s\varepsilon^{-2} u_\varepsilon(s^2/\varepsilon^2, (s^2/\varepsilon^2 + \varrho)\omega) - U(\omega, s, \varrho) \to 0, \quad \text{as } \varepsilon \to 0,$$

locally uniformly in $S^1 \times (0, A) \times \mathbf{R}$; in fact, the difference is locally uniformly $O(\varepsilon^{\frac{1}{2}})$.

When $G(\omega) \equiv 0$ Theorems 6.5.7 and 6.5.8 suggest a much better order of magnitude for the lifespan. In fact, Christodoulou [1] and Klainerman [4] have then proved global existence for small ε when $n = 3$. This will be done in Sections 6.6 and 6.7 here.

We shall now outline an extension of Theorem 6.5.7 to the case where the function f in (6.5.1) is not identically 0. In analogy with (6.5.13), (6.5.14) we introduce

$$(6.5.13)' \qquad f(u') = \sum_{j,k} f^{jk} \partial_j u \partial_k u + O(|u'|^3),$$

$$(6.5.14)' \qquad F(\omega) = \sum_{j,k} f^{jk} \hat{\omega}_j \hat{\omega}_k; \quad \hat{\omega} = (-1, \omega_1, \omega_2, \omega_3).$$

We get new nonlinear terms dominated by $F(\omega)U_q'^2$, and with the notation $u = U_q'$ as in the proof of Lemma 6.5.4 this gives instead of (6.5.15)' the modified Burgers' equation

$$(6.5.32) \qquad 2\partial u(\omega, s, q)/\partial s = G(\omega)u(\omega, s, q)\partial u(\omega, s, q)/\partial q - F(\omega)u(\omega, s, q)^2,$$

which is of the form studied in Lemma 2.3.1. Again we solve this equation with the initial condition $u(\omega, 0, q) = \partial F_0(\omega, q)/\partial q$. The solution exists for $0 \leq s < A$ where (see Lemma 2.3.1)

$$(6.5.33) \qquad A = \left(\max \tfrac{1}{2}(G(\omega)\partial^2 F_0(\omega, \varrho)/\partial \varrho^2 - F(\omega)\partial F_0(\omega, \varrho)/\partial \varrho) \right)^{-1}$$

is finite unless $(G, F) \equiv 0$ or $(u_0, u_1) \equiv 0$. It is no longer true that $\int u(\omega, s, q) \, dq$ vanishes for $s \neq 0$; in fact,

$$2 \frac{\partial}{\partial s} \int u(\omega, s, q) \, dq = - \int F(\omega) u(\omega, s, q)^2 dq,$$

which is hardly ever zero. We define $U(\omega, s, q)$ as the solution of $\partial U / \partial q = u$ which vanishes for $q > M$ if M is an upper bound for $|y|$ when $y \in \text{supp } u_0 \cup \text{supp } u_1$. When $q < -M$ it is a function $U_-(\omega, s)$ independent of q, and we have

(6.5.34)
$$2U_{sq}'' = G(\omega)U_q'U_{qq}'' - F(\omega)U_q'^2.$$

To avoid singularities when $r = 0$ we must cut off by taking a function $\psi(t, x)$ in C^∞ which is homogeneous of degree 0, equal to 1 in a conic neighborhood of the light cone and equal to 0 in a conic neighborhood of the t axis. We keep the definition (6.5.18) of w with U replaced by ψU. When $q < -M$ this is equal to $\psi(t, x)U_-(\omega, \varepsilon \log(\varepsilon t))$, so we have

$$|\partial^\alpha(\psi(t, x)U_-(\omega, \varepsilon \log(\varepsilon t)))| \leq C_{\alpha, B}(1 + t)^{-|\alpha|} \quad \text{if } \varepsilon t \geq 1, \, \varepsilon \log t \leq B.$$

This proves that (6.5.20) remains valid. When $q < -M$ we find that

$$|Z^I \square w| \leq C_{I, B} \varepsilon (1 + t)^{-3} \quad \text{if } \varepsilon t \geq 1, \, \varepsilon \log t \leq B,$$

so (6.5.21) remains valid too, with

$$R = \sum g^{jk}(w')\partial_j \partial_k w - f(w').$$

However, the measure of the support of R is now only $O(1 + t)^3$, so (6.5.23) is replaced by

$$\|Z^I R(t, \cdot)\| \leq C_{I, B} \varepsilon t^{-3/2} \quad \text{if } \varepsilon t \geq 1, \, \varepsilon \log t \leq B,$$

and the integral of this from $1/\varepsilon$ to ∞ is $O(\varepsilon^{3/2})$. In (6.5.1)' we now obtain another term $f(v' + w') - f(w')$ in the right-hand side, which gives another error term

$$R_5 = Z^I(f(v' + w') - f(w'))$$

in the subsequent argument. It is estimated just as the others after writing $f(v' + w') - f(w'))$ as a scalar product with v', so we obtain

$$N_s(t) \leq C\varepsilon^{3/2}, \quad 0 \leq t \leq e^{B/\varepsilon}.$$

Hence we have proved that

Theorem 6.5.9. *The Cauchy problem* (6.5.1), (6.5.2) *with* $u_j \in C_0^\infty(\mathbf{R}^3)$ *has a* C^∞ *solution* u_ε *for* $0 \leq t < T_\varepsilon$ *where*

$$\lim_{\varepsilon \to 0} \varepsilon \log T_\varepsilon \geq A$$

with A now defined by (6.5.33). *Here $\omega \in S^2$ and F, G are defined by* (6.5.14), (6.5.14)' *while εF_0 is the Friedlander radiation field of the solution of the Cauchy problem for the unperturbed wave equation. If U is the solution of* (6.5.34), (6.5.16) *vanishing for $q \gg 0$ then* (6.5.29) *holds locally uniformly in $S^2 \times (0, A) \times \mathbf{R}$; in fact, the difference is locally uniformly $O(\varepsilon^{\frac{1}{2}})$.*

Remark 1. Theorem 6.5.9 was extended in Hörmander [7] to fully non-linear perturbations of the wave equation

(6.5.35)
$$\square u = f(u', u'')$$

where f vanishes of second order at $(0,0)$. The new problem posed by this generalization of (6.5.1) is that (6.5.32) is replaced by a more general equation of the form

$$\partial u/\partial s = a(\omega)(\partial u/\partial q)^2 + 2b(\omega)u\partial u/\partial q + c(\omega)u^2.$$

The lifespan of solutions of the Cauchy problem for this equation can still be determined explicitly, but the formula for it is fairly complicated so the expression for A in the extension of Theorem 6.5.9 is somewhat involved.

Remark 2. If one wants to apply the methods of this section to the equation

(6.5.36) $\Box u = f(u, u', u''),$

where f vanishes of second order at $(0,0,0)$ but may depend on u, there is another difficulty. The energy integral method gives naturally only an estimate of $\|u'(t,\cdot)\|$, and one must expect to lose a factor t when passing to an estimate for $\|u(t,\cdot)\|$. For $n > 5$ this will suffice to prove a global existence theorem for small ε using the proof of Theorem 6.5.2, and the result was extended to $n = 5$ by Li and Chen [1]. In Hörmander [8] it was proved that if there is no u^2 term, that is, $f(u,0,0) = O(u^3)$, then (6.5.36), (6.5.2) has a global solution for small ε when $n \geq 4$, and that the lifespan T_ε is $\geq e^{c/\sqrt{\varepsilon}}$ for some $\varepsilon > 0$ when $n = 3$; this was improved to $T_\varepsilon \geq e^{c/\varepsilon}$ by Lindblad [2]. For arbitrary f it was also proved by Hörmander [8] that $T_\varepsilon \geq e^{c/\varepsilon}$ for some $c > 0$ when $n = 4$; this was improved to $T_\varepsilon \geq e^{c/\varepsilon^2}$ by Li and Zhou [1]. When $n = 3$ it was already proved by John [4] for the equation $\Box u + u^2 = 0$ that $\varepsilon^2 T_\varepsilon$ lies between two positive bounds for small ε. Lindblad [1] proved that the limit as $\varepsilon \to 0$ exists and gave a description of it.

Remark 3. Even if it is hard to doubt that (6.5.28) and (6.5.30) always give the precise asymptotic lifespan of the solutions there are no proofs except for the rotationally symmetric three dimensional case studied by F. John [2] and the recent study by Alinhac [3] of some very special equations and data with two space dimensions. Theorems 6.5.7 and 6.5.8 only show that if the lifespan should be much longer then the asymptotics would be of a different kind. However, Alinhac [1, 2] has given a more precise lower bound for the lifespan in the case discussed in Theorem 6.5.7 and proved very refined results on the asymptotic behavior close to this time which reinforce the belief that Theorem 6.5.7 is fairly precise. (See also the bibliographies in these references.)

6.6. The null condition in three dimensions. In the main results of this section the number n of space variables will be equal to 3, but at first we allow any $n \geq 3$. We shall consider a quasilinear second order equation

(6.6.1) $$\sum_{j,k=0}^{n} g^{jk}(u, u')\partial_j\partial_k u = f(u, u')$$

with small Cauchy data

(6.6.2) $u(0,x) = \varepsilon u_0(x), \quad \partial_0 u(0,x) = \varepsilon u_1(x),$

where $u_j \in C_0^\infty(\mathbf{R}^n)$. We shall often write $t = x_0$ and $\vec{x} = (x_1, \ldots, x_n)$. We assume that $u \equiv 0$ is a solution of (6.6.1) and that the linearisation of (6.6.1) there is the wave equation, that is,

$$\sum g^{jk}(0,0)\partial_j\partial_k = \Box, \quad f = df = 0 \text{ at } (0,0).$$

Definition 6.6.1. The equation (6.6.1) is said to satisfy the *null condition* if

$$g^{jk}(u, u') = \sum_{l=0}^{n} g^{jkl}\partial_l u + O(|u| + |u'|)^2,$$

$$f(u, u') = \sum_{j,k=0}^{n} f^{jk}\partial_j u \partial_k u + O(|u| + |u'|)^3,$$

where

(6.6.3) $\quad \displaystyle\sum_{j,k,l=0}^{n} g^{jkl}\xi_j\xi_k\xi_l = 0, \quad \sum_{j,k=0}^{n} f^{jk}\xi_j\xi_k = 0$ when $\xi_0^2 - \xi_1^2 - \cdots - \xi_n^2 = 0.$

Thus the null condition requires that the quadratic terms are independent of u and that with the notation used in Theorem 6.5.9 we have $G(\omega) \equiv 0$ and $F(\omega) \equiv 0$. We shall improve Theorem 6.5.9 when $A = \infty$ to the following result of Christodoulou [1] and Klainerman [4] (see also Hörmander [6]):

Theorem 6.6.2. *If (6.6.1) satisfies the null condition, $n = 3$, $u_j \in C_0^\infty(\mathbf{R}^3)$, and ε is sufficiently small, then the Cauchy problem (6.6.1), (6.6.2) has a C^∞ solution for $t \geq 0$.*

The proof will occupy the entire section. The main change of the methods used in Section 6.5 is that we shall use energy estimates of the form (6.3.14). To do so we must examine the relation of the null condition to the vector fields (6.2.8), (6.2.9) and the constant vector fields. These suggest using the following norm on covectors

(6.6.4) $\quad |\xi|_x = \left(\displaystyle\sum_{j<k} Z_{jk}(x, \xi)^2 + Z_0(x, \xi)^2 + \sum_0^n \xi_j^2 \right)^{\frac{1}{2}} = (|A'(x) \wedge \xi|^2/4 + \langle x, \xi \rangle^2 + |\xi|^2)^{\frac{1}{2}},$

where the norms are Euclidean and $A(x) = x_0^2 - x_1^2 - \cdots - x_n^2$. The following lemma is perhaps clarified by allowing A to be any nondegenerate real quadratic form.

Lemma 6.6.3. *Let G be a k linear form on $\mathbf{R}^{1+n}$. Then*

(6.6.5) $\quad |G(\xi^1, \ldots, \xi^k)| \leq C(1 + |x|)^{-1}|\xi^1|_x \cdots |\xi^k|_x; \quad \xi^j, x \in \mathbf{R}^{1+n};$

if and only if, with B denoting the dual quadratic form of A,

(6.6.6) $\quad G(\xi, \ldots, \xi) = 0 \quad \text{when } \xi \in \mathbf{R}^{1+n}, \ B(\xi) = 0.$

Proof. Let $\xi^1 = \cdots = \xi^k = A'(x)$ where $A(x) = 0$ but $x \neq 0$. Then it follows from (6.6.4) that $|\xi^j|_{tx} = |A'(x)|$, so the right-hand side of (6.6.5) with x replaced by tx is $O(1/t)$ as $t \to \infty$. This proves that (6.6.5) implies (6.6.6). Now assume that (6.6.6) holds, and let $|\xi^j|_x = 1$ for all j. Then $|\xi^j| \leq 1$ so (6.6.5) is obvious if $|x| \leq 1$. Write

$$\xi^j = r^j + t_j A'(x)$$

where r^j is orthogonal to $A'(x)$ in the Euclidean sense. Then

$$|\xi^j \wedge A'(x)| = |r^j \wedge A'(x)| = |r^j||A'(x)|,$$

so we have a bound for $|r^j|(|x|+1) + |t_j A'(x)| + |t_j A(x)|$. It follows that all terms in the expansion of $G(\xi^1, \ldots, \xi^k)$ containing some r^j can be estimated by $C/(1+|x|)$. It remains to estimate

$$t_1 \cdots t_k G(A'(x), \ldots, A'(x))$$

when $|x| > 1$. By (6.6.6) we have $|G(\xi, \ldots, \xi)| \le C|\xi|^{k-2}|B(\xi)|$, hence

$$|G(A'(x), \ldots, A'(x))| \le C|A'(x)|^{k-2}|B(A'(x))| = 4C|A'(x)|^{k-2}|A(x)|.$$

Since $t_j A(x)$ and $t_j |A'(x)|$ have fixed bounds we obtain

$$|t_1 \cdots t_k G(A'(x), \ldots, A'(x))| \le 4C|t_1 \cdots t_k||A(x)||A'(x)|^{k-2} \le C'/|x|,$$

which completes the proof.

We shall use Lemma 6.6.3 for $k = 2$ and for $k = 3$. For $k = 3$ a closely related estimate of the quadratic term in (6.6.1) is also required:

Lemma 6.6.4. *If g^{jk} satisfy the null condition (6.6.3), then*

$$(6.6.7) \qquad \Big| \sum_{j,k,l=0}^{n} g^{jkl} \partial_j \partial_k u \, \partial_l v \Big| \le C(1+|x|)^{-1}|\partial v|_x \sum_{k=0}^{n} |\partial \partial_k u|_x.$$

Proof. We may assume that $|x| > 1$ and that $\sum_0^n |\partial \partial_k u|_x = 1$, $|\partial v|_x = 1$. Write

$$\partial v(x) = V(x) + t A'(x), \quad \partial \partial_k u(x) = U^k(x) + t_k A'(x),$$

with $V(x), U^k(x)$ orthogonal to $A'(x)$. Then we have a bound for

$$|x||V(x)| + |tx| + |tA(x)| + \sum_{k=0}^{n} |x||U^k(x)| + \sum_{k=0}^{n} |t_k x| + \sum_{k=0}^{n} |t_k A(x)|,$$

as in the proof of Lemma 6.6.3. As there we must estimate

$$M = \sum_{j,k,l=0}^{n} g^{jkl} t_k \partial A/\partial x_j t \partial A/\partial x_l.$$

For $\nu = 0, \ldots, n$ we have

$$t_k \partial A/\partial x_\nu + U_\nu^k = \partial_\nu \partial_k u = t_\nu \partial A/\partial x_k + U_k^\nu, \quad \text{hence}$$

$$M \partial A/\partial x_\nu = t_\nu t \sum_{j,k,l} g^{jkl} \partial A/\partial x_k \partial A/\partial x_j \partial A/\partial x_l + \sum_{j,k,l} g^{jkl}(U_k^\nu - U_\nu^k) \partial A/\partial x_j t \partial A/\partial x_l.$$

The second sum is bounded and the first sum is $\le C|A(x)||x|$ by (6.6.3). Hence $M|x|$ is bounded and the lemma follows.

The next lemma describes how the vector fields (6.2.8), (6.2.9) act on expressions such as that estimated in Lemma 6.6.4.

Lemma 6.6.5. *Let G be a k linear form on $\mathbf{R}^{1+n}$, and let $k = k_1 + \cdots + k_r$ where k_j are positive integers. Then*

$$G(u_1^{(k_1)}, \ldots, u_r^{(k_r)}), \quad u_j \in C^k(\mathbf{R}^{1+n}),$$

is well defined by extension of the k linear form to the tensor products. If Z is a vector field with affine linear coefficients, $Z(x, \partial) = \sum_0^n Z_j(x)\partial_j$, and $u_j \in C^{k+1}(\mathbf{R}^{1+n})$, then

$$ZG(u_1^{(k_1)}, \ldots, u_r^{(k_r)}) = G((Zu_1)^{(k_1)}, \ldots, u_r^{(k_r)}) + \ldots$$
$$+ G(u_1^{(k_1)}, \ldots, (Zu_r)^{(k_r)}) + G_1(u_1^{(k_1)}, \ldots, u_r^{(k_r)}),$$

where

$$G_1(\xi, \ldots, \xi) = \{Z(x, \xi), G(\xi, \ldots, \xi)\} = -\sum_0^n \partial Z/\partial x_j \partial G(\xi, \ldots, \xi)/\partial \xi_j.$$

Here $\{\,,\}$ denotes the Poisson bracket. If $G(\xi, \ldots, \xi) = 0$ when $\xi_0^2 - \cdots - \xi_n^2 = 0$ then this follows for G_1 too if Z is one of the vector fields (6.2.8), (6.2.9) or a constant vector field.

Proof. By Leibniz' rule we must let Z act separately on each of the $u_j^{(k_j)}$, and

$$Z\partial^\alpha = \partial^\alpha Z + Q_\alpha(\partial), \quad \text{where} \quad Q_\alpha(\xi) = -\sum_0^n \partial Z(x, \xi)/\partial x_j \partial \xi^\alpha/\partial \xi_j.$$

The formula for G_1 follows by another application of Leibniz' rule, now for the Poisson bracket. If Z is one of the vector fields (6.2.8), say $Z(x, \xi) = \lambda_j x_j \xi_k - \lambda_k x_k \xi_j$, then

$$\sum_{i=0}^n \partial Z/\partial x_i \partial/\partial \xi_i = \lambda_j \lambda_k (\lambda_k \xi_k \partial/\partial \xi_j - \lambda_j \xi_j \partial/\partial \xi_k),$$

and if Z is the vector field Z_0 in (6.2.9) we obtain the same vector field in the ξ variables. They are all tangent to the light cone, and when Z is a constant vector field we obtain the zero vector field, so $G_1(\xi, \ldots, \xi) = 0$ on the light cone if $G(\xi, \ldots, \xi) = 0$ there. This proves the last statement.

We shall now discuss the modifications of the energy identity $(6.3.3)''$ which are required when the coefficients are variable but close to constants. Thus consider an equation of the form

$$(6.6.8) \qquad \sum_{j,k=0}^n g^{jk}(x)\partial_j \partial_k u = f,$$

where $n \geq 3$ and $\sum_{j,k=0}^n g^{jk}(x)\partial_j \partial_k = \Box + \sum_{j,k=0}^n \gamma^{jk}(x)\partial_j \partial_k$ with γ^{jk} small. We shall use the same vector field K as in Lemma 6.3.4 apart from an addition of $\partial/\partial t$ to get control of the conventional energy form. Thus we set

$$(6.6.9) \qquad Lu = \sum_0^n L^i \partial_i u + (n-1)x_0 u,$$
$$\hat{L} = (L^0, \ldots, L^n) = (1 + x_0^2 + |\bar{x}|^2, 2x_0 x_1, \ldots, 2x_0 x_n),$$

and shall derive a modification of $(6.3.3)''$,

$$(6.6.10) \quad 2Lu \sum_{j,k=0}^{n} g^{jk} \partial_j \partial_k u = \sum_{j=0}^{n} \partial_j \Big(\sum_{i=0}^{n} T_i^j(u) L^i + (n-1)(2x_0 u \sum_{k=0}^{n} g^{jk} \partial_k u - \theta^j u^2) \Big) - R,$$

where $\theta = (1,0,\ldots,0)$ and T_i^j is defined by $(6.3.4)$. We shall write down the error term R later on. The energy integrand, that is, the 0 component of the vector field in the divergence on the right, now becomes

$$\sum_i T_i^0(u) L^i + 2(n-1)x_0 u \sum_k g^{0k} \partial_k u - (n-1)u^2$$
$$= 2 \sum_k g^{0k} \partial_k u \sum_i L^i \partial_i u - \sum_{j,k} g^{jk} \partial_j u \partial_k u L^0 + 2(n-1)x_0 u \sum_k g^{0k} \partial_k u - (n-1)u^2,$$

so for the energy $E(u)$ defined by integration over $\vec{x}$, we have if $E_0(u)$ denotes the same energy for the unperturbed wave operator,

$$(6.6.11) \quad E(u) - E_0(u) = \int \Big(2 \sum_{k=0}^{n} \gamma^{0k} \partial_k u \sum_{i=0}^{n} L^i \partial_i u$$
$$- (1 + x_0^2 + |\vec{x}|^2) \sum_{j,k=0}^{n} \gamma^{jk} \partial_j u \partial_k u + 2(n-1)x_0 u \sum_{k=0}^{n} \gamma^{0k} \partial_k u \Big) \, d\vec{x}.$$

By Lemma 6.3.4

$$(6.6.12) \quad |\langle \hat{L}, \xi \rangle| \le C(1 + |x|)|\xi|_x.$$

If we assume that

$$(6.6.13) \quad \Big| \sum_{j,k=0}^{n} \gamma^{jk}(x) \xi_j r_k \Big| \le \delta(1 + |x|)^{-2} |\xi|_x |r|_x$$

for some small δ, then the integral of the middle sum in $(6.6.11)$ is $O(\delta E_0(u))$, for it follows from Lemma 6.3.5 that

$$(6.6.14) \quad 1/41 \le E_0(u) / (\|Z_0 u\|^2 + \sum_{j<k} \|Z_{jk} u\|^2 + \sum_j \|\partial_j u\|^2 + \|(n-1)u\|^2) \le 2.$$

When $r_i = \delta_{ik}$ it follows from $(6.6.13)$ that

$$(6.6.13)' \quad \Big| \sum_{j=0}^{n} \gamma^{jk}(x) \xi_j \Big| \le C\delta(1 + |x|)^{-1} |\xi|_x,$$

which combined with $(6.6.12)$ proves that also the first and the third sum in $(6.6.11)$ are $O(\delta E_0(u))$. Hence

$$(6.6.15) \quad (1 - C\delta) E_0(u) \le E(u) \le (1 + C\delta) E_0(u)$$

which shows the equivalence of $E(u)$ and $E_0(u)$ when δ is small.

Next we shall prove that the error term R in (6.6.10) is equal to $\sum_1^6 R_j$ where

(6.6.16)

$$R_1 = 2\sum_{j,k,i} \gamma^{jk}\partial_k u(\partial_j L^i)\partial_i u - \sum_i(\partial_i L^i)\sum_{j,k}\gamma^{jk}\partial_j u\partial_k u,$$

$$R_2 = 2\sum_i L^i\partial_i u\sum_{j,k}(\partial_j\gamma^{jk})\partial_k u, \quad R_3 = -\sum_{i,j,k}(L^i\partial_i\gamma^{jk})\partial_j u\partial_k u,$$

$$R_4 = 2(n-1)u\sum_k\gamma^{0k}\partial_k u, \quad R_5 = 2(n-1)x_0\sum_{j,k}\gamma^{jk}\partial_j u\partial_k u,$$

$$R_6 = 2(n-1)x_0 u\sum_{j,k}(\partial_j\gamma^{jk})\partial_k u.$$

To do so we shall consider successively the terms where ∂_j acts on the various terms and factors in the divergence expression in (6.6.10).

i) The terms where ∂_j acts on a derivative of u in $T_i^j(u)$ occur on the left of (6.6.10).

ii) When ∂_j acts on L^i we obtain a quadratic form in ∂u which is the sum of that in the unperturbed case and R_1.

iii) When ∂_j acts on the components of g in $T_i^j(u)$ we obtain the error terms R_2 and R_3.

iv) If ∂_j acts on $\partial_k u$ in the last sum in (6.6.10) we obtain the remaining part of the left-hand side of (6.6.10).

v) If ∂_j acts on x_0 we obtain the term R_4 in addition to the term $2(n-1)u\partial_0 u$ which occurs in the unperturbed case.

vi) If ∂_j acts on g^{jk} we obtain the term R_6.

vii) If ∂_j acts on u we obtain the term R_5 in addition to the quadratic form in ∂u which occurs in the unperturbed case.

viii) If ∂_j acts on the last term $-(n-1)\theta^j u^2$ we obtain the term $-2(n-1)u\partial_0 u$.

In the unperturbed case we know from (6.3.3)″ that the terms not accounted for in the left-hand side of (6.6.10) or in R cancel out, which proves (6.6.10).

The error terms R_1, R_4 and R_5 can be estimated using (6.6.13). From (6.6.9) we obtain $\sum_0^n\partial_i L^i = 2(n+1)x_0$ and

(6.6.17)
$$\sum_{i=0}^n\xi_i\partial_j L^i = \begin{cases} 2\sum_{i=0}^n\xi_i x_i, & \text{when } j=0 \\ 2(\xi_0 x_j + \xi_j x_0), & \text{when } j\neq 0. \end{cases}$$

These sums are bounded by $2|\xi|_x$, so it follows from (6.6.13) and (6.6.13)′ that

(6.6.18)
$$|R_1| + |R_4| + |R_5| \leq C\delta(1+|x|)^{-1}(|\partial u|_x^2 + |u||\partial u|_x).$$

Now we add to (6.6.13) the hypothesis

(6.6.19)
$$\Big|\sum_{j,k=0}^n\xi_k\partial_j\gamma^{jk}(x)\Big| \leq \delta(1+|x|)^{-2}|\xi|_x.$$

By (6.6.12) the estimate (6.6.18) is then also valid for R_6 and R_2. To obtain such a bound for R_3 we also require

(6.6.20)
$$\Big|\sum_{i,j,k=0}^n(L^i\partial_i\gamma^{jk}(x))\xi_j\xi_k\Big| \leq \delta(1+|x|)^{-1}|\xi|_x^2.$$

Then we have the estimate (6.6.18) for all R_j, and we have proved

Proposition 6.6.6. *Let $u \in C^2$ be a solution of (6.6.8) vanishing for large $|\vec{x}|$ when t is bounded, and assume that (6.6.13), (6.6.19) and (6.6.20) hold. If δ is sufficiently small it follows that the energy $E(u; x_0)$ at time x_0 obtained by integrating (6.6.10) with respect to $\vec{x}$ satisfies*

$$(6.6.21) \qquad 1/50 \leq E(u)/(\|Z_0 u\|^2 + \sum \|Z_{jk} u\|^2 + \sum \|\partial_j u\|^2 + \|(n-1)u\|^2) \leq 3.$$

We have

$$
(6.6.22) \qquad
\begin{aligned}
|\partial E(u; x_0)/\partial x_0| \leq C\Big(& \delta(1 + x_0)^{-1} E(u; x_0) \\
& + \big(\int (1 + x_0 + |\vec{x}|)^2 |f(x_0, \vec{x})|^2 \, d\vec{x} \big)^{\frac{1}{2}} E(u; x_0)^{\frac{1}{2}} \Big).
\end{aligned}
$$

Here we have used (6.6.12) when estimating the left-hand side of the energy identity (6.6.10) and of course used that the integrals of the terms on the right with $j \neq 0$ are equal to 0.

After introducing $E(u; x_0)^{\frac{1}{2}}$ as a new unknown in (6.6.22) and multiplication of both sides by $(1 + x_0)^{-C\delta}$ we can integrate and obtain

$$
(6.6.23) \qquad
\begin{aligned}
E(u; x_0)^{\frac{1}{2}} \leq (1 + x_0)^{C\delta} \Big(& E(u; 0)^{\frac{1}{2}} \\
& + C \int_0^{x_0} (1+t)^{-C\delta} dt \big(\int (1 + t + |\vec{x}|)^2 |f(t, \vec{x})|^2 \, d\vec{x} \big)^{\frac{1}{2}} \Big).
\end{aligned}
$$

We shall use (6.6.22) instead of (6.3.6) when we study the equation (6.6.1). The first order terms in g^{jk} will then have the special properties in Definition 6.6.1. To apply Proposition 6.6.6 we need the following

Lemma 6.6.7. *Suppose that*

$$(6.6.24) \qquad \gamma^{jk}(x) = \sum_{l=0}^{n} g^{jkl} \partial_l w(x) + \varrho^{jk}(x),$$

where g^{jkl} are constants satisfying the null condition (6.6.3) and

$$(6.6.25) \qquad (1 + |x|) \sum_{|I| \leq 2} |Z^I w(x)| \leq \delta, \quad (1 + |x|)^2 \sum_{|I| \leq 1} |Z^I \varrho^{jk}(x)| \leq \delta,$$

where Z^I is any product of $|I|$ vector fields of the form (6.2.8), (6.2.9) or $\partial/\partial x_j$. Then (6.6.13), (6.6.19) and (6.6.20) are valid with δ replaced by a constant times δ.

Proof. This is obvious when $\gamma^{jk} = \varrho^{jk}$, if we recall (6.6.12) when verifying (6.6.20). We may therefore assume that $\varrho^{jk} = 0$. To prove (6.6.13) we use that by Lemma 6.6.3

$$\Big| \sum_{j,k,l=0}^{n} g^{jkl} \partial_l w \xi_j r_k \Big| \leq C(1 + |x|)^{-1} |\partial w|_x |\xi|_x |r|_x,$$

and that $|\partial w|_x \leq C\delta(1 + |x|)^{-1}$ by (6.6.25). By Lemma 6.6.4

$$\Big| \sum_{j,k=0}^{n} \xi_k \partial_j \gamma^{jk}(x) \Big| = \Big| \sum_{j,k,l=0}^{n} g^{jkl} \partial^2 w/\partial x_l \partial x_j \xi_k \Big| \leq C(1 + |x|)^{-1} |\xi|_x \sum_{l=0}^{n} |\partial \partial_l w|_x,$$

which proves (6.6.19). To prove (6.6.20) we write, now with $\hat{L} = \sum_0^n L^i \partial/\partial x_i$,

$$\sum_{j,k=0}^n (\hat{L}\gamma^{jk})\xi_j\xi_k = \sum_{j,k,l=0}^n g^{jkl}\xi_j\xi_k\partial_l(\hat{L}w) + \sum_{j,k,l=0}^n g^{jkl}\xi_j\xi_k[\hat{L},\partial_l]w.$$

By Lemma 6.6.3 the first sum can be estimated by

$$C(1+|x|)^{-1}|\xi|_x^2|\partial\hat{L}w|_x,$$

and $|\partial\hat{L}w|_x \le C\delta$ by (6.6.12) and (6.6.25). Since $[\partial_0,\hat{L}] = 2Z_0$ and $[\partial_l,\hat{L}] = 2Z_{0l}$ for $l \ne 0$, we have by (6.6.25)

$$|[\hat{L},\partial_l]w| \le C\delta\{1+|x|\}^{-1},$$

which completes the proof of (6.6.20) also.

We shall need an analogue of Proposition 6.5.1 involving L^1 norms, which also follows from Sobolev's inequality. For the sake of simplicity we assume from now on that $n = 3$. (See also Hörmander [3] for general $n \ge 3$.)

Lemma 6.6.8. *If $u \in C^4$ and $\Box u = F$ in $[0,t] \times \mathbf{R}^3$, and the Cauchy data of u are 0 when $t = 0$, then*

$$(6.6.26) \qquad (1+t+|\vec{x}|)|u(t,\vec{x})| \le C \iint_{0<s<t} \sum_{|I|\le 2} |Z^I F(s,\vec{y})|\, ds\, d\vec{y}/(1+s+|\vec{y}|),$$

where Z^I is any product of $|I|$ vector fields of the form (6.2.8), (6.2.9) or $\partial/\partial x_j$.

Proof. We shall first prove a homogeneous version of (6.6.26): if F vanishes in a neighborhood of the origin then

$$(6.6.26)' \qquad (t+|\vec{x}|)|u(t,\vec{x})| \le C \iint_{0<s<t} \sum_{|I|\le 2}{}' |Z^I F(s,\vec{y})|\, ds\, d\vec{y}/(s+|\vec{y}|),$$

with summation only over products of vector fields of the form (6.2.8), (6.2.9) which preserve homogeneity. A change of scales shows that it suffices to prove (6.6.26)' when $t = 1$.

The solution u of the wave equation is given by the retarded potential

$$u(t,\vec{x}) = (4\pi)^{-1} \int_{|\vec{y}|<t} F(t-|\vec{y}|,\vec{x}-\vec{y})\, d\vec{y}/|\vec{y}|.$$

The difficulties in the proof come from the fact that the integration only takes place over a hypersurface and that the denominator $|\vec{y}|$ vanishes when $\vec{y} = 0$.

i) Let us first assume that supp $F \subset \{(s,\vec{y}); |\vec{y}| < \frac{1}{2}s\}$. In this set the vector fields Z used in (6.6.26)' span all vector fields, and we conclude that

$$(6.6.27) \qquad \iint_{0<s<1} \sum_{|\alpha|\le 2} s^{|\alpha|}|\partial^\alpha F(x,\vec{y})|\, dx\, d\vec{y}/s$$

$$\le C \iint_{0<s<1} \sum_{|I|\le 2}{}' |Z^I F(s,\vec{y})|\, ds\, d\vec{y}/(s+|\vec{y}|).$$

We may assume in the proof of (6.6.26)$'$ with $t = 1$ that $|\vec{x}| \leq 1$ since $u(1, \vec{x}) = 0$ otherwise. When $|\vec{y}| > \frac{1}{2}$ we use the estimate

$$|F(1 - |\vec{y}|, \vec{x} - \vec{y})| \leq \int_0^1 (|F(s, \vec{x} - \vec{y})| + |F_s'(s, \vec{x} - \vec{y})|)\, ds,$$

and $1/|\vec{y}| < 2$ then. When $|\vec{y}| \leq \frac{1}{2}$ we use instead the similar estimate

$$|F(1 - |\vec{y}|, \vec{x} - \vec{y})| \leq \int_{\frac{1}{2}}^1 (2|F(s, \vec{x} - \vec{y})| + |F_s'(s, \vec{x} - \vec{y})|)\, ds,$$

which avoids small values of s, and then we apply the estimate

$$\int |g(\vec{y})|\, d\vec{y}/|\vec{y}| \leq \frac{1}{2} \int |g'(\vec{y})|\, d\vec{y}, \quad g \in C_0^1(\mathbf{R}^3),$$

which follows by introducing polar coordinates and observing that integration by parts gives

$$\int_0^\infty |G(r)| r\, dr \leq \frac{1}{2} \int_0^\infty |G'(r)| r^2\, dr, \quad G \in C_0^1(\mathbf{R}).$$

Summing up, we have proved that

$$4\pi |u(1, \vec{x})| \leq 2 \iint_{0<s<1} (|F(s, \vec{y})| + |F'(s, \vec{y})| + s|F_{s\vec{y}}''(s, \vec{y})|)\, ds\, d\vec{y},$$

which gives (6.6.26)$'$ when combined with (6.6.27).

ii) Assume now that supp $F \subset \{(s, \vec{y}); |\vec{y}| > \frac{1}{3}s\}$. By Sobolev's lemma (see (6.4.13)$'''$)

$$M(t, r) = \sup_{|\omega|=1} |F(t, r\omega)| \leq C \sum_{|I| \leq 2}{}'' \int_{|\omega|=1} |(Z^I F)(t, r\omega)|\, d\omega,$$

where the summation only contains products of the vector fields (6.2.8) corresponding to Euclidean rotations. Hence

$$(6.6.28) \qquad \iint_{0<s<1,\varrho>0} M(s, \varrho)\varrho\, ds\, d\varrho \leq C \iint_{0<s<1} \sum_{|I|\leq 2}{}'' |Z^I F(s, \vec{y})|\, ds\, d\vec{y}/(s + |\vec{y}|).$$

Replacing f by M will increase the retarded potential, so $|u| \leq U$ where $\Box U = M$ and U has Cauchy data zero. It is clear that U is rotationally symmetric in the space variables, and expressing $\Box$ in polar coordinates we have

$$(\partial_t^2 - \partial_r^2) r U(t, r) = r M(t, r),$$

which implies that

$$|\vec{x}||u(1, \vec{x})| \leq r U(1, r) \leq \frac{1}{2} \iint_{0<s<1,\varrho>0} M(s, \varrho)\varrho\, ds\, d\varrho.$$

Combined with (6.6.28) this proves (6.6.26)$'$ when $|\vec{x}| \geq \frac{1}{4}$.

We shall now give a different proof when $|\bar{x}| < \frac{1}{4}$. If $(1 - |\bar{y}|, \bar{x} - \bar{y}) \in \operatorname{supp} F$ we have $3|\bar{x} - \bar{y}| > 1 - |\bar{y}|$, hence $4|\bar{y}| > 1 - 3|\bar{x}| > \frac{1}{4}$ and

$$(6.6.29) \qquad 4\pi|u(1, \bar{x})| \leq 16 \int_{1/16 < |\bar{y}| < 1} |F(1 - |\bar{y}|, \bar{x} - \bar{y})| \, d\bar{y}.$$

We shall now use that we control the radial derivative $Z_0 F$ of F. To do so we write $\varphi(\tau, \bar{y}) = \tau(1 - |\bar{y}|, \bar{x} - \bar{y})$ where $1/16 \leq |\bar{y}| \leq 1$ and $1 \leq \tau \leq 16/15$. The Jacobian $\tau^n(1 - \langle \bar{x}, \bar{y} \rangle/|\bar{y}|)$ is bounded below by $3/4$, and we have

$$\int |F(\varphi(1, \bar{y}))| \, d\bar{y} \leq \int_{1/16 < |\bar{y}| < 1, 1 < \tau < 16/15} (15|F(\varphi(\tau, \bar{y}))| + |\partial F(\varphi(\tau, \bar{y}))/\partial \tau|) \, d\tau \, d\bar{y}.$$

Since $\partial F(\varphi(\tau, \bar{y}))/\partial \tau = (Z_0 F)(\varphi(\tau, \bar{y}))/\tau$, we obtain by changing to the integration variables $\varphi(\tau, \bar{y})$ and combining the estimate with (6.6.29) that

$$|u(1, \bar{x})| \leq C \iint_{0 < s < 1, |\bar{y}| < 2} (|F(s, \bar{y})| + |Z_0 F(s, y)|) \, ds \, d\bar{y}, \qquad |\bar{x}| < \tfrac{1}{4},$$

which completes the proof of (6.6.26)' in this case.

Choose $\psi \in C_0^\infty(\mathbf{R}^3)$ so that $|\bar{y}| < \frac{1}{2}$ in the support of ψ and $|\bar{y}| > 1/3$ in the support of $1 - \psi$. Then $\psi(\bar{y}/s)F(s, \bar{y})$ and $(1 - \psi(\bar{y}/s))F(s, \bar{y})$ satisfy the hypothesis in i) and ii) respectively. Since the operators Z in (6.6.26)' applied to $\psi(\bar{y}/s)$ give a function which is homogeneous of degree 0, hence bounded, the estimate (6.6.26)' follows from the two cases proved above.

If $\operatorname{supp} F \subset \{(s, \bar{y}); s + |\bar{y}| \geq 1\}$, then the same inclusion is true for $\operatorname{supp} u$, so (6.6.26) is valid in that case. On the other hand, if $s + |\bar{y}| \leq 2$ in $\operatorname{supp} F(s, \bar{y})$ we obtain (6.6.26) by applying the case already proved to a translation such as $F(s, y_1 + 3, y_2, y_3)$. (The translation introduces the constant vector fields.) Combining the two cases by a partition of unity yields (6.6.26) in full generality.

Proof of Theorem 6.6.2. Let k be an integer ≥ 5, and assume that we already have a C^∞ solution of (6.6.1), (6.6.2) for $0 \leq x_0 \leq T$ such that for such x_0 and small ε

$$(6.6.30) \qquad (1 + |x|) \sum_{|I| \leq k} |Z^I u(x)| \leq C_0 \varepsilon,$$

$$(6.6.31) \qquad \sum_{|I| \leq k+4} \|Z^I u(x_0, \cdot)\| \leq C_1 (1 + x_0)^{C_2 \varepsilon} \sum_{|I| \leq k+4} \|Z^I u(0, \cdot)\|.$$

Let C_0 be so large that (6.6.30) holds with C_0 replaced by $C_0/3$ if u is replaced by the solution εu^0 of the wave equation with Cauchy data (6.6.2). We shall then prove for ε smaller than some number depending on C_1 and C_2 that
  i) (6.6.30) is valid with C_0 replaced by $C_0/2$;
  ii) (6.6.31) is a consequence of (6.6.30) for suitable C_j.
By the local existence theorem (Theorem 6.4.11) it will follow that a solution exists for all $x_0 \geq 0$ if ε is small enough.
  i) Since the Cauchy data of $Z^I u - Z^I \varepsilon u^0$ are $O(\varepsilon^2)$, it suffices by Lemma 6.6.8 to prove that for small ε

$$\sum_{|I| \leq k} \sum_{|J| \leq 2} \iint_{0 < t < T} |Z^J \square Z^I u(t, \bar{x})| \, dt \, d\bar{x}/(1 + t) \leq C\varepsilon^2.$$

We can write $Z^J \Box Z^I u$ as a sum of terms of the form $Z^K \Box u$ with $|K| \leq |J| + |I| \leq k + 2$, so it suffices to prove that

$$(6.6.32) \qquad \sum_{|K| \leq k+2} \iint_{0 < t < T} |Z^K \Box u(t, \vec{x})| \, dt d\vec{x} / (1 + t) \leq C \varepsilon^2.$$

To do so we write (6.6.1) in the form

$$(6.6.33) \qquad \Box u = - \sum_{j,k=0}^{n} \gamma^{jk}(u, u') \partial_j \partial_k u + f(u, u').$$

When we apply Z^K to (6.6.33) we obtain a number of terms containing products of derivatives of u of order $\leq k + 4$, and two factors can never be of order $> k$ since each such factor requires $k - 1$ factors Z acting on u'' or k factors Z acting on (u, u'), thus altogether $k - 1 + k > k + 2$ factors Z. Apart from the quadratic terms we can factor so that we must have three factors $Z^I u$. We estimate the two factors with highest $|I|$ using the L^2 estimate (6.6.31) and the others by means of (6.6.30). This shows that the L^1 norm can be estimated by $C\varepsilon^3 (1 + t)^{2C_2 \varepsilon - 1}$. For the quadratic terms

$$Z^K \sum_{j,k,l} g^{jkl} \partial_j \partial_k u \, \partial_l u, \quad Z^K \sum_{j,k} f^{jk} \partial_j u \partial_k u$$

we first use Lemma 6.6.5 to obtain a sum of terms of this form with the factors Z next to u. We can then apply Lemma 6.6.3 or Lemma 6.6.4 together with the L^2 estimate (6.6.31). It follows that

$$\sum_{|K| \leq k+2} \int |Z^K \Box u(t, \vec{x})| \, d\vec{x} \leq C \varepsilon^2 (1 + t)^{2C_2 \varepsilon - 1},$$

and this implies (6.6.32) when $2C_2 \varepsilon < 1$.

ii) To prove (6.6.31), assuming (6.6.30) known, we shall apply the energy estimates in Proposition 6.6.6 to the equations obtained when (6.6.1) is multiplied by Z^I for $|I| \leq k + 3$,

$$(6.6.34) \qquad \sum_{j,k=0}^{n} g^{jk}(u, u') Z^I u = \sum_{j=0}^{4} f_I^j,$$

where

$f_I^0 = [\Box, Z^I] u$;

f_I^1 is a sum of terms of the form $\widehat{G}((Z^J u)', (Z^K u)'')$ with $|J| + |K| \leq |I|$, $|K| < |I|$, and a trilinear form $\widehat{G}$ satisfying the null condition (cf. Lemma 6.6.5);

$f_I^2 = \sum c_{IJK}(Z^J \varrho^{jk}(u, u')) Z^K \partial_j \partial_k u$ with the same conditions on J and K, and ϱ^{jk} vanishing of second order at 0;

f_I^3 is a sum of terms of the form $\widehat{F}((Z^K u)', (Z^J u)')$ with $|J| + |K| \leq |I|$ and a bilinear form $\widehat{H}$ satisfying the null condition (cf. Lemma 6.6.5);

$f_I^4 = Z^I R(u, u')$ where R vanishes of third order at 0.

We shall consider $g^{jk}(u, u')$ as a function of x when we apply Proposition 6.6.6. It follows from (6.6.30) and Lemma 6.6.7 that the hypotheses of Proposition 6.6 are then fulfilled with δ equal to a constant times ε. Let

$$E_k(u; x_0) = \sum_{|I| \leq k+3} E(Z^I u; x_0)$$

with E defined as in Proposition 6.6.6; by (6.6.21) this is equivalent to the square of left-hand side of (6.6.31). We shall prove that

$$(6.6.35) \qquad \int (1 + x_0 + |\vec{x}|)^4 |f_I^j(x_0, \vec{x})|^2 d\vec{x} \le C\varepsilon^2 E_k(u; x_0), \quad |I| \le k+3, \, 0 \le j \le 4.$$

To estimate f_I^j we note that no term can contain more than one factor $Z^J u$ with $|J| > k$, for such a factor is only obtained when at least $k-1$ operators Z act on $\partial_j \partial_k u$ or at least k of them act on (u, u'), and this would add up to at least $2k - 1 > k+3$ operators. (This is where we need that $k \ge 5$.) We factor terms vanishing of second or third order using Taylor's formula. In each term we can estimate all factors except one using (6.6.30). For the third order terms f_I^2 and f_I^4 this gives a factor $\le C\varepsilon^2 (1 + x_0 + |\vec{x}|)^{-2}$ in addition to a factor with norm square $O(E_k(u; x_0))$. For the second order terms f_I^1 and f_I^3 we obtain in addition to the factor $\varepsilon/(1 + |x|)$ from (6.6.30) a factor $(1 + |x|)^{-1}$ by using Lemmas 6.6.3 and 6.6.4 as in part i) of the proof. To estimate f_I^0, finally, we write

$$[\square, Z^I] u = \sum_{|J| < |I|} c_{IJ} Z^J \square u$$

and replace $\square u$ by the expression in (6.6.33). The terms then obtained are similar to those already discussed in f_I^j with $j \ne 0$ but they contain one derivative less, which proves (6.6.35).

From (6.6.35) and Proposition 6.6.6 it follows that

$$\partial E_k(u; x_0)/\partial x_0 \le C\varepsilon (1 + x_0)^{-1} E_k(u; x_0),$$

hence that

$$E_k(u; x_0) \le (1 + x_0)^{C\varepsilon} E_k(u; 0).$$

This proves that (6.6.31) follows from (6.6.30) with constants independent of x_0 which completes the proof.

6.7. Global existence theorems by the conformal method. Suppose that in some open subset of Minkowski space we have a C^2 solution of a differential equation

$$(6.7.1) \qquad \square u + f(u, u', u'') = 0$$

where f vanishes of second order at the origin. This means that $u = 0$ is a solution where the linearization is the wave operator in Minkowski space. In Section A.4 of the appendix we have defined a conformal isomorphism Ψ between a bounded part of the Einstein universe $\mathbf{R} \times S^n$ and the Minkowski space $\mathbf{R}^{1+n}$ with the Lorentz metric. (See (A.4.2) and Theorem A.4.1.) On the inverse image in the Einstein universe of the domain of u we define a function $\tilde{u}$ by

$$\tilde{u}(T, X) = \Omega(T, X)^{\frac{1-n}{2}} u(\Psi(T, X)),$$

and obtain using (A.4.3) and (A.3.7), with n replaced by $n + 1$ and $e^\varphi = \cos T + X_0 = \Omega$,

$$(\widetilde{\square} - \widetilde{S}(n-1)/4n)\tilde{u} = \Omega^{-\frac{n+3}{2}} (\square u) \circ \Psi,$$

where $\widetilde{\square}$ is the d'Alembertian on $\mathbf{R} \times S^n$ and $\widetilde{S}$ is the scalar curvature there, that is, $\widetilde{S} = -n(n-1)$ by (A.2.10). Hence (6.7.1) is equivalent to

$$(6.7.2) \qquad (\widetilde{\square} + (n-1)^2/4)\tilde{u} + \Omega^{-\frac{n+3}{2}} f(u, u', u'') \circ \Psi = 0.$$

Here we should interpret $u \circ \Psi$ as $\Omega^{\frac{n-1}{2}} \tilde{u}$ and express the differentiations using (A.4.6), (A.4.6)'. In view of (A.4.7) this means that every component of u' (resp. u'') is the product of $\Omega^{\frac{n-1}{2}}$ and a sum with analytic coefficients of derivatives of $\tilde{u}$ of order ≤ 1 (resp. ≤ 2). By Taylor's formula f can be written as a quadratic form in (u, u', u'') with coefficients which are C^∞ functions of (u, u', u''). Any product $\partial^\alpha u \partial^\beta u$, $0 \leq |\alpha| \leq 2$, $0 \leq |\beta| \leq 2$, gives the product of Ω^{n-1} and a quadratic form with analytic coefficients in the derivatives of $\tilde{u}$. If $n - 1 - (n+3)/2 \geq 0$, that is, $n \geq 5$, and n is odd, it follows that (6.7.1) is equivalent to

$$(6.7.3) \qquad (\tilde{\square} + (n-1)^2/4)\tilde{u} + \tilde{f}(T, X, \tilde{u}, \tilde{u}', \tilde{u}'') = 0,$$

where $\tilde{f}$ is an analytic function of T, X and $U = (\tilde{u}, \tilde{u}', \tilde{u}'')$ for small U, vanishing of second order when $U = 0$.

For $even$ $n \geq 4$ we encounter half integer powers of Ω, and it is easy to see that we get a C^∞ (analytic) function $\tilde{f}$ precisely when f is odd. This is of course far less useful than the case of odd n, so we shall now look in more detail at what happens when $n = 3$. If f vanishes of third order at $(0, 0, 0)$, then $\tilde{f}$ has no singularity for $\Omega = 0$ if $3(n-1)/2 \geq (n+3)/2$, that is, $n \geq 3$. It is therefore sufficient to examine for which quadratic forms f in (u, u', u''),

$$f = \sum_{|\alpha|, |\beta| \leq 2} f_{\alpha\beta} \partial^\alpha_{t,x} u \, \partial^\beta_{t,x} u$$

that the quadratic form $\tilde{f}$ in $(\tilde{u}, \tilde{u}', \tilde{u}'')$ defined by

$$\tilde{f} = \Omega^{-3} \sum_{|\alpha|, |\beta| \leq 2} f_{\alpha\beta} \partial^\alpha_{t,x}(\Omega\tilde{u}) \, \partial^\beta_{t,x}(\Omega\tilde{u})$$

has regular coefficients also when $\Omega = 0$. Expressing the derivatives by (A.4.6), (A.4.6)' and using (A.4.7) we see at once that $G(\tilde{u}, \tilde{u}', \tilde{u}'') = \Omega\tilde{f}$ has analytic coefficients which as functions of T are trigonometric polynomials of degree ≤ 4. Now an analytic function c on $\mathbf{R} \times S^3$ which is a trigonometrical polynomial in T can be written $c = \Omega d$ for some d of the same class if (and only if) $c = 0$ on $\Sigma = \{(T, X); \cos T + X_0 = 0, 0 < T < \pi\}$. In fact, a trigonometrical polynomial is the sum of a polynomial in $\cos T$ and such a polynomial multiplied by $\sin T$, so polynomial division gives

$$c = \Omega d + c_1 \sin T + c_2$$

where c_1 and c_2 are analytic functions on S^3. If $c = 0$ on Σ then

$$c_1(X)^2(\vec{X}, \vec{X}) = c_2(X)^2, X = (X_0, \vec{X}) \in S^n.$$

This remains true for small complex values of $\vec{X}$ when $X_0 = \sqrt{1 - (\vec{X}, \vec{X})}$ which gives a contradiction when $\vec{X}$ is close to a zero $\neq 0$ of $(\vec{X}, \vec{X})$ unless c_1 and c_2 vanish identically on S^3, for $(\vec{X}, \vec{X})$ vanishes simply on a complex hypersurface whereas $c_1(X)^2$ and $c_2(X)^2$ must have zeros of even order or vanish identically.

Thus it is sufficient to determine the conditions for all coefficients of G to vanish at Σ. It follows from (A.4.7) that the restriction of $G(\tilde{u})$ to Σ depends only on the restriction of $\tilde{u}$ to Σ. In particular, we can check this using functions $\tilde{u}$ satisfying the homogeneous conformal d'Alembertian and having Cauchy data vanishing near the pole at infinity, that is, functions $\tilde{u}$ corresponding to solutions u of the wave equation in $\mathbf{R}^4$ with Cauchy data in $C_0^\infty(\mathbf{R}^3)$. It follows from (6.2.15) that apart from a change of variables and a fixed factor the restriction of $\tilde{u}$ to Σ is then the Friedlander radiation field F. Now the radiation field $F(\omega, \varrho)$ can have an arbitrary Taylor expansion of order 2 at a given point, for if $F(\omega, \varrho)$ is a radiation field, then so is $F(\omega, \varrho + \varrho_0)$ for constant ϱ_0, $\partial F(\omega, \varrho)/\partial \varrho$ and $\omega_j \partial F(\omega, \varrho)/\partial \varrho$ for any j. Starting from a radial radiation field this allows us to find another with given Taylor expansion at a chosen point. To check whether $G(\tilde{u}, \tilde{u}', \tilde{u}'')$ vanishes on Σ it is thus sufficient to use such functions. That G vanishes on Σ means that $f(u, u', u'') = o(\Omega^2) = o(t^{-2})$ if $x = (t + \varrho)\omega$ and $t \to \infty$. By Theorem 6.2.1 with $n = 3$

$$\partial_{t,x}^\alpha u(t, x) = t^{-1} \hat{\omega}^\alpha F_{|\alpha|}(\omega, \varrho) + O(t^{-2})$$

where $\hat{\omega} = \varrho' = (-1, \omega)$, and $F_j(\omega, \varrho) = \partial^j F(\omega, \varrho)/\partial \varrho^j$ where F is the Friedlander radiation field of u. Thus we arrive at the condition

$$(6.7.4) \qquad \sum_{|\alpha|,|\beta| \leq 2} f_{\alpha\beta} \hat{\omega}^{\alpha+\beta} F_{|\alpha|}(\omega, \varrho) F_{|\beta|}(\omega, \varrho) = 0.$$

Since $F_{|\alpha|}(\omega, \varrho)$ can be given arbitrary values at a given point and the rays generated by the vectors $\hat{\omega}$ cover the boundary of the light cone, this means that

$$(6.7.4)' \qquad \sum_{|\alpha|=j,|\beta|=k} f_{\alpha\beta}\xi^{\alpha+\beta} = 0, \quad \text{if } \xi_0^2 - \xi_1^2 - \xi_2^2 - \xi_3^2 = 0; \quad 0 \leq j \leq k \leq 2.$$

(We have assumed here that $f_{\alpha\beta} = f_{\beta\alpha}$ which is no restriction. When $j = 0$ this means that u can only occur in a term $c\Box u$, and u is then easily removed from the equation (6.7.1) if we divide by $(1 + cu)$ which we can do for small u.) We sum up the condition encountered in the following definition:

Definition 6.7.1. When $n = 3$ we say that $f(u, u', u'')$ satisfies the *null condition* if f vanishes of second order at $(0, 0, 0)$ and the quadratic part f_2 satisfies the condition

$$(6.7.5) \qquad f_2(a_0, a_1\xi, a_2\xi \otimes \xi) = 0, \quad \text{if } a_j \in \mathbf{R}, \; \xi_0^2 - \xi_1^2 - \xi_2^2 - \xi_3^2 = 0.$$

Here $\xi \otimes \xi = \frac{1}{2}\partial_x^2((x, \xi)^2)$. The condition (6.7.5) is of course just another way of writing (6.7.4), (6.7.4)'. It is an extension of the null condition in Definition 6.6.1, due to Christodoulou [1] and Klainerman [4] who only consider quasilinear equations and required u to be absent in the quadratic terms. As already pointed out, the latter extension is quite trivial though. The following result was proved in Christodoulou [1] in the quasilinear case using the same method as here; an alternative proof using L^2 estimates in Minkowski space due to Klainerman [4] was presented in Section 6.6.

Theorem 6.7.2. *The differential equation (6.7.1) in $\mathbf{R}^{1+n}$ has a global C^∞ solution for arbitrary small Cauchydata in $\mathcal{S}(\mathbf{R}^n)$ if $f \in C^\infty$ vanishes of second order at $(0, 0, 0)$ and n is odd and > 3 or $n = 3$ and f satisfies the null condition. For the solution we have*

$$|u(t, x)| \leq C\left((1 + (|x| - t)^2)(1 + (|x| + t)^2)\right)^{\frac{1-n}{2}}$$

Proof. The Cauchy problem is equivalent to the Cauchy problem for the equation (6.7.2) where the non-linar term is analytic, depending on (T, X) also, and the Cauchy data are small in $C^\infty(S^n)$. The existence of a solution for $0 \leq T \leq \pi$ follows from Theorem 6.4.11 and Remarks 3 and 4 after its proof.

Theorem 6.4.11 yields a much more precise existence theorem. Let the Cauchy data be

$$(6.7.6) \qquad u = u_0, \; \partial_t u = u_1 \quad \text{when } t = 0.$$

For the equation (6.7.3) we then have the Cauchy data

$$(6.7.6)' \qquad\qquad \tilde{u} = \tilde{u}_0, \ \partial_T \tilde{u} = \tilde{u}_1 \quad \text{when } T = 0,$$

where

$$\tilde{u}_0(X) = \Omega(0, X)^{\frac{1-n}{2}} u_0(\Psi(0, X)), \ \tilde{u}_1(X) = \tfrac{1}{2}\Omega(0, X)^{\frac{-1-n}{2}} u_1(\Psi(0, X)), \ X \in S^n.$$

Recall that derivatives in S^n for $T = 0$ can be expressed as linear combinations of products of the operators

$$Z_{jk} = x_j \partial/\partial x_k - x_k \partial/\partial x_j; \ j, k = 1, \dots, n;$$
$$L_k = \tfrac{1}{2}(1 - |x|^2)\partial/\partial x_k + x_k \langle x, \partial/\partial x \rangle.$$

(The operators L_k occur in the right-hand side of (A.4.10).) We have

$$\Omega(0, X)^{\frac{1-n}{2}} = 2^{\frac{1-n}{2}}(1 + |x|^2)^{\frac{n-1}{2}},$$

which is annihilated by all Z_{jk} whereas application of L_k to this factor is equivalent to multiplication by $(n - 1)x_k/2$ since

$$L_k(1 + |x|^2) = (1 - |x|^2)x_k + x_k 2|x|^2 = (1 + |x|^2)x_k.$$

A product of κ operators Z_{jk} and L_i is of the form

$$\sum_{1 \le |\alpha| \le \kappa} a_\alpha(x)\partial^\alpha,$$

where a_α is a polynomial of degree $\le \kappa + |\alpha|$. This follows by induction over κ. If we now recall that the spherical surface measure is $2^n(1 + |x|^2)^{-n}\, dx$ by (A.4.11), which almost compensates for the factor $\Omega(0, X)^{1-n}$, we find that if $u_0 \in H^{\text{loc}}_{(s+1)}(\mathbf{R}^n)$ and $u_1 \in H^{\text{loc}}_{(s)}(\mathbf{R}^n)$ where s is a positive integer, then

$$(6.7.7) \qquad \|\tilde{u}_0\|^2_{(s+1)} \le C \sum_{|\alpha| \le s+1} \int |\partial^\alpha u_0(x)|^2 (1 + |x|^2)^{s+|\alpha|}\, dx,$$

$$(6.7.8) \qquad \|\tilde{u}_1\|^2_{(s)} \le C \sum_{|\alpha| \le s} \int |\partial^\alpha u_1(x)|^2 (1 + |x|^2)^{s+|\alpha|+1}\, dx.$$

In the quasilinear case it follows from Theorem 6.4.11 and Remark 3 after its proof that (6.7.1) has a global solution satisfying (6.7.6) provided that the right-hand sides of (6.7.7) and (6.7.8) are well defined and small enough when $s = (n + 3)/2$. This is a result of Christodoulou [1]. For the fully non-linear case we obtain the same result with $s = (n+5)/2$ by Remark 4 there. Note that the finiteness of these norms requires that

$$|u_j(x)| \le C(1 + |x|)^{-s-j-n/2}, \quad j = 0, 1, \ x \in \mathbf{R}^n.$$

CHAPTER VII

NONLINEAR PERTURBATIONS OF THE
KLEIN-GORDON EQUATION

7.1. Introduction. From the work of Klainerman [5] and Shatah [1] it is known that for nonlinear perturbations of the Klein-Gordon equation in $\mathbf{R}^{1+n}$,

$$(7.1.1) \qquad \Box u + u = F(u, u', u''),$$

where $\Box = \partial_0^2 - \partial_1^2 - \cdots - \partial_n^2$, F vanishes of second order at 0, and F is linear in u'', the Cauchy problem with small data in C_0^∞ has a global solution if $n \geq 3$.

The main purpose of this chapter is to prove this result and to discuss briefly the cases $n = 1, 2$. We shall begin by studying in Section 7.2 the solutions of the unperturbed Klein-Gordon equation $\Box u + u = 0$ in considerable detail for arbitrary n. This covers the estimates of von Wahl [1] and gives in addition a much more precise description of the asymptotic properties to serve as a guide in the study of (7.1.1). In Section 7.3 we discuss L^2 estimates for the inhomogeneous linear Klein-Gordon equation in the spirit of Klainerman [5]. His estimates for the case $n = 3$ were not sharp but sufficient to establish global existence theorems then. Their analogue for $n = 1$ or $n = 2$ would not give a good estimate for the lifespan of the solutions. We shall therefore reexamine the estimates of Klainerman [5] for arbitrary dimension, but some of our estimates may not be optimal when $n > 3$. Using these bounds we outline in Section 7.4 how existence theorems for (7.1.1) follow for $n = 1, 2$ when F vanishes of second or of third order at 0 and the Cauchy data have compact support. In Section 7.5 we pause to discuss the case $n = 0$, that is, the ordinary differential equation

$$u'' + u = F(u, u').$$

In that case one obtains precise estimates for the lifespan by means of an appropriate modification of the definition of the energy. In the original version of these notes the results for $n = 0$ were combined with an approximation of the Klein-Gordon operator by its radial part in terms of hyperbolic polar coordinates to suggest a much longer lifespan for the solutions than guaranteed by the results of Section 7.4, at least if F is independent of u''. The suggested improvements have now been proved by Simon and Taflin [1], Ozawa, Tsutaya and Tsutsumi [1] and Moriyama, Tonegawa and Tsutsumi [1]. The heuristic arguments in Section 7.5 have therefore been omitted and replaced by a presentation in a new Section 7.8 of the method of Shatah [1], which is essential in the papers just quoted. In Sections 7.6 and 7.7, finally, we discuss a simpler approach which works well when $n \geq 2$ and the Cauchy data have compact support.

7.2. Asymptotic behavior of solutions of the Klein-Gordon equation. In this section we shall mainly discuss the solution of the Cauchy problem

$$(7.2.1) \qquad \Box u + u = 0; \quad u(0, x) = u_0(x), \ \partial_t u(0, x) = u_1(x),$$

where u is a function of $(t, x) \in \mathbf{R}^{1+n}$, $\Box = \partial_t^2 - \Delta$, and $u_j \in \mathcal{S}(\mathbf{R}^n)$. It is immediately obtained by Fourier transformation in the x variables, which gives

$$\hat{u}(t, \xi) = \tfrac{1}{2}(\hat{u}_0(\xi) - i\hat{u}_1(\xi)/\langle\xi\rangle)e^{it\langle\xi\rangle} + \tfrac{1}{2}(\hat{u}_0(\xi) + i\hat{u}_1(\xi)/\langle\xi\rangle)e^{-it\langle\xi\rangle}.$$

Here $\langle\xi\rangle = (1 + |\xi|^2)^{\frac{1}{2}}$. This gives a splitting $u = u_+ + u_-$ where

$$(7.2.2) \qquad \partial_t u_\pm = \pm i\langle D_x\rangle u_\pm, \quad u_\pm(0, x) = \varphi_\pm, \quad \varphi_\pm = (u_0 \mp i\langle D_x\rangle^{-1} u_1)/2.$$

The splitting is Lorentz invariant, for the spectrum of $u_\pm$ as a function in $\mathbf{R}^{1+n}$ is contained in $\{(\pm\langle\xi\rangle, \xi), \xi \in \mathbf{R}^n\}$, and these hyperboloids are disjoint. More explicitly, the Fourier transform of $u_\pm$ in all variables is $2\pi\delta(\tau \mp \langle\xi\rangle)\hat{\varphi}_\pm(\xi)$. Since $\varphi_\pm \in \mathcal{S}(\mathbf{R}^n)$, we can as well study the problem

$$(7.2.2)_+ \qquad\qquad \partial_t u = i\langle D_x\rangle u, \quad u(0, x) = \varphi \in \mathcal{S}(\mathbf{R}^n).$$

The solution of $(7.2.2)_+$ is

$$(7.2.3) \qquad\qquad u(t, x) = (2\pi)^{-n} \int e^{i(t\langle\xi\rangle + \langle x, \xi\rangle)} \hat{\varphi}(\xi)\, d\xi.$$

A *formal* application of the method of stationary phase (see e.g. Hörmander [4, Section 7.7]) suggests that one should look for a point ξ where

$$t\xi/\langle\xi\rangle + x = 0,$$

that is,

$$|\xi|^2 = |x|^2/(t^2 - |x|^2), \quad \langle\xi\rangle^2 = t^2/(t^2 - |x|^2).$$

There is a unique solution $\xi = -x\,\mathrm{sgn}\,t/\sqrt{t^2 - |x|^2}$ when $|x| < |t|$ but no real solution otherwise. If $|x| < t$ the critical value of the phase is

$$t\langle\xi\rangle + \langle x, \xi\rangle = t(\langle\xi\rangle - |\xi|^2/\langle\xi\rangle) = t/\langle\xi\rangle = (t^2 - |x|^2)^{\frac{1}{2}},$$

and the Hessian matrix is $t(\delta_{jk}/\langle\xi\rangle - \xi_j\xi_k/\langle\xi\rangle^3)$. The determinant is $t^n\langle\xi\rangle^{-2-n}$ as is immediately seen when $\xi_2 = \cdots = \xi_n = 0$, so we expect that the main contribution to $u(t, x)$ must be

$$e^{i\sqrt{t^2 - |x|^2}}(2\pi)^{-n/2}(\langle\xi\rangle^{n+2}t^{-n})^{\frac{1}{2}}e^{i\pi n/4}\hat{\varphi}(\xi).$$

This suggests that $u(t, x)e^{-i\sqrt{t^2 - |x|^2}}$ behaves when $t > 0$ as a symbol of order $-n/2$ which vanishes for $|x| > t$, hence can be estimated by $(1 + t + |x|)^{-n/2}$ times any power of $(1 + |t - |x||)/(1 + t + |x|)$. We shall prove that this is true apart from an additional term which lies in the Schwartz space $\mathcal{S}(\mathbf{R}^{1+n})$. In order to be able to deduce some estimates for Cauchy data of finite smoothness we shall first prove a weaker result when the Cauchy data just have Fourier transforms in a suitable symbol space.

Theorem 7.2.1. *Assume that the Fourier transform of φ is a C^∞ function $\hat{\varphi}$ such that for every multi-index α*

$$(7.2.4) \qquad\qquad |D^\alpha\hat{\varphi}(\xi)| \le C_\alpha(1 + |\xi|)^{N-|\alpha|}, \quad \xi \in \mathbf{R}^n.$$

If $N < -(n+1)/2$ and u is the solution of $(7.2.2)_+$, it follows that for $|t| + |x| \ge 1$

$$(7.2.5) \qquad |u(t, x)| \le C(|t| + |x|)^{N+1}(1 + (t^2 - |x|^2)_+)^{M_+/2}(1 + (|x|^2 - t^2)_+)^{M_-},$$

where M_- is arbitrary and $M_+ = \max(0, -\frac{n}{2} - N - 1)$; the constant C depends on M_-.

Proof. It suffices to prove that $(7.2.5)$ holds when $\hat{\varphi} \in C_0^\infty$, with a constant depending only on those in $(7.2.4)$, for then an approximation argument can be used to give $(7.2.5)$

for any φ satisfying (7.2.4). First assume that $x = (x_1, 0, \ldots, 0)$ and that $x_1 > 2|t|$, hence $x_1 > \frac{2}{3}$. The derivative of the phase function in (7.2.3) with respect to ξ_1 is

$$f = t\xi_1/\langle\xi\rangle + x_1 \geq x_1/2.$$

Integration by parts gives for any positive integer j

$$(7.2.6) \qquad u(t, x) = (2\pi)^{-n} \int e^{i(t\langle\xi\rangle + \langle x, \xi\rangle)} \left(\frac{\partial}{\partial\xi_1}\frac{i}{f}\right)^j \hat\varphi(\xi)\,d\xi.$$

For $k \neq 0$ we have

$$|\partial_1^k f(\xi)| \leq C_k'|t|\langle\xi\rangle^{-k} \leq \tfrac{1}{2}C_k' x_1 \langle\xi\rangle^{-k} \leq C_k'\langle\xi\rangle^{-k} f(\xi),$$

so we obtain the bound

$$|u(t, x)| \leq C_j'' x_1^{-j} \int \langle\xi\rangle^{N-j}\,d\xi$$

for any j. The integral converges when $j > N + n$, and if $j + 2M_- + N + 1 \geq 0$, the estimate (7.2.5) follows.

The proof is much harder when $|x| \leq 2|t|$, hence $|t| \geq \frac{1}{3}$. It is no restriction to assume that $t > 0$ for otherwise we can consider the complex conjugate of u instead. To start with we shall assume that $n = 1$, $N < -1$. (A reduction of the higher dimensional case to this will be given afterwards.) For the time being we therefore write x instead of x_1.

Let us first consider the case where $t < x < 2t$, for then the phase in (7.2.3) has no critical point. For the derivative f of the phase we obtain

$$f(\xi) = x - t + t(\xi/\langle\xi\rangle + 1) = x - t + t/(\langle\xi\rangle(\langle\xi\rangle - \xi)) \geq x - t + t/2\langle\xi\rangle^2 \geq t/2\langle\xi\rangle^2.$$

If we note that

$$df(\xi)/d\xi = t(1/\langle\xi\rangle - \xi^2/\langle\xi\rangle^3) = t/\langle\xi\rangle^3,$$

it follows that

$$|f^{(j)}(\xi)| \leq C_j t\langle\xi\rangle^{-2-j} \leq 2C_j f(\xi)\langle\xi\rangle^{-j}, \quad j = 1, 2, \ldots.$$

Hence we obtain as above for any j (with new constants C_j)

$$|u(t, x)| \leq C_j \int \langle\xi\rangle^{N-j} f(\xi)^{-j}\,d\xi.$$

Note that

$$\langle\xi\rangle^2 f(\xi)^2 \geq (x - t)t \geq (x^2 - t^2)/3.$$

Replacing j by $j + 2M_-$ we obtain for any integers $j \geq 0$ and $M_- \geq 0$

$$(x^2 - t^2)^{M_-}|u(t, x)| \leq C_{j,M_-} \int (t/\langle\xi\rangle^2)^{-j} \langle\xi\rangle^{N-j}\,d\xi.$$

Suppose first that $\hat\varphi(\xi) = 0$ for $|\xi| > 2t$. When $N + j \geq 0$ the integral can then be estimated by a constant times $t^{N+j+1}t^{-j} = t^{N+1}$, and (7.2.5) follows. On the other hand, if $\hat\varphi(\xi) = 0$ for $|\xi| < t$ then we obtain with $j = 0$

$$(x^2 - t^2)^{M_-}|u(t, x)| \leq C_{M_-} \int_{|\xi|>t} \langle\xi\rangle^N\,d\xi \leq Ct^{N+1},$$

for $N + 1 < 0$ by hypothesis. Now we can always split $\hat{\varphi}$ as a sum $\hat{\varphi}_1 + \hat{\varphi}_2$ where $\hat{\varphi}_1$, $\hat{\varphi}_2$ satisfy (7.2.4) for some other constants and have the preceding properties. In fact, if $\chi \in C_0^\infty(-1/2, 1/2)$ is equal to 1 in $(-1/3, 1/3)$, we can take $\hat{\varphi}_1(\xi) = \hat{\varphi}(\xi)\chi(\xi/4t)$. Hence we have proved (7.2.5) when $x > t$. (Note that we have split the integral into one oscillating part where partial integration pays and one nonoscillating part where it is useless or even counterproductive. It is easy to see that the estimates obtained are optimal.)

Now assume that $|x| < t$. Recall that the phase in (7.2.3) has a critical point η,

$$t\eta/\langle\eta\rangle + x = 0, \quad \langle\eta\rangle^2 = t^2/(t^2 - x^2), \quad \eta = -x(t^2 - x^2)^{-\frac{1}{2}},$$

and that the critical value is $(t^2 - x^2)^{\frac{1}{2}}$. Set $u = v + w$ where

$$v(t, x) = (2\pi)^{-1} \int_{-\infty}^{\infty} e^{i(t\langle\xi\rangle + x\xi)} \chi((\xi - \eta)/\langle\eta\rangle)\hat{\varphi}(\xi)d\xi$$

contains the contributions near the critical point. Writing

$$\Psi(\xi) = (1 - \chi((\xi - \eta)/\langle\eta\rangle))\hat{\varphi}(\xi)$$

we shall first estimate $w(t, x)$. In the support of Ψ we have $|\xi - \eta| > \langle\eta\rangle/3$, and $|\xi - \eta| < \langle\eta\rangle/2$, hence $\langle\xi\rangle \le 3\langle\eta\rangle/2$, in the support of $\chi'((\xi - \eta)/\langle\eta\rangle)$. This shows that Ψ inherits from $\hat{\varphi}$ estimates of the form (7.2.4) with new constants independent of η.

Let

$$f(\xi) = \frac{\partial}{\partial\xi}(t\langle\xi\rangle + x\xi) = t(\xi/\langle\xi\rangle - \eta/\langle\eta\rangle).$$

It is clear that $|f(\xi)/t| \ge 1/\sqrt{37}$ when $|\xi - \eta| > \langle\eta\rangle/3$ and $\xi\eta < 0$, for either $|\xi| > 1/6$ or else $|\eta| > 1/6$. When $\xi\eta \ge 0$ we write

$$f(\xi) = t(\xi^2\langle\eta\rangle^2 - \eta^2\langle\xi\rangle^2)/(\langle\xi\rangle\langle\eta\rangle(\xi\langle\eta\rangle + \eta\langle\xi\rangle)).$$

The numerator is equal to $(\xi - \eta)(\xi + \eta)$. In supp Ψ we have $\langle\eta\rangle < 3|\xi - \eta|$, hence $\langle\xi\rangle + \langle\eta\rangle < 2\langle\eta\rangle + |\xi - \eta| < 7|\xi - \eta| < 7|\xi + \eta|$, and $|\xi\langle\eta\rangle + \eta\langle\xi\rangle| < 2\langle\xi\rangle\langle\eta\rangle$. Thus

$$98|f(\xi)| \ge t(\langle\xi\rangle + \langle\eta\rangle)^2/(\langle\xi\rangle\langle\eta\rangle)^2 = t(1/\langle\xi\rangle + 1/\langle\eta\rangle)^2 \ge t/\langle\xi\rangle^2.$$

As before we obtain

$$|f^{(j)}(\xi)| \le C_j t\langle\xi\rangle^{-j-2} \le 98C_j f(\xi)\langle\xi\rangle^{-j}, \quad \xi \in \text{supp } \Psi,$$

and we can now conclude that $|w(t, x)| \le Ct^{N+1}$, just as when $t < x < 2t$.

What remains is to estimate the contribution v near the critical point. We shall write

$$V(t, x) = 2\pi v(t, x)e^{-i\sqrt{t^2 - x^2}} = \int_{-\infty}^{\infty} e^{it(\langle\xi\rangle - \xi\eta/\langle\eta\rangle - 1/\langle\eta\rangle)}\Phi(\xi)d\xi,$$

where $\Phi(\xi) = \chi((\xi - \eta)/\langle\eta\rangle)\hat{\varphi}(\xi)$. Since $|\xi - \eta| < \langle\eta\rangle/2$, hence $3\langle\eta\rangle/2 \ge \langle\xi\rangle \ge \langle\eta\rangle/2$, in supp Φ, we obtain using (7.2.4), with new C_j independent of η,

$$|\Phi^{(j)}(\xi)| \le C_j\langle\eta\rangle^{N-j}.$$

The phase function is

$$(\langle\xi\rangle\langle\eta\rangle - \eta\xi - 1)/\langle\eta\rangle = (\xi - \eta)^2/((\langle\xi\rangle\langle\eta\rangle + \eta\xi + 1)\langle\eta\rangle).$$

If $|\eta| > \langle\eta\rangle/2$, that is, if $3\eta^2 > 1$, then $\eta\xi \geq 0$ in supp Φ. We shall assume that $3\eta^2 > 1$ and leave for the reader the easy case where $\eta^2 \leq 1/3$. For

$$s = (\xi - \eta)(\langle\xi\rangle\langle\eta\rangle + \eta\xi + 1)^{-\frac{1}{2}}$$

we then have the bound $s^2 \leq 2(\xi - \eta)^2/\langle\eta\rangle^2 \leq 1/2$. Since

$$2sds/d\xi = f(\xi)\langle\eta\rangle/t = (\xi^2 - \eta^2)\langle\eta\rangle/(\langle\xi\rangle\langle\eta\rangle(\xi\langle\eta\rangle + \eta\langle\xi\rangle)),$$

we have

$$2ds/d\xi = (\xi + \eta)(\langle\xi\rangle\langle\eta\rangle + \eta\xi + 1)^{\frac{1}{2}}\langle\eta\rangle/(\langle\xi\rangle\langle\eta\rangle(\xi\langle\eta\rangle + \eta\langle\xi\rangle)))$$

which is bounded above and below by positive constants times $\langle\eta\rangle^{-1}$. Moreover,

$$|d^j s/d\xi^j| \leq C_j\langle\eta\rangle^{-j} \quad \text{in supp } \Phi.$$

With $\Xi = (\xi - \eta)/\langle\eta\rangle$ this means that we have uniform bounds for s and all its derivatives with respect to Ξ; the first derivative is also bounded away from 0. This implies bounds also for the derivatives of the inverse function $\Xi(s)$. From our uniform bounds

$$|d^j \Phi(\eta + \langle\eta\rangle\Xi)/d\Xi^j| \leq C_j\langle\eta\rangle^N,$$

it follows that we have such bounds also for $\Phi(\xi(s))d\xi(s)/ds$, with N replaced by $N + 1$, for $\xi(s) = \eta + \langle\eta\rangle\Xi(s)$ and $d\xi(s)/ds = \langle\eta\rangle d\Xi(s)/ds$. Since

$$V(t, x) = \int_{-\infty}^{\infty} e^{i(t/\langle\eta\rangle)s^2} \Phi(\xi(s))d\xi(s)/ds\, ds,$$

we are now ready to use standard estimates for Gaussian integrals. Note that the frequency $t/\langle\eta\rangle$ is $(t^2 - x^2)^{\frac{1}{2}}$. It is obvious from the expression for V in the original variables that $|V(t, x)| \leq C\langle\eta\rangle^{N+1}$, hence $|V(t, x)| \leq Ct^{N+1}$ if $\langle\eta\rangle > t$, since $N + 1 < 0$. When $t > \langle\eta\rangle$ we obtain using for example Hörmander [4, Lemma 7.7.3]

$$|V(t, x)| \leq C(\langle\eta\rangle/t)^{\frac{1}{2}}\langle\eta\rangle^{N+1}.$$

Hence

$$|u(t, x)| \leq |V(t, x)| + |w(t, x)| \leq Ct^{N+1}(1 + (\langle\eta\rangle/t)^{N+\frac{3}{2}}), \quad |x| < t, \ \langle\eta\rangle < t.$$

Since $M_+ = \max(0, -N - \frac{3}{2})$ we get

$$|u(t, x)| \leq 2Ct^{N+1}(\langle\eta\rangle/t)^{-M_+} = 2Ct^{N+1}(t^2 - x^2)^{M_+/2},$$

which completes the proof of Theorem 7.2.1 when $n = 1$. For later reference we point out that we can of course get an asymptotic series for V with terms $O(\langle\eta\rangle^{N+1}(\langle\eta\rangle/t)^{\frac{1}{2}+j})$, which is useful when N is a large negative number so that $\langle\eta\rangle^{N+1}$ can absorb a high power of $\langle\eta\rangle$.

Assuming still that $x_2 = \cdots = x_n = 0$, we shall complete the proof of Theorem 7.2.1 by the following proposition which reduces the proof to the one dimensional case.

Proposition 7.2.2. Let Φ be a symbol of order N in $\mathbf{R}^n$, that is, assume that for every multi-index α

(7.2.7)
$$|D^\alpha \Phi(\xi)| \leq C_\alpha \langle \xi \rangle^{N-|\alpha|}, \quad \xi \in \mathbf{R}^n.$$

Write $\xi' = (\xi_2, \ldots, \xi_n)$. Then

(7.2.8)
$$\Phi_t(\xi_1) = t^{(n-1)/2} e^{-it\langle \xi_1 \rangle} \int e^{it\langle \xi \rangle} \Phi(\xi) d\xi'$$

is uniformly bounded as a symbol of order $N + (n-1)/2$ in $\mathbf{R}$ when $t > 1$. One can find symbols a_j of order $N + (n-1)/2 - j$ in $\mathbf{R}$ such that for every positive integer J

$$t^J \left(\Phi_t(\xi_1) - \sum_{j<J} a_j(\xi_1) t^{-j} \right)$$

is uniformly bounded as a symbol of order $N + (n-1)/2 - J$ when $t > 1$. The bounds implied by these statements depend only on the constants C_α in (7.2.7).

Proof. The total phase in (7.2.8) is equal to t times

$$\langle \xi \rangle - \langle \xi_1 \rangle = |\xi'|^2 / (\langle \xi \rangle + \langle \xi_1 \rangle).$$

To write (7.2.8) as a Gaussian integral we introduce

$$\theta = \xi'((\langle \xi \rangle + \langle \xi_1 \rangle)\langle \xi_1 \rangle)^{-\frac{1}{2}}.$$

At first we assume that $|\xi'| < n\langle \xi_1 \rangle$ in supp Φ, which implies that $|\theta| < n$ there. Since

$$|\theta|^2 \langle \xi_1 \rangle (\langle \xi \rangle + \langle \xi_1 \rangle) = |\xi'|^2 = \langle \xi \rangle^2 - \langle \xi_1 \rangle^2,$$

cancellation of a factor $\langle \xi \rangle + \langle \xi_1 \rangle$ gives

$$|\theta|^2 \langle \xi_1 \rangle = \langle \xi \rangle - \langle \xi_1 \rangle, \quad (2 + |\theta|^2)\langle \xi_1 \rangle = \langle \xi \rangle + \langle \xi_1 \rangle.$$

Hence

$$\xi' = \langle \xi_1 \rangle \varrho(\theta), \quad \varrho(\theta) = \theta(2 + |\theta|^2)^{\frac{1}{2}},$$

and since $\det \partial \varrho / \partial \theta = (2 + 2|\theta|^2)(2 + |\theta|^2)^{\frac{n-3}{2}}$ it follows that

$$\Phi_t(\xi_1) = t^{\frac{n-1}{2}} \int e^{it\langle \xi_1 \rangle |\theta|^2} \Phi(\xi_1, \langle \xi_1 \rangle \varrho(\theta)) \langle \xi_1 \rangle^{n-1} (2 + 2|\theta|^2)(2 + |\theta|^2)^{\frac{n-3}{2}} d\theta.$$

All θ derivatives of $\Phi(\xi_1, \langle \xi_1 \rangle \varrho(\theta)) / \langle \xi_1 \rangle^N$ are uniformly bounded by (7.2.7), and $|\theta| < n$ in the support of the integrand. Thus Lemma 7.7.3 in Hörmander [4] gives

$$\left| \Phi_t(\xi_1) - \sum_{j<J} a_j(\xi_1) t^{-j} / j! \right| \leq C_J (t\langle \xi_1 \rangle)^{-J} \langle \xi_1 \rangle^{N+(n-1)/2},$$

where

$$a_j(\xi_1) = (\langle \xi_1 \rangle \pi)^{\frac{n-1}{2}} e^{\frac{(n-1)\pi i}{4}} (4i\langle \xi_1 \rangle)^{-j} (-\Delta_\theta)^j \Phi(\xi_1, \langle \xi_1 \rangle \varrho(\theta))(2 + 2|\theta|^2)(2 + |\theta|^2)^{\frac{n-3}{2}},$$

evaluated for $\theta = 0$. This is a symbol of order $N + (n-1)/2 - j$ so Proposition 7.2.2 is an immediate consequence of Hörmander [4, Proposition 18.1.4].

Next assume that $\langle \xi \rangle < n|\xi_j|$ in supp Φ for some $j \neq 1$. Then Φ_t can be interpreted as an oscillatory integral. As at the beginning of the proof of Theorem 7.2.1 we obtain for any integer $k > 0$

$$\Phi_t(\xi_1) = t^{(n-1)/2-k} e^{-it\langle \xi_1 \rangle} \int e^{it\langle \xi \rangle} \left(\frac{\partial}{\partial \xi_j} \frac{i\langle \xi \rangle}{\xi_j} \right)^k \Phi(\xi) d\xi'.$$

The integrand can be estimated by a constant times $\langle \xi \rangle^{N-k}$. When $N - k < 1 - n$ it follows that

$$|\Phi_t(\xi_1)| \leq C_k t^{(n-1)/2-k} \langle \xi_1 \rangle^{N-k+n-1}$$

which implies that Φ_t is of order $-\infty$. The proof is now complete since we can split any Φ into a sum of n terms satisfying one of the preceding assumptions and still satisfying (7.2.7) for some other constants.

Corollary 7.2.3. *For any integer $\nu \geq n/2$ the forward fundamental solution E of the Klein-Gordon equation can be written in the form*

$$(7.2.9) \qquad E = \sum_{|\alpha| \leq \nu} D_x^\alpha E_\alpha$$

where for any M_-

$$(7.2.10) \qquad |E_\alpha(t,x)| \leq C(|t| + |x|)^{-\nu}(1 + (t^2 - |x|^2)_+)^{(2\nu-n)/4}(1 + (|x|^2 - t^2)_+)^{M_-},$$

if $|t| + |x| \geq 1$. We can also write

$$\partial_t E = \sum_{|\alpha| \leq \nu+1} D_x^\alpha E_\alpha$$

for some other E_α satisfying (7.2.10).

Proof. E is the inverse Fourier transform with respect to ξ of

$$\langle\xi\rangle^{-1}\sin(t\langle\xi\rangle) = \langle\xi\rangle^{-1-2\nu}(1 + |\xi|^2)^\nu \sin(t\langle\xi\rangle) = \sum_{|\alpha|\leq\nu} c_\alpha \langle\xi\rangle^{-1-2\nu}\xi^\alpha \sin(t\langle\xi\rangle)\xi^\alpha,$$

where c_α are constants. Let

$$\widehat{E}_\alpha(t,\xi) = c_\alpha\langle\xi\rangle^{-1-2\nu}\xi^\alpha \sin(t\langle\xi\rangle) = c_\alpha\langle\xi\rangle^{-1-2\nu}\xi^\alpha(e^{it\langle\xi\rangle} - e^{-it\langle\xi\rangle})/2i, \quad |\alpha| \leq \nu.$$

Then the hypotheses of Theorem 7.2.1 are fulfilled for each term with $N = -1 - \nu$, and $M_+ = \nu - \frac{n}{2}$. This proves (7.2.10). The decomposition of $\partial_t E$ is obtained in the same way with ν replaced by $\nu + 1$ initially. The proof is complete.

Corollary 7.2.3 contains the estimates of von Wahl [1], which we shall only state in a special case.

Corollary 7.2.4. *For the solution of (7.2.1) with $u_j \in S(\mathbf{R}^n)$ we have the estimates*

$$(7.2.11) \qquad |u(t,x)| \leq C|t|^{-n/2} \sum_{|\alpha|+j\leq(n+3)/2} \int |\partial^\alpha u_j(y)|\, dy, \quad |t| \geq 1,$$

$$(7.2.12) \qquad |u(t,x)| \leq C \sum_{|\alpha|+j\leq n} \int |\partial^\alpha u_j(y)|\, dy, \quad |t| < 1.$$

Proof. Since

$$u(t,\cdot) = E(t,\cdot) * u_1 + \partial_t E(t,\cdot) * u_0$$

we obtain (7.2.11) from Corollary 7.2.3 with ν equal to the smallest integer $\geq n/2$, that is, the integer part of $(n+1)/2$. It is of course not essential that the lower bound for $|t|$ in (7.2.11) is 1, but (7.2.11) cannot hold for $0 < t < 1$ if $n > 3$ and $|t|$ is replaced by 1, for $\sup|u_0|$ cannot be estimated by means of the L^1 norms of the derivatives of order $< n$. Using Fourier transforms as in the proof of Corollary 7.2.3 one obtains (7.2.12) immediately with n replaced by $n+1$, but for the sake of completeness we shall prove the optimal estimate (7.2.12). To do so we denote by E_0 the fundamental solution of the wave operator and set $E_{j+1} = E_j * E_0$ for $j \geq 0$. Thus E_j is a constant times $\chi_+^{(1-n)/2+j}(t^2-|x|^2)$, hence homogeneous of degree $2j+1-n$ and continuous if $2j+1-n > 0$. For any positive integer k we have

$$E = E_0 - E_1 + \cdots + (-1)^k E_k - E * (-1)^k E_k$$

for $\square + 1$ applied to the right hand side is equal to δ. Now the proof of Proposition 3.1' in Hörmander [3] gives

$$|E_j(t, \cdot) * u_1| \leq C_j t^{2j} \sum_{|\alpha|=n-1} \int |D^\alpha u_1| \, dx.$$

E is a sum of derivatives of order $\leq n$ of locally bounded functions with support in the light cone, so $E * E_k$ is locally bounded if k is large enough. This proves that

$$|E(t, \cdot) * u_1| \leq C \sum_{|\alpha|=n-1} \int |D^\alpha u_1(x)| \, dx, \quad |t| \leq 1,$$

and an estimate for $\partial_t E(t, \cdot) * u_0$ is obtained in the same way. The proof is complete.

We shall finally prove the results on the asymptotic behavior of the solution of (7.2.1) when $u_j \in \mathcal{S}(\mathbf{R}^n)$ motivated at the beginning of the section.

Theorem 7.2.5. *If $\varphi \in \mathcal{S}(\mathbf{R}^n)$ then the solution of $(7.2.2)_+$ can be written in the form*

(7.2.13)
$$u(t, x) = U_0(t, x) + U_+(t, x) e^{i\varrho}$$

where $U_0 \in \mathcal{S}(\mathbf{R}^{1+n})$, $\varrho = \operatorname{sgn} t \sqrt{t^2 - |x|^2} = t\sqrt{1 - |x|^2/t^2}$ and U_+ is a polyhomogeneous symbol of order $-n/2$ with support in the full double light cone,

$$U_+(t, x) \sim (+0 - i\varrho)^{-n/2} \sum_0^\infty \varrho^{-j} w_j(t, x)$$

where $w_j(t, x) = w_j(1, x/t)$. The leading term is given by

$$w_0(t, x) = \begin{cases} (2\pi)^{-n/2}(t/\varrho)\hat{\varphi}(-x/\varrho), & \text{if } t^2 > |x|^2, \\ 0, & \text{if } t^2 \leq |x|^2. \end{cases}$$

Proof. In the proof of Theorem 7.2.1 all the terms encountered were rapidly decreasing except the contributions from the critical point. Let us recall how they looked, first when $x_2 = \cdots = x_n = 0$. From Proposition 7.2.2 with $\Phi = (2\pi)^{1-n}\hat{\varphi}$ we get for $\Phi_t(\xi_1)$ an asymptotic expansion when $t > 0$ with the terms

$$(2\pi)^{1-n}(\langle\xi_1\rangle\pi)^{\frac{n-1}{2}} e^{\frac{(n-1)\pi i}{4}}(4it\langle\xi_1\rangle)^{-j}(-\Delta_\theta)^j \hat{\varphi}(\xi_1, \langle\xi_1\rangle\varrho(\theta))(2 + 2|\theta|^2)(2 + |\theta|^2)^{\frac{n-3}{2}}/j!,$$

$j = 0, 1, \ldots,$ evaluated for $\theta = 0$. With the notation in the proof of Theorem 7.2.1 we then obtain

$$V(t, x) \sim t^{-(n-1)/2}(\langle\eta\rangle\pi/t)^{\frac{1}{2}} e^{i\pi/4} \sum_{j=0}^\infty (4it/\langle\eta\rangle)^{-j}(id/ds)^{2j}\Phi_t(\xi(s)) d\xi(s)/ds/j!,$$

evaluated for $s = 0$ where $\xi(0) = \eta$, $\xi'(0) = \langle\eta\rangle\sqrt{2}$. Altogether we obtain for general x with $|x| < t$

$$(7.2.14) \qquad V(t, x)/2\pi \sim (2\pi)^{-\frac{n}{2}}\langle\eta\rangle^{\frac{n}{2}+1}t^{-\frac{n}{2}}e^{i\pi n/4}\sum_{0}^{\infty} A_j(t, x)t^{-j},$$

where $A_0(t, x) = \hat{\varphi}(\eta)$, $\eta = -x/\sqrt{t^2 - |x|^2}$, and $A_j(t, x)$ is a finite sum of derivatives of $\hat{\varphi}$ at η of order $|\alpha| \leq 2j$ multiplied by symbols of η of order $\leq |\alpha| - j$. Since $(+0 - i\varrho)^{-n/2} = \varrho^{-n/2}e^{\pi in/4}$ the leading term is that of U_+ in the statement.

If $\psi \in \mathcal{S}(\mathbf{R}^n)$ then the function Ψ defined by

$$\Psi(t, x) = \begin{cases} \psi(-x/\sqrt{t^2 - |x|^2}) & \text{if } |x| < t, \\ 0 & \text{if } |x| \geq t, \end{cases}$$

is a symbol of order 0 when $t > 1$. In fact, $\partial^\alpha\Psi(t, x)$ is a sum of terms of the form

$$(\partial^\beta\psi)(-x/\sqrt{t^2 - |x|^2})Q(t, x)(t^2 - |x|^2)^{-\gamma/2}, \quad |x| < t,$$

where $Q(t, x)$ is a homogeneous polynomial of degree $\gamma - |\alpha|$, where $3 \leq \gamma \leq 3|\alpha|$. Since $\psi \in \mathcal{S}(\mathbf{R}^n)$ we can estimate this term by a constant times

$$(1 + |x|^2/(t^2 - |x|^2))^{-N}(t + |x|)^{\gamma-|\alpha|}(t^2 - |x|^2)^{-\gamma/2},$$

where N is arbitrary. Choosing $N = 3|\alpha|/2$ we obtain a bound by a constant times $(t + |x|)^{-|\alpha|}$ as claimed.

Thus $\langle\eta\rangle^{\frac{n}{2}+1}A_j(t, x)t^{-j-\frac{n}{2}}$, defined as 0 outside the light cone, is a symbol of order $-j - n/2$ when $t > 1$. Using Proposition 18.1.3 in Hörmander [4] we can find a symbol u_+ of order $-n/2$ with support in the full forward light cone which is an asymptotic sum of the terms in the right-hand side of (7.2.14). The estimates given in the proof of Theorem 7.2.1 show that

$$v_0(t, x) = u(t, x) - v_+(t, x)e^{i\sqrt{t^2-|x|^2}} = O((|t| + |x|)^{-N}) \text{ at } \infty$$

for any N when $t \geq 0$.

All derivatives of $v_+(t, x)e^{i\sqrt{t^2-|x|^2}}$ are $O(t^{-n/2})$ when $t > 1$. In fact, every derivative is a linear combination of terms of the form

$$e^{i\sqrt{t^2-|x|^2}}\partial^\beta v_+(t, x)(t^2 - |x|^2)^{-\nu/2}Q(t, x)$$

where Q is a polynomial of degree $\leq \nu$. Since

$$|\partial^\beta v_+(t, x)| \leq C_{\nu,\beta}t^{-\frac{n}{2}-|\beta|-\nu}(t - |x|)^\nu,$$

we conclude that $\partial^\alpha(v_+(t, x)e^{i\sqrt{t^2-|x|^2}}) = O(t^{-n/2})$. By (7.2.3) it is clear that all derivatives of u are uniformly bounded, hence so are those of v_0. But then it follows from the interpolation inequality (4.2.17)$'$ that v_0 is in $\mathcal{S}$ when $t > 0$.

We can argue in the same way when $t < 0$, or we can apply the result already proved to the complex conjugate of u. The only point which remains to prove is that w_j also when $j > 0$ is fully homogeneous and not only positively homogeneous. This will be more convenient to do after the proof of Theorem 7.2.6.

The decomposition in Theorem 7.2.5 is essentially unique:

Theorem 7.2.6. *If $v_0 \in S(\mathbf{R}^{1+n})$ and v_+, v_- are symbols with support in the full double light cone such that*

$$(7.2.15) \qquad v_0(t,x) + v_+(t,x)e^{i\sqrt{t^2 - |x|^2}} + v_-(t,x)e^{-i\sqrt{t^2 - |x|^2}} = 0$$

then $v_\pm \in S(\mathbf{R}^{1+n})$.

Proof. Assume that $v_\pm$ are of order m, that is,

$$|\partial_{t,x}^\alpha v_\pm(t,x)| \le C_\alpha (1 + |t| + |x|)^{m-|\alpha|}.$$

Differentiation of the equation

$$v_+(t,x)e^{2i\sqrt{t^2 - |x|^2}} = -v_-(t,x) - v_0(t,x)e^{i\sqrt{t^2 - |x|^2}}$$

with respect to t gives since $|t|/\sqrt{t^2 - |x|^2} \ge 1$ that

$$|v_+(t,x)| \le C(1 + |t| + |x|)^{m-1},$$

for Taylor's formula shows that $v_0(t,x)t/\sqrt{t^2 - |x|^2}$ can be estimated by any negative power of $1 + |t| + |x|$. From (4.2.17)′ it follows that v_+ is a symbol of order μ for every $\mu > m - 1$. The same is of course true for v_-. Repeating the argument we find that $v_\pm$ are symbols of order $-\infty$, which completes the proof.

End of proof of Theorem 7.2.5. With polar hyperbolic coordinates $\varrho = t\sqrt{1 - |x|^2/t^2}$, $\omega = (t,x)/\varrho$ inside the light cone we can write (see also Section 7.3 below)

$$\square + 1 = \partial_\varrho^2 + n\varrho^{-1}\partial_\varrho + 1 - \varrho^{-2}\Delta_H,$$

where Δ_H is the Laplacian on the unit hyperboloid $t^2 - |x|^2 = 1$ with respect to the natural Riemannian metric $|dx|^2 - dt^2$. By Theorem 7.2.6 we must have

$$(\partial_\varrho^2 + 2i\partial_\varrho + (n(2-n)/4 - \Delta_H)\varrho^{-2}) \sum_0^\infty \varrho^{-j} w_j \sim 0.$$

Since w_j occurs for the first time in a term of degree $-j-1$, with coefficient $-2ij \ne 0$, we find by induction that all w_j are fully homogeneous since w_0 is. The proof is complete.

Let us now return to the solution of (7.2.1); what we have done so far concerns only the term u_+ in (7.2.2). For u_- we have the same result with φ_+ replaced by φ_- and t replaced by $-t$. Hence

Theorem 7.2.7. *If $u_0, u_1 \in S(\mathbf{R}^n)$, then the solution of (7.2.1) can be written*

$$(7.2.16) \qquad u(t,x) = U_0(t,x) + U_+(t,x)e^{i\varrho} + U_-(t,x)e^{-i\varrho}, \quad \varrho = \operatorname{sgn} t\sqrt{t^2 - |x|^2},$$

where $U_0 \in S(\mathbf{R}^{1+n})$, both U_+ and U_- have their supports in the full double light cone, and $(+0 \mp i\varrho)^{n/2} U_\pm$ are polyhomogeneous symbols of order 0 with fully homogeneous terms and leading terms

$$(2\pi)^{-n/2}(t/\varrho)\hat{\varphi}_\pm(\mp x/\varrho),$$

interpreted as 0 outside the double light cone. Here $\varphi_\pm$ are defined by (7.2.2).

As at the end of the proof of Theorem 7.2.5 it is clear that we can successively compute the complete symbol from the leading symbol, which was given in Theorem 7.2.7. This is a much better way of calculating the asymptotic expansion than by using the method of stationary phase, except for the leading term which must be obtained from it.

So far we have only discussed the Cauchy problem for the homogeneous Klein-Gordon equation. However, replacing (7.2.1) by the inhomogeneous problem

$$(7.2.1)' \qquad \Box u + u = f; \quad u(0,x) = u_0(x), \quad \partial_t u(0,x) = u_1(x),$$

causes no new problems if $f \in \mathcal{S}(\mathbf{R}^{1+n})$. In fact, the Fourier transform in x of the solution can be written

$$\hat{u}(t,\xi) = \hat{\varphi}_+(\xi)e^{it\langle\xi\rangle} + \hat{\varphi}_-(\xi)e^{-it\langle\xi\rangle} + \widehat{R}(t,\xi),$$

where

$$(7.2.17) \qquad 2i\langle\xi\rangle\hat{\varphi}_\pm(\xi) = i\langle\xi\rangle\hat{u}_0(\xi) \pm \hat{u}_1(\xi) \pm i\int_0^\infty \hat{f}(s,\xi)e^{\mp is\langle\xi\rangle}/\langle\xi\rangle\, ds,$$

$$(7.2.18) \qquad \widehat{R}(t,\xi) = -\int_t^\infty \hat{f}(s,\xi)\frac{\sin((t-s)\langle\xi\rangle)}{\langle\xi\rangle}\, ds.$$

It is clear that $R \in \mathcal{S}$ for $t \geq 0$, and the other terms are those already discussed. Thus Theorem 7.2.7 remains true when $t > 0$ with $\varphi_\pm$ defined by (7.2.17) instead of (7.2.2).

7.3. L^2, L^∞ estimates for the Klein-Gordon equation. If u is a solution of the inhomogeneous Klein-Gordon equation

$$(7.3.1) \qquad \Box u + u = f; \quad u(0,\cdot) = u_0, \quad \partial_t u(0,\cdot) = u_1;$$

where $u_j \in C_0^\infty(\mathbf{R}^n)$, then it is well known that

$$(7.3.2) \qquad (\|u'(t,\cdot)\|^2 + \|u(t,\cdot)\|^2)^{\frac{1}{2}} \leq (\|u_1\|^2 + \|u_0'\|^2 + \|u_0\|^2)^{\frac{1}{2}} + \int_0^t \|f(s,\cdot)\|\, ds$$

where the norms are L^2 norms and u' denotes the derivatives with respect to t and x. For the proof it suffices to note that if the left-hand side is denoted by $E(t)$ then

$$2E(t)dE(t)/dt = dE(t)^2/dt = 2\int f(t,\cdot)\partial u(t,\cdot)/\partial t\, dx \leq 2E(t)\|f(t,\cdot)\|.$$

This gives control of u and the first derivatives of u in L^2 for fixed t. As in Sections 6.2 and 6.6, the equation (7.3.1) implies

$$(7.3.1)' \qquad \Box Z^I u + Z^I u = Z^I f$$

if Z^I is any product of the vector fields

$$(7.3.3) \qquad \partial/\partial t, \quad \partial/\partial x_j, \quad j = 1,\ldots n,$$

$$(7.3.4) \quad Z_{0j} = t\partial/\partial x_j + x_j\partial/\partial t, \quad Z_{jk} = x_j\partial/\partial x_k - x_k\partial/\partial x_j, \quad j,k = 1,\ldots,n;$$

generating the Lie algebra of the inhomogeneous Lorentz group. However, *the radial vector field* (6.2.9) *can no longer be used.* Thus one can hope to get estimates of all $Z^I u$ in L^2, for fixed t, even for perturbations of the equation (7.3.1) such that $Z^I f$ in turn can be estimated by some $Z^J u$. The purpose of this section is to examine to what extent one can recover the maximum norm estimates of u established in Section 7.2 if one controls sufficiently many L^2 norms of $Z^I u$ and of $Z^I f$. When $n = 3$ such estimates are the core of the arguments in Klainerman [5]. His estimates are not sharp but sufficient to establish global existence theorems then. However, their analogue for $n = 1$ or $n = 2$ would not give a good estimate for the lifespan of the solutions. We shall therefore reexamine the estimates of Klainerman [5] for an arbitrary dimension n.

We shall begin by proving a simple consequence of Sobolev's lemma which shows how the L^2 norms of $Z^I f$ for products of just the vector fields (7.3.4) give control of f on the hyperboloids $t^2 - |x|^2 = \text{constant}$.

Lemma 7.3.1. *Let $g \in C^\infty(\mathbf{R}^{1+n})$, assume that*

$$(7.3.5) \qquad \operatorname{supp} g \subset \{(t, x); T/2 \leq t \leq 2T, |x| \leq t\},$$

and set

$$M(a) = \sup_{t^2 - |x|^2 = a} |g(t, x)|.$$

Then it follows that

$$(7.3.6) \qquad T^{n-1} \int M(a)^2 \, da \leq C \sum_{|I| \leq (n+2)/2} \int |Z^I g|^2 \, dt dx,$$

where Z^I is a product of $|I|$ vector fields (7.3.4).

Proof. For reasons of homogeneity it is sufficient to prove the estimate when $T = 1$. (Note that $T^{n-1} da$ and $dt dx$ are both of homogeneity $n + 1$ while Z^I is homogeneous of degree 0.) In the support of g we can introduce new coordinates by

$$y_0 = t^2 - |x|^2, \ y_j = x_j \text{ if } j > 0, \quad \text{thus } t = \sqrt{y_0 + |x|^2},$$

for $y_0 + |x|^2 \geq 1/2$. The vector fields in (7.3.4) form a basis for the vector fields tangent to the hyperboloids $y_0 = $ constant so in the new variables they form a basis for the vector fields involving only $\partial/\partial y_j$ for $j \neq 0$. With the notation $G(y) = g(t, x)$ we obtain

$$\int M(y_0)^2 dy_0 \leq C \sum_{|\alpha| \leq (n+2)/2, \alpha_0 = 0} \int |D^\alpha G|^2 \, dy \leq C' \sum_{|I| \leq (n+2)/2} \int |Z^I g|^2 \, dt dx,$$

where the first inequality follows from Sobolev's lemma applied for fixed y_0. (Note that the integer part of $(n + 2)/2$ is the smallest integer larger than $n/2$.) This proves the lemma.

From (7.3.6) we obtain at once

$$(7.3.6)' \qquad T^{n-2} \int M(a)^2 \, da \leq C \sup_t \sum_{|I| \leq (n+2)/2} \int |Z^I g(t, x)|^2 \, dx.$$

In the proof of existence theorems based on standard energy estimates we shall only have control of the supremum in the right-hand side for $t \leq s$, say, and then we need to have an estimate of the form $(7.3.6)'$ with $M(a)$ replaced by

$$M(a, s) = \sup_{t^2 - |x|^2 = a, t \leq s} |g(t, x)|$$

in the left-hand side. This causes no problem if the support of g does not come close to the t axis:

Lemma 7.3.1'. *Let $g \in C^\infty(\mathbf{R}^{1+n})$, assume that*

$$(7.3.5)' \qquad \operatorname{supp} g \subset \{(t, x); T/2 \leq t \leq 2T, t/4 \leq |x| \leq t\}.$$

Then it follows that

$$(7.3.6)'' \qquad T^{n-2} \int M(a, s)^2 \, da \leq C \sup_{t \leq s} \sum_{|I| \leq (n+2)/2} \int |Z^I g|^2 \, dt dx,$$

where Z^I is a product of $|I|$ vector fields (7.3.4).

Proof. Again we may assume that $T = 1$. Since $|x| \geq 1/8$ in supp g we can choose a partition of unity χ_j, $j = 1, \ldots, n$, such that $x_j > 1/9n$ in supp χ_j and $\sum_1^n g_j = 1$ if $g_j = \chi_j g$. In supp g_j we can introduce new coordinates by

$$y_0 = t^2 - |x|^2, \ y_j = t, \ y_k = x_k \text{ if } 0 \neq k \neq j.$$

Then $x_j = \pm(y_j^2 - \sum_{0 \neq k \neq j} y_k^2 - y_0)^{\frac{1}{2}}$ in supp g_j, and if $G_j(y) = g_j(t, x)$ then

$$\int M(y_0, s)^2 \, dy_0 \leq C \sum_{|\alpha| \leq (n+2)/2, \alpha_0 = 0} \int_{y_j \leq s} |D^\alpha G_j(y)|^2 \, dy \leq C' \sum_{|I| \leq (n+2)/2} \int_{t \leq s} |Z^I g_j|^2 \, dt \, dx,$$

where the first inequality follows from Sobolev's lemma in the form (6.4.13)$''$. This implies (7.3.6)$''$.

However, if the support of g comes close to the t axis we cannot take both $t^2 - |x|^2$ and t as new coordinates as in the preceding proof. We shall now prove a related although weaker estimate in this situation.

Lemma 7.3.2. *If $n \leq 3$ and g satisfies (7.3.5), then*

$$(7.3.7) \qquad T^{\frac{n}{2}-2} \int M(a, s) \, da \leq C \sup_{t \leq s} \sum_{|I| \leq n} \|Z^I g(t, \cdot)\|$$

where Z^I is a product of $|I|$ vector fields (7.3.4).

Proof. Since (7.3.7) follows from (7.3.6)$''$, a partition of unity shows that it suffices to prove (7.3.7) assuming that

$$|x| \leq t/2 \quad \text{if } (t, x) \in \text{supp } g.$$

In the support of g we change coordinates as in the proof of Lemma 7.3.1. It is no restriction to assume that $s \leq 2T$. When $t \leq s$ we have

$$y_0 = t^2 - |x|^2 \leq s^2 - |y'|^2, \quad y' = (y_1, \ldots, y_n) = (x_1, \ldots, x_n).$$

Set $R^2 = s^2 - y_0$. With $g(t, x) = G(y)$ we have by Sobolev's lemma as in (6.4.13)$'''$

$$\int M(y_0, s) \, dy_0 \leq C \int_{|y'| < R} R^{-n} \sum_{|\alpha| \leq n, \alpha_0 = 0} |R^{|\alpha|} \partial^\alpha G(y)| \, dy,$$

and $|y'|^2 < R^2$ implies $|x|^2 < s^2 - t^2 + |x|^2$, that is, $t < s$. Now

$$t\partial/\partial y_j = x_j \partial/\partial t + t\partial/\partial x_j = Z_{0j}, \quad j \neq 0,$$

is among the operators (7.3.4), so $(\partial/\partial y)^\alpha$ is when $\alpha_0 = 0$ a linear combination of products Z^I with $|I| \leq |\alpha|$ and coefficients homogeneous in (t, x) of degree $-|\alpha|$. Since $\det \partial y/\partial(t, x) = 2t$, and $R \leq s \leq 2T \leq 4t$ if $(t, x) \in \text{supp } g$, we obtain

$$\int M(y_0, s) \, dy_0 \leq CT \sum_{|I| \leq n} \iint_{t \leq s} (s^2 - t^2 + |x|^2)^{-n/2} |Z^I g(t, x)| \, dt \, dx$$

$$\leq CT \int_0^s dt \left(\int (s^2 - t^2 + |x|^2)^{-n} \, dx \right)^{\frac{1}{2}} \sum_{|I| \leq n} \sup_{t \leq s} \|Z^I g(t, \cdot)\|.$$

The integral with respect to x is a constant times $(s^2 - t^2)^{-n/2}$, and

$$\int_0^s (s^2 - t^2)^{-n/4} dt = s^{1 - \frac{n}{2}} \int_0^1 (1 - t^2)^{-n/4} dt.$$

The integral converges since we have assumed that $n < 4$, which proves the lemma.

When $n \geq 4$ we must use also the operator $\partial/\partial t$ from (7.3.3) since the argument above leads to a divergent integral. However, we shall still use the proof of Lemma 7.3.2 to estimate the integral of $M(y_0, s)$ for $y_0 < s^2 - s$. The integral of $(s^2 - t^2 + |x|^2)^{-n}$ is then restricted to the set where $s^2 - t^2 + |x|^2 > s$. Hence it is equal to

$$(s^2 - t^2)^{-n/2} \int_{1 + |x|^2 > s/(s^2 - t^2)} (1 + |x|^2)^{-n} dx.$$

Since $s/(s^2 - t^2) > 1/(2(s - t))$ we integrate only for $2(1 + |x|^2)(s - t) > 1$, which implies $4|x|^2(s - t) > 1$ if $s - t < 1/4$. The integral over this set is bounded by a constant times $(s - t)^{n/2}$. Now we have

$$\int_{s/4}^{s - 1/4} (s^2 - t^2)^{-n/4} dt + \int_{s - 1/4}^s (s + t)^{-n/4} dt < C s^{-n/4}$$

when $n > 4$, and we have a bound $C s^{-1} \log s$ when $n = 4$, for

$$\int_{s/4}^{s - 1/4} (s - t)^{-n/4} dt = \int_{1/4}^{3s/4} t^{-n/4} dt$$

which is bounded when $n > 4$ and logarithmically divergent when $n = 4$. Altogether we obtain the bound (7.3.7) with the exponent $\frac{n}{2} - 2$ replaced by $\frac{n}{4} - 1$, and an additional factor $\log T$ when $n = 4$. However, it remains to prove such a bound for the integral when $s^2 - s < a < s^2$.

When $t \leq s$ and $t^2 - |x|^2 > s^2 - s$ then $|x|^2 < s$ and $s^2 - t^2 < s$, hence $s - t < 1$. By Sobolev's lemma as in (6.4.13)''' we have

$$\sup_{0 < s - t < 1, |x|^2 < s} |g(t, x)| \leq C s^{-n/2} \int_{s-1}^s dt \int_{|x|^2 < 2s} \sum_{\alpha_0, \dots, \alpha_n \in \{0, 1\}} |\partial_t^{\alpha_0} \partial_x^{\alpha'} g(t, x)| s^{|\alpha'|/2} dx$$

where two differentiations with respect to the same variable never occur in the sum. Since

$$\partial/\partial x_j = t^{-1}(Z_{0j} - x_j \partial/\partial t),$$

we can write ∂^α as a linear combinations of products Z^I of $\partial/\partial t, Z_{01}, \dots, Z_{0n}$ with $|I| \leq n + 1$ and coefficients bounded by a constant times $s^{-|\alpha'|/2}$, for $t > s - 1$ and $|x/t| < \sqrt{2s}/(s - 1)$ if $s > 2$. Using Cauchy-Schwarz' inequality we can estimate the right-hand side by

$$C s^{-n/4} \sup_{t \leq s} \sum_{|I| \leq n+1} \|Z^I g(t, \cdot)\|.$$

Integrating this bound for $M(a, s)$ from $s^2 - s$ to s^2, we complete the proof of the following

Lemma 7.3.3. If $n > 4$ and g satisfies (7.3.5), $T > 1$, then

$$(7.3.7)' \qquad T^{\frac{n}{4}-1} \int M(a, s)\, da \leq C \sup_{t \leq s} \sum_{|I| \leq n+1} \|Z^I g(t, \cdot)\|$$

where Z^I is a product of operators from (7.3.3) and (7.3.4). When $n = 4$ we have the same estimate with a factor $\log(2T)$ in the right-hand side.

After these preparatory lemmas we return to the solutions of the Klein-Gordon equation (7.3.1). As in Klainerman [5] we assume that supp u_j is contained in a ball $\{x; |x| \leq B\}$ with fixed radius. We translate by the distance $2B$ in the time direction, that is, replace (7.3.1) by

$$(7.3.1)'' \qquad \Box u + u = f, \ t \geq 2B; \quad u(2B, \cdot) = u_0, \ \partial_t u(2B, \cdot) = u_1.$$

It follows that

$$(7.3.8) \qquad |x| \leq t - B, \quad \text{when } (t, x) \in \text{supp } u, \quad \text{if } |x| \leq t - B, \quad \text{when } (t, x) \in \text{supp } f,$$

for $t \geq 2B$, as we always assume. This implies that

$$(7.3.9) \qquad t^2 - |x|^2 \geq B(t + |x|) \geq 2B^2, \quad (t, x) \in \text{supp } u.$$

Still following Klainerman [5] we introduce hyperbolic polar coordinates

$$(t, x) = \varrho\omega, \quad \varrho = (t^2 - |x|^2)^{\frac{1}{2}}.$$

Then we can write (7.3.1)'' in the form

$$u''_{\varrho\varrho} + n\varrho^{-1} u'_{\varrho} + u = \varrho^{-2} \Delta_H u + f$$

where Δ_H is the Laplace-Beltrami operator on the hyperboloid $t^2 - |x|^2 = 1$ for the natural metric $|dx|^2 - |dt|^2$,

$$\Delta_H = \sum_1^n (t\partial/\partial x_j + x_j\partial/\partial t)^2 - \sum_{1 \leq j < k \leq n} (x_j\partial/\partial x_k - x_k\partial/\partial x_j)^2,$$

$$\varrho\partial/\partial\varrho = t\partial/\partial t + \sum_1^n x_j\partial/\partial x_j.$$

In fact, with the coordinates x the metric is $|dx|^2 + O(|x|^2)$ at $x = 0$, $t = 1$, so $\Delta_H = \sum_1^n \partial^2/\partial x_j^2$ there, and since

$$\partial u(\sqrt{1 + x^2}, x)/\partial x_j = (\partial u(t, x)/\partial x_j + (x_j/t)\partial u(t, x)/\partial t)|_{t=\sqrt{1+x^2}},$$

it follows that Δ_H is given by this formula when $t = 1$, $x = 0$. The invariance under Lorentz transformations proves that it is valid everywhere. (See also e.g. Hörmander [4, Section 14.7].) With $v = \varrho^{\frac{n}{2}} u$ we can write the equation in the form

$$v''_{\varrho\varrho} + v = \varrho^{\frac{n}{2}}(\varrho^{-2}(\Delta_H u + n(n-2)u/4) + f).$$

Regarding the right-hand side as known we shall split it by a Paley-Littlewood type of partition of unity in t and use the following elementary lemma:

Lemma 7.3.4. *If* $w'' + w = h$ *in* $[a, b] \subset \mathbf{R}$, *then*

$$\sup_{a \leq \varrho \leq b} |w(\varrho)| \leq |w(a)| + |w'(a)| + \int_a^b |h(\varrho)| \, d\varrho.$$

Proof. This is an immediate consequence of the explicit formula

$$w(\varrho) = w(a) \cos(\varrho - a) + w'(a) \sin(\varrho - a) + \int_a^\varrho h(\sigma) \sin(\varrho - \sigma) \, d\sigma.$$

Choose $\chi \in C_0^\infty((1/2, 2))$ such that

$$\sum_{-\infty}^\infty \chi(t/2^k) = 1, \quad t > 0,$$

and set

$$f_k(t, x) = \chi(t/2^k) f(t, x), \quad 2^k \geq B.$$

Clearly $f = \sum f_k$ and more than two supports can never overlap. If M_k is defined as in Lemma 7.3.1 with $g = f_k$, then

$$2^{k(n-1)} \int M_k(\varrho^2)^2 \varrho \, d\varrho \leq C \sum_{|I| \leq (n+2)/2} \int |Z^I f_k|^2 \, dt \, dx.$$

To estimate $u(\bar{t}, \bar{x})$ we shall write $(\bar{t}, \bar{x}) = \bar{\varrho} \omega$ where $\bar{\varrho} = \sqrt{\bar{t}^2 - |\bar{x}|^2}$ and use the estimate

$$\int \varrho^{\frac{n}{2}} |f_k(\varrho\omega)| \, d\varrho \leq \int \varrho^{\frac{n}{2}} M_k(\varrho^2) \, d\varrho \leq \left(\int M_k(\varrho^2)^2 \varrho \, d\varrho \right)^{\frac{1}{2}} \left(\int \varrho^n \, d\varrho/\varrho \right)^{\frac{1}{2}},$$

where the integrals are taken for $\varrho \leq \bar{\varrho}$, and $2^{k-1} \leq t \leq 2^{k+1}$ if we write $\varrho\omega = (t, x)$. Hence we have $\int d\varrho/\varrho \leq \log 4$, and since $\varrho/\bar{\varrho} = t/\bar{t} \leq 2^{k+1}/\bar{t}$ we have $\int \varrho^n \, d\varrho/\varrho \leq (2^{k+1}\bar{\varrho}/\bar{t})^n \log 4$, so the preceding estimates give with a new constant C

$$(\bar{t}/\bar{\varrho})^{\frac{n}{2}} \int \varrho^{\frac{n}{2}} |f_k(\varrho\omega)| \, d\varrho \leq 2^{\frac{k}{2}} C \left(\sum_{|I| \leq (n+2)/2} \int |Z^I f_k|^2 \, dt \, dx \right)^{\frac{1}{2}}$$

$$\leq 2^{k+1} C \sup \sum_{|I| \leq (n+2)/2} \|Z^I f_k(t, \cdot)\|.$$

Put $u_k = \chi(t/2^k)(\Delta_H u + n(n-2)u/4)$, and define $\widetilde{M}_k$ as in Lemma 7.3.1 with $g = u_k$. Then

$$\int \varrho^{\frac{n}{2}-2} |u_k(\varrho\omega)| \, d\varrho \leq \int \varrho^{\frac{n}{2}-2} \widetilde{M}_k(\varrho^2) \, d\varrho \leq \left(\int \widetilde{M}_k(\varrho^2) \varrho \, d\varrho \right)^{\frac{1}{2}} \left(\int \varrho^{n-4} \, d\varrho/\varrho \right)^{\frac{1}{2}}.$$

Since $2^{k-1}/\bar{t} \leq \varrho/\bar{\varrho} \leq 2^{k+1}/\bar{t}$, we obtain as above

$$(\bar{t}/\bar{\varrho})^{\frac{n}{2}} \int \varrho^{\frac{n}{2}-2} |u_k(\varrho\omega)| \, d\varrho \leq (\bar{t}/\bar{\varrho})^2 2^{-k} C \sum_{|I| \leq (n+2)/2} \sup \|Z^I u_k(t, \cdot)\|.$$

If the left-hand side is not 0, then (7.3.9) shows that $\varrho^2 \geq B2^{k-1}$ for some ϱ in the support of the integrand, so

$$(2^{k+1}\bar{\varrho}/\bar{t})^2 \geq B2^{k-1}, \quad \text{that is, } 2^{-k} \leq (8/B)(\bar{\varrho}/\bar{t})^2.$$

The sum of the coefficients in the right-hand side for such k is therefore at most $16C/B$.

We can estimate $Z^I f_k$ by a constant times $\sum_{|J| \leq |I|} |Z^J f|$, and at most two supports overlap for the same t. We can estimate $Z^I u_k$ by a constant times $\sum_{|J| \leq |I|+2} |Z^J u|$. Also note that if the initial data of u_k at $t = 2B$ are not 0 then $\bar{\varrho}/\bar{t} \geq \sqrt{3}/2$ for $\varrho \geq B\sqrt{3}$ in the support of the initial data, so $|\partial u_k/\partial \varrho| \leq 4|u'_k|/\sqrt{3}$. Summing up, we obtain in view of Lemma 7.3.4 with the notation $I_k = (2^{k-1}, 2^{k+1})$

$$\bar{t}^{\frac{n}{2}}|u(\bar{t}, \bar{x})| \leq C \sum_{|I| \leq (n+2)/2} \sum_k 2^k \sup_{2B \leq t \in I_k} \|Z^I f(t, \cdot)\| + C \sum_{|I| \leq (n+6)/2} \sup_{2B \leq t} \|Z^I u(t, \cdot)\|.$$

(Recall that $v(\bar{t}, \bar{x}) = \varrho(\bar{t}, \bar{x})^{\frac{n}{2}}u(\bar{t}, \bar{x})$.) If the energy estimates at the beginning of the section are applied to the equations $(\Box + 1)Z^I u = Z^I f$, it follows that the second sum can be estimated by means of f and the initial data, so we conclude that

$$(7.3.10) \quad \sup t^{\frac{n}{2}}|u(t, x)|$$

$$\leq C\left(\sum_{|I| \leq (n+6)/2} \sum_k 2^k \sup_{2B \leq \tau \in I_k} \|Z^I f(\tau, \cdot)\| + \sum_{|\alpha|+j \leq (n+8)/2} \|\partial^\alpha u_j\| \right).$$

Here we have used that the initial value of $\partial_t^k Z^I u$ can be estimated in terms of the derivatives of u_j of order $\leq k - \dot{} + |I|$ and derivatives of f of order $\leq k + |I| - 2$. We have now proved:

Proposition 7.3.5. *If u satisfies (7.3.1)" and (7.3.9), then the estimate (7.3.10) is valid where Z^I is a product of $|I|$ vector fields of the form (7.3.4).*

In (7.3.10) the supremum is taken for all $\tau \geq 2B$ in the right-hand side. This is acceptable although not convenient when proving global existence theorems for nonlinear perturbations. However, it is not adequate in dimensions 1 and 2 where only a finite lifespan can be proved. In that case we want to be able to take the supremum for $t \leq s$, say, in both sides. As pointed out before the statement of Lemma 7.3.2, this can only cause additional problems when $|x| < |t|/2$. In that case we can use Lemma 7.3.2 instead of Lemma 7.3.1 if $n \leq 3$. Note that if $|x| < |t|/2$ in supp f_k, then we get from Lemma 7.3.2

$$\int \varrho^{\frac{n}{2}}|f_k(\varrho\omega)| \, d\varrho \leq \int \varrho^{\frac{n}{2}-1}M_k(\varrho^2, \bar{t})\varrho \, d\varrho \leq C2^k \sup_{t \leq \bar{t}} \sum_{|I| \leq n} \|Z^I f_k(t, \cdot)\|,$$

for $1/4 \leq \varrho/2^k \leq 4$. The proof then works as before and gives

Proposition 7.3.6. *If $n \leq 3$ and $s \geq 2B$, then*

$$(7.3.11) \quad \sup_{2B \leq t \leq s} t^{\frac{n}{2}}|u(t, x)| \leq C \sum_{|I| \leq n+2} \sum_k 2^k \sup_{I_k \cap (2B, s)} \|Z^I f(t, \cdot)\| + C \sum_{|\alpha|+j \leq n+3} \|\partial^\alpha u_j\|,$$

provided that u satisfies (7.3.1)" and (7.3.9). Here Z^I is a product of $|I|$ vector fields of the form (7.3.3), (7.3.4), and $I_k = (2^{k-1}, 2^{k+1})$.

Using Lemma 7.3.3 instead of Lemma 7.3.2 we also obtain an estimate when $n > 3$ although not with the desired rate of decay when $n \geq 8$:

Proposition 7.3.7. *If $n > 4$ and $s \geq 2B$, then*

(7.3.12)
$$\sup_{t \leq s} t^{\mu + \frac{n}{4}} |u(t, x)| \leq C \Big(\sum_{|I| \leq n+3} \sum_k 2^{\mu k} \sup_{I_k \cap (2B, s)} \|Z^I f(t, \cdot)\|$$
$$+ \sum_{|\alpha| + j \leq n+4} \|\partial^\alpha u_j\|\big)/(2 - \mu), \quad 1 \leq \mu < 2, \ \mu \leq n/4.$$

If $n = 4$ then

(7.3.13)
$$\sup_{t \leq s} t^2 |u(t, x)| \leq C \Big(\sum_{|I| \leq 7} \sum_k (1 + |k|) 2^k \sup_{I_k \cap (2B, s)} \|Z^I f(t, \cdot)\|$$
$$+ \sum_{|\alpha| + j \leq 8} \|\partial^\alpha u_j\| \Big).$$

Here Z^I is a product of $|I|$ vector fields of the form (7.3.3) or (7.3.4).

7.4. Existence theorems. The energy estimates recalled at the beginning of Section 7.3 also work if the coefficients of $\square$ are perturbed:

Lemma 7.4.1. *Let u be a solution of the perturbed Klein-Gordon equation*

(7.4.1)
$$\square u + u + \sum_{j,k=0}^n \gamma^{jk}(x) \partial_j \partial_k u = f, \quad 0 \leq x_0 < T,$$

where $x_0 = t$. If u vanishes for large $|x|$, and if

$$\sum_{j,k=0}^n |\gamma^{jk}| \leq \tfrac{1}{2},$$

then

(7.4.2)
$$(\|u'(t, \cdot)\|^2 + \|u(t, \cdot)\|^2)^{\frac{1}{2}}$$
$$\leq 2 \Big((\|u'(0, \cdot)\|^2 + \|u(0, \cdot)\|^2)^{\frac{1}{2}} + \int_0^t \|f(s, \cdot)\| \, ds \Big) \exp \Big(\int_0^t 2\Gamma(s) \, ds \Big)$$

where

$$\Gamma(s) = \sum_{i,j,k=0}^n \sup |\partial_i \gamma^{jk}(s, \cdot)|.$$

Proof. We multiply (7.4.1) by $2 \partial u / \partial t$ and integrate with respect to x. When $\dot{\jmath}, k \neq 0$ we can write

$$2 \partial_0 u \, \partial_j \partial_k u = \partial_j (\partial_0 u \, \partial_k u) + \partial_k (\partial_0 u \, \partial_j u) - \partial_0 (\partial_j u \, \partial_k u).$$

With the notation $E_0(t; u)$ for the left-hand side of (7.4.2), and $x = (x_1, \ldots, x_n)$,

$$E(t; u)^2 = E_0(t; u)^2 + \int \Big(\gamma^{00} (\partial_0 u)^2 - \sum_{j,k=1}^n \gamma^{jk} \partial_j u \partial_k u \Big) \, dx,$$

we have

$$|E(t; u)^2 - E_0(t; u)^2| \leq E_0(t; u)^2 / 2$$

and obtain (see the proof of (6.3.6))

$$\partial E(t; u)^2/\partial t \leq 2E_0(t; u)\|f(t, \cdot)\| + 2\Gamma(t)E_0(t; u)^2$$
$$\leq 2\sqrt{2}E(t; u)\|f(t, \cdot)\| + 4\Gamma(t)E(t; u)^2.$$

This inequality can be written

$$\partial E(t; u)/\partial t \leq \sqrt{2}\|f(t, \cdot)\| + 2\Gamma(t)E(t; u)$$

which gives (7.4.2) after integration.

This well-known lemma combined with Proposition 7.3.6 allows one to prove along the same lines as in Section 6.5 that if $u_0, u_1 \in C_0^\infty(\mathbf{R}^n)$ then there is some $c > 0$ such that the equation (7.1.1) has a solution with Cauchy data

(7.4.3) $$u(0, \cdot) = \varepsilon u_0, \quad \partial u(0, \cdot)/\partial t = \varepsilon u_1,$$

if ε is small and

$$t \leq \begin{cases} c/\varepsilon^2, & \text{if } n = 1, \\ e^{c/\varepsilon}, & \text{if } n = 2; \end{cases}$$

when $n \geq 3$ one obtains a global solution for small ε by using Proposition 7.3.7 or Proposition 7.3.5, as in Klainerman [5]. (See also Section 7.7 below.) The results for $n \geq 3$ are due to Klainerman [5] and Shatah [1]; the latter paper also treats initial data which do not have compact support. Another proof will be given in Theorem 7.7.1. However, one can do better for $n = 1$ or 2 by using an approximate solution of the Cauchy problem as in Section 6.5. In fact, we shall outline a proof that the constant c above can in fact be chosen arbitrarily.

Theorem 7.4.2. *Assume that the function F in (7.1.1) is in C^∞, vanishes of second order at 0, and is affine linear in u''. Let the number n of space variables be 1 or 2. Then the equation (7.1.1) with Cauchy data (7.4.3) where $u_0, u_1 \in C_0^\infty(\mathbf{R}^n)$ has a C^∞ solution for $0 \leq t \leq T_\varepsilon$ where $\varepsilon\sqrt{T_\varepsilon} \to \infty$ if $n = 1$, and $\varepsilon \log T_\varepsilon \to \infty$ if $n = 2$, as $\varepsilon \to 0$.*

Proof. As in Section 7.3 it is more convenient to put the Cauchy boundary condition at $t = 2B$ instead, where B is an upper bound for $|x|$ in supp $u_0 \cup$ supp u_1. At first we shall just present the arguments of Klainerman [3, 5] to prove the weaker results stated before the theorem. To separate the terms in F involving second order derivatives we write

$$F(u, u', u'') = f(u, u') - \sum_{j,k=0}^{n} \gamma^{jk}(u, u')\partial_j\partial_k u,$$

where f vanishes of second order and γ^{jk} vanishes of first order at the origin. The equation (7.1.1) can then be written

(7.4.4) $$\left(\Box + 1 + \sum_{j,k=0}^{n} \gamma^{jk}(u, u')\partial_j\partial_k\right)u = f(u, u').$$

Choose a positive integer $N \geq 6$, and let A be a positive number. We wish to show that if

(7.4.5) $$t^{n/2}(|Z^I u(t, x)| + |Z^I u'(t, x)|) \leq M\varepsilon, \quad |I| \leq N, \ 2B \leq t \leq T,$$

and if

$$(7.4.6) \qquad \varepsilon \int_{2B}^{T} t^{-n/2} dt \leq A,$$

then there is strict inequality in (7.4.5) for small ε provided that M is large enough and that A is small enough. Here and in what follows Z^I is any product of $|I|$ operators of the form (7.3.3) or (7.3.4). Combined with the local existence theorems this will give the lower bound for the lifespan mentioned before the statement of the theorem, with a fixed $c > 0$. Let $s = 2N$ and apply Z^I to (7.4.4) for all I with $|I| \leq s$. This gives

$$
(7.4.7) \qquad
\begin{aligned}
(\Box + 1 + \sum_{j,k=0}^{n} \gamma^{jk}(u,u')\partial_j\partial_k)Z^I u &= Z^I f(u,u') \\
&+ \sum_{j,k=0}^{n} [\gamma^{jk}, Z^I]\partial_j\partial_k u + \sum_{j,k=0}^{n} \gamma^{jk}[\partial_j\partial_k, Z^I]u.
\end{aligned}
$$

Set

$$(7.4.8) \qquad M_s(t) = \sum_{|I| \leq s} (\|Z^I u(t, \cdot)\| + \|\partial Z^I u(t, \cdot)\|);$$

since $[\partial_j, Z] = \pm\partial_k$ for some k or $[\partial_j, Z] = 0$, an equivalent norm is obtained if ∂Z^I is replaced by $Z^I \partial$. We wish to estimate the L^2 norm of the right-hand side of (7.4.7) by means of $M_s(t)$, noting that (u, u') is small by (7.4.5). We can write $[\gamma^{jk}, Z^I]\partial_j\partial_k u$ as a sum of terms of the form

$$-(Z^J \gamma^{jk}(u,u'))Z^K \partial_j\partial_k u,$$

where $|J| + |K| = |I| \leq s$ and $|J| \neq 0$. Thus $|K| + 1 \leq s$, and $(Z^J \gamma^{jk})Z^K \partial_j\partial_k u$ can be estimated by a sum of products of the form

$$Z^{J_1}(u,u') \cdots Z^{J_r}(u,u')Z^K \partial_j(\partial_k u), \quad |J_1| + \cdots + |J_r| = |J| \leq s, \ |J| + |K| + 1 \leq s+1.$$

Since $s + 1 < 2N + 2$ we can apply (7.4.5) to all factors except one, which we estimate using (7.4.8). We argue similarly for $Z^I f(u, u')$ and $\gamma^{jk}[\partial_j\partial_k, Z^I]u$, regarding $f(u, u')$ as a quadratic form in (u, u') with coefficients depending on (u, u'). The L^2 norm of the right-hand side of (7.4.7) can therefore be estimated by $CM\varepsilon t^{-n/2}M_s(t)$. By Lemma 7.4.1 it follows that for $2B \leq t \leq T$

$$M_s(t) \leq C(\varepsilon + M\varepsilon \int_{2B}^{t} M_s(\tau)\tau^{-n/2} d\tau),$$

hence by Gronwall's lemma and (7.4.6)

$$M_s(t) \leq C\varepsilon e^{CMA}, \quad 2B \leq t \leq T.$$

For $g = (\Box + 1)u$ we also obtain

$$\|Z^I g(t, \cdot)\| \leq CM\varepsilon t^{-n/2}M_s(t), \quad |I| \leq s - 1, \ 2B \leq t \leq T,$$

so it follows from Proposition 7.3.6 that

$$t^{n/2}|Z^I u(t, x)| \leq C'MA\varepsilon e^{CMA} + C''\varepsilon, \quad |I| + 4 \leq s - 1, \ 2B \leq t \leq T.$$

Since $s - 5 = 2N - 5 \geq N + 1$, we confirm (7.4.5) with strict inequality if $M > C''$ and A is small enough. If $N = 7$ we get a maximum norm estimate for one derivative more than needed, and this can be used successively to get bounds for all $\|Z^I u(t, \cdot)\|$ when $t \leq T$, without any further decrease of A. In view of the local existence theorem it follows that (7.4.6) does not hold for the lifespan T_ε of the C^∞ solution of the Cauchy problem, so we have $2\varepsilon T_\varepsilon^{\frac{1}{2}} > A$ if $n = 1$, and $\varepsilon \log(T_\varepsilon/2B) > A$ if $n = 2$.

To get the stronger result in the theorem we must first estimate not u itself but the difference between u and an approximate solution. To construct it, let V be the solution of the equation $(\Box + 1)V = 0$ with Cauchy data u_0, u_1 when $t = 2B$. Then $|x| \leq t - B$ in supp V, and by Theorem 7.2.7 we have a decomposition (7.2.16) of V. Since $U_\pm$ and all derivatives are rapidly decreasing when $|x| > t - 2B$, we can cut them off so that they vanish where $|x| > t - B$, which implies that U_0 vanishes there too. Since $U_0 e^{i\varrho}$ is also in S we can include it in U_+ so we have in fact

$$V = V_+ e^{i\varrho} + V_- e^{-i\varrho}; \quad |x| \leq t - B \text{ if } (t, x) \in \text{supp } V_\pm,$$

where $V_\pm$ are symbols of order $-n/2$ with principal symbols given in Theorem 7.2.7.

εV has the required Cauchy data and $(\Box + 1)\varepsilon V = 0$, but we only know that

$$F(\varepsilon V, \varepsilon V', \varepsilon V'') = O(\varepsilon^2 t^{-n}).$$

Let F_2 be the quadratic part of the Taylor expansion of F, so that $F - F_2$ vanishes of third order at 0. We can write

$$F_2(V, V', V'') = V_2 e^{2i\varrho} + V_0 + V_{-2} e^{-2i\varrho}$$

where V_j are symbols of order $-n$ with $|x| \leq t - B$ in the support. (We are using here that dividing a symbol with such support and order μ by ϱ gives a symbol of order $\mu - 1$. This will be used often in what follows without explicit mention.) If G is such a symbol, of order μ, and a is a real number $\neq \pm 1$, then one can always find another symbol H with the same order and support such that

$$(\Box + 1)(H e^{ia\varrho}) - G e^{ia\varrho} \in S, \quad \text{for } t \geq 2B.$$

In fact, this condition can be written

$$(1 - a^2)H + 2ia\varrho^{-1} \sum_0^n x_j \partial_j H + ianH\varrho^{-1} + \Box H - G \in S.$$

If H is of order μ, then all terms except the first and last are of order $\mu - 1$, so taking $H = G/(1 - a^2) + H_1$ reduces the proof to the statement with μ replaced by $\mu - 1$. Repeating the argument gives the claim by standard symbolic calculus. However, we do not really have to do more than the first step of this argument to conclude that for

$$(7.4.9) \qquad w_\varepsilon = \varepsilon(V_+ e^{i\varrho} + V_- e^{-i\varrho}) + \varepsilon^2(V_2 e^{2i\varrho}/3 - V_0 + V_{-2} e^{-2i\varrho}/3)$$

we have

$$(\Box + 1)w_\varepsilon = F(w_\varepsilon, w_\varepsilon', w_\varepsilon'') + R_\varepsilon,$$

where for all I, if Z^I is a product of the operators (7.3.3), (7.3.4),

$$(7.4.10) \qquad |Z^I w_\varepsilon(t, \cdot)| \leq C_I \varepsilon t^{-n/2},$$

$$(7.4.11) \qquad |Z^I R_\varepsilon(t, \cdot)| \leq C_I(\varepsilon^3 t^{-3n/2} + \varepsilon^2 t^{-n-1}).$$

For the proof we just have to note that the operators (7.3.4) do not act on ϱ and leave the order of symbols unchanged, whereas the operators (7.3.3) decrease the order of symbols but acting on ϱ produce a symbol of order one divided by ϱ. This gives (7.4.10) right away and also gives (7.4.11) since the terms in $F_2(w_\varepsilon, w'_\varepsilon, w''_\varepsilon)$ containing at least a factor ε^3 are of order $\leq -3n/2$ while the error in the preceding solution of a linear problem is of order $-n-1$. The measure of the support of $w_\varepsilon(t, \cdot)$ and of $R_\varepsilon(t, \cdot)$ is $O(t^n)$, so it follows with some new constants that

$$(7.4.12) \qquad \|Z^I w_\varepsilon(t, \cdot)\| \leq C_I \varepsilon, \quad \|Z^I R_\varepsilon(t, \cdot)\| \leq C_I(\varepsilon^3 t^{-n} + \varepsilon^2 t^{-\frac{n}{2}-1}).$$

Also note that for all α

$$(7.4.13) \qquad |\partial^\alpha(u - w_\varepsilon)| \leq C_\alpha \varepsilon^2, \quad t = 2B,$$

if u is the solution of the Cauchy problem (7.1.1), (7.4.3). This is obvious when $\alpha_0 \leq 1$ and follows inductively for larger α_0 if we use the equations satisfied by u and by w_ε.

Write $v = u - w_\varepsilon$ and subtract the equation

$$(\Box + 1 + \sum_{j,k=0}^{n} \gamma^{jk}(w_\varepsilon, w'_\varepsilon)\partial_j \partial_k)w_\varepsilon = f(w_\varepsilon, w'_\varepsilon) + R_\varepsilon$$

from (7.4.4). This gives

$$(7.4.14) \qquad (\Box + 1 + \sum_{j,k=0}^{n} \gamma^{jk}(u, u')\partial_j \partial_k)v = f(u, u') - f(w_\varepsilon, w'_\varepsilon) - R_\varepsilon$$
$$+ \sum_{j,k=0}^{n} (\gamma^{jk}(w_\varepsilon, w'_\varepsilon) - \gamma^{jk}(u, u'))\partial_j \partial_k w_\varepsilon.$$

Set

$$(7.4.15) \quad N_s(t) = \sum_{|I|\leq s} (\|Z^I v(t, \cdot)\| + \|\partial Z^I v(t, \cdot)\|)$$
$$+ \sum_{|I|\leq s-6} t^{n/2} \sup(|Z^I v(t, \cdot)| + |\partial Z^I v(t, \cdot)|),$$

and assume as in the first part of the proof that

$$(7.4.16) \qquad s + 1 < 2(s - 5), \text{ that is, } s \geq 12.$$

When estimating $N_s(t)$ we shall assume that

$$(7.4.17) \qquad N_s(t) \leq \varepsilon, \quad t \leq T,$$

and confirm afterwards that this is true with ε replaced by $\varepsilon/2$ if ε is small, which makes the hypothesis harmless.

Application of Z^I to (7.4.14) gives us even more terms than in (7.4.7),

$$(\Box + 1 + \sum_{j,k=0}^{n} \gamma^{jk}(u, u')\partial_j \partial_k)Z^I v = \sum_{1}^{5} g_j,$$
$$g_1 = -Z^I R_\varepsilon, \quad g_2 = Z^I(f(u, u') - f(w_\varepsilon, w'_\varepsilon)),$$
$$(7.4.18) \qquad g_3 = \sum_{j,k=0}^{n} Z^I(\gamma^{jk}(w_\varepsilon, w'_\varepsilon) - \gamma^{jk}(u, u'))\partial_j \partial_k w_\varepsilon,$$
$$g_4 = \sum_{j,k=0}^{n} [\gamma^{jk}(u, u'), Z^I]\partial_j \partial_k v, \quad g_5 = \sum_{j,k=0}^{n} \gamma^{jk}(u, u')[\partial_j \partial_k, Z^I]v.$$

Recall that $u = v + w_\varepsilon$. We can factor to exploit the zeros assumed; for example, $f(u, u') - f(w_\varepsilon, w'_\varepsilon)$ can be written as a bilinear form in (v, v') and $(v, v', w_\varepsilon, w'_\varepsilon)$ with smooth co-efficients depending on these variables. Similarly we can factor $\gamma^{jk}(u, u') - \gamma^{jk}(w_\varepsilon, w'_\varepsilon)$ as a linear form in (v, v'). Thus we obtain a component of (v, v') as factor in every term except g_1, which we estimated in (7.4.12). Note that when $|I| \leq s$ there are no terms with more than s factors Z applied to $(v, \partial v)$, and that (7.4.16) guarantees that one can use the maximum norm estimate in (7.4.15) in all factors except one. This yields, if we use (7.4.10), (7.4.12), (7.4.17) in (7.4.18),

$$(7.4.19) \qquad \|g_1(t, \cdot)\| \leq C(\varepsilon^3 t^{-n} + \varepsilon^2 t^{-\frac{n}{2}-1}), \quad \|g_j(t, \cdot)\| \leq C\varepsilon t^{-\frac{n}{2}} N_s(t), \text{ if } j > 1.$$

In view of (7.4.13) we conclude that

$$\|Z^I v(t, \cdot)\| \leq C\left(\varepsilon^2 + \varepsilon \int_{2B}^t \tau^{-\frac{n}{2}} N_s(\tau) \, d\tau\right), \quad |I| \leq s.$$

We get an estimate similar to (7.4.19) for $\|Z^I(\square + 1)v(t, \cdot)\|$ when $|I| \leq s - 1$; the only difference is that we also have to estimate the terms $\gamma^{jk}(u, u')\partial_j \partial_k Z^I v$. Hence an application of Proposition 7.3.6 gives

$$t^{\frac{n}{2}} \sup |Z^I v(t, \cdot)| \leq C\left(\varepsilon^2 + \varepsilon \int_{2B}^t \tau^{-\frac{n}{2}} N_s(\tau) \, d\tau\right), \quad |I| \leq s - 5.$$

Summing up, we have proved that

$$N_s(t) \leq C\left(\varepsilon^2 + \varepsilon \int_{2B}^t \tau^{-\frac{n}{2}} N_s(\tau) \, d\tau\right),$$

and by Gronwall's lemma it follows that

$$(7.4.20) \qquad N_s(t) \leq C\varepsilon^2 \exp(CA), \quad 2B \leq t \leq T$$

if (7.4.6) holds. The estimate (7.4.17) with ε replaced by $\varepsilon/2$ is a consequence of (7.4.20) for small enough ε, no matter how large A is. Starting from $s = 13$ we can now derive estimates for higher derivatives of u when $t \leq T$ as in the first part of the proof. By the local existence theorem it follows that (7.4.6) is not true for the life span T_ε. Since A is arbitrary now, the theorem is proved.

We have in fact also proved that locally uniformly in $t \in (0, \infty)$ we have for the solution u_ε of the Cauchy problem (7.1.1), (7.4.3)

$$u_\varepsilon(t/\varepsilon^2, x)/\varepsilon - V(t/\varepsilon^2, x) = O(\varepsilon(t/\varepsilon^2)^{-\frac{n}{2}})$$

as $\varepsilon \to 0$, if $n = 1$. When $n = 2$ we have a similar result with t/ε^2 replaced by $e^{t/\varepsilon}$ in the left-hand side. Thus the effect of the nonlinearity is weak during the time for which we have proved the existence of the solution. This suggests strongly that the lifespan is actually much longer than stated in Theorem 7.4.2. We shall discuss this question further in Sections 7.5 and 7.8.

If the perturbation F vanishes of third order at the origin then another factor $\varepsilon t^{-\frac{n}{2}}$ appears in the estimates made in the first part of the proof. This gives global existence for small ε when $n = 2$, since t^{-n} is integrable in $(2B, \infty)$ then. For $n = 1$ we get existence when $\varepsilon^2 \log t \leq c$, for some $c > 0$.

7.5. Remarks on the lowest space dimensions. We shall begin by discussing the ordinary differential equation which is the analogue of the nonlinear Klein-Gordon equation with no space variables present,

$$(7.5.1) \qquad u'' + u = f(u, u'); \quad u(0) = \varepsilon u_0, \ u'(0) = \varepsilon u_1;$$

where $f \in C^\infty$ vanishes of second order at 0 and $u_0^2 + u_1^2 = E_0$ is independent of ε. Since the solution of the unperturbed equation does not decay at infinity, one might expect that the lifespan T_ε should only be of the order $1/\varepsilon$ in general. However, it is much longer than that. Set

$$(7.5.2) \qquad 8\gamma = -(\partial_1^3 + \partial_0^2 \partial_1) f(u_0, u_1) - \partial_1 \partial_0 f(u_0, u_1)(\partial_0^2 + \partial_1^2) f(u_0, u_1)\Big|_{u_0 = u_1 = 0}.$$

Theorem 7.5.1. *If T_ε is the lifespan of the solution u_ε of (7.5.1), we have*

$$\lim_{\varepsilon \to 0} \varepsilon^2 T_\varepsilon = \infty, \ \text{if } \gamma \geq 0, \quad (-\gamma E_0) \lim_{\varepsilon \to 0} \varepsilon^2 T_\varepsilon \geq 1 \text{ if } \gamma < 0.$$

Moreover,

$$(7.5.3) \qquad (u_\varepsilon(t/\varepsilon^2)^2 + u_\varepsilon'(t/\varepsilon^2)^2)/\varepsilon^2 \to E_0/(1 + E_0 \gamma t)$$

uniformly on $[0, t_0]$ if $1 + E_0 \gamma t_0 > 0$.

Proof. If we multiply (7.5.1) by $2u'$ we obtain

$$d(u'^2 + u^2)/dt = 2u' f(u, u').$$

More generally, if $g(u_0, u_1)$ is a C^1 function in $\mathbf{R}^2$ then

$$\frac{d}{dt} g(u, u') = u_1 \partial_0 g(u_0, u_1) - (u_0 - f(u_0, u_1)) \partial_1 g(u_0, u_1); \quad u_0 = u, \ u_1 = u';$$

so we have

$$\frac{d}{dt}(u'^2 + u^2 - g(u, u')) = 2u_1 f - (u_1 \partial_0 g - (u_0 - f) \partial_1 g).$$

Here $u_1 \partial_0 - u_0 \partial_1$ is differentiation with respect to the polar angle in the plane, and the term $f \partial_1 g$ is of higher order. The quadratic part of f,

$$f_2(u_0, u_1) = a u_0^2 + b u_0 u_1 + c u_1^2$$

is eliminated by taking g equal to

$$g_3 = (4c + 2a) u_0^3/3 + 2c u_0 u_1^2 - 2b u_1^3/3,$$

which leaves us with the nonlinear terms

$$2u_1 f_3(u_0, u_1) - f_2(u_0, u_1)(4c u_0 u_1 - 2b u_1^2) + O(|u_0|^5 + |u_1|^5).$$

All fourth order terms cannot be eliminated if the mean value around the unit circle does not vanish. Since

$$(u_1 \partial_0 - u_0 \partial_1)(\alpha u_1^3 u_0 - \beta u_1 u_0^3) = \alpha u_1^4 - 3(\alpha + \beta) u_0^2 u_1^2 + \beta u_0^4,$$

a fourth order polynomial is of the form

$$(u_1 \partial_0 - u_0 \partial_1) g_4$$

where g_4 is also of fourth order, if and only if the coefficient of $u_1^2 u_0^2$ plus three times the sum of those of u_1^4 and of u_0^4 is equal to 0. This condition is fulfilled if we add $\gamma (u_1^2 + u_0^2)^2$ to the fourth order terms, so for some g_4 and $g = g_3 + g_4$ we have

$$(7.5.4) \qquad dE(t)/dt + \gamma E(t)^2 = O(|u(t)|^5 + |u'(t)|^5),$$

if $E(t) = u'(t)^2 + u(t)^2 - g(u(t), u'(t))$. Note that $E(0) = \varepsilon^2 E_0 + O(\varepsilon^3)$ and that

$$E(t)/2 \le u'(t)^2 + u(t)^2 \le 2E(t)$$

when $u'(t)^2 + u(t)^2$ is small. Assume now that we already know that there is a solution for $0 \le t \le T$ and that

$$(7.5.5) \qquad u'(t)^2 + u(t)^2 \le M^2 \varepsilon^2, \quad 0 \le t \le T,$$

for some M to be specified later on. If we divide (7.5.4) by $E(t)^2$ and integrate it follows then that for small ε

$$|1/E(0) - 1/E(t) + \gamma t| \le CM\varepsilon t, \quad 0 \le t \le T,$$

where C does not depend on M. Hence, with $E(0) = \varepsilon^2 E_0 + O(\varepsilon^3)$,

$$(7.5.6) \qquad (1 + E(0)(\gamma - CM\varepsilon)t)/E(0) \le 1/E(t) \le (1 + E(0)(\gamma + CM\varepsilon)t)/E(0).$$

If $t_0 > 0$ and $1 + \gamma E_0 t_0 > 0$, $\varepsilon^2 T \le t_0$, it follows that for $0 \le t \le T$ there is strict inequality in (7.5.5) provided that

$$M^2 > 2E_0/(1 + \gamma E_0 t), \quad 0 \le t \le t_0,$$

and ε is sufficiently small. This means that a solution satisfying (7.5.5) will exist for $T = t_0/\varepsilon^2$ when ε is small enough, and (7.5.3) follows from (7.5.6).

If $f(u, u') = f(u)$ is independent of u', then the number γ defined by (7.5.2) is equal to 0. If ε is so small that $|f(u)| \le |u|/2$ when $|u| < 2\varepsilon$, and $u_0^2 + u_1^2 = 1$, then the solution of (7.5.1) exists for all t, and it is a periodic function with

$$(7.5.7) \qquad \tfrac{1}{3}\varepsilon^2 \le u'(t)^2 + u(t)^2 \le 3\varepsilon^2, \quad t \in \mathbf{R}.$$

In fact, if $g(u) = \int_0^u 2f(v)\, dv$, then $|g(u)| \le u^2/2$ when $|u| < 2\varepsilon$, and

$$d(u'(t)^2 + u(t)^2 - g(u(t)))/dt = 2u'(t)(u''(t) + u(t) - f(u(t))) = 0.$$

As long as the solution exists and $|u(t)| < 2\varepsilon$ it follows that

$$u'(t)^2 + u(t)^2 - g(u(t)) = \varepsilon^2 - g(\varepsilon u_0),$$

and since $|g(\varepsilon u_0)| \le \varepsilon^2/2$, $|g(u(t))| \le u(t)^2/2$, we obtain

$$u'(t)^2 + u(t)^2/2 \le 3\varepsilon^2/2, \quad u'(t)^2 + 3u(t)^2/2 \ge \varepsilon^2/2.$$

This implies $|u(t)| \le \varepsilon\sqrt{3} < 2\varepsilon$ and (7.5.7), so $(u(t), u'(t))$ moves for all t periodically around a closed curve in the phase plane defined by $u'(t)^2 + u(t)^2 - g(u(t)) = u_1^2 + u_0^2 - g(u_0)$.

There is a similar result for $n = 1$ (see Delort [1]):

Theorem 7.5.2. *If $f \in C^\infty(\mathbf{R})$ vanishes of second order at 0 then the Cauchy problem*

$$(\partial_t^2 - \partial_x^2 + 1)u = f(u); \quad u(0, \cdot) = \varepsilon u_0, \; \partial_t u(0, \cdot) = \varepsilon u_1,$$

where $u_j \in \mathcal{S}(\mathbf{R})$, has a solution in $C^\infty(\mathbf{R}^2)$ if ε is sufficiently small.

Proof. With $g(u) = \int_0^u 2f(v)\, dv$ as above, it follows that

$$I(t) = \int \left((\partial u(t,x)/\partial t)^2 + (\partial u(t,x)/\partial x)^2 + u(t,x)^2 - g(u(t,x)) \right) dx$$

is constant for $0 \le t \le T$ if a rapidly decreasing C^∞ solution exists then. Choose $c > 0$ so that $|f(u)| \le |u|/2$, hence $|g(u)| \le u^2/2$, when $|u| \le c$. If $|u(t,x)| \le c$ when $0 \le t \le T$, it follows that

$$\int \left((\partial u(t,x)/\partial t)^2 + (\partial u(t,x)/\partial x)^2 + \tfrac{1}{2}u(t,x)^2 \right) dx \le I(0) \le C\varepsilon^2, \quad 0 \le t \le T.$$

Since

$$\sup_x u(t,x)^2 \le 2 \int |u(t,x)\partial u(t,x)/\partial x|\, dx \le \int \left(u(t,x)^2 + (\partial u(t,x)/\partial x)^2 \right) dx \le C\varepsilon^2,$$

we conclude that $|u(t,x)| \le c/2$ if $4C\varepsilon^2 \le c^2$.

Differentiation of the equation with respect to x gives the equation

$$(\partial_t^2 - \partial_x^2 + 1)v = f'(u)v,$$

if v is a first order derivative of u. If $M = \sup_{|u| \le c} |f'(u)|$, then $\|f'(u)v\| \le M\sqrt{C}\varepsilon$, so the energy estimate (7.3.2) gives

$$(\|v'(t,\cdot)\|^2 + \|v(t,\cdot)\|^2)^{\frac{1}{2}} \le C_1\varepsilon(1+t), \quad \text{hence} \quad \sup |v(t,\cdot)| \le C_1\varepsilon(1+t).$$

Inductively it follows that

$$(7.5.8) \quad \sum_0^1 \|\partial_j \partial^\alpha u(t,\cdot)\|^2 + \|\partial^\alpha u(t,\cdot)\|^2)^{\frac{1}{2}} \le C_\alpha \varepsilon(1+t)^{|\alpha|}, \quad \sup |\partial^\alpha u(t,\cdot)| \le C_\alpha \varepsilon(1+t)^{|\alpha|},$$

for if $v_\alpha = \partial^\alpha u$ then

$$(\partial_t^2 - \partial_x^2 + 1)v_\alpha = f'(u)v_\alpha + h_\alpha$$

where h_α for $|\alpha| \ge 2$ is a sum of products of at least two derivatives of u of order $< |\alpha|$. If (7.5.8) is already proved for smaller values of α, then $\|f'(u)v_\alpha(t,\cdot)\| \le C_\alpha\varepsilon(1+t)^{|\alpha|-1}$, and $\|h_\alpha(t,\cdot)\| \le C_\alpha\varepsilon(1+t)^{|\alpha|-1}$ since we can use the L^2 estimate (7.5.8) in one of the factors and the L^∞ estimate in all the others. Hence the energy estimates give (7.5.8). This proves that there is a C^∞ solution with all derivatives bounded when t is bounded. Since the solution at (t,x) only depends on the data in $[x-t, x+t]$, it is rapidly decreasing which completes the proof.

When $n = 2$ there is a global solution for small data when the non-linear term in (7.1.1) only depends on (u, u'). (See the end of Section 7.8.)

7.6. Alternative energy and Sobolev estimates. Let u be a solution of the Klein-Gordon equation in $\mathbf{R}^{1+n}$

$$\text{(7.6.1)} \qquad \Box u + u = f,$$

with Cauchy data vanishing for $|x| > B$, say. As before we pose the Cauchy data at time $2B$,

$$\text{(7.6.2)} \qquad u(2B, \cdot) = u_0, \quad \partial_t u(2B, \cdot) = u_1.$$

If $f = 0$ it follows that

$$\text{(7.6.3)} \qquad |x| \leq t - B, \quad (t, x) \in \operatorname{supp} u,$$

when $t \geq 2B$ as we always assume. This is also true for the inhomogeneous equation if

$$\text{(7.6.4)} \qquad |x| \leq t - B, \quad (t, x) \in \operatorname{supp} f.$$

In the application to the Cauchy problem for (7.1.1) we have $f = F(u, u', u'')$, and (7.6.3), (7.6.4) are then consequences of (7.6.1), (7.6.2) and the support condition on the Cauchy data. Note that (7.6.3) implies

$$\varrho^2 = t^2 - |x|^2 \geq B(t + |x|) \geq 2B^2, \quad (t, x) \in \operatorname{supp} u.$$

As in the standard energy integral method we start by rewriting the product of (7.6.1) by $2\partial u/\partial t$,

$$2f \partial u/\partial t = \partial(|u'|^2 + |u|^2)/\partial t - 2 \sum_1^n \partial(\partial u/\partial t \partial u/\partial x_j)/\partial x_j.$$

However, for some $T \geq 2B$ we shall now integrate over

$$\text{(7.6.5)} \qquad G_T = \{(t, x); t \geq 2B, \, t^2 - |x|^2 \leq T^2\}.$$

Note that $t + |x| \leq T^2/B$ when $(t, x) \in G_T \cap \operatorname{supp} u$. With the hyperboloid $H_T = \{(t, x); t^2 - |x|^2 = T^2\}$ parametrized by x, we get by integration

$$\text{(7.6.6)} \qquad \begin{aligned} E(T; u) &= \int_{H_T} (|u'|^2 + 2\partial u/\partial t \langle \partial u/\partial x, x/t \rangle + |u|^2) \, dx \\ &= \int_{G_T} 2f \partial u/\partial t \, dt dx + E_0(2B; u), \end{aligned}$$

where

$$E_0(T; u) = \int (|u'(T, x)|^2 + |u(T, x)|^2) \, dx$$

is the standard energy associated with the Klein-Gordon equation. The quadratic form in the first integral in (7.6.6) is positive since H_T is spacelike; completing squares we can write it in the form

$$\text{(7.6.7)} \qquad |\partial u/\partial x + x/t \partial u/\partial t|^2 + (\partial u/\partial t)^2 \varrho^2/t^2 + |u|^2, \quad \varrho^2 = t^2 - |x|^2.$$

We shall differentiate (7.6.6) with respect to T. If H is the Heaviside function then

$$\frac{\partial}{\partial T} H((T^2 + |x|^2)^{\frac{1}{2}} - t) = T(T^2 + |x|^2)^{-\frac{1}{2}} \delta((T^2 + |x|^2)^{\frac{1}{2}} - t),$$

so we obtain in view of (7.6.6) and (7.6.7), writing $E(T; u) = E(T)$,

$$\partial E(T)/\partial T = 2 \int_{H_T} f \partial u/\partial t \, T/t \, dx$$

$$\leq 2 \left(\int_{H_T} f^2 dx \right)^{\frac{1}{2}} \left(\int_{H_T} (\partial u/\partial t)^2 \varrho^2 /t^2 \, dx \right)^{\frac{1}{2}} \leq 2 \left(\int_{H_T} f^2 \, dx \right)^{\frac{1}{2}} E(T)^{\frac{1}{2}}.$$

Introducing $E(T) = (E(T)^{\frac{1}{2}})^2$ in the left-hand side we conclude that

$$(7.6.8) \qquad E(T)^{\frac{1}{2}} \leq E(2B)^{\frac{1}{2}} + \int_{2B}^{T} ds \left(\int_{H_s} f^2 dx \right)^{\frac{1}{2}}.$$

In $G_{2B} \cap \operatorname{supp} u$ we have $t \leq 5B/2$, so the standard energy estimate

$$E_0(t)^{\frac{1}{2}} \leq E_0(2B)^{\frac{1}{2}} + \int_{2B}^{t} \|f(s, \cdot)\| \, ds$$

implies that

$$(7.6.9) \qquad \begin{aligned} E(2B) &\leq 2 \int_{2B}^{5B/2} \|f(s, \cdot)\| \, ds \left(E_0(2B)^{\frac{1}{2}} + \int_{2B}^{5B/2} \|f(s, \cdot)\| \, ds \right) + E_0(2B) \\ &\leq 2 \left(E_0(2B)^{\frac{1}{2}} + \int_{2B}^{5B/2} \|f(s, \cdot)\| \, ds \right)^2. \end{aligned}$$

(7.6.8) and (7.6.9) together give good control of $E(T)$. In what follows we shall usually ignore the simple and standard estimate in G_{2B}.

With Z^I still denoting any product of the vector fields (7.3.3), (7.3.4), we can get estimates for

$$(7.6.10) \qquad \int_{H_T} |Z^I u|^2 \, dx, \quad T \geq 2B,$$

in terms of $Z^I f$ and the Cauchy data of $Z^I u$. To pass from L^2 estimates to maximum norm estimates we can now use a much simpler version of Sobolev's lemma than in Section 7.3:

Lemma 7.6.1. *If ν is the smallest integer $> n/2$ there is a constant C such that*

$$(7.6.11) \qquad \sup_{H_T} t^n |u(t, x)|^2 \leq C \sum_{|I| \leq \nu} \int_{H_T} |Z^I u|^2 \, dx,$$

when $u \in C^\nu$ and $T \geq 2B$.

Proof. With x as coordinate on H_T the vector field $t \partial/\partial x_j + x_j \partial/\partial t$ acts just as $t \partial/\partial x_j$ with $t = (T^2 + |x|^2)^{\frac{1}{2}}$, for $x_j = t \partial t/\partial x_j$. Since $\partial^\alpha t = O(t^{1-|\alpha|})$ for every α, it follows that the sum in the right-hand side of (7.6.11) bounds a constant times

$$(7.6.12) \qquad \sum_{|\alpha| \leq \nu} \int (T^2 + |x|^2)^{|\alpha|} |\partial^\alpha U(x)|^2 \, dx, \quad U(x) = u((T^2 + |x|^2)^{\frac{1}{2}}, x).$$

Now we can apply the standard Sobolev lemma in a ball with arbitrary center x_0 and radius $t_0 = \frac{1}{2}(T^2 + |x_0|^2)^{\frac{1}{2}}$. Since

$$(T^2 + |x + x_0|^2)^{\frac{1}{2}} \geq 2t_0 - |x| \geq t_0, \quad |x| < t_0,$$

the estimate (7.6.11) follows at once.

Combination of Lemma 7.6.1 with (7.6.8) gives if u satisfies (7.6.1) and (7.6.3)

$$(7.6.13) \qquad \sup_{H_T} t^{n/2}|u(t,x)| \leq C \sum_{|I| \leq \nu} \left(E(2B; Z^I u)^{\frac{1}{2}} + \int_{2B}^T \left(\int_{H_s} |Z^I f|^2 \, dx \right)^{\frac{1}{2}} ds \right).$$

In our application to the nonlinear equation (7.1.1) and the equations obtained by applying an operator Z^I to it, we shall estimate all factors in a nonlinear term except one using (7.6.13). For the highest order derivatives we shall have to appeal to an energy estimate for a linear perturbation of (7.6.1). Before discussing such equations we shall make a remark on the quadratic form (7.6.7) in the first derivatives occurring in the definition (7.6.6) of $E(T)$. To shorten notation we write $\tau = \partial u/\partial t$, $\xi = \partial u/\partial x$. The form becomes

$$\tau^2 + |\xi|^2 + 2\tau\langle x/t, \xi\rangle = |\xi + \tau x/t|^2 + (1 - |x|^2/t^2)\tau^2.$$

The vector fields (7.3.4) correspond to

$$t(\xi_j + \tau x_j/t), \quad x_j\xi_k - x_k\xi_j, \quad j,k = 1,\ldots,n.$$

Since

$$x_j\xi_k - x_k\xi_j = x_j(\xi_k + \tau x_k/t) - x_k(\xi_j + \tau x_j/t),$$

we conclude that $|Zu|^2/t^2$ can be estimated by (7.6.7) if Z is of the form (7.3.4). When Z is of the form (7.3.3) we shall use instead that

$$(7.6.14) \qquad \tau^2 + |\xi|^2 + 2\tau\langle x/t, \xi\rangle \geq (\tau^2 + |\xi|^2)(1 - |x|/t) \geq (\tau^2 + |\xi|^2)\varrho^2/2t^2.$$

We must finally discuss how the energy estimate (7.6.9) is modified for a linear perturbation of $\square$. Suppose that (7.6.1) is replaced by

$$(7.6.15) \qquad \square u + u + \sum_{j,k=0}^n \gamma^{jk}(t,x)\partial_j\partial_k u + \sum_{j=0}^n \gamma^j(t,x)\partial_j u = f$$

where γ is small. As before we multiply by $2\partial_t u = 2\partial_0 u$ and integrate by parts. Since

$$2\partial_0 u \sum_{j,k=0}^n \gamma^{jk}\partial_j\partial_k u = 2\sum_{j=0}^n \partial_j \sum_{k=0}^n \gamma^{jk}\partial_k u\partial_0 u - \partial_0 \sum_{j,k=0}^n \gamma^{jk}\partial_j u\partial_k u$$

$$- 2\sum_{j,k=0}^n (\partial_j\gamma^{jk})\partial_k u\partial_0 u + \sum_{j,k=0}^n (\partial_0\gamma^{jk})\partial_j u\partial_k u$$

the surface integral in (7.6.6) is modified to

$$\tilde{E}(T;u) = E(T;u) + \int_{H_T} (2\sum_{j,k=0}^n \nu_j\gamma^{jk}\partial_k u\partial_0 u - \sum_{j,k=0}^n \gamma^{jk}\partial_j u\partial_k u)dx,$$

where $\nu_0 = 1$ and $\nu_j = -x_j/t$ for $j \neq 0$. To the volume integral in (7.6.6) we must also add

$$\int_{G_T} (2\sum_{j,k=0}^n (\partial_j\gamma^{jk})\partial_k u\partial_0 u - \sum_{j,k=0}^n (\partial_0\gamma^{jk})\partial_j u\partial_k u - 2\sum_{j=0}^n \gamma^j(t,x)\partial_j u\partial_0 u)dt dx.$$

Recalling (7.6.14) we conclude that

$$E(T; u)/2 \leq \tilde{E}(T; u) \leq 3E(T; u)/2$$

if

(7.6.16)
$$\sup \sum_{j,k=0}^{n} |\gamma^{jk}(t,x)| t^2/\varrho^2 \leq c_n$$

where c_n is a positive constant depending only on the dimension and the supremum is taken over supp u. Thus $t^2/\varrho^2 \leq t/B$, so (7.6.16) is valid if

(7.6.16)'
$$\sup \sum_{j,k=0}^{n} |\gamma^{jk}(t,x)| t \leq c_n B.$$

When differentiating the modified formula (7.6.6) with respect to T we get a new term

$$\int_{H_T} \Big(2\sum_{j,k} (\partial_j \gamma^{jk}) \partial_k u \partial_0 u - \sum_{j,k} (\partial_0 \gamma^{jk}) \partial_j u \partial_k u - 2\sum_{j} \gamma^j(t,x) \partial_j u \partial_0 u \Big) T/t \, dx.$$

If

(7.6.17)
$$\sum_{i,j,k=0}^{n} |\partial_i \gamma^{jk}(t,x)| + \sum_{j=0}^{n} |\gamma^j(t,x)| \leq \kappa t^{-\mu}$$

for some $\mu \geq 1$, then the integrand can be estimated by

$$C_n \kappa t^{-\mu} |u'|^2 T/t = C_n \kappa (T/t)^2 |u'|^2 T^{-1} t^{1-\mu} \leq C_n \kappa (T/t)^2 |u'|^2 T^{-\mu},$$

so we obtain using (7.6.14), with another C_n,

$$\partial \tilde{E}/\partial T \leq 4 \Big(\int_{H_T} |f|^2 \, dx \Big)^{\frac{1}{2}} \tilde{E}(T)^{\frac{1}{2}} + C_n \kappa T^{-\mu} \tilde{E}(T).$$

Integration of this inequality gives by the equivalence of E and $\tilde{E}$ above

(7.6.18)
$$E(T)^{\frac{1}{2}} \leq \Big(3E(2B)^{\frac{1}{2}} + 4\int_{2B}^{T} ds \big(\int_{H_s} f^2 dx \big)^{\frac{1}{2}} \Big) \exp \Big(\int_{2B}^{T} C_n \kappa s^{-\mu} ds \Big),$$

so we have proved

Lemma 7.6.2. *If u is a solution of (7.6.15) satisfying (7.6.3), and if γ satisfies (7.6.16), (7.6.17) for some $\mu \geq 1$, then (7.6.18) is valid.*

Remark. The proof works with no essential change if u and f take values in $\mathbf{R}^N$ and γ^j are $N \times N$ matrices.

7.7. Existence theorems for Cauchy data of compact support. Using Lemmas 7.6.1 and 7.6.2 it is easy to recover the results of Klainerman [6] and Shatah [1] when the data have compact support:

Theorem 7.7.1. Let u_0, $u_1 \in C_0^\infty(\{x \in \mathbf{R}^n; |x| \le B\})$, and consider the Cauchy problem for (7.1.1), where $F \in C^\infty$ vanishes of second order at 0, with Cauchy data (7.6.2) for u/ε. If $n \ge 3$ and ε is small enough there is a unique C^∞ solution for $t \ge 2B$.

Proof. By the local existence theorems there is a C^∞ solution which is $O(\varepsilon)$ with all derivatives when $\varrho^2 = t^2 - |x|^2 \le (2B)^2$. Let s be an integer $> n+4$, and set

$$
(7.7.1) \qquad M_s(t) = \sum_{|I| \le s} E(t; Z^I u)^{\frac{1}{2}},
$$

where E is defined by (7.6.6) and Z^I stands for the product of $|I|$ vector fields of the form (7.3.3) or (7.3.4). We shall prove that there is a constant M such that if the solution exists for $\varrho < T$ then

$$
(7.7.2) \qquad M_s(t) \le M\varepsilon, \quad 2B \le t < T.
$$

By the local existence theory this proves that the solution exists for $\varrho < T'$ with some $T' > T$, so global existence follows. To prove (7.7.2) it suffices to show that M can be chosen such that (7.7.2) implies the same estimate with M replaced by $M/2$.

For the proof we first observe that by Lemma 7.6.1 it follows from (7.7.1) that

$$
(7.7.3) \qquad \sum_{|I| < s - n/2} \sup_{H_\tau} t^{\frac{n}{2}} |Z^I u(t,x)| \le C M_s(\tau),
$$

and (7.7.1) implies

$$
(7.7.1)' \qquad \sum_{|I| \le s} \left(\int_{H_\tau} |Z^I u|^2 dx \right)^{\frac{1}{2}} \le M_s(\tau).
$$

Let $|I| \le s$ and apply Z^I to the equation (7.1.1). With $\gamma^{jk} = -\partial F/\partial u''_{jk}$ we obtain

$$
(7.7.4) \qquad (\Box + 1 + \sum \gamma^{jk} \partial_j \partial_k) Z^I u + \sum_j \sum_J \gamma^j_{I,J} \partial_j Z^J u = f_I,
$$

where $\gamma^j_{I,J} = 0$ unless $|I| = |J| = s$, and

$$
(7.7.5) \qquad \sum_{j,k=0}^n |\gamma^{jk}(t,x)| + \sum_{i,j,k=0}^n |\partial_i \gamma^{jk}(t,x)| + \sum_{i=0}^n \sum_{|I|=|J|=s} |\gamma^j_{I,J}(t,x)| \le C M \varepsilon t^{-\frac{n}{2}},
$$

$$
(7.7.6) \qquad \left(\int_{H_t} |f_I|^2 dx \right)^{\frac{1}{2}} \le C M \varepsilon t^{-\frac{n}{2}} M_s(t), \quad 2B \le t < T.
$$

(7.7.3) is valid for $|I| \le 3$ since $s - n/2 > 3 + n/2$ by hypothesis, so the estimates of γ^{jk} are obvious. If $|I| = 0$ we define $\gamma^j_{I,J} = 0$ and

$$
f_I = F(u, u', u'') - \sum_{j,k=0}^n \gamma^{jk} \partial_j \partial_k u
$$

and obtain (7.7.6) in that case using (7.7.3) and (7.7.1)'. Let $|I| \ne 0$ now. Since Z^I commutes with $\Box$ we only have to discuss the terms obtained when Z^I acts on $F(u, u', u'')$.

No terms obtained involve derivatives of order $|I| + 2$ except those displayed in (7.7.4). In no term can there be two factors which cannot be estimated using (7.7.3), for such a factor requires that $\geq s - 2 - n/2$ factors Z fall on an argument in F, and to obtain two factors one needs $\geq 2s - 4 - n > s$ factors when $s > n + 4$. There are no terms in f_I of order $s + 1$ if $|I| < s$. If $|I| = s$, we have first of all terms from the commutator $\gamma^{jk}[Z^I, \partial_j \partial_k] u$, and then there are terms where all factors in Z^I fall on an argument u' of F or at least $s - 1$ of them fall on an argument u''. Since $[Z, \partial_j]$ is 0 or $\pm \partial_i$ for some i, it is clear that all terms involving derivatives of order $s + 1$ have the form listed in (7.7.4) with $\gamma^j_{I,J}$ satisfying (7.7.5). From (7.7.1)' and (7.7.3) it follows that (7.7.6) is valid for the sum f_I of the other terms. (Functions of (u, u', u'') vanishing at $(0, 0, 0)$ can be estimated by $CM\varepsilon$.) If we now apply Lemma 7.6.2 to the system (7.7.4) with $|I| \leq s$, recalling the remark following the statement, it follows that

$$M_s(t) \leq C\Big(M_s(2B) + \int_{2B}^t M\varepsilon \tau^{-\frac{n}{2}} M_s(\tau) \, d\tau \Big) \exp \Big(\int_{2B}^t C_n C M\varepsilon \tau^{-\frac{n}{2}} \, d\tau \Big), \quad t < T.$$

When ε is small enough the exponential factor is ≤ 2 no matter how large M is, and we obtain using Gronwall's lemma for small ε

$$M_s(t) \leq 2C M_s(2B) \exp \Big(\int_{2B}^t 2CM\varepsilon \tau^{-\frac{n}{2}} \, d\tau \Big) \leq 4C M_s(2B).$$

If $M > 8C M_s(2B)/\varepsilon$ this proves (7.7.2) with M replaced by $M/2$, which completes the proof.

When $n = 2$ the proof works with no change except that we must impose the condition that $\varepsilon \log T$ is small. Hence we obtain a crude version of Theorem 7.4.2:

Theorem 7.7.2. *Let the hypotheses be as in Theorem 7.7.1 except that $n = 2$. Then there is a constant $c > 0$ such that the Cauchy problem has a solution for $t \leq e^{c/\varepsilon}$ provided that ε is small enough.*

Unfortunately, the method does not seem to work at all when $n = 1$ because we only have (7.6.17) with $\mu = \frac{1}{2}$ then.

7.8. The method of Shatah. Existence theorems were proved by Shatah [1] for $n \geq 3$ space dimensions with a method which does not require the fairly complicated a priori estimates of Section 7.3. It is also applicable when the initial data do not have compact support; in fact, it is necessary to allow such data. A combination with the methods of Klainerman has recently been used by a number of authors to improve the existence theorems of Section 7.4 for $n = 1$ and $n = 2$, and we shall discuss their results also at the end of the section.

Instead of the *a priori* estimates in Section 7.3 the original arguments of Shatah [1] used only the following easy consequence of Corollary 7.2.4:

Lemma 7.8.1. *If $u \in C^\infty(\{(t, x) \in \mathbf{R}^{1+n}; 0 \leq t \leq T\})$ is rapidly decreasing as $x \to \infty$ and is a solution of the Cauchy problem*

$$(7.8.1) \qquad \Box u + u = f; \quad u(0, \cdot) = u_0, \quad \partial_t u(0, \cdot) = u_1,$$

then

$$(7.8.2) \quad \sup_x |u(t, x)| \leq C\Big((1 + t)^{-n/2} \sum_{j=0}^1 \sum_{|\alpha| + j \leq n+1} \int |\partial^\alpha u_j(x)| \, dx$$
$$+ \sum_{|\alpha| \leq n} \iint_{0 < s < t} (1 + t - s)^{-n/2} |\partial_x^\alpha f(s, x)| \, ds \, dx \Big), \quad 0 \leq t \leq T.$$

Proof. With E denoting the fundamental solution of $\Box + 1$, we have by d'Alembert's principle

$$(7.8.3) \qquad u(t,x) = u_0 * \partial_t E(t,\cdot) + u_1 * E(t,\cdot) + \int_0^t f(s,\cdot) * E(t-s,\cdot)\, ds.$$

Corollary 7.2.4 gives the desired estimate of the first two terms, and the estimate for $u_1 * E(t,\cdot)$ gives an estimate of $f(s,\cdot) * E(t-s,\cdot)$ which completes the proof.

Using Lemma 7.8.1 and the energy estimate Lemma 7.4.1 we shall now prove a crude existence theorem as a motivation for the main technical points in this section.

Proposition 7.8.2. *Assume that the function F in (7.1.1) is in C^∞ and vanishes of third order at 0, and that the number n of space variables is at least equal to 3. Then the equation (7.1.1) has a solution in $C^\infty(\mathbf{R}^{1+n})$ with Cauchy data (7.4.3) if $u_0, u_1 \in \mathcal{S}(\mathbf{R}^n)$ and ε is sufficiently small.*

Proof. It suffices to prove that there is a solution for $t \geq 0$. Let s be an integer $\geq 2n+6$. We shall prove that there is a number M independent of ε and T but depending on u_0, u_1 such that if u is a solution for $0 \leq t \leq T$ with the bounds

$$(7.8.4) \qquad \sum_{|\alpha| \leq s+1} \|\partial_{t,x}^\alpha u(t,\cdot)\| \leq M\varepsilon, \quad 0 \leq t \leq T,$$

$$(7.8.5) \qquad \sum_{|\alpha| \leq s/2+2} \sup_x |\partial_{t,x}^\alpha u(t,x)| \leq M\varepsilon(1+t)^{-n/2}, \quad 0 \leq t \leq T,$$

then (7.8.4), (7.8.5) are in fact valid with M replaced by $M/2$ if ε is small enough. (As usual $\|\cdot\|$ denotes the L^2 norm.) By the local existence theorems this implies that the set of T for which a solution satisfying (7.8.5) exists for $0 \leq t \leq T$ is open as well as closed, so T is arbitrary which will prove the statement.

To verify (7.8.4) with M replaced by $M/2$ we shall apply Lemma 7.4.1 to the equation (7.1.1) and the equations obtained by applying ∂^α to (7.1.1) when $0 < |\alpha| \leq s$. They can be written in the form

$$(7.8.6) \qquad \left(\Box + \sum_{j,k=0}^n \gamma^{jk} \partial_j \partial_k + 1\right)\partial^\alpha u = f_\alpha,$$

where $\gamma^{jk} = -\partial F/\partial u_{jk}''$ and f_α only depends on derivatives of u of order $\leq |\alpha| + 1$. Set $f_0 = F$. Since F vanishes of third order it follows from (7.8.5) that

$$|\gamma^{jk}(t,x)| \leq C(M\varepsilon)^2(1+t)^{-n}, \quad |\partial_{t,x}\gamma^{jk}(t,x)| \leq C(M\varepsilon)^2(1+t)^{-n}, \quad 0 \leq t \leq T.$$

Hence Lemma 7.4.2 applied to all the equations (7.8.6) and the analogue with $\alpha = 0$ show that for small ε

$$(7.8.7) \qquad \sum_{|\beta| \leq s+1} \|\partial^\beta u(t,\cdot)\| \leq C\left(\varepsilon + \sum_{|\alpha| \leq s} \int_0^t \|f_\alpha(s,\cdot)\|\, ds\right).$$

Using Taylor's formula we can write every f_α as a finite sum of products of at least three derivatives of u of order $\leq s+1$ with bounded coefficients, for (u, u', u'') is small by (7.8.5). At most one of the derivatives is of order $> s/2 + 2$ for producing a derivative of order μ requires at least $\mu - 2$ differentiations, so two such factors can only occur if $2(\mu - 2) \leq s$, that is, $\mu \leq s/2 + 2$. Using (7.8.4) and (7.8.5) we can therefore estimate the sum in the right-hand side of (7.8.7) and obtain since $\int_0^\infty (1+t)^{-n}\, dt < \infty$

$$\sum_{|\alpha|\leq s+1} \|\partial^\alpha u(t,\cdot)\| \leq C\varepsilon + C'(\varepsilon M)^3, \quad 0 \leq t < T.$$

If $M > 2C$ it follows for small ε that (7.8.4) is valid with M replaced by $M/2$.

It remains to prove a corresponding improvement of (7.8.5) using Lemma 7.8.1. To do so we write the differentiated equation in the form

$$(7.8.6)' \qquad\qquad (\Box + 1)\partial^\alpha u = F_\alpha$$

where F_α is of third order and contains derivatives of order $\leq |\alpha|+2$. Let $|\alpha|+2+n \leq s+1$, which is true if $|\alpha| \leq s/2+2$ and $s \geq 2n+6$. Every term in $\partial_x^\beta F_\alpha$, $|\beta| \leq n$, can be estimated by the product of two factors for which (7.8.4) is applicable and one for which (7.8.5) is valid, which proves that $\int |\partial_x^\beta F_\alpha(t,x)|\, dx \leq C(M\varepsilon)^3(1+t)^{-n/2}$. We can now apply (7.8.2) noting that

$$\int_0^t (1+t-s)^{-n/2}(1+s)^{-n/2}\, ds = 2\int_0^{t/2} (1+t-s)^{-n/2}(1+s)^{-n/2}\, ds$$

$$\leq 2^{1+n/2}(1+t)^{-n/2}\int_0^{t/2} (1+s)^{-n/2}\, ds.$$

The integral is bounded since $n \geq 3$ by hypothesis. Thus it follows from (7.8.2) that

$$\sum_{|\alpha|\leq s/2+2} \sup_x |\partial_{t,x}^\alpha u(t,x)| \leq C(\varepsilon + (M\varepsilon)^3)(1+t)^{-n/2}, \quad 0 \leq t \leq T,$$

where C is independent of M and ε. If $M > 2C$ it follows that (7.8.5) is valid with M replaced by $M/2$ when ε is small. This completes the proof.

Shatah [1] proved that Theorem 7.8.2 is also valid when F just vanishes of second order at 0. The first part of the proof above where we strengthened (7.8.4) works with no change in that case, but for the improvement of (7.8.5) by means of (7.8.2) it was essential that for the terms in $\partial^\alpha F$ we could bound two factors using (7.8.4) and one using (7.8.5). The essential idea of Shatah [1] is to split u into a sum of a term which is quite explicitly determined by u and a term which satisfies a Klein-Gordon equation with right-hand side of third order.

It is no restriction to assume that F is independent of u_{tt}'', for by the implicit function theorem the equation $\Box u + u = F(u, u', u'')$ with the derivatives of order ≤ 2 considered as independent variables can be solved for u_{tt}'' in a neighborhood of 0. This gives an equation of the same form where u_{tt}'' does not occur in the non-linear term. By Taylor's formula we can then write

$$(7.8.8) \qquad F(u, u', u'') = \sum_{j,k=0}^{1} \mathcal{A}_{jk}(\partial', \partial'')[\partial_t^j u][\partial_t^k u] + R(u, u', u'').$$

Here R vanishes of third order at 0 and $\mathcal{A}_{jk}$ is a real polynomial of degree $\leq 2 - j$ in ∂' which operates on the first factor $\partial_t^j u$ and of degree $\leq 2 - k$ in ∂'' which operates on the second factor $\partial_t^k u$, in both cases with respect to the space variables x. The Fourier transform with respect to x of an expression of the form $\mathcal{B}(D', D'')[u][v]$, where $D' = -i\partial'$, $D'' = -i\partial''$ and $u, v \in \mathcal{S}(\mathbf{R}^n)$, is

$$\xi \mapsto (2\pi)^{-n} \int \mathcal{B}(\xi - \eta, \eta) \hat{u}(\xi - \eta) \hat{v}(\eta) \, d\eta.$$

With this as definition we shall look for functions $\mathcal{B}_{jk}$, no longer polynomials, such that if u is a solution of (7.1.1) and

$$(7.8.9) \qquad U = \sum_{j,k=0}^{1} \mathcal{B}_{jk}(D', D'')[\partial_t^j u][\partial_t^k u], \quad V = u - U,$$

then $(\Box + 1)V = F(u, u', u'') - (\Box + 1)U$ is of third order in u. By Leibniz' rule

$$(1 - \Delta)U = \sum_{j,k=0}^{1} ((1 + |D'|^2) + (2\langle D', D'' \rangle - 1) + (1 + |D''|^2)) \mathcal{B}_{jk}(D', D'')[\partial_t^j u][\partial_t^k u],$$

$$\partial_t^2 U = \sum_{j,k=0}^{1} \mathcal{B}_{jk}(D', D'')([\partial_t^{j+2} u][\partial_t^k u] + 2[\partial_t^{j+1} u][\partial_t^{k+1} u] + [\partial_t^j u][\partial_t^{k+2} u]).$$

In the second equation we introduce

$$(7.8.10) \qquad \begin{aligned} \partial_t^{j+2} u &= \partial_t^j (\Delta u - u + F(u, u', u'')) = -(|D_x|^2 + 1)\partial_t^j u + \partial_t^j F(u, u', u''), \\ \partial_t^{j+1} u &= \partial_t^2 u = -(|D_x|^2 + 1)u + F(u, u', u''), \quad \text{if } \dot{\cdot} = 1, \end{aligned}$$

and similarly with $\dot{\cdot}$ replaced by k. This shows that $(\Box + 1)U$ apart from terms involving F or a derivative of F is equal to

$$\mathcal{B}_{00}(D', D'')(2\langle D', D'' \rangle - 1)[u][u] + 2\mathcal{B}_{00}(D', D'')[\partial_t u][\partial_t u]$$
$$+\mathcal{B}_{01}(D', D'')(2\langle D', D'' \rangle - 1)[u][\partial_t u] - 2\mathcal{B}_{01}(D', D'')(1 + |D''|^2)[\partial_t u][u]$$
$$+\mathcal{B}_{10}(D', D'')(2\langle D', D'' \rangle - 1)[\partial_t u][u] - 2\mathcal{B}_{10}(D', D'')(1 + |D'|^2)[u][\partial_t u]$$
$$+\mathcal{B}_{11}(D', D'')(2\langle D', D'' \rangle - 1)[\partial_t u][\partial_t u] + 2\mathcal{B}_{11}(D', D'')(1 + |D'|^2)(1 + |D''|^2)[u][u],$$

which is equal to the sum in (7.8.8) if

$$\mathcal{B}_{00}(\xi, \eta)(2\langle \xi, \eta \rangle - 1) + 2\mathcal{B}_{11}(\xi, \eta)(1 + |\xi|^2)(1 + |\eta|^2) = \mathcal{A}_{00}(i\xi, i\eta),$$
$$\mathcal{B}_{11}(\xi, \eta)(2\langle \xi, \eta \rangle - 1) + 2\mathcal{B}_{00}(\xi, \eta) = \mathcal{A}_{11}(i\xi, i\eta),$$
$$\mathcal{B}_{01}(\xi, \eta)(2\langle \xi, \eta \rangle - 1) - 2\mathcal{B}_{10}(\xi, \eta)(1 + |\xi|^2) = \mathcal{A}_{01}(i\xi, i\eta),$$
$$\mathcal{B}_{10}(\xi, \eta)(2\langle \xi, \eta \rangle - 1) - 2\mathcal{B}_{01}(\xi, \eta)(1 + |\eta|^2) = \mathcal{A}_{10}(i\xi, i\eta).$$

The solution is given by

$$\mathcal{B}_{00}(\xi, \eta) = ((1 - 2\langle \xi, \eta \rangle)\mathcal{A}_{00}(i\xi, i\eta) + 2(1 + |\xi|^2)(1 + |\eta|^2)\mathcal{A}_{11}(i\xi, i\eta))\mathcal{K}(\xi, \eta),$$
$$\mathcal{B}_{11}(\xi, \eta) = ((1 - 2\langle \xi, \eta \rangle)\mathcal{A}_{11}(i\xi, i\eta) + 2\mathcal{A}_{00}(i\xi, i\eta))\mathcal{K}(\xi, \eta),$$
$$(7.8.11) \quad \mathcal{B}_{01}(\xi, \eta) = ((1 - 2\langle \xi, \eta \rangle)\mathcal{A}_{01}(i\xi, i\eta) - 2(1 + |\xi|^2)\mathcal{A}_{10}(i\xi, i\eta))\mathcal{K}(\xi, \eta),$$
$$\mathcal{B}_{10}(\xi, \eta) = ((1 - 2\langle \xi, \eta \rangle)\mathcal{A}_{10}(i\xi, i\eta) - 2(1 + |\eta|^2)\mathcal{A}_{01}(i\xi, i\eta))\mathcal{K}(\xi, \eta),$$
$$\mathcal{K}(\xi, \eta) = \left(4(|\xi|^2|\eta|^2 - \langle \xi, \eta \rangle^2 + |\xi|^2 + |\eta|^2 + \langle \xi, \eta \rangle) + 3\right)^{-1}.$$

It is fairly obvious and will be made explicit later on (see (7.8.17)) that this implies that $(\Box + 1)V = F(u, u', u'') - (\Box + 1)U$ will be of third order in u. However, before discussing this point we must study the continuity properties of the bilinear operators $\mathcal{B}_{jk}$.

Let $P(x, y)$ be the inverse Fourier transform of a continuous bounded function $\mathcal{P}(\xi, \eta)$ in $\mathbf{R}^{n+n}$. We have $P \in \mathcal{S}'(\mathbf{R}^{n+n})$, and if $u, v \in \mathcal{S}(\mathbf{R}^n)$ then

$$(7.8.12) \quad \mathcal{P}(D', D'')[u][v](x) = \iint P(x - y, x - z)u(y)v(z)\, dy\, dz$$
$$= \iint P(y, z)u(x - y)v(x - z)\, dy\, dz,$$

which should be interpreted in the sense of distribution theory, as P acting on $u(x - \cdot) \otimes v(x - \cdot)$. The proof follows at once from Fourier's inversion formula if $\mathcal{P} \in \mathcal{S}$ and the definitions of $\hat{u}$ and $\hat{v}$ are introduced in the left-hand side, and the formula then follows in general by continuity.

We expect the bilinear operators $(u, v) \mapsto \mathcal{B}_{jk}(D', D'')[u][v]$ obtained to have estimates similar to those of products, with a possible loss of differentiability. This will be proved by the following analogue of Hölder's inequality.

Lemma 7.8.3. *If $P \in L^1(\mathbf{R}^{n+n})$ then*

$$\|\widehat{P}(D', D'')[u][v]\|_{L^r} \leq \|P\|_{L^1} \|u\|_{L^p} \|v\|_{L^q}, \quad u, v \in \mathcal{S}(\mathbf{R}^n),$$

if $1 \leq p, q, r \leq \infty$ and $1/r = 1/p + 1/q$.

Note that this becomes the standard Hölder inequality when P approaches the Dirac measure at 0.

Proof. By the Riesz convexity theorem it suffices to prove the statement in the extreme cases when $(1/p, 1/q, 1/r)$ equal to $(1, 0, 1)$, $(0, 1, 1)$ or $(0, 0, 0)$. The estimate is then a trivial consequence of (7.8.12).

In general it is of course difficult to decide for a given bounded $\mathcal{P}$ if the inverse Fourier transform is in L^1, and we shall only use the following simple sufficient condition:

Lemma 7.8.4. *If $\partial_\xi^\alpha \partial_\eta^\beta \mathcal{P}(\xi, \eta) \in L^2(\mathbf{R}^{n+n})$ when $|\alpha|, |\beta| \leq (n + 2)/2$, then the inverse Fourier transform P of $\mathcal{P}$ is in $L^1(\mathbf{R}^{n+n})$.*

Proof. Let k be the largest integer $\leq (n + 2)/2$, thus $2k \geq n + 1$. By Parseval's formula $P \in L^2$ and

$$\sum_{|\alpha|, |\beta| \leq k} \iint |x^\alpha y^\beta P(x, y)|^2\, dx\, dy < \infty.$$

Since $\iint (1 + |x|^2)^{-k}(1 + |y|^2)^{-k}\, dx\, dy < \infty$, the lemma follows from Cauchy-Schwarz' inequality.

If $n = 1$ then the function $\mathcal{K}$ defined by (7.8.11) is equal to $\left(4(|\xi|^2 + |\eta|^2 + \langle \xi, \eta \rangle) + 3\right)^{-1}$, and since $4(|\xi|^2 + |\eta|^2 + \langle \xi, \eta \rangle) + 3 \geq 2(|\xi|^2 + |\eta|^2) + 3$ the inverse Fourier transform K is the fundamental solution of an elliptic operator in $\mathbf{R}^2$ with only a logarithmic singularity at the origin, and it is exponentially decreasing at infinity by the Paley-Wiener theorem, so it is in $L^1(\mathbf{R}^2)$. We exclude this simple case in the following lemma.

Lemma 7.8.5. *If $\mathcal{K}$ is defined by (7.8.11) and $n \geq 2$, then*

$$(7.8.13) \qquad \mathcal{P}_{a,b}(\xi, \eta) = (1 + |\xi|^2)^{a/2}(1 + |\eta|^2)^{b/2}\mathcal{K}(\xi, \eta)$$

is in $L^2(\mathbf{R}^{n+n})$ if and only if $2\max(a,b)+n < 4$ and $2(a+b)+n < 3$; the second condition follows from the first if $n \geq 5$. These conditions also imply that $\partial_\xi^\alpha \partial_\eta^\beta \mathcal{P}_{a,b}(\xi,\eta) \in L^2(\mathbf{R}^{n+n})$ for all α, β.

Proof. If h is a nonnegative continuous function in $\mathbf{R}^3$ then

$$\iint_{\mathbf{R}^n \times \mathbf{R}^n} h(|\xi|^2, \langle \xi, \eta \rangle, |\eta|^2)\, d\xi\, d\eta$$

$$= C_n \iiint_{X,Y>0,|Z|<1} h(X^2, XYZ, Y^2)(XY)^{n-1}(1 - Z^2)^{(n-3)/2}\, dX\, dY\, dZ.$$

In fact, by the orthogonal invariance the integral with respect to ξ only depends on $Y = |\eta|$, so taking $\eta = (Y, 0, \ldots, 0)$ we find that the integral is equal to

$$C_n \int_0^\infty Y^{n-1}\, dY \iint_{r>0} h(r^2 + \xi_1^2, \xi_1 Y, Y^2)\, r^{n-2}\, dr\, d\xi_1.$$

If we introduce $X = \sqrt{r^2 + \xi_1^2}$ and $Z = \xi_1/\sqrt{r^2 + \xi_1^2}$ as new variables, the assertion follows. The problem is therefore to decide when

$$\iiint_{X,Y>0,|Z|<1} \frac{(1 + X^2)^a(1 + Y^2)^b(XY)^{n-1}(1 - Z^2)^{(n-3)/2}}{(X^2Y^2(1 - Z^2) + X^2 + Y^2 + 1)^2}\, dX\, dY\, dZ < \infty.$$

The integral when $\max(X, Y) < 1$ is convergent since $n \geq 2$, so it suffices to estimate the integral when $X < Y$ and $Y > 1$ or $Y < X$ and $X > 1$. In the first case we can replace $1 + Y^2$ by Y^2 and drop $X^2 + 1$ in the denominator which leads to the equivalent integral

$$\iiint_{|Z|<1<Y,0<X<Y} (1 + X^2)^a X^{n-1} Y^{2b+n-5}(1 - Z^2)^{(n-3)/2}(X^2(1 - Z^2) + 1)^{-2}\, dX\, dY\, dZ.$$

The integral with respect to Y diverges unless $2b+n < 4$, and when X, Y are interchanged we get the necessary condition $2a + n < 4$. When these are fulfilled, the integral when $X < 1$ is obviously convergent. The remaining integral when $1 < X < Y$ is convergent if and only if

$$\iint_{|Z|<1,X>1} X^{2a+2b+2n-5}(1 - Z^2)^{(n-3)/2}(X^2(1 - Z^2) + 1)^{-2}\, dX\, dZ < \infty.$$

The integral when $|Z| < 1/2$ converges since $2a + 2b + 2n < 8$ by the necessary conditions already obtained. When $1/2 < Z < 1$ we can take $1 - Z^2$ as a new variable, so what remains is to determine when

$$\iint_{X>1,0<t<1} X^{2a+2b+2n-5} t^{(n-3)/2}(X^2 t + 1)^{-2}\, dX\, dt < \infty.$$

Here

$$\int_0^1 t^{(n-3)/2}(X^2 t + 1)^{-2}\, dt = X^{1-n} \int_0^{X^2} t^{(n-3)/2}(t + 1)^{-2}\, dt.$$

The integral is bounded when $X \to \infty$ if $n \leq 4$, it is $O(\log X)$ if $n = 5$ and it is $O(X^{n-5})$ if $n > 5$. When $n \geq 5$ the convergence of the integral is therefore a consequence of the necessary conditions already given, but when $n \leq 4$ it requires in addition that $2(a+b)+n < 3$, which completes the proof of the first part of the lemma.

With $\mathcal{I}(\xi,\eta) = 1/\mathcal{K}(\xi,\eta)$ we have

$$\mathcal{I}(\xi,\eta) \geq 2(|\xi|^2 + |\eta|^2) + 3 \geq 4\sqrt{|\xi|^2 + |\eta|^2},$$
$$|\partial(|\xi|^2|\eta|^2 - \langle\xi,\eta\rangle^2)/\partial(\xi,\eta)|^2 = 2(|\xi|^2 + |\eta|^2)(|\xi|^2|\eta|^2 - \langle\xi,\eta\rangle^2) \leq \mathcal{I}(\xi,\eta)^2/4.$$

The first inequality follows from the inequality between geometric and arithmetic means. By the orthogonal invariance it suffices to prove the second (in)equality when $\xi_j = \eta_j = 0$ for $j > 2$. Since $|\xi|^2|\eta|^2 - \langle\xi,\eta\rangle^2 = (\xi_1\eta_2 - \xi_2\eta_1)^2 + O((|\xi|^2 + |\eta|^2)\sum_{j>2}(|\xi_j|^2 + |\eta_j|^2))$ it is obvious then. Hence we conclude that $|\partial\mathcal{I}(\xi,\eta)/\partial(\xi,\eta)| \leq 5\mathcal{I}(\xi,\eta)$ which proves that

$$|\partial_\xi^\alpha \partial_\eta^\beta \mathcal{I}(\xi,\eta)| \leq C_{\alpha\beta}\mathcal{I}(\xi,\eta)$$

for all α and β, for this is obvious when $|\alpha| + |\beta| \geq 2$. Since $\mathcal{K}(\xi,\eta)\mathcal{I}(\xi,\eta) = 1$ it follows inductively by differentiation of this equation that

$$|\partial_\xi^\alpha \partial_\eta^\beta \mathcal{K}(\xi,\eta)| \leq C'_{\alpha\beta}\mathcal{K}(\xi,\eta).$$

Hence

$$|\partial_\xi^\alpha \partial_\eta^\beta \mathcal{P}_{a,b}(\xi,\eta)| \leq C''_{\alpha\beta}\mathcal{P}_{a,b}(\xi,\eta),$$

which concludes the proof.

Lemma 7.8.6. *If $1 \leq p,q,r \leq \infty$ and $1/r = 1/p + 1/q$ then*

$$(7.8.14) \qquad \|\mathcal{K}(D',D'')[u][v]\|_{L^r} \leq C_n \sum_{|\alpha|\leq\nu} \|\partial^\alpha u\|_{L^p} \sum_{|\beta|\leq\nu} \|\partial^\beta v\|_{L^q}, \quad u,v \in \mathcal{S}(\mathbf{R}^n),$$

if ν is the smallest integer $> (n-4)/2$ when $n > 3$, $\nu = 1$ when $n = 3$ and $\nu = 0$ when $n = 1, 2$.

Proof. When $n = 1$ we observed before stating Lemma 7.8.5 that the inverse Fourier transform of $\mathcal{K}$ is in L^1, so (7.8.14) follows from Lemma 7.8.3 in that case. If $n \geq 2$ we can apply Lemma 7.8.5 with $a = b = -\nu$. We can write

$$\mathcal{K}(\xi,\eta) = (1+|\xi|^2)^\nu(1+|\eta|^2)^\nu(1+|\xi|^2)^{-\nu}(1+|\eta|^2)^{-\nu}\mathcal{K}(\xi,\eta) = \sum_{|\alpha|\leq\nu,|\beta|\leq\nu} c_{\alpha\beta}\xi^\alpha\eta^\beta\mathcal{K}_{\alpha\beta}(\xi,\eta)$$

where $c_{\alpha\beta}$ are constants and

$$\mathcal{K}_{\alpha\beta}(\xi,\eta) = \xi^\alpha(1+|\xi|^2)^{-\nu/2}\eta^\beta(1+|\eta|^2)^{-\nu/2}\mathcal{P}_{-\nu,-\nu}(\xi,\eta).$$

It follows from Lemma 7.8.5 that $\mathcal{K}_{\alpha\beta}$ and all its derivatives are in L^2. Hence the inverse Fourier transform is in L^1 by Lemma 8.5.4, and (7.8.14) follows from Lemma 7.8.3 since

$$\mathcal{K}(D',D'')[u][v] = \sum_{|\alpha|\leq\nu}\sum_{|\beta|\leq\nu} c_{\alpha\beta}\mathcal{K}_{\alpha\beta}(D',D'')[D'^\alpha u][D''^\beta v].$$

The proof is complete.

The inverse Fourier transform K of $\mathcal{K}$ is real valued since $\mathcal{K}$ is real valued and even. This implies that $\mathcal{B}_{jk}(D',D'')[u][v]$ is real valued when u and v are, for $\mathcal{B}_{jk}(D',D'')$ differs from $\mathcal{K}$ only by a differential operator with real coefficients, of order $\leq 3 - j$ when acting on the first argument and order $\leq 3 - k$ when acting on the second argument.

We are now ready to prove the full result of Shatah [1]:

Theorem 7.8.7. *Assume that the function F in (7.1.1) is in C^∞ and vanishes of second order at 0, and that the number n of space variables is at least equal to 3. Then the equation (7.1.1) has a solution in $C^\infty(\mathbf{R}^{1+n})$ with Cauchy data (7.4.3) if $u_0, u_1 \in \mathcal{S}(\mathbf{R}^n)$ and ε is sufficiently small.*

Proof. To a large extent we can repeat the proof of Proposition 7.8.2. Let s be an integer $\geq 3n + 11$. It suffices to prove that there is a number M independent of ε and T but depending on u_0, u_1 such that if u is a solution for $0 \leq t \leq T$ with the bounds (7.8.4) and (7.8.5) then (7.8.4) and (7.8.5) are in fact valid with M replaced by $M/2$ if ε is small enough. As before we obtain

$$\sum_{|\alpha| \leq s+1} \|\partial^\alpha u(t, \cdot)\| \leq C\varepsilon + C'(\varepsilon M)^2, \quad 0 \leq t \leq T;$$

the only difference is that $(\varepsilon M)^3$ has now been replaced by $(\varepsilon M)^2$. However, if $M > 2C$ it still follows for small ε that (7.8.4) is valid with M replaced by $M/2$.

To prove (7.8.5) with M replaced by $M/2$ we write $u = U + V$ as in (7.8.9). To estimate $\partial^\gamma U$ when $|\gamma| \leq \mu = s/2 + 2$ we note that

$$\partial^\gamma U = \sum_{\gamma' + \gamma'' = \gamma} \frac{\gamma!}{\gamma'!\gamma''!} \sum_{j,k=0}^{1} \mathcal{B}_{jk}(D', D'')[\partial^{\gamma'} \partial_t^j u][\partial^{\gamma''} \partial_t^k u],$$

since Leibniz' rule is valid for the bilinear maps $\mathcal{B}_{jk}(D', D'')$. From (7.8.14) with $p = q = r = \infty$ it follows that

$$(7.8.15) \qquad \sum_{|\gamma| \leq \mu} |\partial^\gamma U(t, \cdot)| \leq C \sum_{\substack{|\gamma' + \gamma''| \leq \mu \\ |\alpha| \leq \nu+3, |\beta| \leq \nu+3}} \sup |\partial^{\gamma'+\alpha} u(t, \cdot)| \sup |\partial^{\gamma''+\beta} u(t, \cdot)|.$$

Assuming as we may that $|\gamma'| \leq |\gamma''|$ we have $|\gamma'| \leq \mu/2$ and $|\gamma'| + |\alpha| \leq \mu/2 + \nu + 3 \leq \mu$ since $2\nu + 6 \leq n + 5 \leq \mu$. The first factor in the sum in the right-hand side of (7.8.15) can therefore be estimated using (7.8.5). By Sobolev's lemma and (7.8.4) the second factor is $\leq CM\varepsilon$, for $|\gamma''| + |\beta| + (n+2)/2 \leq s/2 + 2 + \nu + 3 + (n+2)/2 \leq s + 1$. Thus

$$(7.8.16) \qquad |\partial^\gamma U(t, \cdot)| \leq C(M\varepsilon)^2 (1+t)^{-n/2}, \quad 0 \leq t \leq T.$$

What remains is to show that the other term V in (7.8.9) can be estimated essentially as in the proof of Proposition 7.8.2. With the notation R in (7.8.8) for the nonlinear perturbation apart from second order terms, we have

$$(7.8.17) \qquad (\square + 1)\partial^\alpha V = \partial^\alpha \left(R(u, u', u'') - \sum_{1}^{4} S_j(u, u', u'') \right) = F_\alpha, \quad \text{where}$$

$$S_1(u, u', u'') = \sum_{j,k=0}^{1} \left(\mathcal{B}_{jk}(D', D'')[\partial_t^j F(u, u', u'')][\partial_t^k u] + \mathcal{B}_{jk}(D', D'')[\partial_t^j u][\partial_t^k F(u, u', u'')] \right),$$

$$S_2(u, u', u'') = \mathcal{B}_{10}(D', D'')[F(u, u', u'')][\partial_t u] + \mathcal{B}_{01}(D', D'')[\partial_t u][F(u, u', u'')],$$

$$S_3(u, u', u'') = \mathcal{B}_{11}(D', D'')[F(u, u', u'')][F(u, u', u'')],$$

$$S_4(u, u', u'') = \mathcal{B}_{11}(D', D'')[\Delta u - u][F(u, u', u'')] + \mathcal{B}_{11}(D', D'')[F(u, u', u'')][\Delta u - u].$$

An estimate of the L^1 norm of $\partial^\alpha R(u, u', u'')$ was given in the proof of Proposition 7.8.2. By Lemma 7.8.6 with $r = 1$ and $p = q = 2$ we have

$$\|\partial^\gamma S_1(t, \cdot)\|_{L^1} \leq C \sum_{\substack{\gamma'+\gamma''=\gamma \\ |\alpha|\leq 3+\nu, |\beta|\leq 3+\nu}} \|\partial^{\gamma'+\alpha} F(u, u', u'')(t, \cdot)\| \|\partial^{\gamma''+\beta} u(t, \cdot)\|.$$

If $|\gamma| \leq \mu + n$ then $|\gamma| + |\alpha| + 2 \leq \mu + n + 5 + \nu \leq s + 1$, for $2\nu \leq n - 1$ and $s \geq 3n + 11$. Hence we have

$$\|\partial^{\gamma'+\alpha} F(u, u', u'')(t, \cdot)\| \leq C(M\varepsilon)^2(1 + t)^{-n/2}, \quad \|\partial^{\gamma''+\beta} u(t, \cdot)\| \leq CM\varepsilon,$$

which gives $\|\partial^\gamma S_1(t, \cdot)\|_{L^1} \leq C(M\varepsilon)^3(1+t)^{-n/2}$. Since S_2, S_3 and S_4 can be estimated in the same way it follows that

$$\int |F_\gamma(t, x)| \, dx \leq C(M\varepsilon)^3(1+t)^{-n/2}, \quad 0 \leq t \leq T, \ |\gamma| \leq \mu + n.$$

Now an application of Lemma 7.8.1 as in the proof of Proposition 7.8.2 proves that

$$(7.8.18) \qquad \sum_{|\gamma|\leq s/2+2} \sup_x |\partial^\gamma V(t, x)| \leq C(\varepsilon + (M\varepsilon)^3)(1+t)^{-n/2}, \quad 0 \leq t \leq T.$$

If $M > 2C$ it follows from (7.8.16) and (7.8.18) that (7.8.5) is valid with M replaced by $M/2$ when ε is small enough. This completes the proof.

There are two reasons why the preceding proof is not applicable when $n = 2$. The first is that in (7.8.6) we would only obtain $|\gamma^{jk}(t, x)| \leq CM\varepsilon(1+t)^{-1}$, which is not integrable with respect to t. This problem disappears if F is independent of u'', or at least the quadratic terms in F only depend on u and on u'. The second reason is that

$$(2 + t)\int_0^t (1 + t - s)^{-1}(1 + s)^{-1} = 2\log(1 + t)$$

is not bounded. This obstacle was overcome in Ozawa, Tsutaya and Tsutsumi [1] by combining the methods used to prove Theorem 7.8.7 with the methods used to prove Theorem 6.4.2 here. Thus (7.8.4), (7.8.5) are modified to conditions on the action on u of operators commuting with the unperturbed operator. This program requires the use of improvements of the a priori estimates of Section 7.3 due to Georgiev [1] which are also applicable when the Cauchy data do not have compact support. The result is that Theorem 7.8.7 remains valid when $n = 2$ for perturbations of the form $F(u, u')$ vanishing of second order at the origin. This was proved earlier with another method by Simon and Taflin [1]. When $n = 1$ it was proved in Moriyama, Tonegawa and Tsutsumi [1] with methods similar to those of Ozawa, Tsutaya and Tsutsumi [1] that the lifespan of the solution is $\geq A\exp(B\varepsilon^{-2})$ for small ε if $F = F(u, u')$. We refer to these papers for the details of proof. Some of them have already been given here, for Lemmas 7.8.3–7.8.6 are also valid for $n = 1$ or 2.

Yordanov [1] has proved that in general there is no better bound for the lifespan when $n = 1$ than those just described, if F depends on u':

Proposition 7.8.8. *For the solution u of the differential equation*

$$u''_{tt} - u''_{xx} + u = u'^2_t u'_x$$

with Cauchy data $u(0, x) = \varepsilon u_0(x)$, $u_t'(0, x) = \varepsilon u_1(x)$ where $u_0, u_1 \in C_0^\infty([-R, R])$, the lifespan is $\leq R(\exp(2/\sigma\varepsilon^2) - 1)$ if $\sigma = \int_{\mathbf{R}} u_0'(x)u_1(x)\, dx > 0$.

Proof. If a solution exists for $0 \leq t \leq T$ and $I(t) = \int_{\mathbf{R}} u_t'(t, x)u_x'(t, x)\, dx$, then

$$I'(t) = \int (u_{tt}'' u_x' + u_t' u_{xt}'')\, dx = \int (u_{xx}'' - u + u_t'^2 u_x')u_x'\, dx = \int (u_t' u_x')^2\, dx \geq 0$$

when $0 \leq t \leq T$. Since $u = 0$ when $|x| > t + R$ it follows that

$$I(t)^2 \leq 2(t + R) \int (u_t' u_x')^2\, dx = 2(t + R)I'(t),$$

which means that $-d(1/I(t))/dt = I'(t)/I(t)^2 \geq 1/(2(t + R))$. Hence

$$2/I(0) - 2/I(t) \geq \log((t + R)/R), \quad 0 \leq t \leq T,$$

and since $I(t) \geq I(0) = \sigma\varepsilon^2$, it follows that $T \leq R(\exp(2/\sigma\varepsilon^2) - 1)$.

The existence theorems just mentioned for $n = 1$ or $n = 2$ assume a fairly rapid decay of the initial data at infinity. When $n = 1$ and the data just decay as $H_{(s)}$ functions then Delort [1] has proved that the lifespan can be bounded from below by $c\varepsilon^{-4}|\log\varepsilon|^{-6}$.

CHAPTER VIII

MICROLOCAL ANALYSIS

8.1. Introduction. The study of singularities of solutions of a (hyperbolic) differential equation is simplified and the results are improved by taking what is now known as a microlocal point of view. To motivate it we consider the solution of the Cauchy problem in $\mathbf{R}^{1+n}$,

$$\Box u = 0; \qquad u = u_0, \quad \partial_t u = u_1 \quad \text{when } t = 0;$$

where u_0 and u_1 are smooth except at the origin. The solution is for $t > 0$

$$(8.1.1) \qquad\qquad u = u_0 * \partial_t E + u_1 * E,$$

where $E = c_n \chi_+^{\frac{1-n}{2}}(t^2 - |x|^2)$, $t > 0$, and the convolution is taken in the space variables for fixed t. It is clear that the solution is smooth except when $t = |x|$. If we now solve the Cauchy problem for $t > T$ with the Cauchy data of u for $t = T$, we get of course the same solution u. However, if we apply (8.1.1) we might expect to have singularities at $(t, y + x)$ for all y, x with $|y| = T$ and $|x| = t - T$, whether x is proportional to y or not. Thus the Cauchy data of u at (T, y) must have some special property which causes the singularity to propagate only in the direction (T, y) from that point. Now a look at (8.1.1) shows that the Cauchy data are essentially functions (or rather distributions) of $T^2 - |x|^2$ there. In a small neighborhood this is close to the linear function $x \mapsto -2\langle x - y, y\rangle$ so the Fourier transform of the Cauchy data localized to a small neighborhood should be small at infinity except near the direction y.

The preceding observation leads to the definition in Section 8.2 of the wave front set $WF(u)$ of a distribution u in $\mathbf{R}^n$. It is a subset of $\mathbf{R}^n \times (\mathbf{R}^n \setminus 0)$; one can think of it as the complement of the set of all (x, ξ) such that u is smooth at x as far as oscillations with frequency in the direction ξ are concerned. This notion allows one to define for example that $u \in H_{(s)}$ (microlocally) at (x, ξ) if $(x, \xi) \notin WF(u - v)$ for some $v \in H_{(s)}$ in the usual (global) sense.

In Section 8.3 we use the basic definitions to discuss directly the microlocal regularity of products. However, to work efficiently microlocally one needs some basic facts on pseudo-differential operators. These are summed up without proofs in Section 8.4. Pseudo-differential operators allow one to give an alternative and very natural definition of the wave front set. They were designed for linear differential operators with smooth coefficients though. To motivate what one needs in the study of nonlinear operators we reexamine in Section 8.5 the results on composite functions proved in Section 6.4 and find that one must be able to work with pseudo-differential operators having much more general symbols than those considered in Section 8.4. These will be the subject of Chapters 9 and 10, and the techniques developed there will be important in Chapter XI. Section 8.6, finally, contains the facts on Hölder and Zygmund spaces which are needed in Chapter 10.

8.2. The wave front set. Let $v \in \mathcal{E}'(\mathbf{R}^n)$ and let $\hat{v}(\xi) = v(e^{-i\langle x, \xi\rangle})$ be the Fourier transform of v. In the introduction we motivated studying the set $\Sigma(v) \subset \mathbf{R}^n \setminus 0$ such that $\eta \notin \Sigma(v)$ is equivalent to

$$(8.2.1) \qquad\qquad |\hat{v}(\xi)| \le C_N(1 + |\xi|)^{-N}, \qquad \xi \in V,$$

for all integers N and a conic neighborhood V of η independent of N. It is clear that $v \in C_0^\infty$ if and only if $\Sigma(v) = \emptyset$. The following elementary result is basic:

Lemma 8.2.1. If $\varphi \in C_0^\infty(\mathbf{R}^n)$ and $v \in \mathcal{E}'(\mathbf{R}^n)$ then

$$(8.2.2) \qquad \Sigma(\varphi v) \subset \Sigma(v).$$

Proof. Let Γ be a closed cone disjoint with $\Sigma(v)$. We must show that $\widehat{\varphi v}$ is rapidly decreasing in Γ. To do so we write $\hat{v} = V_1 + V_2$ where V_1 is rapidly decreasing and supp $V_2 \subset \Gamma_2$ where Γ_2 is a closed cone meeting Γ only at 0, thus

$$(8.2.3) \qquad |\xi - \eta| \geq \delta|\xi|, \qquad \text{if } \xi \in \Gamma, \quad \eta \in \Gamma_2,$$

for some $\delta > 0$. Then

$$(2\pi)^n \widehat{\varphi v}(\xi) = \sum_1^2 \int \hat{\varphi}(\xi - \eta) V_j(\eta)\, d\eta,$$

where the term with $j = 1$ is rapidly decreasing since $\hat{\varphi} \in \mathcal{S}(\mathbf{R}^n)$. If M is the order of v, we have $|V_2(\eta)| \leq C(1 + |\eta|)^M$. Hence

$$(1 + |\xi|)^N \left| \int \hat{\varphi}(\xi - \eta) V_2(\eta)\, d\eta \right| \leq C \int_{\Gamma_2} (1 + |\xi|)^N (1 + |\eta|)^M |\hat{\varphi}(\xi - \eta)|\, d\eta,$$

and since (8.2.3) implies that

$$1 + |\xi| \leq 1 + |\xi - \eta|/\delta, \qquad 1 + |\eta| \leq 1 + (1 + 1/\delta)|\xi - \eta|, \qquad \text{if } \xi \in \Gamma, \quad \eta \in \Gamma_2,$$

we obtain a bound for the right-hand side when $\xi \in \Gamma$. The proof is complete.

Remark. Later on we shall use that the proof is valid for any $v \in \mathcal{S}'$ such that $\hat{v}$ has a polynomial bound.

If X is an open set in $\mathbf{R}^n$ and $u \in \mathcal{D}'(X)$, we set for $x \in X$

$$(8.2.4) \qquad \Sigma_x(u) = \bigcap_{\varphi \in C_0^\infty(X), \varphi(x) \neq 0} \Sigma(\varphi u).$$

Lemma 8.2.1 implies

$$(8.2.5) \qquad \Sigma(\varphi u) \to \Sigma_x(u) \qquad \text{if } \varphi \in C_0^\infty(X), \quad \varphi(x) \neq 0, \quad \text{and supp } \varphi \to \{x\},$$

for $\Sigma(\psi u) \subset \Sigma(\varphi u)$ if $\varphi \neq 0$ in supp ψ. Thus $\Sigma_x(u)$ only depends on the local behavior of u at x, and $\Sigma_x(u) = \emptyset$ if and only if $x \notin$ sing supp u. The sets Σ_x are collected in the wave front set:

Definition 8.2.2. If $u \in \mathcal{D}'(X)$, where $X \subset \mathbf{R}^n$ is an open set, then the set

$$WF(u) = \{(x, \xi) \in X \times (\mathbf{R}^n \setminus 0); \xi \in \Sigma_x(u)\}$$

is called the wave front set of u.

It is clear that the wave front set is closed in $X \times (\mathbf{R}^n \setminus 0)$ and that the projection in X is equal to sing supp u. If T is a nonsingular linear transformation and $V = v \circ T, v \in \mathcal{E}'$, then $\widehat{V} = |\det T|^{-1} \hat{v} \circ {}^t T^{-1}$ which shows that $\Sigma(V) = {}^t T(\Sigma(v))$. This proves the following theorem for linear maps:

Theorem 8.2.3. *Let X and Y be open subsets of $\mathbf{R}^n$ and let $f : X \to Y$ be a diffeomorphism. Then it follows that*

$$WF(u \circ f) = \{(x, {}^t f'(x)\xi); (f(x), \xi) \in WF(u)\}.$$

We refer to Hörmander [4, Theorem 8.2.4] for a general proof since we shall only work in $\mathbf{R}^n$ here and do not need a general change of variables.

One can find distributions with arbitrarily prescribed wave front sets. The following example gives a construction for a ray, which will be needed later on; for a more general but less explicit construction we refer to Hörmander [4, Theorem 8.1.4].

Example 8.2.4. Let $\chi \in C^\infty(\mathbf{R})$ be equal to 1 in $(-\infty, \frac{1}{2})$ and equal to 0 in $(1, \infty)$, and define $u \in \mathcal{S}'(\mathbf{R}^n)$ by

$$\hat{u}(\xi) = (1 - \chi(\xi_1))\xi_1^{-s}\chi((\xi_2^2 + \cdots + \xi_n^2)/\xi_1^{2\varrho}),$$

where $0 < \varrho < 1$ and $s \in \mathbf{R}$. Then

$$WF(u) = \{(0, \xi); \xi_2 = \cdots = \xi_n = 0, \ \xi_1 > 0\},$$

and u coincides with a function in $\mathcal{S}(\mathbf{R}^n)$ outside a neighborhood of the origin.

Proof. Since $\Sigma(u) \subset \{\xi; \xi_2 = \cdots = \xi_n = 0, \xi_1 > 0\}$ it suffices by the remark after Lemma 8.2.1 to prove the last statement. The Fourier transform of $x^\alpha D^\beta u$ is $(-D)^\alpha \xi^\beta \hat{u}(\xi)$. If a product $\prod \xi_j^{a_j}$ is given the weight $a_1 + \varrho \sum_2^n a_j$ then differentiation with respect to ξ_j decreases the weight by 1 if $j = 1$ and by ϱ if $j \neq 1$. Since $(\xi_2^2 + \cdots + \xi_n^2)/\xi_1^{2\varrho}$ is of weight 0 it follows that $(-D)^\alpha \xi^\beta \hat{u}(\xi)$, apart from a term in C_0^∞ obtained when $1 - \chi(\xi_1)$ is differentiated, is a sum with terms of weight $\beta_1 - s - \alpha_1 + \varrho(\beta_2 - \alpha_2 + \cdots + \beta_n - \alpha_n)$, so it is integrable if

$$\beta_1 - s - \alpha_1 + 1 + \varrho(\beta_2 - \alpha_2 + \cdots + \beta_n - \alpha_n + n - 1) < 0.$$

Hence $x^\alpha D^\beta u$ is a bounded continuous function for any β if $|\alpha|$ is large enough, and this completes the proof.

It is useful to think of the statement $(x, \xi) \notin WF(u)$ as meaning that u belongs to C^∞ at (x, ξ); indeed, the condition (8.2.1) means that the Fourier transform of u, appropriately localized at x, behaves in the direction ξ as if u were really in C^∞. This suggests a definition of other regularity properties at (x, ξ), such as:

Definition 8.2.5. If $u \in \mathcal{D}'(X)$ and $(x, \xi) \in X \times (\mathbf{R}^n \setminus 0)$, then u is said to be in $H_{(s)}^{\mathrm{loc}}$ at (x, ξ) if $(x, \xi) \notin WF(u - v)$ for some $v \in H_{(s)}(\mathbf{R}^n)$.

We recall that $v \in H_{(s)}(\mathbf{R}^n)$ if $v \in \mathcal{S}'(\mathbf{R}^n)$ and $(1 + |\xi|^2)^{s/2}\hat{v}(\xi) \in L^2(\mathbf{R}^n)$. The L^2 norm of this function is $(2\pi)^{n/2}$ times the norm in $H_{(s)}(\mathbf{R}^n)$. If $\varphi \in \mathcal{S}(\mathbf{R}^n)$ then $\varphi v \in H_{(s)}(\mathbf{R}^n)$ too; if $(x, \xi) \notin WF(u - v)$ and supp φ is sufficiently close to x, then the Fourier transform of $\varphi(u - v)$ is rapidly decreasing in a conic neighborhood Γ of ξ, hence

$$(8.2.6) \qquad\qquad \int_\Gamma |\widehat{\varphi u}(\xi)|^2 (1 + |\xi|^2)^s \, d\xi < \infty.$$

Conversely, if (8.2.6) holds for some $\varphi \in C_0^\infty(X)$ with $\varphi(x) \neq 0$, then $u \in H_{(s)}^{\mathrm{loc}}$ at (x, ξ). In fact, we can define $v \in H_{(s)}(\mathbf{R}^n)$ by

$$\hat{v}(\xi) = \begin{cases} \widehat{\varphi u}(\xi), & \text{if } \xi \in \Gamma \\ 0, & \text{if } \xi \notin \Gamma. \end{cases}$$

Then $\hat{w} = \widehat{\varphi u} - \hat{v}$ vanishes in Γ, and since $\hat{w}$ has a polynomial bound the remark following the proof of Lemma 8.2.1 gives that $(x, \xi) \notin WF(w)$. We can choose $\psi \in C_0^\infty$ so that $\psi \varphi = 1$ in a neighborhood of x. Then $\psi v \in H_{(s)}$, $(x, \xi) \notin WF(\psi w)$, and $u - \psi v = (1 - \psi \varphi)u + \psi w$, so $(x, \xi) \notin WF(u - \psi v)$. We have proved:

Proposition 8.2.6. *If $u \in \mathcal{D}'(X)$ and (8.2.6) is valid for some $\varphi \in C_0^\infty(X)$ with $\varphi(x) \neq 0$, then $u \in H_{(s)}^{loc}$ at (x, ξ). Conversely, if this is true then (8.2.6) holds for all $\varphi \in C_0^\infty$ with supp φ sufficiently close to x. Thus $u \in H_{(s)}^{loc}$ at x_0 if $u \in H_{(s)}^{loc}$ at (x_0, ξ_0) for every $\xi_0 \neq 0$.*

As an example we note for later reference that the distribution u in Example 8.2.4 is in $H_{(t)}^{loc}$ at $(0, (1, 0, \ldots, 0))$ if and only if

$$(8.2.7) \qquad\qquad t < s - (1 + \varrho(n - 1))/2.$$

8.3. Microlocal regularity of a product. If $u_j \in H_{(s_j)}(\mathbf{R}^n)$ for $j = 1, 2$, then the product $u_1 u_2$ is a well defined distribution in $\mathcal{S}'(\mathbf{R}^n)$ if $s_1 + s_2 \geq 0$; if $\varphi \in \mathcal{S}(\mathbf{R}^n)$ then

$$|\langle u_1 u_2, \varphi \rangle| = |\langle u_1, u_2 \varphi \rangle| \leq \|u_1\|_{(s_1)} \|u_2 \varphi\|_{(s_2)} \leq C(\varphi) \|u\|_{(s_1)} \|u_2\|_{(s_2)},$$

for the dual space of $H_{(s_1)}(\mathbf{R}^n)$ is $H_{(-s_1)}(\mathbf{R}^n) \supset H_{(s_2)}(\mathbf{R}^n)$. We shall prove

Theorem 8.3.1. *If $u_j \in H_{(s_j)}(\mathbf{R}^n)$, $j = 1, 2$, and $s_1 + s_2 \geq 0$, then $u_1 u_2 \in H_{(s)}(\mathbf{R}^n)$ if*

$$(8.3.1) \qquad\qquad s \leq s_j, \quad j = 1, 2, \quad \text{and } s \leq s_1 + s_2 - n/2,$$

with the second inequality strict if s_1 or s_2 or $-s$ is equal to $n/2$.

For the proof we need an elementary lemma:

Lemma 8.3.2. *Let F be a piecewise continuous function in $\mathbf{R}^n \times \mathbf{R}^n$ and set*

$$T_F(f, g)(\xi) = \int F(\xi, \eta) f(\eta) g(\xi - \eta) \, d\eta$$

when $f, g \in C_0(\mathbf{R}^n)$. Then $\|T_F(f, g)\| \leq M \|f\| \|g\|$, with L^2 norms, if

$$(8.3.2) \qquad \int |F(\xi, \eta)|^2 \, d\eta \leq M^2 \quad \forall \xi \quad \text{or} \quad \int |F(\xi, \eta)|^2 \, d\xi \leq M^2 \quad \forall \eta.$$

Proof. If the first condition in (8.3.2) is fulfilled, then

$$|T_F(f, g)(\xi)|^2 \leq M^2 \int |f(\eta) g(\xi - \eta)|^2 \, d\eta,$$

by Cauchy-Schwarz' inequality which proves the statement then. If the second condition is fulfilled we note that when $h \in C_0(\mathbf{R}^n)$ then

$$\left| \int T_F(f, g)(\xi) h(\xi) \, d\xi \right| = \left| \int f(\eta) \, d\eta \int F(\xi, \eta) g(\xi - \eta) h(\xi) \, d\xi \right| \leq M \|f\| \|g\| \|h\|$$

by the estimate already proved, which completes the proof.

Remark. Replacing η by $\xi - \eta$ in the integral we find that

$$(8.3.2)' \qquad\qquad \int |F(\xi, \xi - \eta)|^2 \, d\xi \leq M^2 \quad \forall \eta$$

also implies the same bound for the bilinear map.

Proof of Theorem 8.3.1. We have

$$\langle\xi\rangle^s \widehat{u_1 u_2}(\xi) = (2\pi)^{-n} \int F(\xi,\eta) f_1(\xi-\eta) f_2(\eta)\, d\eta$$

where $f_j(\xi) = \langle\xi\rangle^{s_j} \hat{u}_j(\xi)$, $\langle\xi\rangle = (1+|\xi|^2)^{\frac{1}{2}}$ and

$$F(\xi,\eta) = \langle\xi\rangle^s \langle\xi-\eta\rangle^{-s_1} \langle\eta\rangle^{-s_2}.$$

Let us first assume that $\langle\eta\rangle < \langle\xi\rangle/2$, that is, multiply F by the characteristic function of $\{(\xi,\eta); \langle\eta\rangle < \langle\xi\rangle/2\}$. Then we have $|F(\xi,\eta)| \leq C\langle\xi\rangle^{s-s_1}\langle\eta\rangle^{-s_2}$. If $s_2 < n/2$ then $\int |F(\xi,\eta)|^2 d\eta$ is bounded since $2(s-s_1) + n - 2s_2 \leq 0$; the same conclusion holds if $s_2 > n/2$ since $s \leq s_1$ and if $s_2 = n/2$ since $s < s_1$ then by hypothesis. Thus the first condition (8.3.2) is fulfilled. Similarly we can handle the set where $\langle\xi-\eta\rangle < \langle\xi\rangle/2$, this time using (8.3.2)'. It remains to estimate

$$\int |F(\xi,\eta)|^2\, d\xi,$$

taken over the set where $\langle\xi\rangle < 2\langle\eta\rangle$ and $\langle\xi\rangle < 2\langle\xi-\eta\rangle$. There is a uniform bound for the integral when $\langle\xi\rangle \leq 4$, since $s_1 + s_2 \geq 0$. Thus we let $\langle\xi\rangle > 4$, hence $\langle\eta\rangle > 2$, $\langle\xi-\eta\rangle > 2$. After introducing $\xi/|\eta|$ as a new integration variable, we can then estimate the integral by

$$C|\eta|^{n+2(s-s_1-s_2)} \int\limits_{2/|\eta|<|\xi|<4, |\xi-\eta/|\eta||>|\xi|/4} |\xi|^{2s}|\xi - \eta/|\eta||^{-2s_1}\, d\xi.$$

The integral is bounded as $\eta \to \infty$ unless $s \leq -n/2$. If $s < -n/2$ then the integral is $\leq C|\eta|^{-n-2s}$, which gives the desired bound since $s_1 + s_2 \geq 0$. If $s = -n/2$ the integral is $O(\log|\eta|)$, and the bound follows since $s_1+s_2 > 0$ then by assumption. Another application of Lemma 8.3.2 completes the proof.

We shall next prove some additions to Theorem 8.3.1 when microlocal information is available concerning the smoothness of u_j.

Theorem 8.3.3. *Let $u_j \in H_{(s_j)}(\mathbf{R}^n)$, $j = 1, 2$. Then*

  (i) $u_1 u_2 \in H_{(s_2)}^{\text{loc}}$ *outside* $WF(u_1)$ *if* $s_1 > n/2$ *and* $s_1 + s_2 > n/2$.
  (ii) $u_1 u_2 \in H_{(s)}^{\text{loc}}$ *outside* $WF(u_1)$ *if* $s_1 < n/2$ *and* $s_1 + s_2 - n/2 > s \geq 0$.
  (iii) $u_1 u_2 \in H_{(s_1+s_2-n/2)}^{\text{loc}}$ *outside* $WF(u_1) \cup WF(u_2)$ *if* $s_1 + s_2 > 0$.

Proof. Multiplication of u_j by suitable cutoff functions reduces the proof of (i) to proving that if $u_j \in \mathcal{S}'(\mathbf{R}^n)$,

$$\int |\hat{u}_j(\xi)|^2 (1+|\xi|^2)^{s_j}\, d\xi < \infty, \quad j = 1, 2,$$

and $\hat{u}_1$ is rapidly decreasing in a conic neighborhood Γ_1 of the closed cone Γ, then

$$(8.3.3) \qquad \int_\Gamma |\hat{u}(\xi)|^2 (1+|\xi|^2)^{s_2}\, d\xi < \infty,$$

where $\hat{u} = (2\pi)^{-n}\hat{u}_1 * \hat{u}_2$. Since $\langle a+b\rangle \leq \langle a\rangle + |b| \leq \langle a\rangle(1+|b|)$ when $a, b \in \mathbf{R}^n$, we have

$$\langle\xi\rangle^{s_2} \left| \int_{\Gamma_1} \hat{u}_1(\eta)\hat{u}_2(\xi-\eta)\, d\eta \right| \leq \int_{\Gamma_1} |\hat{u}_1(\eta)|(1+|\eta|)^{|s_2|}\langle\xi-\eta\rangle^{s_2}|\hat{u}_2(\xi-\eta)|\, d\eta.$$

This is the convolution of one function in L^1 and one in L^2, hence in L^2. When $\xi \in \Gamma$ and $\eta \notin \Gamma_1$ we have $\langle \xi \rangle \leq C \langle \xi - \eta \rangle$, and $\langle \xi \rangle^{-1} \leq \langle \xi - \eta \rangle^{-1}(1 + |\eta|)$ for arbitrary ξ, η, so we have

$$\langle \xi \rangle^{s_2} \left| \int_{\eta \notin \Gamma_1} \hat{u}_1(\eta) \hat{u}_2(\xi - \eta) \, d\eta \right| \leq \begin{cases} \int C^{s_2} |\hat{u}_1(\eta)| \langle \xi - \eta \rangle^{s_2} |\hat{u}_2(\xi - \eta)| \, d\eta, & \text{if } s_2 \geq 0 \\ \int (1 + |\eta|)^{-s_2} |\hat{u}_1(\eta)| \langle \xi - \eta \rangle^{s_2} |\hat{u}_2(\xi - \eta)| \, d\eta, & \text{if } s_2 \leq 0. \end{cases}$$

We have $\hat{u}_1 \in L^1$ since $s_1 > n/2$, and

$$\int \langle \eta \rangle^{-s_2} |\hat{u}_1(\eta)| \, d\eta \leq \left(\int |\hat{u}_1(\eta)|^2 \langle \eta \rangle^{2s_1} \, d\eta \right)^{\frac{1}{2}} \left(\int \langle \eta \rangle^{-2s_1 - 2s_2} \, d\eta \right)^{\frac{1}{2}} < \infty$$

since $s_1 + s_2 > n/2$, which proves (8.3.3).

To prove (ii) we observe that since $s > 0$ we have

$$\langle \xi \rangle^s \left| \int_{\eta \notin \Gamma_1} \hat{u}_1(\eta) \hat{u}_2(\xi - \eta) \, d\eta \right| \leq \int F(\xi, \eta) \langle \eta \rangle^{s_1} |\hat{u}_1(\eta)| \langle \xi - \eta \rangle^{s_2} |\hat{u}_2(\xi - \eta)| \, d\eta,$$

where

$$F(\xi, \eta) = \begin{cases} C^s \langle \eta \rangle^{-s_1} \langle \xi - \eta \rangle^{s - s_2}, & \text{if } \eta \notin \Gamma_1, \\ 0, & \text{if } \eta \in \Gamma_1. \end{cases}$$

The statement follows from Lemma 8.3.2 if we show that there is a uniform bound for $\int |F(\xi, \eta)|^2 \, d\eta$ when $\xi \in \Gamma$. This is clear when $|\xi|$ is bounded, for $2(s - s_1 - s_2) < -n$. Since $s - s_2 < 0$ there is also a bound for the integral when $|\eta| < 2$, say. When $|\xi| > 2C$ we have $\langle \xi - \eta \rangle \geq 2$, hence $|\xi - \eta| \geq 1$ and $|\xi| \leq 2C|\xi - \eta|$. Hence the integral when $|\eta| > 2$ can be estimated by a constant times

$$|\xi|^{n + 2(s - s_1 - s_2)} \int_{|\eta - \xi/|\xi|| \geq 1/2C} |\eta|^{-2s_1} |\eta - \xi/|\xi||^{2(s - s_2)} \, d\eta.$$

The integral converges since $2s_1 < n$ and $2(s - s_1 - s_2) < -n$, which proves the second statement in the theorem.

When proving (iii) we may assume that $\hat{u}_1$ and $\hat{u}_2$ are both rapidly decreasing in Γ_1. Choose $\delta \in (0, \frac{1}{2})$ such that $\xi - \eta \in \Gamma_1$ if $\xi \in \Gamma$ and $|\eta| < \delta|\xi|$. Since $|\xi - \eta| > |\xi|/2$ we have for arbitrary N a uniform bound for $|\xi|^N \hat{u}_2(\xi - \eta)$ then, hence

$$\int_{|\eta| < \delta|\xi|} |\hat{u}_1(\eta) \hat{u}_2(\xi - \eta)| \, d\eta \leq C_N |\xi|^{-N} \left(\int |\hat{u}_1(\eta)|^2 \langle \eta \rangle^{2s_1} d\eta \right)^{\frac{1}{2}} \left(\int_{|\eta| < \delta|\xi|} \langle \eta \rangle^{-2s_1} \, d\eta \right)^{\frac{1}{2}},$$

so this is rapidly decreasing as $\xi \to \infty$ in Γ. The same is true for the integral where $|\xi - \eta| < \delta|\xi|$, so it suffices to prove that there is a bound for the L^2 norm of

$$\langle \xi \rangle^{s_1 + s_2 - n/2} \int_{|\eta| > \delta|\xi|, |\xi - \eta| > \delta|\xi|} \hat{u}_1(\eta) \hat{u}_2(\xi - \eta) \, d\eta.$$

To do so we apply Lemma 8.2.2 with

$$F(\xi, \eta) = \begin{cases} \langle \xi \rangle^{s_1 + s_2 - n/2} \langle \eta \rangle^{-s_1} \langle \xi - \eta \rangle^{-s_2}, & \text{when } \delta|\xi| \leq \min(|\eta|, |\xi - \eta|), \\ 0, & \text{otherwise.} \end{cases}$$

Then $\int_{|\xi|<1} F(\xi,\eta)^2\,d\xi$ is uniformly bounded since $s_1 + s_2 \geq 0$, and the integral when $|\xi| > 1$ can be estimated by a constant times

$$|\eta|^{-2s_1} \int\limits_{\delta|\xi|<\min(|\eta|,|\xi-\eta|)} |\xi|^{2(s_1+s_2)-n}|\xi-\eta|^{-2s_2}\,d\xi$$

$$= \int\limits_{\delta|\xi|<1,\delta|\xi|<|\xi-\eta/|\eta||} |\xi|^{2(s_1+s_2)-n}|\xi-\eta/|\eta||^{-2s_2}\,d\xi.$$

The only singularity is at 0, and since $s_1 + s_2 > 0$ by hypothesis, the integral is convergent and independent of η, which completes the proof.

An improvement of Theorem 8.3.3 will be given in Theorem 10.2.10.

Corollary 8.3.4. *For $j = 1,\ldots,N$ let $u_j \in H_{(s)}^{loc}$ at $x_0 \in \mathbf{R}^n$, where $s > n/2$, and let $u \in H_{(\sigma)}^{loc}$ at (x_0,ξ_0), where $s \leq \sigma \leq 2s - n/2$. Then it follows that $F(u_1,\ldots,u_N) \in H_{(s)}^{loc}$ at x_0 and that $F(u_1,\ldots,u_N) \in H_{(\sigma)}^{loc}$ at (x_0,ξ_0) if F is a polynomial.*

Proof. It is enough to prove this when $N = 2$ and $F(u_1,u_2) = u_1u_2$. The first statement is then an immediate consequence of Theorem 8.3.1. To prove the second one we write $u_j = v_j + w_j$ where $w_j \in H_{(\sigma)}$ and $(x_0,\xi_0) \notin WF(v_j)$; we may achieve that $v_j \in H_{(s)}$ by multiplication with a cutoff function. Then it follows from Theorem 8.3.1 that $w_1w_2 \in H_{(\sigma)}$ while Theorem 8.3.3 gives

$$v_1v_2 \in H_{(2s-n/2)}^{loc}, \qquad w_1v_2 + v_1w_2 \in H_{(\sigma)}^{loc} \qquad \text{at } (x_0,\xi_0).$$

This completes the proof.

Corollary 6.4.5 contains the first part of Corollary 8.3.4 for arbitrary smooth F when s is an integer. Using Theorem 8.3.3 one can also prove the complete statement for general F (cf. Beals [1]):

Corollary 8.3.4'. *Corollary 8.3.4 remains valid for every $F \in C^\infty$ if $s \geq \nu$ where ν is an integer $> n/2$.*

The result is actually valid when $s > n/2$. This will be clear later on when we give another less elementary but more illuminating proof using more elaborate techniques due to Bony.

Proof. We may assume that $F(0) = 0$ and that $u_j \in H_{(s)}(\mathbf{R}^n)$. By Corollary 6.4.5 we have $F(u) \in H_{(t)}(\mathbf{R}^n)$ then, if $t = \nu$. We shall prove the first statement by a bootstrap argument with this starting point; the restriction $s \geq \nu$ just allows us to begin the argument. Suppose that we have already proved for some $t \geq \nu$ that for arbitrary smooth F we have $F(u) \in H_{(t)}$. Then $F'(u) \in H_{(t)}$, and by Theorem 8.3.1

$$D_kF(u) = F'(u)D_ku \in H_{(r)}, \quad k = 1,\ldots,n, \quad F(u) \in H_{(r)},$$

if $r \leq t, r \leq s-1$ and $r < t+s-1-n/2$. Hence $F(u) \in H_{(r+1)}$ if $r+1 \leq \min(t+1,s)$ and $r+1 < t+s-n/2$. We can take $r+1 = s$ if $t+1 \geq s$. Otherwise $s > t+1 > n/2+1$, and we obtain $F(u) \in H_{(t+1)}$. After a finite number of iterations of the argument we obtain $F(u) \in H_{(s)}$.

The microlocal statement is proved similarly. There is nothing to prove unless $\sigma > s$. Assume that we already know some $\tau \geq s$ such that that $F(u) \in H_{(\tau)}^{loc}$ at (x_0,ξ_0) for arbitrary F; this is true when $\tau = s$. We can write

$$D_ku = v_k + w_k, \quad w_k \in H_{(\sigma-1)}, \quad v_k \in H_{(s-1)}, \quad (x_0,\xi_0) \notin WF(v_k),$$
$$F'(u) = v + w, \quad w \in H_{(\tau)}, \quad v \in H_{(s)}, \quad (x_0,\xi_0) \notin WF(v).$$

Then it follows from Theorem 8.3.1 that $ww_k \in H_{(\varrho)}$, if $\varrho \leq \tau$ and $\varrho \leq \sigma - 1$, for $\varrho < \tau + \sigma - 1 - n/2$ then. Theorem 8.3.3 gives at (x_0, ξ_0) that $vv_k \in H^{\mathrm{loc}}_{(2s-1-n/2)} \subset H^{\mathrm{loc}}_{(\sigma-1)}$, that $vw_k \in H^{\mathrm{loc}}_{(\varrho)}$, and that $v_k w \in H^{\mathrm{loc}}_{(\varrho)}$ if in addition $\varrho < \tau + s - 1 - n/2$. Since $2s - 1 - n/2 \geq \sigma - 1 \geq \varrho$, this condition is fulfilled if $\tau > s$ or $\varrho < \sigma - 1$. When $\tau = s$ we can take ϱ so that $\varrho + 1$ is any number $< \min(\sigma, s + 1)$, and when $\tau > s$ we can take $\varrho + 1 = \min(\sigma, \tau + 1)$ and conclude that $F(u) \in H^{\mathrm{loc}}_{(\varrho+1)}$ at (x_0, ξ_0). The assertion follows after a finite number of iterations of the argument.

8.4. Pseudo-differential operators. A linear differential operator of order m in an open set $X \subset \mathbf{R}^n$ can be written in the form

$$a(x, D) = \sum_{|\alpha| \leq m} a_\alpha(x) D^\alpha,$$

where $D = -i\partial = -i\partial/\partial x$, and $\alpha = (\alpha_1, \ldots, \alpha_n)$ is a multi-index, $|\alpha| = \sum_1^n \alpha_j$. We shall also use the notation $\alpha! = \prod_1^n \alpha_j!$. For $u \in C_0^\infty(X)$ we have by Fourier's inversion formula

$$(8.4.1) \qquad a(x, D)u(x) = (2\pi)^{-n} \int e^{i\langle x, \xi \rangle} a(x, \xi) \hat{u}(\xi)\, d\xi,$$

when $x \in X$. Here $\hat{u}(\xi) = \int e^{-i\langle x, \xi \rangle} u(x)\, dx$ is the Fourier transform of u. If the coefficients of $a(x, D)$ are in $C^\infty(X)$ then

$$(8.4.2) \qquad |a^{(\alpha)}_{(\beta)}(x, \xi)| \leq C_{\alpha\beta}(1 + |\xi|)^{m-|\alpha|}$$

for x in a compact subset of X; here $a^{(\alpha)}_{(\beta)}(x, \xi) = \partial_\xi^\alpha \partial_x^\beta a(x, \xi)$. The reason for writing $a(x, D)$ in the form (8.4.1) is that in the constant coefficient case composition of operators will just correspond to multiplication of functions then, so one might hope to invert the operator (8.4.1) at least approximately by substituting $a(x, \xi)^{-1}$ for $a(x, \xi)$ in (8.4.1), if a satisfies suitable hypotheses.

To make these vague ideas precise one first introduces the space S^m consisting of all $a \in C^\infty(\mathbf{R}^n \times \mathbf{R}^n)$ satisfying (8.4.2) in $\mathbf{R}^n \times \mathbf{R}^n$ for some $C_{\alpha\beta}$ and arbitrary α, β. With the best constants $C_{\alpha\beta}$ in (8.4.2) as seminorms, S^m is a Fréchet space. Clearly $a \in S^m$, $b \in S^{m'}$ implies $ab \in S^{m+m'}$, and it is elementary to verify that if $a \in S^m$ and $|a(x, \xi)| \geq c(1 + |\xi|)^m$ for some $c > 0$, then $1/a \in S^{-m}$. We write $S^{-\infty} = \cap_m S^m$.

The operator (8.4.1) with $a \in S^m$ is obviously continuous from $\mathcal{S}(\mathbf{R}^n)$ to $L^\infty(\mathbf{R}^n) \cap C(\mathbf{R}^n)$. We shall sometimes denote it by $\mathrm{Op}\, a$. Since the commutators with D_j and with the multiplication operator x_j are of the same form,

$$(8.4.3) \qquad [a(x, D), D_j] = ia_{(j)}(x, D), \qquad [a(x, D), x_j] = -ia^{(j)}(x, D),$$

it follows that $x^\alpha D^\beta a(x, D)$ is also continuous in these spaces. Hence we have proved:

Theorem 8.4.1. *If $a \in S^m$ then $a(x, D)$ is continuous from $\mathcal{S}(\mathbf{R}^n)$ to $\mathcal{S}(\mathbf{R}^n)$.*

The Schwartz kernel $K \in \mathcal{S}'(\mathbf{R}^n \times \mathbf{R}^n)$ of $a(x, D)$ is obtained by introducing the definition of the Fourier transform $\hat{u}$ in (8.4.1),

$$K(x, y) = (2\pi)^{-n} \int e^{i\langle x-y, \xi \rangle} a(x, \xi)\, d\xi,$$

which should be interpreted as $A(x, x - y)$ where $A(x, y)$ is the inverse Fourier transform in the sense of Schwartz of $a(x, \xi)$ with respect to ξ. Note that $a \mapsto K$ is an isomorphism of the space $\mathcal{S}'(\mathbf{R}^n \times \mathbf{R}^n)$ of temperate distributions. By the Schwartz kernel theorem we conclude that (8.4.1) can be interpreted as a continuous map from $\mathcal{S}(\mathbf{R}^n)$ to $\mathcal{S}'(\mathbf{R}^n)$ for arbitrary $a \in \mathcal{S}'(\mathbf{R}^n \times \mathbf{R}^n)$. Conversely, every continuous linear map from $\mathcal{S}(\mathbf{R}^n)$ to $\mathcal{S}'(\mathbf{R}^n)$ can be so defined. By Theorem 8.4.1 it maps $\mathcal{S}(\mathbf{R}^n)$ into $\mathcal{S}(\mathbf{R}^n)$ when $a \in S^m$. It will be useful in Chapter 9 to remember that the proof only used that every derivative of $a(x, \xi)$ is bounded by a polynomial in ξ.

To develop a calculus for the set $\operatorname{Op} S^m$ of *pseudo-differential operators* (8.4.1) with symbol in S^m we start with studying the adjoint $a(x, D)^*$ which is always well defined as a map from $\mathcal{S}(\mathbf{R}^n)$ to $\mathcal{S}'(\mathbf{R}^n)$. The kernel is $\overline{K(y, x)}$. If $a \in \mathcal{S}(\mathbf{R}^n \times \mathbf{R}^n)$ then a direct calculation gives $a(x, D)^* = b(x, D)$ where

$$(8.4.4) \qquad b(x, \xi) = (2\pi)^{-n} \int e^{-i\langle y, \eta \rangle} \overline{a(x - y, \xi - \eta)} \, dy \, d\eta = e^{i\langle D_x, D_\xi \rangle} \overline{a(x, \xi)}.$$

Here the expression on the right is defined according to (8.4.1), that is, the Fourier transform of b with respect to all $2n$ variables is equal to that of $\bar{a}$ multiplied by $e^{i\langle \cdot, \cdot \rangle}$. For reasons of continuity (8.4.4) remains valid for arbitrary $a \in \mathcal{S}'(\mathbf{R}^n \times \mathbf{R}^n)$.

If one would make a formal power series expansion of $e^{i\langle D_x, D_\xi \rangle}$ in (8.4.4) one would obtain a series with the terms

$$(i\langle D_x, D_\xi \rangle)^k \overline{a(x, \xi)}/k! \in S^{m-k},$$

so the order decreases to $-\infty$. This suggests the following theorem:

Theorem 8.4.2. *If $a \in S^m$ then (8.4.4) defines a function $b \in S^m$,*

$$(8.4.5) \qquad (a(x, D)u, v) = (u, b(x, D)v), \quad \text{if } u, v \in \mathcal{S}(\mathbf{R}^n),$$

$$(8.4.6) \qquad b(x, \xi) \sim \sum_0^\infty (i\langle D_x, D_\xi \rangle)^k \overline{a(x, \xi)}/k! = \sum_\alpha \partial_\xi^\alpha D_x^\alpha \overline{a(x, \xi)}/\alpha!.$$

This means that the difference between b and a partial sum is in S^μ if all the terms left out are in S^μ.

Thus $b - \bar{a} \in S^{m-1}$. Note that the theorem shows at once that $a(x, D)$ extends to a continuous map from $\mathcal{S}'(\mathbf{R}^n)$ to $\mathcal{S}'(\mathbf{R}^n)$ such that (8.4.5) remains valid when $u \in \mathcal{S}'(\mathbf{R}^n)$.

For a proof of Theorem 8.4.2 we refer to the proof of Hörmander [4, Theorem 8.1.7], but we shall prove a more precise form of (8.4.6) which will be useful in Chapter IX.

Theorem 8.4.2'. *If $a \in S^m$ and b is defined by (8.4.4), then every seminorm of the remainder term*

$$b(x, \xi) - \sum_0^{N-1} (i\langle D_x, D_\xi \rangle)^k \overline{a(x, \xi)}/k!$$

in S^{m-N} can be estimated by a seminorm of $\langle D_x, D_\xi \rangle^N a(x, \xi)$ in S^{m-N}.

Proof. We shall prove this using only the fact that $a \mapsto e^{i\langle D_x, D_\xi \rangle} \overline{a(x, \xi)}$ is a continuous map $S^m \to S^m$. If $a \in S^m$, then $A_t(x, \xi) = a(tx, \xi)$ is bounded in S^m for $0 \le t \le 1$, so

$$b_t(x, \xi) = e^{it\langle D_x, D_\xi \rangle} \overline{a(x, \xi)} = e^{i\langle D_x, D_\xi \rangle} \overline{A_t(z, \xi)}|_{tz=x}$$

has a bound for $0 \le t \le 1$ of the form $|b_t(x, \xi)| \le C_m(a)(1+|\xi|)^m$ where C_m is a seminorm in S^m. Now

$$d^k b_t(x, \xi)/dt^k = e^{it\langle D_x, D_\xi \rangle} \overline{a_k(x, \xi)} \quad \text{where} \quad \overline{a_k(x, \xi)} = (i\langle D_x, D_\xi \rangle)^k \overline{a(x, \xi)},$$

so Taylor's formula gives

$$\left| b(x,\xi) - \sum_0^{N-1} \overline{a_k(x,\xi)}/k! \right| \le C_{m-N}(a_N)(1+|\xi|)^{m-N}$$

where C_{m-N} is a seminorm in S^{m-N}. Application of this estimate to $a_{(\beta)}^{(\alpha)}$ for all α, β proves the theorem.

For a proof of the following closely related multiplicative property we refer to that of Hörmander [4, Theorem 18.1.8].

Theorem 8.4.3. *If $a_j \in S^{m_j}$, $j = 1, 2$, then as operators in $\mathcal{S}(\mathbf{R}^n)$ or in $\mathcal{S}'(\mathbf{R}^n)$*

$$(8.4.7) \qquad\qquad a_1(x,D)a_2(x,D) = b(x,D)$$

where $b \in S^{m_1+m_2}$ is given by

$$(8.4.8) \qquad\qquad b(x,\xi) = e^{i\langle D_y, D_\eta\rangle} a_1(x,\eta)a_2(y,\xi)|_{\eta=\xi, y=x}$$

and has the asymptotic expansion

$$(8.4.9) \qquad \begin{aligned} b(x,\xi) &\sim \sum_0^\infty (i\langle D_y, D_\eta\rangle)^j a_1(x,\eta)a_2(y,\xi)|_{\eta=\xi, y=x}/j! \\ &= \sum_\alpha a_1^{(\alpha)}(x,\xi)D_x^\alpha a_2(x,\xi)/\alpha!. \end{aligned}$$

There is a refinement of Theorem 8.4.3 which is analogous to Theorem 8.4.2':

Theorem 8.4.3'. *If $a_j \in S^{m_j}$, $j = 1, 2$, and $b \in S^{m_1+m_2}$ is defined by (8.4.8) then every seminorm of the remainder term*

$$b(x,\xi) - \sum_{|\alpha|<N} a_1^{(\alpha)}(x,\xi)D_x^\alpha a_2(x,\xi)/\alpha!$$

in $S^{m_1+m_2-N}$ can be estimated by a sum of products of seminorms of $a_1^{(\alpha)}(x,\xi)$ in $S^{m-|\alpha|}$ and $a_{2(\alpha)}(x,\xi)$ in S^{m_1} when $|\alpha| = N$.

To prove the theorem one just has to apply Taylor's formula to

$$e^{it\langle D_y, D_\eta\rangle} a_1(x,\eta)a_2(y,\xi)|_{\eta=\xi, y=x} = e^{i\langle D_z, D_\eta\rangle} a_1(x,\eta)a_2(tz,\xi)|_{\eta=\xi, tz=x},$$

as in the proof of Theorem 8.4.2'. This is left as an exercise for the reader.

The multiplicative properties easily give continuity properties:

Theorem 8.4.4. *If $a \in S^0$ then $a(x,D)$ is continuous in $L^2(\mathbf{R}^n)$.*

Proof. If $a \in S^m$ and $m < -n$ then the kernel $K(x,y)$ is a bounded continuous function, and so is the kernel $(x-y)^\alpha K(x,y)$ of the commutator

$$[x_{j_1}, [x_{j_2}, \ldots, [x_{j_{|\alpha|}}, a(x,D)]\ldots]] = i^{|\alpha|}a^{(\alpha)}(x,D)$$

with α_j factors x_j. Hence

$$|K(x,y)| \le C(1+|x-y|)^{-n-1}$$

which implies L^2 continuity since $\|k * u\|_{L^2} \le \|k\|_{L^1}\|u\|_{L^2}$. If $a \in S^m$ and $u \in \mathcal{S}(\mathbf{R}^n)$, then

$$\|a(x,D)u\|^2 = (a(x,D)^*a(x,D)u, u),$$

where $a(x,D)^*a(x,D) \in \operatorname{Op} S^{2m}$. Hence we conclude successively that $a(x,D)$ is L^2 continuous if $m < -n$, if $m < -n/2$, if $m < -n/4, \ldots$. If we just know that $a \in S^0$, we choose $M > \sup|a|$, and note that

$$b(x,\xi) = (M^2 - |a(x,\xi)|^2)^{\frac{1}{2}} \in S^0,$$

so the calculus gives $M^2 - a(x,D)^*a(x,D) - b(x,D)^*b(x,D) \in \operatorname{Op} S^{-1}$ which implies L^2 continuity. Hence $a(x,D)$ is L^2 continuous, which proves the theorem.

We have defined the Sobolev space $H_{(s)}(\mathbf{R}^n)$ as the set of all $u \in \mathcal{S}'(\mathbf{R}^n)$ such that $E_s(\xi)\hat{u}(\xi) \in L^2(\mathbf{R}^n)$, where $E_s(\xi) = (1 + |\xi|^2)^{s/2}$, that is, $E_s(D)u \in L^2(\mathbf{R}^n)$; the norm is

(8.4.10) $$\|u\|_{(s)} = \|E_s(D)u\|_{L^2} = \left((2\pi)^{-n} \int E_s(\xi)^2 |\hat{u}(\xi)|^2 \, d\xi\right)^{\frac{1}{2}}.$$

Note that $\mathcal{S}(\mathbf{R}^n)$ is dense in $H_{(s)}(\mathbf{R}^n)$ and that $E_s \in S^s$.

Theorem 8.4.5. If $a \in S^m$ then $a(x,D)$ extends for every $s \in \mathbf{R}$ to a continuous map $H_{(s)}(\mathbf{R}^n) \to H_{(s-m)}(\mathbf{R}^n)$.

Proof. We can write

$$E_{s-m}(D)a(x,D)u = b(x,D)E_s(D)u$$

where $b(x,D) = E_{s-m}(D)a(x,D)E_{-s}(D) \in \operatorname{Op} S^0$ is L^2 continuous.

Theorem 8.4.3 allows us to discuss approximate inversion of operators in $\operatorname{Op} S^m$:

Theorem 8.4.6. Let $a \in S^m$ and $b \in S^{-m}$. Then the conditions

(i) $$a(x,D)b(x,D) - \operatorname{Id} \in \operatorname{Op} S^{-\infty}$$

(ii) $$b(x,D)a(x,D) - \operatorname{Id} \in \operatorname{Op} S^{-\infty}$$

are equivalent, and a determines b (mod $S^{-\infty}$). Here Id is the identity operator $\operatorname{Op} 1$. They imply

(iii) $$a(x,\xi)b(x,\xi) - 1 \in S^{-1}, \qquad b(x,\xi)a(x,\xi) - 1 \in S^{-1},$$

which in turn imply that for some positive constants c and C

(iv) $$|a(x,\xi)| > c|\xi|^m, \qquad if \ |\xi| > C.$$

Conversely, if (iv) is fulfilled one can find $b \in S^{-m}$ satisfying (i), (ii), (iii).

Proof. (i) and (ii) both imply (iii) by Theorem 8.4.3. From (iii) it follows that we have $|a(x,\xi)b(x,\xi) - 1| < \frac{1}{2}$ for $|\xi| > C$, hence

$$\tfrac{1}{2} < |a(x,\xi)b(x,\xi)| < C'|a(x,\xi)||\xi|^{-m}, \qquad |\xi| > C,$$

which proves (iv). From (iii) it also follows that

$$a(x,D)b(x,D) = \operatorname{Id} - r(x,D), \qquad r \in S^{-1}.$$

We want to invert $\text{Id} - r(x, D)$ by the Neumann series, so we set

$$b(x, D)r(x, D)^k = b_k(x, D) \in \text{Op } S^{-m-k}.$$

We can find $b' \in S^{-m}$ with $b' \sim \sum_0^\infty b_j$; to do so we just have to cut off b_j with a function $(1 - \varphi(\varepsilon_j \xi))$, where $\varphi \in C_0^\infty$ is equal to 1 in a neighborhood of 0 and $\varepsilon_j \to 0$ sufficiently rapidly. The standard sum b' will then have the required property. (See Hörmander [4, Prop. 18.1.3] for details.) Now we obtain (i) with b replaced by b'. In the same way we can find $b'' \in S^{-m}$ such that (ii) is fulfilled with b replaced by b''. Hence

$$b''(x, D) - b'(x, D) = b''(x, D)(\text{Id} - a(x, D)b'(x, D)) + (b''(x, D)a(x, D) - \text{Id})b'(x, D),$$

is in $\text{Op } S^{-\infty}$, so b' and b'' satisfy both (i) and (ii). Since it is elementary to see that (iv) implies that $b(x, \xi) = (1 - \varphi(\varepsilon\xi))/a(x, \xi) \in S^{-m}$ for sufficiently small $\varepsilon > 0$, we have (iv) $\implies$ (iii), and the proof is complete.

Remark. The condition (iv) in Theorem 8.4.6 is called *ellipticity*. One says that a is *noncharacteristic* at $(x_0, \xi_0) \in \mathbf{R}^n \times (\mathbf{R}^n \setminus 0)$ if (iv) is valid at infinity in a conic neighborhood, that is, a product $U \times \Gamma$ where U is a neighborhood of x_0 in $\mathbf{R}^n$ and Γ is a conic neighborhood of ξ_0 in $\mathbf{R}^n$. One can then find $b \in S^{-m}$ such that the symbols of $a(x, D)b(x, D) - \text{Id}$ and $b(x, D)a(x, D) - \text{Id}$ are of order $-\infty$ in a conic neighborhood of (x_0, ξ_0). The proof is the same except that one cuts off r inside such a small neighborhood.

Corollary 8.4.7. *If $a \in S^m$ satisfies the conditions in Theorem 8.4.6, say the ellipticity condition (iv), and if $u \in H_{(-\infty)}(\mathbf{R}^n) = \cup_t H_{(t)}(\mathbf{R}^n)$, then $a(x, D)u \in H_{(s)}(\mathbf{R}^n)$ implies $u \in H_{(s+m)}(\mathbf{R}^n)$, and for every t*

$$\|u\|_{(s+m)} \leq C_{s,t}(\|a(x, D)u\|_{(s)} + \|u\|_{(t)}).$$

Proof. With $b(x, D)$ as in Theorem 8.4.5 we have

$$u = b(x, D)a(x, D)u + r(x, D)u$$

where $b \in S^{-m}$ and $r \in S^{-\infty}$. Hence $b(x, D)$ is continuous from $H_{(s)}$ to $H_{(s+m)}$ and $r(x, D)$ is continuous from $H_{(t)}$ to $H_{(s+m)}$.

So far we have only discussed pseudo-differential operators globally in $\mathbf{R}^n$. In the applications it is essential to localize them. First we observe that if $K \in \mathcal{S}'(\mathbf{R}^n \times \mathbf{R}^n)$ is the Schwartz kernel of $a(x, D) \in \text{Op } S^m$, then $K \in C^\infty(\mathbf{R}^n \times \mathbf{R}^n \setminus \Delta)$ where Δ denotes the diagonal $\{(x, x) ; x \in \mathbf{R}^n\}$. This is an immediate consequence of the fact that the Schwartz kernel $(x - y)^\alpha K(x, y)$ of $i^{-|\alpha|}a^{(\alpha)}(x, D)$ is in C^k if $|\alpha| > m + k + n$. (Note that we just need here that $a_{(\beta)}^{(\alpha)}(x, \xi)(1 + |\xi|)^k$ is integrable for $|\alpha| > C_{k,\beta}$, so the argument is valid for much more general symbol classes encountered in Chapter 9.) On the other hand, every operator with kernel in $\mathcal{S}(\mathbf{R}^n \times \mathbf{R}^n) \supset C_0^\infty(\mathbf{R}^n \times \mathbf{R}^n)$ is in $\text{Op } S^{-\infty}$, for the symbol is also in $\mathcal{S}(\mathbf{R}^n \times \mathbf{R}^n)$. Thus the only local conditions satisfied by the kernels of pseudo-differential operators concern the singularities at the diagonal.

If $X \subset \mathbf{R}^n$ is open then a continuous linear map $A : C_0^\infty(X) \to C^\infty(X)$ is said to be a pseudo-differential operator of order m in X if for arbitrary $\varphi, \psi \in C_0^\infty(X)$ the operator

$$\mathcal{S}(\mathbf{R}^n) \ni u \mapsto \varphi A(\psi u)$$

is in $\text{Op } S^m$. The set of such operators is denoted by $\Psi^m(X)$. In particular, the restriction of $a(x, D) \in \text{Op } S^m$ to X is a pseudo-differential operator in X. More precisely, every pseudo-differential operator in X is a sum $a(x, D) + R$ where $a \in S^m(X \times \mathbf{R}^n)$ in the sense that (8.4.2) is valid for x in any compact subset K of X, with $C_{\alpha\beta}$ depending also on K of course, and the kernel of R is in $C^\infty(X \times X)$. (See Hörmander [4, Proposition 18.1.14].) The decomposition is not unique, but a is determined mod $S^{-\infty}(X \times \mathbf{R}^n)$ since an operator with kernel in C_0^∞ is in $\text{Op}\, S^{-\infty}$.

If $u \in \mathcal{E}'(X)$ and $A \in \Psi^m(X)$, then $Au \in \mathcal{D}'(X)$ is well defined, and we have the *pseudo-local property*

$$(8.4.11) \qquad\qquad\qquad \text{sing supp}\, Au \subset \text{sing supp}\, u$$

since the kernel K of A is in $C^\infty(X \times X \setminus \Delta_X)$. In fact,

$$Au(x) = u_y(K(x, y)), \qquad x \notin \text{sing supp}\, u.$$

It is usually convenient to require pseudo-differential operators to be *properly supported*, which means that the projections

$$\text{supp}\, K \ni (x, y) \mapsto x \in X \quad \text{and} \quad \text{supp}\, K \ni (x, y) \mapsto y \in X$$

are proper (that is, the inverse image of any compact set in X is a compact set in $\text{supp}\, K \subset X \times X$). This guarantees that $a(x, D)\mathcal{E}'(X) \subset \mathcal{E}'(X)$ and that $a(x, D)u$ vanishes in any relatively compact subset of X if u vanishes in a corresponding sufficiently large relatively compact subset of X; this implies that there is a unique extension of $a(x, D)$ to a map in $\mathcal{D}'(X)$ with the same property. (Here we have tacitly used the fact that every properly supported pseudo-differential operator is of the form $a(x, D)$ without any additional C^∞ term.)

Just as a distribution $u \in \mathcal{D}'(X)$ is in $H_{(s)}^{\text{loc}}$ at x_0 if and only if one can find $\varphi \in C_0^\infty(X)$ with $\varphi(x_0) \neq 0$ and $\varphi u \in H_{(s)}$, we can use pseudo-differential operators to give a concise form to the corresponding microlocal condition (Definition 8.2.5):

Theorem 8.4.8. *If $u \in \mathcal{D}'(X)$ and $(x_0, \xi_0) \in X \times (\mathbf{R}^n \setminus 0)$ then the following conditions are equivalent:*

(i) *$u \in H_{(s)}^{\text{loc}}$ at (x_0, ξ_0);*

(ii) *$Au \in L_{\text{loc}}^2(X)$ for some properly supported $A \in \Psi^s(X)$ which is noncharacteristic at (x_0, ξ_0);*

(iii) *There is a conic neighborhood Γ_0 of (x_0, ξ_0) in $X \times (\mathbf{R}^n \setminus 0)$ such that $Bu \in H_{(s-m)}^{\text{loc}}(X)$ for every properly supported $B \in \Psi^m(X)$ with symbol of order $-\infty$ outside Γ_0.*

Proof. If (i) is true we can choose $\varphi \in C_0^\infty(X)$ equal to 1 in a neighborhood of x_0 such that (8.2.6) holds for some conic neighborhood Γ of ξ_0. Choose $q(\xi) \in C^\infty$, homogeneous of degree s for $|\xi| \geq 1$, with support in Γ, and let $A = \varphi q(D)\varphi \in \Psi^s(X)$. Then

$$\|Au\|_{(0)} \leq C\|q(D)\varphi u\|_{(0)} < \infty,$$

and the symbol of A is $q(\xi)$ (mod $S^{-\infty}$) near x_0, which proves (ii). If (iii) is valid we choose φ, ψ in $C_0^\infty(X)$ and $q(\xi)$ homogeneous of degree s for $|\xi| \geq 1$ with $\psi = 1$ in a neighborhood of $\text{supp}\, \varphi$ and $\text{supp}\, \varphi \times \text{supp}\, q \subset \Gamma_0$. Then we obtain

$$q(D)\varphi u = \psi q(D)\varphi u + (1 - \psi)q(D)\varphi(\psi u) \in L^2$$

since $(1 - \psi)q(D)\varphi$ is of order $-\infty$ and $\psi q(D)\varphi u \in L^2$ by (iii) with $B = \psi q(D)\varphi$. Hence (i) holds. Finally assume that (ii) is valid, and choose a closed conic neighborhood Γ_0 of (x_0, ξ_0) such that the symbol of A satisfies (iv) in Theorem 8.4.6 in Γ_0, with m replaced by s. In view of the remark following the proof of Theorem 8.4.6 we can for every B as in condition (iii) find a properly supported $C \in \Psi^{m-s}(X)$ such that $B - CA \in \Psi^{-\infty}(X)$. Hence $Bu - CAu \in C^\infty(X)$. Since $CAu \in H^{loc}_{(s-m)}(X)$ by Theorem 8.4.5, the condition (iii) follows and the proof is complete.

Corollary 8.4.9. *If $u \in \mathcal{D}'(X)$ is in $H^{loc}_{(s)}$ at (x_0, ξ_0), then $Au \in H^{loc}_{(s-m)}$ at (x_0, ξ_0) for every properly supported $A \in \Psi^m(X)$.*

Proof. If $B \in \Psi^{s-m}(X)$ is properly supported, noncharacteristic at (x_0, ξ_0), and of order $-\infty$ outside the set Γ_0 in (iii), then $BAu \in H^{loc}_{(0)}(X) = L^2_{loc}(X)$, for $BA \in \Psi^s(X)$. Hence the implication (ii) $\Longrightarrow$ (i) in Theorem 8.4.8 proves the statement.

Corollary 8.4.10. *If $u \in \mathcal{D}'(X)$ and $Au \in H^{loc}_{(s-m)}$ at (x_0, ξ_0) for some properly supported $A \in \Psi^m$ which is noncharacteristic at (x_0, ξ_0), then $u \in H^{loc}_{(s)}$ at (x_0, ξ_0).*

Proof. Choose $C \in \Psi^{s-m}(X)$ properly supported and noncharacteristic at (x_0, ξ_0) such that $CAu \in H^{loc}_{(0)}$. Then $CA \in \Psi^s(X)$ is noncharacteristic at (x_0, ξ_0), so the statement follows from the implication (ii) $\Longrightarrow$ (i) in Theorem 8.4.8.

Corollary 8.4.11. *If $(x_0, \xi_0) \notin WF(u)$, then there is a conic neighborhood Γ_0 of (x_0, ξ_0) in $X \times (\mathbf{R}^n \setminus 0)$ such that $u \in H^{loc}_{(s)}$ at every point in Γ_0 for every $s \in \mathbf{R}$. Conversely, this implies that $(x_0, \xi_0) \notin WF(u)$.*

Proof. If $(x_0, \xi_0) \notin WF(u)$ it follows at once from Definitions 8.2.2 and 8.2.5 that there exists such a Γ_0. Now assume that Γ_0 exists with the stated properties. Choose $\varphi \in C^\infty_0$ and $q(\xi)$ homogeneous of degree 0 for $|\xi| > 1$, so that

$$\varphi(x_0) = q(\xi_0/|\xi_0|) = 1, \qquad \operatorname{supp} \varphi \times \operatorname{supp} q \subset \Gamma_0.$$

Then $q(D)\varphi(x)u \in H^{loc}_{(s)}$ for every s by condition (iii) in Theorem 8.4.8, for we can cover $\operatorname{supp} \varphi \times \operatorname{supp} q$ by a finite number of conic neighborhoods with the properties required there and decompose the symbol correspondingly. Hence $q(D)\varphi(x)u \in C^\infty$. Since the kernel of $q(D)$ is of the form $K(x - y)$ with K in C^∞ except at 0 and rapidly decreasing, it follows that $q(D)\varphi u \in \mathcal{S}$, hence that $q(\xi)\widehat{\varphi u}(\xi)$ is rapidly decreasing. The proof is complete.

Corollary 8.4.12. *If $A \in \Psi^m(X)$ is properly supported, then we have the microlocal property*

$$(8.4.12) \qquad WF(Au) \subset WF(u), \qquad u \in \mathcal{D}'(X).$$

Proof. This is a consequence of Corollaries 8.4.11 and 8.4.9.

So far we have just discussed linear equations. However, using the results on microlocal regularity of nonlinear functions established in Section 8.3, we can discuss *semi-linear* equations, following Rauch [1]. The more general results of Bony [1] will be discussed in Chapter X. We shall use the notation

$$(8.4.13) \qquad J_k u = (\partial^\alpha u)_{|\alpha| \leq k}$$

for the *k-jet* of u, that is, the array of all derivatives of order $\leq k$.

Theorem 8.4.13. *Let $u \in H^{\mathrm{loc}}_{(s+k)}(X)$, where X is an open set in $\mathbf{R}^n$ and $s > n/2$, be a solution of the semi-linear differential equation*

$$P(x, D)u = f(x, J_k u),$$

where f and the coefficients of P are in C^∞ and k is smaller than the order m of $P(x, D)$. If P is noncharacteristic at $(x_0, \xi_0) \in X \times (\mathbf{R}^n \setminus 0)$, it follows that $u \in H^{\mathrm{loc}}_{(2s+m-n/2)}$ at (x_0, ξ_0).

Proof. Assume that we already know that $u \in H^{\mathrm{loc}}_{(\sigma+k)}$ at (x_0, ξ_0) for some $\sigma \geq s$. Then $J_k u \in H^{\mathrm{loc}}_{(\sigma)}$ at (x_0, ξ_0), and since $J_k u \in H^{\mathrm{loc}}_{(s)}$ at x_0, it follows from Corollary 8.3.4' that $P(x, D)u = f(x, J_k u) \in H^{\mathrm{loc}}_{(\sigma)}$ at (x_0, ξ_0) if $\sigma \leq 2s - n/2$. Hence $u \in H^{\mathrm{loc}}_{(\sigma+m)}$ at (x_0, ξ_0) by Corollary 8.4.10. Since $m > k$ we have gained in regularity. Repeating the argument we obtain that $u \in H^{\mathrm{loc}}_{(2s+m-n/2)}$ at (x_0, ξ_0) as claimed.

Remark. We have anticipated here that Corollary 8.3.4' will be extended to all $s > n/2$ in Corollary 10.3.2. Without this extension the proof is just valid if $s \geq \nu$ where ν is the smallest integer $> n/2$.

In this section we have only considered the simplest types of symbols. It is easy to replace S^m by the space $S^m_{\varrho, \delta}$ of symbols of order m and type ϱ, δ defined by

$$(8.4.2)' \qquad\qquad |a^{(\alpha)}_{(\beta)}(x, \xi)| \leq C_{\alpha\beta}(1 + |\xi|)^{m - \varrho|\alpha| + \delta|\beta|},$$

provided that $0 \leq \delta < \varrho \leq 1$. Thus S^m is the special case $S^m_{1,0}$. Since $\delta < \varrho$ the asymptotic expansions in Theorems 8.4.2 and 8.4.3 make sense, and the proofs of the subsequent results are essentially unchanged. If $\varrho = \delta < 1$ the asymptotic expansions in Theorems 8.4.2 and 8.4.3 break down but everything else remains true although the proofs must be changed. When $\varrho = \delta = 1$, however, there are operators in $\mathrm{Op}\, S^0_{1,1}$ which are not L^2 continuous. We shall see in Section 8.5 that they are important in nonlinear analysis, and Chapters 9 and 10 will then be devoted to their study.

8.5. Composite functions. We shall now discuss another approach to results such as Corollary 6.4.5 on the regularity of composite functions which is better suited to the study of microlocal regularity. Choose $\varphi \in C^\infty_0(\mathbf{R}^n)$ real valued, equal to 1 when $|\xi| < \frac{1}{2}$, equal to 0 when $|\xi| > 1$, and symmetric with respect to the origin. Then we have

$$d\varphi(\xi/t)/dt = \psi(\xi/t)/t; \qquad \psi(\xi) = -\langle \xi, \partial\varphi/\partial\xi \rangle,$$

which yields a continuous partition of unity

$$(8.5.1) \qquad\qquad 1 = \varphi(\xi) + \int_1^\infty \psi(\xi/t)\, dt/t.$$

Note that $t/2 \leq |\xi| \leq t$ in the support of $\psi(\xi/t)$. If $u \in L^\infty(\mathbf{R}^n)$ then

$$u = \varphi(D)u + \int_1^\infty \dot{u}_t\, dt/t, \qquad \dot{u}_t = \psi(D/t)u,$$

with convergence almost everywhere and a uniform bound for

$$(8.5.2) \qquad\qquad u_T = \varphi(D)u + \int_1^T \dot{u}_t\, dt/t = \varphi(D/T)u = T^n \Phi(T\cdot) * u,$$

where $\Phi \in S$ has Fourier transform φ. Note that u_T will in fact be uniformly bounded in any Banach space with translation invariant norm containing u.

If u is real valued and $F \in C^\infty(\mathbf{R})$ we have

$$F(u) = \lim F(u_T) = F(u_1) + \int_1^\infty F'(u_t)\dot{u}_t \, dt/t.$$

It is obvious that $F(u_1) - F(0) \in H_{(\infty)} = \cap_t H_{(t)}$ if $u \in H_{(-\infty)} = \cup_t H_{(t)}$, for $u_1 \in H_{(\infty)}$ then. To study the second term we introduce the linear operator

$$(8.5.3) \qquad L_u(g) = \int_1^\infty F'(u_t)\psi(D/t)g \, dt/t, \qquad g \in S(\mathbf{R}^n);$$

we are interested in estimating $L_u(u)$. Note that

$$(8.5.4) \qquad\qquad L_u(g) = a(x, D)g,$$

$$(8.5.5) \qquad\qquad a(x, \xi) = \int_1^\infty F'(u_t(x))\psi(\xi/t) \, dt/t.$$

By (8.5.2) we have for every $u \in L^\infty$

$$|D^\alpha u_t| \le C_\alpha t^{|\alpha|} \sup |u|.$$

In (8.5.5) the integrand vanishes unless $\frac{1}{2} \le |\xi/t| \le 1$, and the integral of dt/t over this interval is $\log 2$. Hence

$$(8.5.6) \qquad\qquad |a_{(\beta)}^{(\alpha)}(x, \xi)| \le C_{\alpha\beta}(1 + |\xi|)^{|\beta|-|\alpha|},$$

that is, $a \in S_{1,1}^0$. In Chapter 9 we shall study such operators and prove in particular that they are continuous in $H_{(s)}$ when $s > 0$, although not always when $s \le 0$. This will prove

Theorem 8.5.1. *If $u \in L^\infty \cap H_{(s)}$ is real valued and $s \ge 0$, then*

$$F(u) \in L^\infty \cap H_{(s)} \qquad \text{if } F \in C^\infty(\mathbf{R}) \text{ and } F(0) = 0.$$

When $s = 0$ it is of course enough to note that $|F(u)| \le C|u|$. The proof is not quite satisfactory when $s \to 0$ since the estimates obtained are not uniformly bounded then. However, it is elementary to see that

$$\|F(u)\|_{(s)} \le \sup |F'| \, \|u\|_{(s)} \qquad \text{for } 0 \le s \le 1.$$

In the preceding discussion we could equally well allow the values of u to belong to $\mathbf{R}^N$. We shall use Theorem 8.5.1 in that generality later on. However, the real advantage of the proof given here compared to that of Corollary 6.4.5 is that additional smoothness of u gives better properties of the operator L_u which lead in particular to good control of commutators with pseudo-differential operators with symbols of type $1,0$ and therefore to microlocal information. We shall return to this point in Chapter 10.

8.6. Hölder and Zygmund classes. To describe regularity properties we shall in addition to Sobolev spaces need to use Hölder classes and some modifications of them called Zygmund classes. Let us first recall that if $\varrho > 0$ is not an integer, and k is the integer with $k < \varrho < k + 1$, then the Hölder class $C^\varrho(\mathbf{R}^n)$ is the set of all $u \in C^k(\mathbf{R}^n)$ with the norm

$$|u|_\varrho = \sum_{|\alpha| \le k} \sup_x |D^\alpha u(x)| + \sum_{|\alpha|=k} \sup_{x \ne y} |D^\alpha u(x) - D^\alpha u(y)||x - y|^{k-\varrho} < \infty.$$

The limiting case where $\varrho = k + 1$ will be denoted by $C^{\varrho-0}$ to distinguish it from the space C^{k+1}.

Let φ and ψ be defined as in Section 8.5. Then we have

Proposition 8.6.1. *If $u \in C^\varrho$, $\varrho > 0$, then*

$$(8.6.1) \qquad\qquad \sup |\varphi(D)u| + \sup_{t>1} t^\varrho \sup |\psi(D/t)u| \le K_\varrho |u|_\varrho,$$

where K_ϱ is bounded when ϱ is bounded.

Proof. We have $\varphi = \widehat{\Phi}$, $\psi = \widehat{\Psi}$, with Φ and Ψ in S and $\int x^\alpha \Psi(x)\, dx = 0$ for every α. Hence

$$|\varphi(D)u| = |\Phi * u| \le \sup |u| \int |\Phi(x)|\, dx \le |u|_\varrho \int |\Phi(x)|\, dx.$$

Since

$$\psi(D/t)u(x) = \int \Psi(y)u(x - y/t)\, dy = \int \Psi(y)\Big(u(x - y/t) - \sum_{|\alpha| \le k} \partial^\alpha u(x)(-y/t)^\alpha/\alpha!\Big)\, dy,$$

we have by Taylor's formula

$$|\psi(D/t)u(x)| \le C_k |u|_\varrho \int |\Psi(y)| |y/t|^\varrho\, dy = C_k' |u|_\varrho t^{-\varrho},$$

which proves the proposition.

Since $|\xi| \le 1$ if ξ is in the spectrum of $\varphi(D)u$, and $|\xi| \le t$ if ξ is in the spectrum of $\psi(D/t)u$, the estimate (8.6.1) is much stronger than it may seem:

Lemma 8.6.2 (Bernstein's inequality). *If $v \in L^\infty(\mathbf{R}^n)$ and $|\xi| \le R$ if $\xi \in \operatorname{supp} \hat{v}$, then*

$$(8.6.2) \qquad\qquad \sup |v'| \le CR \sup |v|,$$

where C is independent of v and n. (The best constant is $C = 1$.)

Proof. The spectral condition is satisfied for the restriction of v to any straight line, so it suffices to prove the statement when $n = 1$, and we can take $R = 1$, since $v(x/R)$ can be introduced otherwise. Choose $\chi \in S$ with $\hat{\chi} = 1$ in $(-1, 1)$. Then $\hat{\chi}\hat{v} = \hat{v}$, so $v = \chi * v$ and

$$\sup |v'| \le \int |\chi'|\, dx \sup |v|.$$

This proves (8.6.2). That $C \ge 1$ is clear if we take $v(x) = e^{ix_1}$, but it takes somewhat longer to prove that $C = 1$ will do. (See e.g. Hörmander [4, Exercise 7.3.3].)

Proposition 8.6.3. *If $u \in S'(\mathbf{R}^n)$ and the left-hand side of (8.6.1) is finite, and if $k < \varrho < k + 1$ for some integer $k \ge 0$, then $u \in C^\varrho(\mathbf{R}^n)$ and*

$$(8.6.3) \quad |u|_\varrho \le K_k\big(\sup |\varphi(D)u| + \big((\varrho - k)^{-1} + (k + 1 - \varrho)^{-1}\big) \sup_{t>1} t^\varrho \sup |\psi(D/t)u|\big).$$

Proof. Since $|D^\alpha \varphi(D)u| \le \sup |\varphi(D)u|$ for every α it suffices to estimate the Hölder norm of

$$(8.6.4) \qquad\qquad v = \int_1^\infty \psi(D/t)u\, dt/t.$$

Set

$$M = \sup_{t>1} t^\varrho \sup |\psi(D/t)u|.$$

By Lemma 8.6.2 we have for every α

$$|D^\alpha \psi(D/t)u| \leq Mt^{|\alpha|-\varrho},$$

which implies that when $|\alpha| = k$

$$|D^\alpha \psi(D/t)u(x) - D^\alpha \psi(D/t)u(y)| \leq t^{k+1-\varrho}|x-y|M.$$

Hence

$$|D^\alpha v| \leq M \int_1^\infty t^{k-\varrho}\, dt/t, \qquad |\alpha| \leq k,$$

$$|D^\alpha v(x) - D^\alpha v(y)| \leq M\left(|x-y| \int_1^T t^{k+1-\varrho}\, dt/t + 2 \int_T^\infty t^{k-\varrho}\, dt/t\right), \qquad |\alpha| = k,$$

if $T \geq 1$. When $|x - y| \geq 1$ we choose $T = 1$, and otherwise we choose $T = 1/|x - y|$ to complete the proof.

When ϱ is a positive integer the proof breaks down and the space of functions such that the left-hand side of (8.6.1) is finite is actually larger than $C^{\varrho-0}$.

Definition 8.6.4. If ϱ is any real number we define the *Zygmund class* $C_*^\varrho(\mathbf{R}^n)$ as the set of all $u \in \mathcal{S}'(\mathbf{R}^n)$ with the norm

$$(8.6.5) \qquad |u|_\varrho^* = \sup |\varphi(D)u| + \sup_{t>1} t^\varrho \sup |\psi(D/t)u| < \infty.$$

Here φ and ψ are chosen as in Section 8.5.

By Propositions 8.6.3 and 8.6.4 we have $C_*^\varrho(\mathbf{R}^n) = C^\varrho(\mathbf{R}^n)$ if $\varrho > 0$ and ϱ is not an integer. Up to equivalence of the norm the definition is always independent of the choice of φ and of ψ. This follows from the following lemma where $|u|_\varrho^*$ is defined by (8.6.5) for some fixed choice of φ and ψ.

Lemma 8.6.5. *Let χ belong to a bounded subset of $C_0^\infty(\mathbf{R}^n \setminus 0)$ and let $s \in \mathbf{R}$. Then*

$$(8.6.6) \qquad t^{\varrho-s}|\chi(D/t)(1+|D|^2)^{s/2}u| \leq K_{\varrho,s}|u|_\varrho^*, \qquad t \geq 1, \quad u \in C_*^\varrho(\mathbf{R}^n).$$

For all χ_1 in a bounded subset of $C_0^\infty(\mathbf{R}^n)$ we have

$$(8.6.7) \qquad |\chi_1(D)u| \leq K_\varrho|u|_\varrho^*, \qquad u \in C_*^\varrho(\mathbf{R}^n),$$

$$(8.6.8)\ |D^\alpha \chi_1(D/t)u| \leq \begin{cases} K_{\varrho,\alpha} t^{|\alpha|-\varrho}|u|_\varrho^*, & \text{if } |\alpha| > \varrho, \\ K_\varrho(1+\log t)|u|_\varrho^*, & \text{if } |\alpha| = \varrho, \quad \text{for } t \geq 1, \ u \in C_*^\varrho(\mathbf{R}^n). \\ K_{\varrho,\alpha}|u|_\varrho^*, & \text{if } |\alpha| < \varrho, \end{cases}$$

If in addition $\chi_1 = 1$ in a fixed neighborhood of 0 and $\sigma \leq \varrho$, then

$$(8.6.9) \qquad |(1 - \chi_1(D/t))u|_\sigma^* \leq K_{\varrho,\sigma} t^{\sigma-\varrho}|u|_\varrho^*, \qquad \text{if } t \geq 1, \quad u \in C_*^\varrho(\mathbf{R}^n).$$

Proof. After splitting χ by a partition of unity we may assume that there is some $T \neq 0$ such that $\psi(\xi/T) \neq 0$ if $\xi \in \operatorname{supp} \chi$. Then

$$\chi(\xi) = \psi(\xi/T)G(\xi) \qquad \text{where} \quad G \in C_0^\infty(\mathbf{R}^n \setminus 0).$$

Hence
$$\chi(\xi/t)(1+|\xi|^2)^{s/2} = t^s \psi(\xi/tT) G_t(\xi/t)$$
with $G_t(\xi) = G(\xi)(t^{-2}+|\xi|^2)^{s/2}$ bounded in $C_0^\infty(\mathbf{R}^n)$ when $t \geq 1$, so the inverse Fourier transform g_t is bounded in $\mathcal{S}(\mathbf{R}^n)$. Hence
$$t^{\varrho-s}|\chi(D/t)(1+|D|^2)^{s/2}u| = t^\varrho|t^n g_t(t\cdot) * \psi(D/tT)u| \leq K|u|_\varrho^*/T^\varrho, \qquad \text{if } t \geq 1/T.$$
Since
$$|\varphi(D/t)u| = |\varphi(D)u + \int_1^t \psi(D/\tau)u \, d\tau/\tau| \leq |u|_\varrho^*(1+\int_1^t \tau^{-\varrho-1}\,d\tau), \qquad t \geq 1,$$
and $\varphi(\xi/t) = 1$ when $|\xi| \leq t/2$, we obtain (8.6.7) and hence the remaining part of (8.6.6), for $1 \leq t \leq 1/T$.

To prove (8.6.8) we write, with φ and ψ as in Section 8.5,
$$\chi_1(\xi/t) = \chi_1(\xi/t)\varphi(\xi) + \int_1^\infty \chi_1(\xi/t)\psi(\xi/\tau)\,d\tau/\tau.$$
The first term lies in a bounded subset of C_0^∞ when $t \geq 1$, hence
$$|D^\alpha \chi_1(D/t)\varphi(D)u| \leq K_{\varrho,\alpha}|u|_\varrho^*$$
by (8.6.7). If $|\xi| \leq A$ when $\xi \in \operatorname{supp}\chi_1$, then the integrand vanishes unless $\tau/2 \leq |\xi| \leq At$. Now $\chi(\xi) = \xi^\alpha \chi_1(\tau\xi/t)\psi(\xi)$ belongs for fixed α to a bounded subset of $C_0^\infty(\mathbf{R}^n \setminus 0)$ when $\tau \leq 2At$, so
$$|(D/\tau)^\alpha \chi_1(D/t)\psi(D/\tau)u| \leq K_{\varrho,\alpha}\tau^{-\varrho}|u|_\varrho^*, \qquad 1 \leq \tau \leq 2At,$$
by (8.6.6) with $s = 0$ and t replaced by τ. Since
$$\int_1^{2At} \tau^{|\alpha|-\varrho}\,d\tau/\tau \leq \begin{cases} (2At)^{|\alpha|-\varrho}/(|\alpha|-\varrho), & \text{if } |\alpha| > \varrho, \\ \log(2At), & \text{if } |\alpha| = \varrho, \\ 1/(\varrho-|\alpha|), & \text{if } |\alpha| < \varrho, \end{cases}$$
we have now proved (8.6.8). If $\chi_1(\xi) = 1$ for $|\xi| < r$ and $\chi_1(\xi) = 0$ for $|\xi| > R$, then
$$\psi(\xi/\tau)(1-\chi_1(\xi/t)) = \begin{cases} 0, & \text{if } \tau/t < r, \\ \psi(\xi/\tau), & \text{if } \tau/t > 2R. \end{cases}$$
(8.6.9) is therefore an immediate consequence of (8.6.6) with $s = 0$. The proof is complete.

As already observed, it follows from the lemma that $C_*^\varrho(\mathbf{R}^n)$ is independent of the choice of φ and ψ up to equivalence of the norm; this follows already from (8.6.6) with $s = 0$. In addition we obtain:

Proposition 8.6.6. *If $E_s(\xi) = (1+|\xi|^2)^{s/2}$ then $E_s(D)$ is an isomorphism of $C_*^\varrho(\mathbf{R}^n)$ on $C_*^{\varrho-s}(\mathbf{R}^n)$ for arbitrary ϱ and s in $\mathbf{R}$. If $u \in C_*^\varrho$ then $D_j u \in C_*^{\varrho-1}$, $j = 1,\ldots,n$. Conversely, $u \in C_*^\varrho$ if $D^\alpha u \in C_*^{\varrho-1}$ when $|\alpha| \leq 1$.*

Proof. From (8.6.6) it follows that $E_s(D)$ is continuous from $C_*^\varrho(\mathbf{R}^n)$ to $C_*^{\varrho-s}(\mathbf{R}^n)$. Since the inverse $E_{-s}(D)$ is continuous from $C_*^{\varrho-s}(\mathbf{R}^n)$ to $C_*^\varrho(\mathbf{R}^n)$, it is an isomorphism. That $D_j u \in C_*^{\varrho-1}(\mathbf{R}^n)$ if $u \in C_*^\varrho(\mathbf{R}^n)$ is an obvious consequence of (8.6.5) and Lemma 8.6.2. If $D^\alpha u \in C_*^{\varrho-1}(\mathbf{R}^n)$ when $|\alpha| \leq 1$, then $E_2(D)u = (1+\sum_1^n D_j^2)u \in C_*^{\varrho-2}(\mathbf{R}^n)$ so $u \in C_*^\varrho(\mathbf{R}^n)$.

This result shows that Definition 8.6.4 is the only way to extend the definition of the Hölder classes $C^\varrho(\mathbf{R}^n)$ to integers ϱ and to $\varrho \leq 0$ so that $E_s(D)$ operates in a natural way. If ϱ is a positive integer we have $u \in C_*^\varrho(\mathbf{R}^n)$ if and only if $D^\alpha u \in C_*^1(\mathbf{R}^n)$ when $|\alpha| \leq \varrho - 1$. This gives a concrete description of these spaces when combined with the following result, which goes back to work of Zygmund and motivates the terminology used.

Proposition 8.6.7. $C_*^1(\mathbf{R}^n)$ *consists of all bounded continuous functions* u *such that*

(8.6.10)
$$\sup |u(x)| + \sup_{y \neq 0} |u(x+y) + u(x-y) - 2u(x)|/|y| < \infty,$$

and the norm is equivalent to the left-hand side.

Proof. Assume that $u \in C_*^1(\mathbf{R}^n)$. With $M = |u|_1^*$ we have $|\psi(D/t)u| \leq M/t$, hence $|D^2\psi(D/t)u| \leq tM$ by Lemma 8.6.2, and

$$|\psi(D_x/t)(u(x+y) + u(x-y) - 2u(x))| \leq Mt|y|^2, \quad t \geq 1,$$

by Taylor's formula. Since $|\psi(D/t)u| \leq M/t$, we obtain for every $T \geq 1$ with v defined by (8.6.4)

$$|v(x+y) + v(x-y) - 2v(x)| \leq M|y|^2 \int_1^T t\,dt/t + 4M \int_T^\infty t^{-1}\,dt/t.$$

When $|y| > 2$ we choose $T = 1$ and when $|y| \leq 2$ we take $T = 2/|y|$, which yields

$$|v(x+y) + v(x-y) - 2v(x)| \leq 4M|y|.$$

Conversely, assume that the left-hand side M_1 of (8.6.10) is finite. Then

$$\psi(D/t)u(x) = \int u(x - y/t)\Psi(y)\,dy = \int (u(x-y/t) + u(x+y/t) - 2u(x))\Psi(y)\,dy/2,$$

where Ψ is the inverse Fourier transform of ψ which we have assumed to be even. Then

$$|\psi(D/t)u| \leq M_1 \int |y/t||\Psi(y)|\,dy = KM_1/t.$$

Since $|\varphi(D)u| \leq C \sup |u| \leq C'|u|_1^*$, the proof is complete.

We shall need some multiplicative properties of Zygmund spaces analogous to Theorem 8.3.1:

Proposition 8.6.8. *If* $u_j \in C_*^{\varrho_j}(\mathbf{R}^n)$, $j = 1, 2$ *and* $\varrho_1 + \varrho_2 > 0$, *then the product* $u = u_1 \cdot u_2$ *can be defined in a unique way so that on bounded sets in* $C_*^{\varrho_j}(\mathbf{R}^n)$ *it is a sequentially continuous function* $\mathcal{D}'(\mathbf{R}^n) \times \mathcal{D}'(\mathbf{R}^n) \to \mathcal{D}'(\mathbf{R}^n)$ *with the weak topologies and is standard multiplication if* u_j *are continuous. We have* $u \in C_*^{\varrho}(\mathbf{R}^n)$ *if* $\varrho = \min(\varrho_1, \varrho_2)$, *and*

$$|u_1 u_2|_\varrho^* \leq K_{\varrho_1, \varrho_2} |u_1|_{\varrho_1}^* |u_2|_{\varrho_2}^*.$$

Proof. With φ, ψ defined as in Section 8.5 we have $\varphi(D/t)u_j \to u_j$ in $\mathcal{S}'(\mathbf{R}^n)$ as $t \to \infty$, boundedly in $C_*^{\varrho_j}(\mathbf{R}^n)$. This proves uniqueness, and existence will follow easily from our representation of $u_1 \cdot u_2$ as the limit of $u_1^t \cdot u_2^t$ where

$$u_j^t = \varphi(D/t)u_j = \varphi(D)u_j + \int_1^t \psi(D/s)u_j\,ds/s, \quad t \geq 1.$$

The spectrum of $\varphi(D)u_k \cdot \psi(D/t)u_j$ lies outside the unit ball when $t > 4$, which gives for $t > 4$

$$\varphi(D)(u_1^t \cdot u_2^t) = \varphi(D)\Big(\varphi(D)u_1 \cdot \varphi(D/4)u_2 + \varphi(D/4)u_1 \cdot \varphi(D)u_2 - \varphi(D)u_1 \cdot \varphi(D)u_2$$

$$+ \iint_{1 \leq s_j \leq t} \psi(D/s_1)u_1 \cdot \psi(D/s_2)u_2\,ds_1\,ds_2/s_1 s_2\Big).$$

If $s_1 > 2(1+s_2)$ then the spectrum of the product in the integrand lies outside the unit ball which gives no contribution. Thus we may restrict the integration to $s_2 \leq s_1 \leq 2(1 + s_2)$ or $s_1 \leq s_2 \leq 2(1 + s_1)$. Since $|\psi(D/s_j)u_j| \leq s_j^{-\varrho_j}|u|^*_{\varrho_j}$ and $\varrho_1 + \varrho_2 > 0$ by assumption this gives a bound for $\varphi(D)(u_1^t \cdot u_2^t)$ which is independent of t. We must also estimate $\psi(D/T)(u_1^t \cdot u_2^t)$ when $T \geq 1$. Nothing needs to be changed above when T is bounded, so we may assume that $T \geq 4$, and write for $t > T$

$$u_j^t = \varphi(4D/T)u_j + \int_{T/4}^t \psi(D/s)u_j \, ds/s.$$

Arguing as before we obtain that

$$\psi(D/T)(u_1^t \cdot u_2^t) = \psi(D/T)\Big(\varphi(4D/T)u_1 \cdot (\varphi(2D/5T)u_2 - \varphi(4D/T)u_2)$$
$$+ (\varphi(2D/5T)u_1 - \varphi(4D/T)u_1) \cdot \varphi(4D/T)u_2$$
$$+ \iint_{T/4 \leq s_j \leq t} \psi(D/s_1)u_1 \cdot \psi(D/s_2)u_2 \, ds_1 \, ds_2/s_1 s_2 \Big).$$

There is no contribution from the double integral unless $s_2 \leq s_1 \leq 2(s_2 + T)$ or $s_1 \leq s_2 \leq 2(s_1 + T)$. Hence the double integral only contributes to the estimate of $\psi(D/T)(u_1^t \cdot u_2^t)$ a constant times $T^{-\varrho_1-\varrho_2}|u_1|^*_{\varrho_1}|u_2|^*_{\varrho_2}$. If $\varrho_1 < 0$ the same estimate holds for the first term on the right-hand side. When $\varrho_1 = 0$ we can estimate it by $T^\varrho|u_1|^*_{\varrho_1}|u_2|^*_{\varrho_2}$ for any $\varrho < \varrho_2$, hence for $\varrho = \varrho_1 = 0$, and when $\varrho_1 > 0$ the estimate contains a factor $T^{-\varrho_2}$ instead. Since $\varrho < \varrho_1 + \varrho_2$, because some ϱ_j is positive, the proof is complete for the second term has the same estimate for reasons of symmetry.

In Chapter 10 we shall need to change the order of multiplication and regularization using the following result:

Proposition 8.6.9. *Let χ and χ_1 belong to a bounded set in $C_0^\infty(\mathbf{R}^n)$, and assume that $\chi = 1$ in a fixed neighborhood of 0 and that $\chi_1 = 1$ in $\{x + y; x, y \in \operatorname{supp}\chi\}$. If $0 \leq \varrho_2 \leq \varrho_1$ and $\varrho_1 > 0$, it follows that*

$$(8.6.11) \quad |\chi_1(D/t)(u_1 \cdot u_2) - \chi(D/t)u_1 \cdot \chi(D/t)u_2|$$
$$\leq K_{\varrho_1,\varrho_2}t^{-\varrho_2}|u_1|^*_{\varrho_1}|u_2|^*_{\varrho_2}, \quad t > 1, \; u_j \in C_*^{\varrho_j}(\mathbf{R}^n).$$

Proof. Write $u_j = v_j + w_j$ where $v_j = \chi(D/t)u_j$. Then

$$u_1 \cdot u_2 = v_1 \cdot v_2 + v_1 \cdot w_2 + w_1 \cdot v_2 + w_1 \cdot w_2, \qquad \chi_1(D/t)(v_1 \cdot v_2) = v_1 \cdot v_2,$$

so (8.6.11) can be written

$$|\chi_1(D/t)(v_1 \cdot w_2 + w_1 \cdot v_2 + w_1 \cdot w_2)| \leq K_{\varrho_1,\varrho_2}t^{-\varrho_2}|u_1|^*_{\varrho_1}|u_2|^*_{\varrho_2}, \quad t > 1, \; u_j \in C_*^{\varrho_j}(\mathbf{R}^n).$$

Let $|\xi| \leq R$ when $\xi \in \operatorname{supp}\chi$ and $|\xi| \leq R_1$ when $\xi \in \operatorname{supp}\chi_1$. Then the spectrum of w_2 where $|\xi| > (R + R_1)t$ does not influence $\chi_1(D/t)(v_1 \cdot w_2)$, so we have

$$(8.6.12) \qquad \chi_1(D/t)(v_1 \cdot w_2) = \chi_1(D/t)(v_1 \cdot \varphi(D/(2Rt + 2R_1t))(1 - \chi(D/t))u_2),$$

if φ is chosen as in Section 8.5. By (8.6.8) and (8.6.6)

$$|v_1| \leq K|u_1|^*_{\varrho_1}, \qquad |\varphi(D/(2Rt + 2R_1t))(1 - \chi(D/t))u_2| \leq Kt^{-\varrho_2}|u_2|^*_{\varrho_2},$$

which gives the desired estimate of (8.6.12). For the analogous term with indices interchanged we can argue in the same way if $\varrho_2 > 0$. If $\varrho_2 = 0$ we have by (8.6.8)

$$|v_2| \le K t^{\varrho_1} |u_2|^*_{-\varrho_1} \le K t^{\varrho_1} |u_2|^*_{\varrho_2}$$

and the desired estimate follows again.

What remains is to estimate

$$\chi_1(D/t)(w_1 \cdot w_2).$$

If $\varrho_2 > 0$ we have $|w_j| \le K |u_j|^*_{\varrho_j} t^{-\varrho_j}$, which gives the desired estimate at once. If $\varrho_2 = 0$ and $0 < \varrho < \varrho_1$, we have

$$|w_1|^*_\varrho \le K t^{\varrho - \varrho_1} |u_1|^*_{\varrho_1}$$

in view of (8.6.9), so it follows from Proposition 8.6.8 that

$$|w_1 \cdot w_2|^*_0 \le K t^{\varrho - \varrho_1} |u_1|^*_{\varrho_1} |u_2|^*_0.$$

If $0 < \varepsilon \le \varrho_1 - \varrho$ another application of (8.6.8) gives

$$|\chi_1(D/t)(w_1 \cdot w_2)| \le K' t^\varepsilon |w_1 \cdot w_2|^*_{-\varepsilon} \le K' K t^{\varepsilon + \varrho - \varrho_1} |u_1|^*_{\varrho_1} |u_2|^*_{\varrho_2},$$

which completes the proof.

Using Zygmund spaces we can give an improved version of Sobolev's lemma (see Proposition 6.4.8) in the L^2 case:

Proposition 8.6.10. *For any $s \in \mathbf{R}$ we have $H_{(s)}(\mathbf{R}^n) \subset C_*^{s-n/2}(\mathbf{R}^n)$.*

Proof. Let $u \in H_{(s)}(\mathbf{R}^n)$. With φ, ψ as in (8.6.5) we have

$$|\varphi(D)u| \le (2\pi)^{-n} \int |\varphi(\xi)\hat{u}(\xi)| \, d\xi \le \|u\|_{(s)} \left((2\pi)^{-n} \int |\varphi(\xi)|^2 (1 + |\xi|^2)^{-s} \, d\xi \right)^{\frac{1}{2}},$$

$$|\psi(D/t)u| \le \|u\|_{(s)} \left((2\pi)^{-n} \int |\psi(\xi/t)|^2 (1 + |\xi|^2)^{-s} \, d\xi \right)^{\frac{1}{2}}$$

$$\le t^{n/2-s} \|u\|_{(s)} \left((2\pi)^{-n} \int |\psi(\xi)|^2 (t^{-2} + |\xi|^2)^{-s} \, d\xi \right)^{\frac{1}{2}}.$$

The last integral is bounded for $t \ge 1$, which proves that $|u|^*_{s-n/2} \le C_s \|u\|_{(s)}$. The proof is complete.

So far we have not emphasized that the Zygmund spaces are interpolation spaces. This is done in the following lemma, which could also be used to prove that Definition 8.6.4 does not depend on the choice of φ and ψ when $\varrho > 0$.

Lemma 8.6.11. *Assume that U_t is a continuous function of $t \ge 1$ with values in $C^N(\mathbf{R}^n)$ and that*

$$(8.6.13) \qquad |D^\alpha U_t| \le M t^{|\alpha|-\varrho} \qquad \text{when } \alpha = 0 \text{ or } |\alpha| = N,$$

where $0 < \varrho < N$. Then

$$(8.6.14) \qquad U = \int_1^\infty U_t \, dt/t \in C_*^\varrho, \qquad |U|^*_\varrho \le C_{\varrho,N} M.$$

Note that on the other hand, if $u \in C_*^\ell$ then $u - \varphi(D)u$ is of the form (8.6.14) with $U_t = \psi(D/t)u$ satisfying (8.6.13) in view of Definition 8.6.4 and Bernstein's inequality.

Proof. Choose φ and ψ as in Section 8.5. Since $\varphi(D)$ is convolution by a function in $\mathcal{S}$, it follows from (8.6.13) with $\alpha = 0$ that

$$|\varphi(D)U_t| \le C_1 M t^{-\ell}, \qquad \text{hence } |\varphi(D)U| \le C_1 M/\varrho.$$

Similarly $\psi(D/s)$ is convolution by a kernel in $\mathcal{S}$ with L^1 norm independent of s, so we have

$$|\psi(D/s)U_t| \le C_2 M t^{-\ell}.$$

Writing

$$\psi(\xi) = |\xi|^{2N} \psi(\xi)|\xi|^{-2N} = \sum_{|\alpha|=N} \xi^\alpha \psi_\alpha(\xi)$$

where $\psi_\alpha \in C_0^\infty$, we obtain

$$\psi(D/s) = \sum_{|\alpha|=N} \psi_\alpha(D/s)(D/s)^\alpha,$$

which gives in view of (8.6.13) with $|\alpha| = N$

$$|\psi(D/s)U_t| \le C_3 M t^{N-\ell} s^{-N}.$$

Using this estimate when $t \le s$ and the earlier one when $t > s$, we obtain for $s \ge 1$

$$|\psi(D/s)U| \le M\left(C_3 \int_1^s t^{N-\ell-1} s^{-N}\, dt + C_2 \int_s^\infty t^{-\ell-1}\, dt\right) \le (C_2/\varrho + C_3/(N-\varrho))M s^{-\ell}.$$

This completes the proof.

The lemma allows us in particular to deal with composite functions:

Proposition 8.6.12. *Let* $F \in C^{N+1}(\mathbf{R})$ *and let* u *be a real valued function in* C_*^ℓ *for some* $\varrho \in (0, N)$. *Then* $F(u) \in C_*^\ell$, *and if* $F(0) = 0$, $|u| \le M$ *we have*

$$(8.6.15) \qquad |F(u)|_\varrho^* \le C_{\varrho,N} \sum_0^N M^j \sup |F^{(j+1)}|\, |u|_\varrho^*.$$

Note that $F(u)$ does not change if we change F outside $(-M, M)$, so the supremum in (8.6.15) may be restricted to that interval.

Proof. With φ, ψ chosen as usual we set $u_t = \varphi(D/t)u$ and note that (cf. Section 8.5)

$$F(u) = F(u_1) + \int_1^\infty U_t\, dt/t, \qquad U_t = t\partial_t F(u_t) = (\psi(D/t)u)F'(u_t).$$

For all α we have

$$|D^\alpha \psi(D/t)u| \le C|u|_\varrho^* t^{|\alpha|-\varrho}, \quad t \ge 1,$$

and since

$$(8.6.16) \qquad |D^\alpha u_t| \le CM t^{|\alpha|}, \quad t \ge 1,$$

we obtain

$$|D^\alpha U_t| \leq C|u|_\varrho^* t^{|\alpha|-\varrho} \sum_0^N M^j \sup |F^{(j+1)}|, \qquad |\alpha| \leq N,$$

if we note that terms where F' is differentiated j times also have j derivatives of u_t as factors. By Lemma 8.6.11 we now get the desired estimate for $F(u) - F(u_1)$. Since $F(0) = 0$ we have

$$|F(u_1)| \leq \sup |F'| \, |u_1| \leq \sup |F'| \, |u|_\varrho^*.$$

Differentiation gives

$$|D^\alpha F(u_1)| \leq C|u|_\varrho^* \sum_0^{N-1} \sup |F^{(j+1)}| M^j, \qquad |\alpha| \leq N,$$

if we estimate one derivative of u_1 by

$$|D^\alpha u_1| \leq C|u|_\varrho^*$$

and the others by means of (8.6.16). The proof is complete.

Proposition 8.6.13. *Let $F \in C^{N+1}(\mathbf{R})$ and let $u \in C_*^\varrho$ for some $\varrho \in (0, N)$, $|u| \leq M$. Let χ_1, χ_2 by real valued even functions in $C_0^\infty(\mathbf{R}^n)$ which are equal to 1 in a neighborhood of 0. Then we have for $0 \leq |\alpha| \leq N$*

$$(8.6.17) \qquad |D^\alpha(\chi_1(D/t)F(u) - F(\chi_2(D/t)u))| \leq C(F, M, |u|_\varrho^*) t^{|\alpha|-\varrho}, \qquad t \geq 1.$$

Proof. By the interpolation inequalities $(4.2.17)'$ it suffices to prove the proposition when $\alpha = 0$ and when $|\alpha| = N$. A constant term in F does not contribute so we may assume that $F(0) = 0$. Since

$$|\chi_1(D/t)F(u) - F(u)| \leq Ct^{-\varrho}|F(u)|_\varrho^*, \qquad |\chi_2(D/t)u - u| \leq Ct^{-\varrho}|u|_\varrho^*$$

by (8.6.9), the estimate (8.6.17) follows from Proposition 8.6.12 when $\alpha = 0$. By (8.6.8) we have

$$|D^\alpha \chi_1(D/t)F(u)| \leq Ct^{|\alpha|-\varrho}|F(u)|_\varrho^*, \qquad |\alpha| = N,$$

so by Proposition 8.6.12 it just remains to prove that

$$|D^\alpha F(\chi_2(D/t)u)| \leq C(F, M, |u|_\varrho^*) t^{|\alpha|-\varrho}, \qquad |\alpha| = N.$$

If a term in the derivative on the left contains a factor $D^\beta \chi_2(D/t)u$ with $|\beta| > \varrho$, we can estimate that factor by $C|u|_\varrho^* t^{|\beta|-\varrho}$ and the others by $C|u|_\varrho^* t^{|\beta|}$, in view of (8.6.8). If there is no such factor with $|\beta| > \varrho$, then all of the factors can be estimated by $C(1 + \log t)|u|_\varrho^*$, which gives the desired estimate since $N > \varrho$. The proof is complete.

Finally we shall discuss the continuity of pseudo-differential operators in the Zygmund spaces. The following is an analogue of Theorem 8.4.5.

Theorem 8.6.14. *If $a \in S^m$ then $a(x, D)$ defines a continuous map from $C_*^\varrho(\mathbf{R}^n)$ to $C_*^{\varrho-m}(\mathbf{R}^n)$ for every $\varrho \in \mathbf{R}$.*

Proof. It follows from Proposition 8.6.6 as in the proof of Theorem 8.4.5 that it suffices to prove Theorem 8.6.14 when $m = 0$, $\varrho = 0$. The proof of Theorem 8.4.4 also shows that every operator in $\mathrm{Op}\, S^{-n-1}$ is bounded in L^∞. Choose φ, ψ as in Definition 8.6.4, and let $u \in C_*^0$, $|u|_0^* = 1$. Then $v = (1 + |D|^2)^{-1}u$ is bounded in C_*^2, so v and v' have fixed bounds, by Lemma 8.6.5. Since $\varphi(D)a(x, D)(1 + |D|^2) \in \text{Op } S^{-\infty}$, we have

$$|\varphi(D)a(x, D)u| = |\varphi(D)a(x, D)(1 + |D|^2)v| \leq C \sup |v| \leq C'.$$

To estimate $\psi(D/t)a(x, D)u$ we choose $\chi \in C_0^\infty(\mathbf{R}^n \setminus 0)$ so that $\chi = 1$ in a neighborhood of supp ψ. Then

$$|\psi(D/t)a(x, D)u| \leq |\psi(D/t)a(x, D)\chi(D/t)^2 u| + |\psi(D/t)a(x, D)(1 - \chi(D/t)^2)(1 + |D|^2)v|.$$

From the calculus it follows that $\psi(D/t)a(x, D)(1 - \chi(D/t)^2)(1 + |D|^2)$ is bounded in $\text{Op } S^{-\infty}$ when $t \geq 1$, for $\psi(\cdot/t)$ and $\chi(\cdot/t)$ are bounded in S^0 and $1 - \chi^2 = 0$ in a neighborhood of supp ψ, so all terms in the asymptotic expansion of the product will vanish. This gives a bound for the last term. By definition of C_*^0 we have a fixed bound for $\chi(D/t)u$. The operator $a(x, D)\chi(D/t)$ becomes $a(x/t, tD)\chi(D)$ if we introduce tx as a new variable instead of x. Thus the symbol is bounded in $S^{-\infty}$ when $t \geq 1$, so the operator has a uniform bound as an operator in L^∞. The same is obviously true for the operator $\psi(D/t)$, so we obtain a bound for $\psi(D/t)a(x, D)u$ independent of t and u when $t \geq 1$ and $|u|_0^* = 1$, which completes the proof.

We leave for the reader to verify that the results following Theorem 8.4.5 remain valid with $H_{(s)}$ spaces replaced by Zygmund spaces. In particular, this leads to the fact that if $u \in C_*^\varrho$ at (x_0, ξ) for every $\xi \neq 0$, in a sense analogous to Definition 8.2.5, then $u \in C_*^\varrho$ locally at x_0 in the sense that $\varphi u \in C_*^\varrho$ for every $\varphi \in C_0^\infty$ with support sufficiently near x_0. This is not as obvious as for $H_{(s)}$ spaces, since Zygmund spaces are *not* defined by inequalities on the Fourier transform side.

CHAPTER IX

PSEUDO-DIFFERENTIAL OPERATORS OF TYPE 1,1

9.1. Introduction. As observed in Section 8.4 there is a good calculus for the class $\operatorname{Op} S_{\varrho,\delta}^m$ of pseudo-differential operators $a(x, D)$ with $a \in S_{\varrho,\delta}^m$, that is,

$$(9.1.1) \qquad |\partial_\eta^\alpha \partial_x^\beta a(x, \eta)| \leq C_{\alpha\beta}(1 + |\eta|)^{m-\varrho|\alpha|+\delta|\beta|}; \quad x, \eta \in \mathbf{R}^n;$$

provided that $0 \leq \delta < \varrho \leq 1$. When $0 \leq \delta = \varrho < 1$ everything except the asymptotic expansions remains valid. However, there are operators in $\operatorname{Op} S_{1,1}^0$ which are not L^2 continuous, as was first proved by Ching [1]. The reputation of operators of type 1,1 was somewhat restored when E.M. Stein proved (unpublished; see also Meyer [1, 2], Bourdaud [1, 2]) that they are continuous in the Sobolev spaces $H_{(s)}$ for every $s > 0$. However, combination of these facts also implies lack of the desirable multiplicative properties. That operators of type 1,1 still deserve serious attention is due to their important role in the paradifferential calculus of Bony [1]. (See Section 8.5 and Chapter X.)

Bourdaud [1, 2] has shown that all pathologies of operators of type 1,1 can be attributed to the fact that they are not closed under taking adjoints: The operators in $\operatorname{Op} S_{1,1}^0 \cap (\operatorname{Op} S_{1,1}^0)^*$ are L^2 continuous and form an algebra. His proofs rely on a characterization of $\operatorname{Op} S_{1,1}^0$ in terms of the Schwartz kernel and the operation on certain test functions, but here we shall work directly with the symbols. Thus the set of $a \in S_{1,1}^0$ such that $a(x, D)^* = b(x, D)$ for some $b \in S_{1,1}^0$ is characterized by means of the behavior of the Fourier transform of $a(x, \eta)$ with respect to x at the *twisted diagonal* $\{(\xi, -\xi); \xi \in \mathbf{R}^n\}$. When carrying out this program we shall treat operators of arbitrary order. On the other hand we shall not discuss L^p estimates but only estimates in L^2, or rather in $H_{(s)}$, and in C_*^ϱ. We refer to Meyer [1, 2], Bourdaud [1, 2] for methods applicable for arbitrary $p \in (1, \infty)$.

The plan of the chapter is as follows. In Section 9.2 we recall some basic facts on the calculus of pseudo-differential operators, including expressions for the adjoint and the composition in terms of the Fourier transforms of the symbols. In Section 9.3 we introduce for symbols $a \in S_{1,1}^m$ a condition, (9.3.12), depending on a real number σ which implies that $a(x, D)$ is continuous from $H_{(s+m)}(\mathbf{R}^n)$ to $H_{(s)}(\mathbf{R}^n)$ when $s + \sigma > 0$ and follows if the operator is continuous for some s with $s + \sigma < 0$. The condition is automatically fulfilled when $\sigma = 0$ which gives the theorem of Stein mentioned above. The condition for continuity for all s coincides with the characterization of

$$(9.1.2) \qquad \tilde{\Psi}_{1,1}^m = \operatorname{Op} S_{1,1}^m \cap (\operatorname{Op} S_{1,1}^m)^*$$

given in Section 9.4. The multiplicative property $\tilde{\Psi}_{1,1}^m \tilde{\Psi}_{1,1}^\mu \subset \tilde{\Psi}_{1,1}^{m+\mu}$ is established in Section 9.5, and we show using the results of Section 9.3 that it is essential that the factor to the right in a composition is in $\tilde{\Psi}_{1,1}$. In Section 9.6 we extend the results of Sections 9.4 and 9.5 to the full calculus of paradifferential operators as exposed in Meyer [2]. No discussion is made of variable changes, for the proof of Hörmander [4, Theorem 18.1.17] is valid for operators of type 1,1. However, in Section 9.7 we give an extension of the "sharp Gårding inequality" to paradifferential operators using a modification of the proof of Hörmander [4, Theorem 18.1.14].

9.2. Some basic facts on pseudo-differential operators. We recall from Section 8.4 that if a belongs to the Schwartz space $S(\mathbf{R}^n \times \mathbf{R}^n)$, the pseudo-differential operator $a(x, D)$ is defined by

$$(9.2.1) \qquad a(x,D)u(x) = (2\pi)^{-n} \int e^{i\langle x,\eta\rangle} a(x,\eta)\hat{u}(\eta)\, d\eta, \quad u \in S(\mathbf{R}^n),$$

where $\hat{u}$ denotes the Fourier transform

$$\hat{u}(\eta) = \int e^{-i\langle x,\eta\rangle} u(x)\, dx.$$

Thus the Fourier transform of $a(x, D)u(x)$ is given by

$$(9.2.2) \qquad \mathbf{R}^n \ni \xi \mapsto (2\pi)^{-n} \int \hat{a}(\xi - \eta, \eta)\hat{u}(\eta)\, d\eta,$$

where $\hat{a}(\xi, \eta)$ is the Fourier transform of $a(x, \eta)$ with respect to x. By the Schwartz kernel theorem the right-hand side defines a continuous map from $S(\mathbf{R}^n)$ to $S'(\mathbf{R}^n)$ for any $a \in S'(\mathbf{R}^n \times \mathbf{R}^n)$, and conversely every such map can be so defined. Thus $a(x, D)$ is well defined as a continuous map from $S(\mathbf{R}^n)$ to $S'(\mathbf{R}^n)$ for all $a \in S'(\mathbf{R}^n \times \mathbf{R}^n)$. The adjoint is also a continuous map from $S(\mathbf{R}^n)$ to $S'(\mathbf{R}^n)$, so it is equal to $b(x, D)$ for some $b \in S'(\mathbf{R}^n \times \mathbf{R}^n)$. If $u, v \in S(\mathbf{R}^n)$ then, in the sense of distribution theory,

$$(a(x,D)u, v) = (2\pi)^{-2n} \iint \hat{a}(\xi - \eta, \eta)\hat{u}(\eta)\overline{\hat{v}(\xi)}\, d\xi d\eta;$$

$$(b(x,D)v, u) = (2\pi)^{-2n} \iint \hat{b}(\eta - \xi, \xi)\hat{v}(\xi)\overline{\hat{u}(\eta)}\, d\xi d\eta.$$

This means that

$$(9.2.3) \qquad \hat{b}(\eta - \xi, \xi) = \overline{\hat{a}(\xi - \eta, \eta)}, \quad \xi, \eta \in \mathbf{R}^n,$$

or equivalently

$$(9.2.4) \qquad \hat{b} = \overline{\hat{a}} \circ T; \quad T(\xi, \eta) = (-\xi, \xi + \eta),$$

when $a, b \in S'(\mathbf{R}^n \times \mathbf{R}^n)$ define adjoint operators. Since the adjoint operation is an involution this is also the case for T. The eigenspace of T with eigenvalue 1 is $\{(0, \eta); \eta \in \mathbf{R}^n\}$, and the eigenspace with eigenvalue -1 is $\{(-2\eta, \eta); \eta \in \mathbf{R}^n\}$, so T consists of reflection in the first space along the second one. It will be particularly significant here that T maps the *horizontal space* $\{(\xi, 0); \xi \in \mathbf{R}^n\}$ to the *twisted diagonal* $\{(-\xi, \xi); \xi \in \mathbf{R}^n\}$, hence maps the twisted diagonal to the horizontal space.

Direct calculation gives the commutation identities (see (8.4.3))

$$(9.2.5) \qquad [D_j, a(x, D)] = -ia_{(j)}(x, D), \quad [x_j, a(x, D)] = ia^{(j)}(x, D),$$

with the standard notation $a^{(\alpha)}_{(\beta)}(x, \xi) = \partial_\xi^\alpha \partial_x^\beta a(x, \xi)$.

Proposition 9.2.1. *If $a \in C^\infty$ and for arbitrary α, β*

$$|a_{(\beta)}^{(\alpha)}(x, \xi)| \le C_{\alpha\beta}(1 + |\xi|)^{m_{\alpha\beta}}, \quad x, \xi \in \mathbf{R}^n,$$

for some constants $C_{\alpha\beta}$ and $m_{\alpha\beta}$, then $a(x, D)$ maps $\mathcal{S}(\mathbf{R}^n)$ into $\mathcal{S}(\mathbf{R}^n)$.

Proof. This is actually what the proof of Theorem 8.4.1 gave, as remarked there.

If $b \in \mathcal{S}'(\mathbf{R}^n \times \mathbf{R}^n)$ and a satisfies the hypotheses of Proposition 9.2.1, then $b(x, D)a(x, D)$ is a continuous map from $\mathcal{S}(\mathbf{R}^n)$ to $\mathcal{S}'(\mathbf{R}^n)$ so it has a kernel in $\mathcal{S}'(\mathbf{R}^n \times \mathbf{R}^n)$. We shall give a convenient expression for it when $a, b \in \mathcal{S}(\mathbf{R}^n \times \mathbf{R}^n)$. If $u \in \mathcal{S}(\mathbf{R}^n)$ then

$$b(x, D)a(x, D)u = (2\pi)^{-2n} \iint e^{i\langle x, \theta \rangle} b(x, \theta) \hat{a}(\theta - \eta, \eta) \hat{u}(\eta) \, d\theta d\eta,$$

which is equal to $c(x, D)u$ if

$$(9.2.6) \qquad \begin{aligned} c(x, \eta) &= (2\pi)^{-n} \int e^{i\langle x, \theta - \eta \rangle} b(x, \theta) \hat{a}(\theta - \eta, \eta) \, d\theta \\ &= (2\pi)^{-n} \int e^{i\langle x, \theta \rangle} b(x, \theta + \eta) \hat{a}(\theta, \eta) \, d\theta. \end{aligned}$$

Hence

$$(9.2.7) \qquad \hat{c}(\xi, \eta) = (2\pi)^{-n} \int \hat{b}(\xi - \theta, \theta + \eta) \hat{a}(\theta, \eta) \, d\theta.$$

Proposition 9.2.1 proves in particular that $a(x, D)$ is a continuous map in $\mathcal{S}(\mathbf{R}^n)$ when $a \in S_{1,1}^m$. We shall use standard Littlewood-Paley partitions of unity to reduce the study of operators of type $1, 1$ to operators such that $|\eta|$ is bounded in the support of the symbol so that the symbol is also contained in the symbol classes discussed in Chapter VIII. To define the partitions we choose a fixed $\varphi \in C_0^\infty(\mathbf{R}^n)$ such that $\varphi(\xi) = 1$ when $|\xi| < 1/2$ and $\varphi(\xi) = 0$ when $|\xi| > 1$. Set

$$(9.2.8) \qquad \varphi_\nu(\xi) = \varphi(\xi/2^\nu), \ \psi_\nu(\xi) = \varphi_{\nu+1}(\xi) - \varphi_\nu(\xi); \quad \nu = 0, \pm 1, \pm 2, \dots$$

Then

$$(9.2.9) \qquad 1 = \varphi_\mu(\xi) + \sum_\mu^\infty \psi_\nu(\xi)$$

for every integer μ, and

$$(9.2.10) \qquad |\xi| \le 2^\nu \text{ if } \xi \in \operatorname{supp} \varphi_\nu; \quad 2^{\nu-1} \le |\xi| \le 2^{\nu+1} \text{ if } \xi \in \operatorname{supp} \psi_\nu.$$

The partition of unity (9.2.9) is the discrete analogue of the one used in Sections 8.5 and 8.6. With the notation used there we have $\psi_0(\xi) = \int_1^2 \psi(\xi/t) \, dt/t$.

The spaces $S_{\varrho, \delta}^m$ are all Fréchet spaces, with the topology defined by taking the best constants $C_{\alpha\beta}$ in (9.1.1) as semi-norms.

9.3. Continuity in $H_{(s)}$ and in C_*^s. As a first step in the proof of $H_{(s)}$ estimates we shall verify that all difficulties stem from the behavior at the twisted diagonal of the Fourier transform $\hat{a}(\xi, \eta)$ of the symbol $a(x, \eta)$ with respect to x.

Proposition 9.3.1. *If $a(x,\eta) \in S^m_{1,1}$ and*

$$(9.3.1) \qquad \hat{a}(\xi,\eta) = 0 \quad when \ |\xi + \eta| + 1 < |\eta|/B,$$

for some constant B, then $a(x,D)$ is continuous from $H_{(s+m)}(\mathbf{R}^n)$ to $H_{(s)}(\mathbf{R}^n)$ for every $s \in \mathbf{R}$, with norm depending on s, B and seminorms of a.

Proof. Since $a(x,D)u = a(x,D)(1 + |D|^2)^{-m/2}(1 + |D|^2)^{m/2}u$, it is no restriction to assume in the proof that $m = 0$. With φ_0 and ψ_ν defined in (9.2.8) and $u \in \mathcal{S}(\mathbf{R}^n)$ we have

$$u = \varphi_0(D)u + \sum_0^\infty \psi_\nu(D)u.$$

The product $a(x,D)\varphi_0(D)$ is continuous from $H_{(s)}(\mathbf{R}^n)$ to $H_{(t)}(\mathbf{R}^n)$ for arbitrary s and t since $a(x,\eta)\varphi_0(\eta) \in S^\mu_{1,0}$ for every μ. Set for $\nu \geq 0$

$$h_\nu(x) = a(x,D)\psi_\nu(D)u = a_\nu(x,D)\psi_\nu(D)u,$$
$$a_\nu(x,\eta) = (\psi_{\nu+1}(\eta) + \psi_\nu(\eta) + \psi_{\nu-1}(\eta))a(x,\eta).$$

Changing scales we obtain the symbol

$$b_\nu(x,\eta) = a_\nu(2^{-\nu}x, 2^\nu\eta) = (\psi_1(\eta) + \psi_0(\eta) + \psi_{-1}(\eta))a(2^{-\nu}x, 2^\nu\eta)$$

with the estimate

$$|b^{(\alpha)}_{\nu(\beta)}(x,\eta)| \leq C_{\alpha\beta}(a), \quad \nu \geq 0,$$

where $C_{\alpha\beta}$ is a semi-norm in $S^0_{1,1}$. This is clear since $1/4 \leq |\eta| \leq 4$ in the support. Thus b_ν is bounded in $S^\mu_{1,0}$ for every μ, and it follows that

$$\|b_\nu(x,D)v\|_{(s)} \leq C_s(a)\|v\|_{(0)}, \quad v \in \mathcal{S}(\mathbf{R}^n),$$

where the semi-norm $C_s(a)$ is an increasing function of s. Replacing x by $2^\nu x$ we conclude that

$$\|(1 + |D/2^\nu|^2)^{s/2}a_\nu(x,D)v\| \leq C_s(a)\|v\|,$$

where the norms are L^2 norms, and with $v = 2^{\nu s}\psi_\nu(D)u$ we obtain

$$\|(2^{2\nu} + |D|^2)^{s/2}h_\nu\| \leq C_s(a)\|2^{\nu s}\psi_\nu(D)u\|.$$

We have

$$(9.3.2) \qquad \sum_{\nu \geq 0} \|2^{\nu s}\psi_\nu(D)u\|^2 \leq 5^{|s|}\|u\|^2_{(s)},$$

for $\sum_{\nu \geq 0} 2^{2\nu s}\psi_\nu(\eta)^2 \leq \sum_{\nu \geq 0} 2^{2\nu s}\psi_\nu(\eta) \leq 5^{|s|}(1 + |\eta|^2)^s$ since $2^{2\nu s} \leq 5^{|s|}(1 + |\eta|^2)^s$ if $2^{\nu-1} \leq |\eta| \leq 2^{\nu+1}$ and $\nu \geq 0$, and $\sum_{\nu \geq 0} \psi_\nu \leq 1$. Since

$$\hat{h}_\nu(\xi) = (2\pi)^{-n} \int \hat{a}(\xi - \eta, \eta)\psi_\nu(\eta)\hat{u}(\eta)\,d\eta$$

we have $\hat{h}_\nu(\xi) = 0$ if $|\xi| + 1 < |\eta|/B$ when $\eta \in \operatorname{supp} \psi_\nu$, hence if $|\xi| + 1 \leq 2^{\nu-1}/B$. By the Cauchy-Schwarz inequality

$$\left|\sum_0^\infty \hat{h}_\nu(\xi)\right|^2 \leq \sum_0^\infty 2^{-2\nu}(2^{2\nu} + |\xi|^2)^{s+1}|\hat{h}_\nu(\xi)|^2 \sum_{1 \leq 2^\nu < 2B(|\xi|+1)} 2^{2\nu}(2^{2\nu} + |\xi|^2)^{-s-1}.$$

In the last sum we have $1 + |\xi|^2 \leq 2^{2\nu} + |\xi|^2 \leq (8B^2 + 1)(|\xi|^2 + 1)$, and $\sum 2^{2\nu}$ when $1 \leq 2^\nu < 2B(|\xi| + 1)$ is at most $8B^2(|\xi| + 1)^2 \leq 16B^2(|\xi|^2 + 1)$, so we obtain

$$\int (1 + |\xi|^2)^s \left| \sum_0^\infty \hat{h}_\nu(\xi) \right|^2 d\xi \leq 16B^2(8B^2 + 1)^{|s|+1} \sum_0^\infty 2^{-2\nu} \|(2^{2\nu} + |\xi|^2)^{(s+1)/2} \hat{h}_\nu(\xi)\|^2$$

$$\leq 16B^2(8B^2 + 1)^{|s|+1} C_{s+1}(a)^2 \sum_0^\infty (2\pi)^n \|2^{\nu s} \psi_\nu(D)u\|^2$$

which completes the proof in view of (9.3.2).

In order to apply Proposition 9.3.1 to an arbitrary $a \in S_{1,1}^m$ we must first remove a part in order to satisfy (9.3.1). To do so we choose a cutoff function $\chi \in C^\infty(\mathbf{R}^n \times \mathbf{R}^n)$ such that

$$(9.3.3) \qquad \begin{aligned} &\chi(t\xi, t\eta) = \chi(\xi, \eta), \quad \text{if } t \geq 1, \ |\eta| \geq 2, \\ &\operatorname{supp} \chi \subset \{(\xi, \eta); |\xi| \leq |\eta|, \ |\eta| \geq 1\}, \\ &\chi = 1 \text{ in } \{(\xi, \eta); 2|\xi| \leq |\eta|, \ |\eta| \geq 2\}, \end{aligned}$$

and define $a_{\chi,\varepsilon}$ by

$$(9.3.4) \qquad \hat{a}_{\chi,\varepsilon}(\xi, \eta) = \chi(\xi + \eta, \varepsilon\eta)\hat{a}(\xi, \eta),$$

with Fourier transforms taken with respect to the first variable.

Lemma 9.3.2. *If $a \in S_{1,1}^m$ and $0 < \varepsilon \leq 1$ then $a_{\chi,\varepsilon} \in C^\infty$ and*

$$(9.3.5) \qquad |\partial_x^\beta \partial_\eta^\alpha a_{\chi,\varepsilon}(x, \eta)| \leq C_{\alpha\beta} \varepsilon^{-|\alpha|}(1 + |\eta|)^{m+|\beta|-|\alpha|},$$

$$(9.3.6) \qquad \left(\int_{R \leq |\eta| \leq 2R} |\partial_\eta^\alpha a_{\chi,\varepsilon}(x, \eta)|^2 \, d\eta \right)^{1/2} \leq C_\alpha R^m (\varepsilon R)^{n/2 - |\alpha|}.$$

Proof. Let $\check{\chi}(x, \eta) = (2\pi)^{-n} \int \chi(\xi, \eta) e^{i\langle x, \xi \rangle} \, d\xi$ be the inverse Fourier transform of χ with respect to the first variable. It is a continuous function of η with values in $\mathcal{S}(\mathbf{R}^n)$, and the Fourier transform of $x \mapsto \check{\chi}(x, \varepsilon\eta)e^{-i\langle x, \eta \rangle}$ is $\xi \mapsto \chi(\xi + \eta, \varepsilon\eta)$, so we have

$$(9.3.7) \qquad a_{\chi,\varepsilon}(x, \eta) = \int e^{-i\langle y, \eta \rangle} \check{\chi}(y, \varepsilon\eta) a(x - y, \eta) \, dy.$$

When $|\varepsilon\eta| \geq 2$ we have $\chi(\xi, \varepsilon\eta) = \chi(2\xi/|\varepsilon\eta|, 2\eta/|\eta|)$, so $\int |\check{\chi}(y, \varepsilon\eta)| \, dy = \int |\check{\chi}(y, 2\eta/|\eta|)| \, dy$ is bounded then. Hence it follows from (9.3.7) that

$$|a_{\chi,\varepsilon}(x, \eta)| \leq C(1 + |\eta|)^m.$$

Differentation of (9.3.7) with respect to x acts directly on a, so (9.3.5) follows when $\alpha = 0$. However, when we apply ∂_η^α some derivatives ∂_η^γ may act on the factor $e^{-i\langle y, \eta \rangle} \check{\chi}(y, \varepsilon\eta)$ instead, which gives a new function with the Fourier transform $\partial_\eta^\gamma \chi(\xi + \eta, \varepsilon\eta)$. This is a sum of terms each of which is a nonnegative power of ε times a derivative of χ of order $|\gamma|$ evaluated at $(\xi + \eta, \varepsilon\eta)$. By (9.3.3) we have now homogeneity of degree $-|\gamma|$, so the L^1 norm of the inverse Fourier transform can be estimated by a constant times $(1 + \varepsilon|\eta|)^{-|\gamma|} \leq \varepsilon^{-|\gamma|}(1 + |\eta|)^{-|\gamma|}$, which completes the proof of (9.3.5).

The proof of (9.3.6) is similar. Since $\psi_0(\eta) + \psi_1(\eta) = 1$ when $1 \leq |\eta| \leq 2$, we have by (9.3.7) for $R \leq |\eta| \leq 2R$

(9.3.8)
$$a_{\chi,\varepsilon}(x,\eta) = \int e^{-i\langle y,\eta\rangle} K(y,\eta) a(x-y,\eta)\, dy,$$
$$K(y,\eta) = \check{\chi}(y,\varepsilon\eta)(\psi_0(\eta/R) + \psi_1(\eta/R)).$$

To prove (9.3.6) we shall estimate the L^2 norm of the right-hand side and its derivatives with respect to η for fixed x. Equivalently, we can estimate the L^2 norm with respect to z of the Fourier transform

$$F(x,z) = \iint a(x-y,\eta)K(y,\eta)e^{-i\langle y+z,\eta\rangle}\, dy\, d\eta$$

and the product by monomials in z. For fixed η the support of

$$\widehat{K}(\xi,\eta) = \chi(\xi,\varepsilon\eta)(\psi_0(\eta/R) + \psi_1(\eta/R))$$

is contained in a ball of radius $4\varepsilon R$, and

$$|\partial_\eta^\alpha \partial_\xi^\beta \widehat{K}(\xi,\eta)| \leq C_{\alpha\beta}\varepsilon^{-|\beta|}R^{-|\alpha|-|\beta|},$$

which gives for arbitrary N

$$|\partial_\eta^\alpha K(y,\eta)| \leq C_{\alpha N}(\varepsilon R)^n(1 + \varepsilon R|y|)^{-N}R^{-|\alpha|}.$$

The measure of the support of $K(y,\eta)$ for fixed y is $\leq CR^n$, hence integration by parts gives

$$\left| \int a(x-y,\eta)K(y,\eta)e^{-i\langle y+z,\eta\rangle}\, d\eta \right| \leq C_N R^{m+n}(\varepsilon R)^n(1+\varepsilon R|y|)^{-N}(1+R|y+z|)^{-N}.$$

Since $|z| \leq (\varepsilon R|y| + R|y+z|)/(\varepsilon R)$, it follows that for any M

$$|z|^M|F(x,z)| \leq C_M R^{m+n}(\varepsilon R)^{n-M}\int(1+\varepsilon R|y|)^{-n}(1+R|y+z|)^{-n-1}\, dy.$$

In this convolution one factor has L^2 norm $O((\varepsilon R)^{-n/2})$ and the other has L^1 norm $O(R^{-n})$, so the L^2 norm of $|z|^M F(x,z)$ with respect to z is $O(R^m(\varepsilon R)^{n/2-M})$. This completes the proof of (9.3.6).

For a general $a \in S_{1,1}^m$ the estimate (9.3.6) cannot be improved. We shall give an example which is closely related to counterexamples due to Ching [1] and Bourdaud [2].

Example 9.3.3. Let θ be a fixed vector in $\mathbf{R}^n$, let $A \in C_0^\infty(\{\eta; 1 < |\eta| < 2\})$, and set

$$a(x,\eta) = \sum_0^\infty e^{-i\langle x,2^\nu\theta\rangle}A(\eta/2^\nu).$$

Since the terms have disjoint supports it is clear that $a \in S_{1,1}^0$. We have $\hat{a}(\xi,\eta) = (2\pi)^n\sum_0^\infty \delta(\xi + 2^\nu\theta)A(\eta/2^\nu)$, hence

$$a_{\chi,\varepsilon}(x,\eta) = \sum_0^\infty e^{-i\langle x,2^\nu\theta\rangle}\chi(\eta - 2^\nu\theta,\varepsilon\eta)A(\eta/2^\nu).$$

When $R = 2^\nu$ and $R < |\eta| < 2R$ the sum reduces to the νth term, and $\chi(\eta - 2^\nu \theta, \varepsilon\eta) = 1$ if $|\eta - 2^\nu \theta| < \varepsilon R/2$ and $\varepsilon R \geq 2$. If $1 < |\theta| < 2$ we obtain for small ε and $\varepsilon R \geq 2$

$$\int_{R \leq |\eta| \leq 2R} |a_{\chi,\varepsilon}(x,\eta)|^2 \, d\eta \geq \int_{|\eta - R\theta| < \varepsilon R/2} |A(\eta/R)|^2 \, d\eta = R^n \int_{|\eta - \theta| < \varepsilon/2} |A(\eta)|^2 \, d\eta$$

which is asymptotic to $(\varepsilon R/2)^n |A(\theta)|^2$ times the volume of the unit ball as $\varepsilon \to 0$, if $A(\theta) \neq 0$. This proves that (9.3.6) cannot be improved in general. However, if A vanishes of order k at θ we obtain an improvement of the estimate (9.3.6) by a factor ε^k in the right-hand side, and we shall see that this has a decisive importance for the $H_{(s)}$ continuity.

If $a \in S_{1,1}^m$, it follows from (9.3.5) and Proposition 9.3.1 that $a(x, D) - a_{\chi,1}(x, D)$ is continuous from $H_{(s+m)}(\mathbf{R}^n)$ to $H_{(s)}(\mathbf{R}^n)$ for every $s \in \mathbf{R}$. To prove $H_{(s)}$ continuity of $a_{\chi,1}(x, D)$ under suitable hypotheses on the behavior at the twisted diagonal of the Fourier transform $\hat{a}(\xi, \eta)$ in the first variable of the symbol $a(x, \eta)$, we shall use the following result.

Lemma 9.3.4. Let $a \in L^2_{\mathrm{loc}}(\mathbf{R}^n \times \mathbf{R}^n)$, and assume that the derivatives $\partial_\eta^\alpha a(x, \eta)$ also are in L^2_{loc} when $|\alpha| \leq n + 1$,

$$(9.3.9) \qquad \int_{\mathbf{R}^n} |\partial_\eta^\alpha a(x, \eta)|^2 \, d\eta \leq 1, \quad |\alpha| \leq n + 1.$$

Then it follows that $a(x, D)$ is continuous in $L^2(\mathbf{R}^n)$ and that $\|a(x, D)\| \leq C$, where C is independent of a.

Proof. Choose $\chi_0 \in C_0^\infty(\mathbf{R}^n)$ such that

$$(9.3.10) \qquad \sum_{g \in \mathbf{Z}^n} \chi_0(\eta - g)^2 \equiv 1.$$

If $\hat{u} \in C_0^\infty(\mathbf{R}^n)$ we have then

$$a(x, D)u = \sum_{g \in \mathbf{Z}^n} a(x, D)\chi_0(D - g)u_g, \quad u_g = \chi_0(D - g)u,$$

and $u_g \neq 0$ for finitely many g. By Parseval's formula

$$(9.3.11) \qquad \sum_{g \in \mathbf{Z}^n} \|u_g\|^2 = \|u\|^2,$$

and the kernel K_g of $a(x, D)\chi_0(D - g)$ is given by

$$K_g(x, y) = (2\pi)^{-n} \int e^{i\langle x - y, \eta\rangle} a(x, \eta)\chi_0(\eta - g) \, d\eta.$$

Since

$$i^{|\alpha|}(y - x)^\alpha K_g(x, y) = (2\pi)^{-n} \int e^{i\langle x - y, \eta\rangle} \partial_\eta^\alpha(a(x, \eta)\chi_0(\eta - g)) \, d\eta, \quad |\alpha| \leq n + 1,$$

and χ_0 has compact support, we obtain by Cauchy-Schwarz' inequality and (9.3.9)

$$\sum_{|\alpha| \leq n+1} \sum_{g \in \mathbf{Z}^n} |(y - x)^\alpha K_g(x, y)|^2 \leq C \sum_{|\alpha| \leq n+1} \sum_{g \in \mathbf{Z}^n} \int |\partial_\eta^\alpha(a(x, \eta)\chi_0(\eta - g))|^2 \, d\eta \leq C'.$$

By Cauchy-Schwarz' inequality for sequences we have

$$|a(x, D)u| \leq \int \Big(\sum_{g \in \mathbf{Z}^n} |K_g(x, y)|^2 \Big)^{\frac{1}{2}} \Big(\sum_{g \in \mathbf{Z}^n} |u_g(y)|^2 \Big)^{\frac{1}{2}} \, dy$$

$$\leq C'' \int (1 + |x - y|)^{-n-1} \Big(\sum_{g \in \mathbf{Z}^n} |u_g(y)|^2 \Big)^{\frac{1}{2}} \, dy.$$

Since convolution by an integrable function is a bounded operator in L^2, we conclude using (9.3.11) that $\|a(x, D)u\| \leq C'''\|u\|$. The proof is complete.

Theorem 9.3.5. *Let $a \in S_{1,1}^m$ and assume that for some $\sigma \in \mathbf{R}$*

$$(9.3.12) \qquad \left(\int_{R \leq |\eta| \leq 2R} |\partial_\eta^\alpha a_{\chi,\varepsilon}(x,\eta)|^2 \, d\eta \right)^{1/2} \leq C_\alpha \varepsilon^\sigma R^m (\varepsilon R)^{n/2 - |\alpha|}, \quad R > 0,$$

for all $\varepsilon \in (0,1)$ and all multi-indices α, where $a_{\chi,\varepsilon}$ is defined by (9.3.4) with χ satisfying (9.3.3). Then it follows that $a(x,D)$ is continuous from $H_{(s+m)}(\mathbf{R}^n)$ to $H_{(s)}(\mathbf{R}^n)$ when $s + \sigma > 0$.

By (9.3.6) we obtain as a corollary the result of Stein mentioned in the introduction:

Corollary 9.3.6. *If $a \in S_{1,1}^m$ then $a(x,D)$ is continuous from $H_{(s+m)}(\mathbf{R}^n)$ to $H_{(s)}(\mathbf{R}^n)$ for every $s > 0$.*

Proof of Theorem 9.3.5. As in the proof of Proposition 9.3.1 we may assume that $m = 0$. From Proposition 9.3.1 we know already that $a(x,D) - a_{\chi,1}(x,D)$ is continuous in $H_{(s)}(\mathbf{R}^n)$ for every $s \in \mathbf{R}$. Since

$$(9.3.13) \qquad a_{\chi,1}(x,D) = \sum_0^\infty (a_{\chi,2^{-\nu}}(x,D) - a_{\chi,2^{-\nu-1}}(x,D))$$

we shall be able to conclude that $a(x,D)$ is continuous in $H_{(s)}$ if we can prove that the norm of $a_{\chi,\varepsilon}(x,D) - a_{\chi,\varepsilon/2}(x,D)$ in $H_{(s)}$ is $O(\varepsilon^c)$ for some $c > 0$ when $\varepsilon \to 0$. Set

$$e_\varepsilon(x,\eta) = a_{\chi,\varepsilon}(x,\eta) - a_{\chi,\varepsilon/2}(x,\eta), \qquad \text{that is,}$$
$$\hat{e}_\varepsilon(\xi,\eta) = (\chi(\xi + \eta, \varepsilon\eta) - \chi(\xi + \eta, \varepsilon\eta/2))\hat{a}(\xi,\eta).$$

By (9.3.3)

$$\operatorname{supp} \hat{e}_\varepsilon \subset \{(\xi,\eta); \varepsilon|\eta|/4 \leq \max(1, |\xi + \eta|) \leq \varepsilon|\eta|\}.$$

With $\hat{u} \in C_0^\infty(\mathbf{R}^n)$ and the notation (9.2.8), we write

$$e_\varepsilon(x,D)u = h = \sum h_\nu; \quad h_\nu = e_\varepsilon(x,D)\psi_0(D/2^\nu)u.$$

Since

$$\hat{h}_\nu(\xi) = (2\pi)^{-n} \int \hat{e}_\varepsilon(\xi - \eta, \eta)\psi_0(\eta/2^\nu)\hat{u}(\eta) \, d\eta,$$

we have

$$\varepsilon 2^{\nu-3} \leq \max(1, |\xi|) \leq \varepsilon 2^{\nu+1}, \quad \text{if } \xi \in \operatorname{supp} \hat{h}_\nu.$$

In particular, $h_\nu = 0$ when $\varepsilon 2^{\nu+1} < 1$, and since at most 8 supports $\hat{h}_\nu$ can overlap we have $|\hat{h}(\xi)|^2 \leq 8 \sum |h_\nu(\xi)|^2$, hence

$$(9.3.14) \qquad \|h\|_{(s)}^2 \leq 8 \sum \|h_\nu\|_{(s)}^2 \leq 2^{3+6|s|} \sum (\varepsilon 2^\nu)^{2s} \|h_\nu\|^2,$$

where $\| \cdot \|$ is the L^2 norm.

To estimate the L^2 norm of h_ν we change scales and write, with θ denoting the characteristic function of $\{\eta; 1/2 \leq |\eta| \leq 2\}$,

$$h_\nu(x/\varepsilon 2^\nu) = c_{\varepsilon,\nu}(x,D)\theta(\varepsilon D)u(x/\varepsilon 2^\nu), \quad c_{\varepsilon,\nu}(x,\eta) = e_\varepsilon(x/\varepsilon 2^\nu, 2^\nu \varepsilon \eta)\psi_0(\varepsilon \eta).$$

Since (9.3.12) with $m = 0$ implies

$$\left(\int\limits_{\frac{1}{2}R \leq |\eta| \leq 2R} |\partial_\eta^\alpha e_\varepsilon(x, \eta)|^2 \, d\eta \right)^{1/2} \leq C_\alpha \varepsilon^\sigma (\varepsilon R)^{n/2 - |\alpha|},$$

we obtain by taking $R = 2^\nu$ and replacing η by $\varepsilon R \eta$

$$\left(\int |\partial_\eta^\alpha c_{\varepsilon, \nu}(x, \eta)|^2 \, d\eta \right)^{1/2} \leq C_\alpha' \varepsilon^\sigma.$$

By Lemma 9.3.4 it follows that

$$\|c_{\varepsilon, \nu}(x, D)\| \leq C \varepsilon^\sigma,$$

and we conclude that

$$\|h_\nu\| \leq C \varepsilon^\sigma \|\theta(D/2^\nu) u\|.$$

Combined with (9.3.14) this proves that

$$\|h\|_{(s)}^2 \leq C_s \varepsilon^{2(\sigma+s)} \sum_0^\infty 2^{2\nu s} \|\theta(D/2^\nu) u\|^2 \leq C_s' \varepsilon^{2(\sigma+s)} \|u\|_{(s)}^2.$$

Recalling that $h = e_\varepsilon(x, D) u$ we conclude that

$$\sum_0^\infty \|e_{2^{-\mu}}(x, D) u\|_{(s)} \leq \sqrt{C_s'} \sum_0^\infty 2^{-\mu(\sigma+s)} \|u\|_{(s)} = \sqrt{C_s'} \|u\|_{(s)} / (1 - 2^{-(\sigma+s)}).$$

This completes the proof.

The condition (9.3.12) may seem complicated and unnatural, but the following theorem shows that it is quite close to giving a necessary and sufficient condition for $H_{(s)}$ continuity.

Theorem 9.3.7. *Suppose that $a \in S_{1,1}^m$ and that $a(x, D)$ is bounded from $H_{(s+m)}(\mathbf{R}^n)$ to $H_{(s)}(\mathbf{R}^n)$. Then it follows that (9.3.12) is valid for every σ with $s + \sigma < 0$.*

By Lemma 9.3.2 the conclusion is only interesting when $s < 0$, and by Corollary 9.3.6 the hypothesis on $H_{(s)}$ continuity is always fulfilled if $s > 0$. Together Theorems 9.3.5 and 9.3.7 prove that the set of all s such that $a(x, D)$ is continuous from $H_{(s+m)}(\mathbf{R}^n)$ to $H_{(s)}(\mathbf{R}^n)$ is an interval containing the positive half axis, and they describe the left end point of the interval but do not decide if it is a member of the interval. In Example 9.3.3 above it is easy to see that when A vanishes precisely of order k at θ then the interval is $(-k, \infty)$, but Theorem 9.3.7 itself does not exclude the point $-k$.

To prove Theorem 9.3.7 we need a result which interpolates between information on the norm of a pseudo-differential operator and bounds for the derivatives of its symbol.

Lemma 9.3.8. *Let $a \in C^N(\mathbf{R}^n \times \mathbf{R}^n)$, and assume that*

$$(9.3.15) \qquad |\partial^\alpha a| \leq 1, \quad |\alpha| = N.$$

If the norm M of $a(x, D)$ as operator in $L^2(\mathbf{R}^n)$ is ≤ 1, then

$$(9.3.16) \qquad |\partial^\alpha a| \leq C_N M^{(N - |\alpha|)/(N+n)}, \quad |\alpha| < N.$$

Proof. The hypothesis on the norm means that

$$\left|(2\pi)^{-n}\int e^{i\langle x,\xi\rangle}a(x,\xi)\hat{v}(\xi)u(x)\,dx\,d\xi\right|\leq M\|u\|\|v\|,\quad u,\hat{v}\in C_0^\infty(\mathbf{R}^n),$$

where the norms are L^2 norms. Replacing x,ξ by $\varepsilon x,\varepsilon\xi$ for some $\varepsilon\in(0,1)$ and writing $a_\varepsilon(x,\xi)=a(\varepsilon x,\varepsilon\xi)$, $U(x)=u(\varepsilon x)$, $V(\xi)=\hat{v}(\varepsilon\xi)$ gives

$$\varepsilon^n\left|\iint e^{i\varepsilon^2\langle x,\xi\rangle}a_\varepsilon(x,\xi)U(x)V(\xi)\,dx\,d\xi\right|\leq M(2\pi)^{n/2}\|U\|\|V\|,\quad U,V\in C_0^\infty(\mathbf{R}^n).$$

Using a standard proof of the Schwartz kernel theorem we can conclude that for every $\Psi\in C_0^\infty(\mathbf{R}^{2n})$ there is an estimate

$$(9.3.17)\qquad\varepsilon^n\left|\iint a_\varepsilon(x,\xi)\Psi(x,\xi)\,dx\,d\xi\right|\leq C_\Psi M,\quad 0<\varepsilon<1.$$

In fact, if $\Phi\in C_0^\infty(\mathbf{R}^n)$ is chosen so that $\Phi(x)\Phi(\xi)=1$ when $(x,\xi)\in\operatorname{supp}\Psi$, then

$$e^{-i\varepsilon^2\langle x,\xi\rangle}\Psi(x,\xi)=(2\pi)^{-2n}\iint e^{i\langle x,\hat{x}\rangle}\Phi(x)e^{i\langle\xi,\hat{\xi}\rangle}\Phi(\xi)f_\varepsilon(\hat{x},\hat{\xi})\,d\hat{x}\,d\hat{\xi}$$

by Fourier's inversion formula, and the Fourier transform f_ε of $e^{-i\varepsilon^2\langle\cdot,\cdot\rangle}\Psi$ is bounded in $\mathcal{S}(\mathbf{R}^{2n})$ when $0<\varepsilon<1$. Since

$$\varepsilon^n\left|\iint e^{i\varepsilon^2\langle x,\xi\rangle}a_\varepsilon(x,\xi)e^{i\langle x,\hat{x}\rangle}\Phi(x)e^{i\langle\xi,\hat{\xi}\rangle}\Phi(\xi)\,dx\,d\xi\right|\leq M(2\pi)^{n/2}\|\Phi\|^2,$$

we obtain (9.3.17) with $C_\Psi=(2\pi)^{n/2}\|\Phi\|^2\sup_{0<\varepsilon<1}\iint|f_\varepsilon(\hat{x},\hat{\xi})|\,d\hat{x}\,d\hat{\xi}$.

It suffices to prove (9.3.16) at $(0,0)$. Choose $\Phi_1\in C_0^\infty(\mathbf{R}^{2n})$ equal to 1 in a neighborhood of the origin. With polar coordinates r,ω in $\mathbf{R}^{2n}$ we have for $|\alpha|<N$

$$\partial^\alpha a_\varepsilon(0)=(-1)^{N-|\alpha|}\iint_{r>0}((\partial/\partial r)^{N-|\alpha|}\Phi_1\partial^\alpha a_\varepsilon)r^{N-|\alpha|-1}dr\,d\omega/(N-|\alpha|-1)!$$

where $d\omega$ is the normalized surface measure on the unit sphere in $\mathbf{R}^{2n}$. Since

$$(\partial/\partial r)^{N-|\alpha|}(\Phi_1\partial^\alpha a_\varepsilon)-\Phi_1(\partial/\partial r)^{N-|\alpha|}\partial^\alpha a_\varepsilon$$

is a differential operator with coefficients in $C_0^\infty(\mathbf{R}^{2n}\setminus\{0\})$ acting on a_ε, we get after an integration by parts with $\Psi_\alpha\in C_0^\infty(\mathbf{R}^{2n}\setminus\{0\})$

$$\partial^\alpha a_\varepsilon(0)=(-1)^{N-|\alpha|}\iint_{r>0}\Phi_1((\partial/\partial r)^{N-|\alpha|}\partial^\alpha a_\varepsilon)r^{N-|\alpha|-1}\,dr\,d\omega/(N-|\alpha|-1)!$$
$$+\iint a_\varepsilon(x,\xi)\Psi_\alpha(x,\xi)\,dx\,d\xi.$$

Estimating the first term on the right by (9.3.15) and the other by (9.3.17), we obtain

$$\varepsilon^{|\alpha|}|\partial^\alpha a(0)|=|\partial^\alpha a_\varepsilon(0)|\leq C_N(\varepsilon^N+\varepsilon^{-n}M),\quad 0<\varepsilon<1.$$

The estimate (9.3.16) follows by taking $\varepsilon=M^{1/(N+n)}$.

Actually we shall need a closely related lemma which is closer in spirit to Lemma 9.3.4:

Lemma 9.3.9. *Let* $a \in L_{\mathrm{loc}}^2(\mathbf{R}^n \times \mathbf{R}^n)$, *and assume that the derivatives* $a^{(\alpha)}(x, \eta) = \partial_\eta^\alpha a(x, \eta)$ *also are in* L_{loc}^2 *when* $|\alpha| \le N$,

(9.3.18)
$$\int |\partial_\eta^\alpha a(x, \eta)|^2 \, d\eta \le 1, \qquad |\alpha| \le N,$$

where $N > n$. *Further assume that*

(9.3.19)
$$|\xi + \eta| \le 1, \quad \text{if } (\xi, \eta) \in \operatorname{supp} \hat{a}.$$

If M *is the norm of* $a(x, D)$ *as operator in* L^2, *it follows that*

(9.3.20)
$$\int |\partial_\eta^\alpha a(x, \eta)|^2 \, d\eta \le C_N M^{2(1-n/N)(1-|\alpha|/N)}, \qquad |\alpha| \le N.$$

Proof. In this lemma we have a global hypothesis on the η derivatives and no assumption on the x derivatives. However, we shall attain the situation studied in Lemma 9.3.8 by a localization similar to that used in the proof of Lemma 9.3.4.

The estimate (9.3.20) follows from (9.3.18) when $|\alpha| = N$, and it follows in general by Parseval's formula and Hölder's inequality if we can prove it for $|\alpha| = 0$. Choose $\chi_0 \in C_0^\infty(\mathbf{R}^n)$ as in the proof of Lemma 9.3.4, and write for $v \in C_0^\infty(\mathbf{R}^n)$ and $g \in \mathbf{Z}^n$

$$a(x, D)e^{i\langle x, g \rangle}\chi_0(D)v = a_g(x, D)v, \quad a_g(x, \eta) = e^{i\langle x, g \rangle}a(x, \eta + g)\chi_0(\eta).$$

Let $\{c_g\} \in l^2$, $\sum_{\mathbf{Z}^n} |c_g|^2 = 1$, with only finitely many $c_g \ne 0$, and set

$$A(x, \eta) = \sum_{\mathbf{Z}^n} c_g a_g(x, \eta).$$

We shall prove that

(9.3.21)
$$|A(x, \eta)| \le C_N M^{(N-n)/N}.$$

This implies that

$$\sum_{g \in \mathbf{Z}^n} |a_g(x, \eta)|^2 \le C_N^2 M^{2(N-n)/N},$$

and since $a_g(x, \eta) = 0$ when $\eta \notin \operatorname{supp} \chi_0$, it follows from (9.3.10) that

$$\int |a(x, \eta)|^2 \, d\eta = \sum_{g \in \mathbf{Z}^n} \int |a_g(x, \eta)|^2 \, d\eta \le C C_N^2 M^{2(N-n)/N},$$

which is (9.3.20) with $\alpha = 0$.

Since $\hat{a}_g(\xi, \eta) = \hat{a}(\xi - g, \eta + g)\chi_0(\eta)$, we have $|\eta| \le C$ and $|\xi + \eta| \le 1$ in supp $\hat{a}_g$, hence $|\xi| \le C + 1$, and this is also true for $\hat{A}$ which by Bernstein's inequality will lead to control of the x derivatives of $A(x, \xi)$. However, before discussing that we observe that $\|A(x, D)\| \le M$, because for $v \in C_0^\infty(\mathbf{R}^n)$

$$A(x, D)v = a(x, D)w, \quad w = \sum_{g \in \mathbf{Z}^n} c_g e^{i\langle x, g \rangle}\chi_0(D)v,$$

$$\hat{w}(\xi) = \sum_{g \in \mathbf{Z}^n} c_g \chi_0(\xi - g)\hat{v}(\xi - g), \quad |\hat{w}(\xi)|^2 \le \sum_{g \in \mathbf{Z}^n} |c_g|^2 |\hat{v}(\xi - g)|^2,$$

which implies $\|w\| \leq \|v\|$, hence $\|A(x, D)v\| \leq M\|v\|$. We have

$$\sum_{|\alpha| \leq N} |\partial_\eta^\alpha A(x, \eta)|^2 \leq \sum_{g \in \mathbf{Z}^n} \sum_{|\alpha| \leq N} |\partial_\eta^\alpha a_g(x, \eta)|^2 \leq C \sum_{g \in \mathbf{Z}^n} \sum_{|\alpha| \leq N} |\partial_\eta^\alpha a(x, \eta + g)|^2,$$

and integration over $\operatorname{supp} \chi_0$ gives by (9.3.18)

$$\sum_{|\alpha| \leq N} \int |\partial_\eta^\alpha A(x, \eta)|^2 \, d\eta \leq C'.$$

By Sobolev's lemma it follows that

$$|\partial_\eta^\alpha A(x, \eta)| \leq C'', \quad |\alpha| \leq N - n,$$

and since $|\xi| \leq C + 1$ in $\operatorname{supp} \widehat{A}(\xi, \eta)$ we also have by Bernstein's theorem

$$|\partial_x^\beta \partial_\eta^\alpha A(x, \eta)| \leq C''', \quad |\alpha| + |\beta| = N - n.$$

This means that A/C''' satisfies the hypotheses of Lemma 9.3.8 with N replaced by $N - n$ and M replaced by M/C'''. With the notation of Lemma 9.3.8 this gives

$$|A(x, \eta)| \leq C_{N-n} C'''^{n/N} M^{(N-n)/N}$$

if $M \leq C'''$, and (9.3.20) follows from (9.3.18) if $M \geq C'''$. This completes the proof of (9.3.21) and of the lemma.

Proof of Theorem 9.3.7. We may assume that $m = 0$. Let M be the norm of $a(x, D)$ in $H_{(s)}$ for some $s < 0$. Then

$$\|a(x, D)\psi_0(D/R)u\|_{(s)} \leq M(R/2)^s \|u\|, \quad u \in \mathcal{S}(\mathbf{R}^n), \ R \geq 1,$$

where $\|\cdot\|$ is the L^2 norm, for $\|\psi_0(D/R)u\|_{(s)} \leq (R/2)^s \|u\|$. The left-hand side is equal to $\|(1 + |D|^2)^{s/2} a(x, D)\psi_0(D/R)u\|$, so changing scales we obtain if $u \in \mathcal{S}(\mathbf{R}^n)$

$$\|(R^{-2} + |D|^2)^{s/2} A_R(x, D)u\| \leq 2^{-s} M \|u\|, \quad A_R(x, \eta) = a(x/R, R\eta)\psi_0(\eta).$$

If $1 \leq B \leq R$ then $R^{-2} + |\xi|^2 \leq R^{-2} + B^{-2} \leq 2B^{-2}$ when $\varphi_0(B\xi) \neq 0$, so we conclude that

$$\|b_{R,B}(x, D)u\| \leq M 2^{-3s/2} B^s \|u\|, \quad u \in \mathcal{S}(\mathbf{R}^n),$$
$$b_{R,B}(x, D) = \varphi_0(BD)A_R(x, D), \quad \hat{b}_{R,B}(\xi, \eta) = \varphi_0(B(\xi + \eta))\widehat{A}_R(\xi, \eta).$$

(Note that $\varphi_0(BD)$ extracts the frequencies which have been much lowered.) $A_R(x, \eta)$ is bounded in $S_{1,0}^{-\infty}$. If $\Phi \in \mathcal{S}(\mathbf{R}^n)$ is the inverse Fourier transform of φ_0, then $\varphi_0(BD)$ is convolution by $B^{-n}\Phi(\cdot/B)$, hence

$$e^{i\langle x, \eta \rangle} b_{R,B}(x, \eta) = \int B^{-n}\Phi(y/B)e^{i\langle x-y, \eta \rangle} A_R(x - y, \eta) \, dy,$$

that is,

$$b_{R,B}(x, \eta) = \int B^{-n}\Phi(y/B)e^{-i\langle y, \eta \rangle} A_R(x - y, \eta) \, dy.$$

The derivatives with respect to x are uniformly bounded, but differentiation with respect to η also produces a factor y, so we only obtain

$$(9.3.22) \qquad |\partial_\eta^\alpha \partial_x^\beta b_{R,B}(x,\eta)| \le C_{\alpha\beta} B^{|\alpha|}.$$

Changing scales again we set $c_{R,B}(x,\eta) = b_{R,B}(Bx, \eta/B)$, and obtain since $\frac{1}{2} \le |\eta| \le 2$ if $(x,\eta) \in \operatorname{supp} b_{R,B}$

$$(9.3.23) \qquad \hat{c}_{R,B}(\xi, \eta) = \hat{b}_{R,B}(\xi/B, \eta/B)/B^n = 0, \quad \text{when } |\xi + \eta| > 1,$$

$$(9.3.24) \qquad |\partial_\eta^\alpha c_{R,B}(x,\eta)| \le C_\alpha, \quad B/2 \le |\eta| \le 2B \text{ if } (x,\eta) \in \operatorname{supp} c_{R,B},$$

$$(9.3.25) \qquad \|c_{R,B}(x,D)u\| \le M 2^{-3s/2} B^s \|u\|, \quad u \in \mathcal{S}(\mathbf{R}^n).$$

From (9.3.24) it follows that

$$\left(\int |\partial_\eta^\alpha c_{R,B}(x,\eta)|^2 \, d\eta \right)^{1/2} \le C_N B^{n/2}, \quad |\alpha| \le N,$$

so we can apply Lemma 9.3.9 to $c_{R,B}/(C_N B^{n/2})$ with M replaced by a constant times $B^{s-n/2}$. For any $\sigma < -s$ and α we have

$$n/2 + (s - n/2)(1 - n/N)(1 - |\alpha|/N) < -\sigma$$

if N is large enough, so we obtain

$$\left(\int |\partial_\eta^\alpha c_{R,B}(x,\eta)|^2 \, d\eta \right)^{1/2} \le C_{\alpha,\sigma} B^{-\sigma}.$$

Thus

$$(9.3.26) \qquad \left(\int |\partial_\eta^\alpha b_{R,B}(x,\eta)|^2 \, d\eta \right)^{1/2} \le C_{\alpha,\sigma} B^{-\sigma + |\alpha| - n/2}.$$

We now introduce $c_B(x,\eta)$ defined by

$$\hat{c}_B(\xi, \eta) = \sum_{2^\nu \ge B} \varphi_0(B(\xi + \eta)/2^\nu) \psi_0(\eta/2^\nu) \hat{a}(\xi, \eta),$$

which is equal to $\hat{a}(\xi, \eta)$ in the conic neighborhood

$$\{(\xi, \eta); |\eta| > 2B, \ 4B|\xi + \eta| \le |\eta|\},$$

of ∞ in the twisted diagonal. We have

$$c_B(x, \eta) = \sum_{2^\nu \ge B} b_{2^\nu, B}(2^\nu x, 2^{-\nu}\eta)$$

and conclude using (9.3.26) that

$$(9.3.27) \qquad \left(\int_{\varrho \le |\eta| \le 2\varrho} |\partial_\eta^\alpha c_B(x,\eta)|^2 \, d\eta \right)^{1/2} \le C_{\alpha,\sigma} B^{-\sigma} (B/\varrho)^{|\alpha| - n/2},$$

for at most three terms, with $\varrho/2 < 2^\nu < 4\varrho$, are not identically 0 when $\varrho \le |\eta| \le 2\varrho$. As in the proof of Lemma 9.3.2 we conclude with $B \sim 1/\varepsilon$ that the estimate (9.3.27) is also valid for the cleaner cutoff $a_{\chi,\varepsilon}$ in (9.3.12). This completes the proof.

Remark. By Sobolev's lemma it follows from (9.3.12) that

$$(9.3.28) \qquad |\partial_\eta^\alpha a_{\chi,\varepsilon}(x,\eta)| \le C_\alpha \varepsilon^{\sigma - |\alpha|} |\eta|^{m - |\alpha|}, \quad 0 < \varepsilon < 1,$$

so this is necessary for continuity of $a(x,D)$ from $H_{(s+m)}(\mathbf{R}^n)$ to $H_{(s)}(\mathbf{R}^n)$ for some s with $s + \sigma < 0$. On the other hand, (9.3.28) only implies (9.3.12) with σ replaced by $\sigma - n/2$. From (9.3.28) we can therefore only conclude continuity when $s + \sigma > n/2$, which shows the importance of using the L^2 condition (9.3.12) rather than conditions of the form (9.3.28).

We shall finally discuss briefly the analogous continuity results for the Zygmund spaces C_*^ϱ introduced in Section 8.6.

Theorem 9.3.10. *If $a \in S_{1,1}^m$ then $a(x, D)$ is continuous from $C_*^{\varrho+m}(\mathbf{R}^n)$ to $C_*^{\varrho}(\mathbf{R}^n)$ for every $\varrho > 0$.*

Proof. In view of Proposition 8.6.6 we may assume that $m = 0$. We keep the notation in the proof of Proposition 9.3.1. Since $a(x, \xi)\varphi_0(\xi) \in S_{1,0}^0$, it follows from Theorem 8.6.14 that

$$|a(x, D)\varphi_0(D)u|_\varrho^* \leq C|u|_\varrho^*, \quad u \in \mathcal{S}(\mathbf{R}^n).$$

We have $a(x, D)u = a(x, D)\varphi_0(D)u + \sum_0^\infty h_\nu(x)$ where

$$h_\nu(x) = a(x, D)\psi_\nu(D)u = a_\nu(x, D)\psi_\nu(D)u,$$
$$a_\nu(x, \eta) = (\psi_{\nu+1}(\eta) + \psi_\nu(\eta) + \psi_{\nu-1}(\eta))a(x, \eta).$$

Changing scales we obtain the symbol $b_\nu(x, \eta) = a_\nu(2^{-\nu}x, 2^\nu\eta)$ for which $|\eta|$ is bounded in the support and all derivatives have uniform bounds. Hence b_ν is uniformly bounded in $S_{1,0}^{-N}$, so $|b_\nu(x, D)v|_N^* \leq C_N \sup|v|$, thus

$$|b_\nu(x, D)v| \leq C \sup|v|, \quad |\psi_0(D/t)b_\nu(x, D)v| \leq C_N t^{-N} \sup|v|, \quad t \geq 1,$$

for every N and $v \in \mathcal{S}(\mathbf{R}^n)$. When the scales are changed back this means that

$$|a_\nu(x, D)v| \leq C \sup|v|, \quad |\psi_0(D/t)a_\nu(x, D)v| \leq C_N(2^\nu/t)^N \sup|v|, \quad t \geq 2^\nu.$$

Hence

$$|h_\nu| \leq C2^{-\nu\varrho}|u|_\varrho^*, \quad |\psi_0(D/t)h_\nu| \leq C_N(2^\nu/t)^N 2^{-\nu\varrho}|u|_\varrho^*, \quad t \geq 2^\nu.$$

Since $\varphi_0(D)$ and $\psi_0(D/t)$ are convolutions by functions in L^1 with norm independent of t, we conclude that for $h = \sum_0^\infty h_\nu$

$$|\varphi_0(D)h| \leq C' \sum_0^\infty 2^{-\nu\varrho}|u|_\varrho^* \leq C''|u|_\varrho^*,$$
$$|\psi_0(D/t)h| \leq C'\Big(\sum_{2^\nu > t} 2^{-\nu\varrho} + \sum_{2^\nu \leq t}(2^\nu/t)^N 2^{-\nu\varrho}\Big)|u|_\varrho^*.$$

When $N > \varrho$ summation of the geometrical series gives

$$t^\varrho|\psi_0(D/t)h| \leq C|u|_\varrho^*,$$

which completes the proof. (Corollary 9.3.6 can be proved in essentially the same way.)

9.4. Adjoints. If $a \in S_{1,1}^m$ and the adjoint $a(x, D)^*$ is in $\operatorname{Op} S_{1,1}^m$ then $a(x, D)$ and $a(x, D)^*$ are continuous from $H_{(s+m)}(\mathbf{R}^n)$ to $H_{(s)}(\mathbf{R}^n)$ for every $s > 0$. By duality it follows that $a(x, D)$ is then continuous from $H_{(s)}(\mathbf{R}^n)$ to $H_{(s-m)}(\mathbf{R}^n)$ for every $s < 0$, and continuity follows for every s by interpolation. (See e.g. Hörmander [4, III: Section B.1].) Hence it follows from Theorem 9.3.7 that a must be small at the twisted diagonal in the sense that (9.3.12) (and (9.3.28)) must be valid for every $\sigma \in \mathbf{R}$. When discussing the adjoint we shall therefore as in Proposition 9.3.1 begin with the case of symbols with Fourier transform vanishing at infinity in a conic neighborhood of the twisted diagonal.

Lemma 9.4.1. *If* $a(x, \eta) \in S^m_{1,1}$ *and*

$$(9.4.1) \qquad \hat{a}(\xi, \eta) = 0 \quad when \ |\xi + \eta| + 1 < |\eta|/B,$$

where $B \geq 1$, *then* $a(x, D)^* = b(x, D)$ *where* $b \in S^m_{1,1}$ *and*

$$(9.4.2) \qquad \hat{b}(\xi, \eta) = 0 \quad when \ |\xi + \eta| > B(|\eta| + 1).$$

With semi-norms $C_{\alpha\beta}$ *in* $S^m_{1,1}$ *we have*

$$(9.4.3) \qquad |b^{(\alpha)}_{(\beta)}(x, \eta)| \leq C_{\alpha\beta}(a)B(B^{m+|\beta|-|\alpha|} + 1)(1 + |\eta|)^{m+|\beta|-|\alpha|}.$$

Proof. We split a into a sum

$$a(x, \eta) = r(x, \eta) + \sum_0^\infty a_\nu(x, \eta), \quad r(x, \eta) = a(x, \eta)\varphi_0(\eta), \ a_\nu(x, \eta) = a(x, \eta)\psi_\nu(\eta).$$

(See (9.2.8) for the notation.) Then $r(x, \eta)$ and

$$A_\nu(x, \eta) = a_\nu(2^{-\nu}x, 2^\nu\eta)2^{-m\nu} = a(2^{-\nu}x, 2^\nu\eta)\psi_0(\eta)2^{-m\nu}$$

have uniformly bounded derivatives and a uniform bound for $|\eta|$ in the support. This means that they are uniformly bounded in $S^{-M}_{1,0}$ for every M. Thus $r(x, D)^* = s(x, D)$ and $A_\nu(x, D)^* = B_\nu(x, D)$ where by Theorem 8.4.2 for every $M \geq 0$,

$$|s(x, \eta)| \leq C_M(1 + |\eta|)^{-M}, \quad |B_\nu(x, \eta)| \leq C_M(1 + |\eta|)^{-M}.$$

Here C_M is a semi-norm of a. Changing scales gives $a_\nu(x, D)^* = b_\nu(x, D)$ where $b_\nu(x, \eta) = 2^{m\nu}B_\nu(2^\nu x, 2^{-\nu}\eta)$, hence

$$(9.4.4) \qquad |b_\nu(x, \eta)| \leq 2^{m\nu}C_M(1 + |2^{-\nu}\eta|)^{-M}.$$

Since $2^{\nu-1} \leq |\eta| \leq 2^{\nu+1}$ in supp $\hat{a}_\nu$, we have by (9.2.4)

$$2^{\nu-1} \leq |\eta + \xi| \leq 2^{\nu+1} \quad \text{in supp } \hat{b}_\nu.$$

From (9.2.4) we also obtain (9.4.2), hence

$$(9.4.5) \qquad B(|\eta| + 1) \geq 2^{\nu-1} \quad \text{in supp } b_\nu.$$

If $|\eta| \leq 1$ it follows from (9.4.5) that $\sum_0^\infty b_\nu(x, \eta)$ has at most $\log_2(8B) \leq 3B$ terms different from 0, and by (9.4.4) with $M = 0$ each of them can be estimated by $B^m + 1$. If $|\eta| > 1$ we choose the smallest $M \geq 0$ such that $M + m \geq 1$ and obtain from (9.4.4)

$$\sum_0^\infty |b_\nu(x, \eta)| \leq C_M \sum_{1 \leq 2^\nu \leq 2B(|\eta|+1)} 2^{(m+M)\nu}|\eta|^{-M}$$

$$\leq 2C_M(2B(|\eta| + 1))^{m+M}|\eta|^{-M} \leq C'_M B^{m+M}(1 + |\eta|)^m.$$

This proves (9.4.3) when $\alpha = \beta = 0$, for $B^{m+M} \leq B^{m+1} + B$. We obtain $b^{(\alpha)}_{(\beta)}$ when a is replaced by $a^{(\alpha)}_{(\beta)}$ since $b(x, \eta) = \exp\langle iD_x, D_\eta\rangle\overline{a(x, \eta)}$, so the general statement follows at once.

Theorem 9.4.2. *If $a \in S_{1,1}^m$ then the following three conditions are equivalent:*

(i) $a(x,D)^* \in \mathrm{Op}\, S_{1,1}^m$.

(ii) *With $a_{\chi,\varepsilon}$ defined by (9.3.4) and χ as in (9.3.3) there is an estimate*

$$(9.4.6) \qquad |\partial_\eta^\alpha \partial_x^\beta a_{\chi,\varepsilon}(x,\eta)| \le C_{\alpha\beta N}\varepsilon^N (1+|\eta|)^{m+|\beta|-|\alpha|}, \quad 0 < \varepsilon < 1,$$

for arbitrary N, α, β.

(iii) *The operator $a(x,D)$ is continuous from $H_{(s+m)}(\mathbf{R}^n)$ to $H_{(s)}(\mathbf{R}^n)$ for every $s \in \mathbf{R}$.*

Proof. At the beginning of this section we saw that (i) $\Longrightarrow$ (iii), and (iii) $\Longrightarrow$ (ii) by Theorem 9.3.7 (see (9.3.28)). Assume now that condition (ii) is fulfilled. By (9.4.6) and Bernstein's inequality $e_{-1}(x,D) = a(x,D) - a_{\chi,1}(x,D)$ and all the terms

$$e_\nu(x,D) = a_{\chi,2^{-\nu}}(x,D) - a_{\chi,2^{-\nu-1}}(x,D), \quad \nu = 0, 1, \ldots,$$

in the decomposition (9.3.13) of $a_{\chi,1}(x,D)$ are in $\mathrm{Op}\, S_{1,1}^m$, and it follows from Lemma 9.4.1 that $e_\nu(x,D)^* = b_\nu(x,D)$ where $b_\nu \in S_{1,1}^m$. We can estimate b_ν using (9.4.3) with $B = 2^{\nu+2}$. Since $C_{\alpha\beta}(e_\nu) = O(2^{-N\nu})$ for every N by (9.4.6), it follows that $b = \sum_{-1}^\infty b_\nu$ is in $S_{1,1}^m$. Hence $a(x,D)^* \in \mathrm{Op}\, S_{1,1}^m$, which proves (i) and completes the proof.

Remark. It is easy to prove that (i) $\Longrightarrow$ (ii) without passing through condition (iii) which requires the results of Section 9.3. To do so one observes that if $a(x,D)^* = b(x,D)$ with $b \in S_{1,1}^m$, then $\hat{a}_{\chi,\varepsilon}(\xi,\eta) = \chi(\xi+\eta, \varepsilon\eta)\overline{\hat{b}}(-\xi,\xi+\eta)$. When $\chi(\xi+\eta,\varepsilon\eta) \ne 0$ we have $|\xi+\eta| \le \varepsilon|\eta| \le \varepsilon|\xi+\eta| + \varepsilon|\xi|$, hence $|\xi+\eta| \le \varepsilon|\xi|/(1-\varepsilon)$, so $\hat{a}_{\chi,\varepsilon}$ is determined by $\hat{b}$ near the horizontal space, and (9.4.6) follows by elementary estimates of convolutions. (See Hörmander [9, p. 1101].)

As in the introduction we shall define

$$(9.4.7) \qquad \tilde{\Psi}_{1,1}^m = \mathrm{Op}\, S_{1,1}^m \cap (\mathrm{Op}\, S_{1,1}^m)^* = \mathrm{Op}\, \tilde{S}_{1,1}^m.$$

By Theorem 9.4.2 $\tilde{S}_{1,1}^m$ is thus the set of symbols in $S_{1,1}^m$ satisfying (9.4.6).

9.5. Composition. Let $A \in \mathrm{Op}\, S_{1,1}^m$ and assume that for some $\mu \in \mathbf{R}$

$$(9.5.1) \qquad (1+|D|^2)^{\mu/2} A \in \mathrm{Op}\, S_{1,1}^{m+\mu}.$$

Then it follows from Corollary 9.3.6 that $(1+|D|^2)^{\mu/2}A$ is continuous from $H_{(m+\mu+s)}(\mathbf{R}^n)$ to $H_{(s)}(\mathbf{R}^n)$ for every $s > 0$, that is, A is continuous from $H_{(m+s)}(\mathbf{R}^n)$ to $H_{(s)}(\mathbf{R}^n)$ for every $s > \mu$. If this is true for every negative μ it follows from Theorem 9.4.2 that $A \in \tilde{\Psi}_{1,1}^m$, defined by (9.1.2) or (9.4.7). When studying composition of operators of type $1,1$ it is therefore essential to assume that the factor to the right is in $\tilde{\Psi}_{1,1}^m$. The main step in proving multiplicative properties is then an analogue of Proposition 9.3.1.

Lemma 9.5.1. *Let $a \in S_{1,1}^m$ satisfy (9.3.1) for some $B \ge 1$, and let $b \in S_{1,1}^\mu$. Then $c(x,D) = b(x,D)a(x,D) \in \mathrm{Op}\, S_{1,1}^{m+\mu}$. Every semi-norm in $S_{1,1}^{m+\mu}$ of the symbol can be estimated by a power of B times a product of a semi-norm of a in $S_{1,1}^m$ and a seminorm of b in $S_{1,1}^\mu$.*

Proof. In the proof we may assume that $a, b \in \mathcal{S}(\mathbf{R}^n \times \mathbf{R}^n)$ when proving estimates which only depend on the seminorms of a and b allowed in the statement. We can then use (9.2.6). Differentiation with respect to η just acts on a or on b, lowering the degree by one. Since $\theta_j \hat{a}(\theta,\eta)$ is the Fourier transform of $D_{x_j}a(x,\eta)$, differentiation with respect to

x either acts on a or on b, raising the degree by one. It is therefore sufficient to prove an estimate for $c(x, \eta)(1 + |\eta|)^{-m-\mu}$.

By hypothesis we have $\hat{a}(\theta - \eta, \eta) = 0$ unless $|\theta| + 1 \geq |\eta|/B$, which implies $|\theta| \geq |\eta|/2B$ if $|\eta| \geq 2B$. Hence $c(x, \eta)$ is not changed when $|\eta| = T \geq 2B$ if we replace $a(x, \xi)$ by $a(x, \xi)\psi_0(\xi/T)$ and $b(x, \xi)$ by $b(x, \xi)(1 - \varphi(2B\xi/T))$. Changing scales we introduce

$$a_T(x, \xi) = a(x/T, T\xi)\psi_0(\xi)/T^m, \quad b_T(x, \xi) = b(x/T, T\xi)(1 - \varphi_0(2B\xi))/T^\mu.$$

The seminorms of a_T in $S^0_{1,0}$ and those of $b_T(x^0, \cdot)$ in $S^\mu_{1,1}$, with x^0 fixed and $T \geq 2B$, are bounded by admissible seminorms of a and b multiplied by a power of B, for $4B|\xi| \geq 1$ in supp b_T which implies $T/(1 + |T\xi|) \leq 1/|\xi| \leq (4B + 1)/(1 + |\xi|)$. If $c_T(x, D) = b_T(x, D)a_T(x, D)$ we conclude from the standard calculus (see Theorem 8.4.3) that $c_T(x, \eta)$ for $|\eta| = 1$ is bounded by a power of B times a product of a seminorm of a in $S^m_{1,1}$ and one of b in $S^\mu_{1,1}$. Since $c(x, \eta) = T^{m+\mu}c_T(xT, \eta/T)$ when $|\eta| = T$, we have proved the desired bound when $|\eta| \geq 2B$. When $|\eta| \leq 2B$ we note that $c(x, \eta)$ is equal to the symbol of $b(x, D)a(x, D)\varphi_0(D/4B)$, and since $a(x, D)\varphi_0(D/4B)$ is bounded as the operator a_T above, we obtain as before an estimate for $c(x, \eta)$ when $|\eta| \leq 2B$, which completes the proof of the lemma.

As in the proof of Theorem 9.4.2 we can weaken the hypothesis (9.3.1) to (9.4.6) and obtain the first part of the following theorem:

Theorem 9.5.2. *For arbitrary $m, \mu \in \mathbf{R}$ we have*

$$(9.5.2) \qquad \operatorname{Op} S^\mu_{1,1} \tilde{\Psi}^m_{1,1} \subset \operatorname{Op} S^{m+\mu}_{1,1}, \quad \tilde{\Psi}^\mu_{1,1} \tilde{\Psi}^m_{1,1} \subset \tilde{\Psi}^{m+\mu}_{1,1}.$$

Proof. It remains to prove the second part of (9.5.2). If $A \in \tilde{\Psi}^m_{1,1}$ and $B \in \tilde{\Psi}^\mu_{1,1}$ then A^* and B^* are in the same spaces by Theorem 9.4.2, and it follows from the first part of (9.5.2) that BA and the adjoint A^*B^* are both in $\operatorname{Op} S^{m+\mu}_{1,1}$. By definition this means that $BA \in \tilde{\Psi}^{m+\mu}_{1,1}$, which completes the proof.

We shall end this section by emphasizing the continuity properties of elements in $\operatorname{Op} \tilde{S}^m_{1,1}$.

Theorem 9.5.3. *If $a \in \tilde{S}^m_{1,1}$ then $a(x, D)$ defines a continuous map from $H_{(s+m)}(\mathbf{R}^n)$ to $H_{(s)}(\mathbf{R}^n)$ and from $C^{s+m}_*(\mathbf{R}^n)$ to $C^s_*(\mathbf{R}^n)$ for every $s \in \mathbf{R}$.*

Proof. We know from Corollary 9.3.6 and Theorem 9.3.10 that this is true for an arbitrary $a \in S^m_{1,1}$ when $s > 0$. If $s \leq 0$ we use that by Theorem 9.5.2

$$a(x, D) = (1 + |D|^2)^{(1-s)/2}b(x, D); \quad b(x, D) = (1 + |D|^2)^{(s-1)/2}a(x, D) \in \operatorname{Op} \tilde{S}^{m+s-1}_{1,1}.$$

Thus $b(x, D)$ maps $H_{(s+m)}$ to $H_{(1)}$ and C^{s+m}_* to C^1_*, which proves the statement in view of Proposition 8.6.6.

The statement about $H_{(s)}$ continuity is of course contained already in Theorem 9.4.2. However, we needed the composition theorem to get the continuity in C^s_*.

9.6. Symbols with additional smoothness.

The calculus of paradifferential operators as exposed by Meyer [2] gives leading terms for the symbol of the adjoint and for the composition of operators of type 1,1, provided that one has additional information on the derivatives of the symbol with respect to x. To prove such results we begin with an extension of Lemma 9.4.1.

Lemma 9.6.1. Let $a(x,\eta) \in S_{1,1}^m$ and $a_{(\beta)}(x,\eta) \in S_{1,1}^{m_N+|\beta|}$ when $|\beta| = N$, a positive integer, and assume that (9.4.1) holds with $B \geq 1$. Then $a(x,D)^* = b(x,D)$ where b satisfies (9.4.2) and

$$|\partial_\eta^\beta \partial_\eta^\alpha (b(x,\eta) - \sum_{j<N} \langle iD_x, D_\eta \rangle^j \overline{a(x,\eta)}/j!)| \leq C_{\alpha\beta} B(B^{m_N+|\beta|-|\alpha|} + 1)(1 + |\eta|)^{m_N+|\beta|-|\alpha|},$$

with $C_{\alpha\beta}$ denoting a sum of semi-norms of $a_{(\gamma)}$ in $S_{1,1}^{m_N+|\gamma|}$ when $|\gamma| = N$.

Proof. As in the proof of Lemma 9.4.1 the general statement follows from the case $\alpha = \beta = 0$. With the notation used there we have

$$|\partial_x^\beta \partial_\xi^\alpha r(x,\eta)| \leq C_{\alpha\beta}(a), \quad |\beta| \geq N,$$

$$|\partial_x^\beta \partial_\xi^\alpha A_\nu(x,\eta)| \leq C_{\alpha\beta}(a) 2^{(m_N-m)\nu}, \quad |\beta| \geq N,$$

with a semi-norm $C_{\alpha\beta}$ as in Lemma 9.6.1. Hence it follows from Theorem 8.4.2' that

$$(1 + |\eta|)^M |s(x,\eta) - \sum_{j<N} \langle iD_x, D_\eta \rangle^j \overline{r(x,\eta)}/j!| \leq C_{M,N}(a),$$

$$(1 + |\eta|)^M |B_\nu(x,\eta) - \sum_{j<N} \langle iD_x, D_\eta \rangle^j \overline{A_\nu(x,\eta)}/j!| \leq C_{M,N}(a) 2^{(m_N-m)\nu},$$

for arbitrary M. The second estimate means that

$$|b_\nu(x,\eta) - \sum_{j<N} \langle iD_x, D_\eta \rangle^j \overline{a_\nu(x,\eta)}/j!| \leq C_{M,N}(a) 2^{m_N\nu}(1 + |2^{-\nu}\eta|)^{-M},$$

and the proof proceeds just as before with m replaced by m_N.

Theorem 9.6.2. Let $a(x,\eta) \in S_{1,1}^m$ and assume that $a_{(\beta)}(x,\eta) \in \tilde{S}_{1,1}^{m_N+|\beta|}$ when $|\beta| = N$. Then it follows that $a(x,D)^* = b(x,D)$ where

$$b(x,\eta) - \sum_{j<N} \langle iD_x, D_\eta \rangle^j \overline{a(x,\eta)}/j! \in S_{1,1}^{m_N}.$$

Proof. Define $a_{\chi,\varepsilon}$ by (9.3.4), (9.3.3), and note that $(a_{\chi,\varepsilon})_{(\beta)} = (a_{(\beta)})_{\chi,\varepsilon}$. The proof then proceeds just as the proof that (ii) $\Longrightarrow$ (i) in Theorem 9.4.2, with the reference to Lemma 9.4.1 replaced by a reference to Lemma 9.6.1.

Next we prove an analogue of Lemma 9.5.1:

Lemma 9.6.3. Let $a \in S_{1,1}^m$ and $a_{(\beta)} \in S_{1,1}^{m_N+|\beta|}$ when $|\beta| = N$, a positive integer; assume that (9.3.1) holds for some $B \geq 1$, and let $b \in S_{1,1}^\mu$. Then $b(x,D)a(x,D) = c(x,D)$ where

$$|\partial_x^\beta \partial_\eta^\alpha (c(x,\eta) - \sum_{j<N} \langle iD_y, D_\xi \rangle^j b(x,\xi)a(y,\eta)/j!|_{y=x,\xi=\eta})|$$

$$\leq C_{\alpha\beta} B^{\kappa(\alpha,\beta,N,m_N,\mu)}(1 + |\eta|)^{m_N+\mu+|\beta|-|\alpha|},$$

where $C_{\alpha\beta}$ is the product of a semi-norm of b in $S_{1,1}^\mu$ and a sum of semi-norms of $a_{(\gamma)}$ in $S_{1,1}^{m_N+|\gamma|}$ when $|\gamma| = N$.

Proof. As usual it suffices to prove the estimate when $\alpha = \beta = 0$. Then we just repeat the proof of Lemma 9.5.1 with a reference to Theorem 8.4.3' in the application of the standard calculus. The details are left for the reader.

The proofs of Theorems 9.5.2 and 9.6.2 give, if one uses Lemma 9.6.3 instead of Lemma 9.5.1:

Theorem 9.6.4. Let $a(x, \eta) \in S_{1,1}^m$, $a_{(\beta)} \in \tilde{S}_{1,1}^{m_N + |\beta|}$ when $|\beta| = N$, and $b(x, \eta) \in S_{1,1}^\mu$. Then $b(x, D)a(x, D) = c(x, D)$ where

$$c(x, \eta) - \sum_{j < N} \langle iD_y, D_\xi \rangle^j b(x, \xi) a(y, \eta)/j!|_{x=y, \xi=\eta} \in S_{1,1}^{m_N + \mu}.$$

We shall now prove that the results of this section contain the standard calculus of operators in Op $S_{1,\delta}^m$, $0 \le \delta < 1$.

Theorem 9.6.5. When $\delta < 1$ we have $S_{1,\delta}^\mu \subset \tilde{S}_{1,1}^\mu$; if $b \in S_{1,\delta}^\mu$ and $b_{\chi,\varepsilon}$ is defined by (9.3.3), (9.3.4), then

(9.6.1) $$|\partial_x^\beta \partial_\xi^\alpha b_{\chi,\varepsilon}(x, \eta)| \le C_{N,\alpha,\beta} \varepsilon^N (1 + |\eta|)^{-N}, \qquad \forall N, \alpha, \beta, \quad 0 < 2\varepsilon < 1.$$

Moreover, $S_{1,\delta}^\mu \tilde{S}_{1,1}^m \subset \tilde{S}_{1,1}^{\mu+m}$.

Proof. Let $b \in S_{1,\delta}^\mu$ and set

$$\hat{B}(\xi, \eta) = \chi(\xi, \eta/2)\hat{b}(\xi, \eta), \qquad C(x, \eta) = b(x, \eta) - B(x, \eta),$$

where χ satisfies (9.3.3). If $\check{\chi}(x, \eta)$ is the inverse Fourier transform of $\chi(\xi, \eta)$ with respect to ξ, then

$$B(x, \eta) = \int b(x - y, \eta)\check{\chi}(y, \eta/2)\, dy.$$

When $|\eta| > 4$ we have

$$\int y^\alpha \check{\chi}(y, \eta/2)\, dy = \begin{cases} 1, & \text{if } \alpha = 0 \\ 0, & \text{if } \alpha \ne 0, \end{cases}$$

which proves that for every positive integer N

$$C(x, \eta) = -\int \Big(b(x - y, \eta) - \sum_{|\beta| < 2N} b_{(\beta)}(x, \eta)(-y)^\beta/\beta!\Big)\check{\chi}(y, \eta/2)\, dy.$$

By Taylor's formula the integrand can be estimated by $(1 + |\eta|)^{\mu + 2N\delta}|y|^{2N}|\check{\chi}(y, \eta/2)|$. Now the Fourier transform of $|y|^{2N}\check{\chi}(y, \eta/2)$ is $(-\Delta_\xi)^N \chi(\xi, \eta/2)$ which is homogeneous of degree $-2N$ when $|\eta| > 4$. Hence the L^1 norm of the inverse Fourier transform is a constant times $|\eta|^{-2N}$, which proves that

$$|C(x, \eta)| \le C_N (1 + |\eta|)^{\mu + 2N(\delta-1)}$$

for any N. Derivatives of C with respect to x can be placed on b while derivatives with respect to η give two terms, one where b is differentiated and one where $\check{\chi}$ is differentiated. The latter is also rapidly decreasing and it follows that $C \in S_{1,0}^{-\infty}$, hence that $B \in S_{1,\delta}^\mu \subset S_{1,1}^\mu$. Now $\hat{B}$ vanishes in a conic neighborhood of the twisted diagonal, so $B \in \tilde{S}_{1,1}^\mu$, and it is clear that $C \in \tilde{S}_{1,1}^\mu$ and that (9.6.1) holds since only C contributes to $b_{\chi,\varepsilon}$. (Note that $|\eta| > 1/\varepsilon$ in supp $b_{\chi,\varepsilon}$.)

The proof of the second (stronger) statement is an elaboration of the preceding argument. First we note that if $0 < \varepsilon < 1/2$ then

$$\hat{b}_\varepsilon(\xi, \eta) = \chi(\xi, \varepsilon\eta)\hat{b}(\xi, \eta).$$

has support in $\{(\xi, \eta); |\xi| \leq \varepsilon|\eta|,\ \varepsilon|\eta| \geq 1\}$, and for $c_\varepsilon = b - b_\varepsilon$ we have for every positive integer N

$$|c_{\varepsilon(\beta)}^{(\alpha)}(x, \eta)| \leq C_{N\alpha\beta}(1 + |\eta|)^{\mu+|\beta|+N\delta}(1 + \varepsilon|\eta|)^{-N-|\alpha|}.$$

This is proved just as the estimate of b_1 above. Now we decompose a as in (9.3.13)

$$a(x, \eta) = (a(x, \eta) - a_{\chi,1}(x, \eta)) + \sum_0^\sigma (a_{\chi,2^{-\nu}}(x, \eta) - a_{\chi,2^{-\nu-1}}(x, \eta)) + a_{\chi,2^{-\sigma-1}}(x, \eta).$$

The Fourier transform of $a(x,\eta) - a_{\chi,1}(x,\eta)$ vanishes when $|\eta| > 2$ and $|\xi+\eta| < |\eta|/2$, so the Fourier transform of the product by $b_{1/4}(x,\eta)$ vanishes when $|\eta| > 2$ and $|\xi + \eta| < |\eta|/4$ while the product by $c_{1/4}(x,\eta)$ is of order $-\infty$. With $\kappa = 2^{-\nu}$ the Fourier transform of $a_{\chi,\kappa}(x,\eta) - a_{\chi,\kappa/2}(x,\eta)$ vanishes when $\kappa|\eta| \geq 4$ and $4|\xi + \eta| \leq \kappa|\eta|$, hence the Fourier transform of $b_{\kappa/8}(x,\eta)(a_{\chi,\kappa}(x,\eta) - a_{\chi,\kappa/2}(x,\eta))$ vanishes when $8|\xi+\eta| \leq \kappa|\eta|$ and $\kappa|\eta| \geq 4$, while for arbitrary N

$$|\partial_\xi^\alpha \partial_x^\beta c_{\kappa/8}(x,\eta)(a_{\chi,\kappa}(x,\eta) - a_{\chi,\kappa/2}(x,\eta))| \leq C_{\alpha\beta N}(1 + |\eta|)^{-N}\kappa^N.$$

It is here that we use that a satisfies (9.4.6). The last term $a_{\chi,2^{-\sigma-1}}(x,\eta)$ in the decomposition of $a(x,\eta)$ we just multiply by $b(x,\eta)$. It follows that $b(x,\eta)a(x,\eta)$ is the sum of a function with Fourier transform vanishing when $8|\xi + \eta| \leq 2^{-\sigma}|\eta|$ and $|\eta| > 2^{\sigma+2}$ and another function $r_\sigma(x,\eta)$ such that

$$|\partial_\xi^\alpha \partial_x^\beta r_\sigma(x,\eta)| \leq C_{\alpha\beta N}2^{-\sigma N}(1 + |\eta|)^{-N}$$

when $|\eta| > 2^{\sigma+2}$. For small ε we let σ be the largest integer such that $2^{\sigma+2}\varepsilon < 1$. Then

$$\chi(\xi + \eta, \varepsilon\eta)\widehat{ba}(\xi, \eta) = \chi(\xi + \eta, \varepsilon\eta)\hat{r}_\sigma(\xi, \eta),$$

so we conclude as in the proof of Lemma 9.3.2 that $ba \in \tilde{S}_{1,1}^{\mu+m}$. The proof is complete.

From Theorem 9.4.2 it follows easily that

$$(9.6.2) \qquad\qquad a \in \tilde{S}_{1,1}^m \implies a_{(\beta)}^{(\alpha)} \in \tilde{S}_{1,1}^{m+|\beta|-|\alpha|}.$$

It suffices to verify (9.6.2) when $|\alpha| + |\beta| = 1$, and then it follows from the fact that

$$a^{(j)}(x, D) = i[a(x, D), x_j] \in \text{Op}\,S_{1,1}^{m-1}, \quad a^{(j)}(x, D)^* = i[a(x, D)^*, x_j] \in \text{Op}\,S_{1,1}^{m-1}$$

and the corresponding commutation identities with D_j.

We shall now show that Theorem 9.6.4 can be improved when the factor b to the left is in a better class which makes Theorem 9.6.5 applicable.

Theorem 9.6.4′. *Assume in addition to the hypotheses in Theorem 9.6.4 that $b \in S_{1,\delta}^\mu$ for some $\delta < 1$ and that $a \in \tilde{S}_{1,1}^m$. Then $b(x, D)a(x, D) = c(x, D)$ where*

$$r(x, \eta) = c(x, \eta) - \sum_{j < N} \langle iD_y, D_\xi \rangle^j b(x, \xi)a(y, \eta)/j!|_{x=y, \xi=\eta} \in \tilde{S}_{1,1}^{\mu+m_N}.$$

Proof. By Theorem 9.6.5 we know that the terms in the sum are in $\tilde{S}_{1,1}^{\mu+m}$. To prove the theorem it suffices to show, as remarked at the beginning of Section 9.5, that if $E(\eta) = (1 + |\eta|^2)^{\kappa/2}$ then $E(D)r(x, D) \in \text{Op}\,S_{1,1}^{\kappa+\mu+m_N}$, for arbitrary $\kappa \in \mathbf{R}$. Now

$$E(D)r(x, D) = (E(D)b(x, D))a(x, D) - \sum_{|\gamma| < N} i^{-|\gamma|}E(D)\,\text{Op}(b^{(\gamma)}(x, \eta)a_{(\gamma)}(x, \eta)/\gamma!).$$

If $M(1 - \delta) \geq m - m_N$ then the symbol of $E(D)b(x, D)$ is equal to

$$\sum_{|\beta| < M} i^{-|\beta|} E^{(\beta)}(\eta) b_{(\beta)}(x, \eta) / \beta!$$

modulo $S_{1,1}^{\kappa + \mu + m_N - m}$. We have $\partial_x^\beta (b^{(\gamma)}(x, \eta) a_{(\gamma)}(x, \eta) \in S_{1,1}^{\mu + m_N + |\beta|}$ if $|\beta| \geq N + M$, for if $\partial_x^{\beta'}$ acts on $b^{(\gamma)}$ and $\partial_x^{\beta''}$ acts on $a_{(\gamma)}$, we have

$$\partial_x^{\beta'} b^{(\gamma)}(x, \eta) \in S_{1,1}^{\mu + |\beta'| - |\gamma|}, \quad \partial_x^{\beta''} a_{(\gamma)}(x, \eta) \in \tilde{S}_{1,1}^{m + |\beta''| + |\gamma|},$$

and can use one of the estimates

$$\partial_x^{\beta'} b^{(\gamma)}(x, \eta) \in S_{1,1}^{\mu + m_N - m + |\beta'| - |\gamma|}, \ |\beta'| \geq M; \quad \partial_x^{\beta''} a_{(\gamma)}(x, \eta) \in \tilde{S}_{1,1}^{m_N + |\beta''| + |\gamma|}, \ |\beta''| \geq N.$$

Using Theorem 9.6.4 we now conclude that modulo $S_{1,1}^{\kappa + \mu + m_N}$ the symbol of $E(D)r(x, D)$ is

$$\sum_{|\beta| < M, |\alpha| < N} i^{-|\alpha| - |\beta|} \partial_\eta^\alpha (E^{(\beta)}(\eta) b_{(\beta)}(x, \eta)) a_{(\alpha)}(x, \eta) / \alpha! \beta!$$

$$- \sum_{|\alpha| < M + N, |\gamma| < N} i^{-|\alpha| - |\gamma|} E^{(\alpha)}(\eta) \partial_x^\alpha (b^{(\gamma)}(x, \eta) a_{(\gamma)}(x, \eta)) / \alpha! \gamma!.$$

The residue class does not change if we increase N and M. Both sums can be written in the form

$$\sum i^{-|\alpha| - |\beta| - |\gamma|} E^{(\beta + \alpha)}(\eta) b_{(\beta)}^{(\gamma)}(x, \eta) a_{(\alpha + \gamma)}(x, \eta) / \alpha! \beta! \gamma!$$

with various ranges for the sum. Now the order of such a term is at most

$$\leq \kappa - |\beta| - |\alpha| + \mu + \delta |\beta| - |\gamma| + m + |\alpha| + |\gamma| = \kappa + \mu + m + (\delta - 1) |\beta| \leq \kappa + \mu + m_N$$

if $|\beta| \geq (m - m_N)/(1 - \delta)$, and this inequality is also valid if $|\alpha| + |\gamma| \geq N$. The terms with $|\alpha| + |\gamma| < N$ and $|\beta| < (m - m_N)/(1 - \delta)$ occur in both sums which completes the proof of the theorem.

It is important to keep in mind that operators of type 1, 1 are capable of changing the frequencies in the wave front set very much, so they are local but not microlocal. However, we can prove some limited analogues of Corollaries 8.4.9 and 8.4.10 for operators satisfying the hypotheses of Theorem 9.6.2:

Theorem 9.6.6. *If a satisfies the hypotheses of Theorem 9.6.2 and if*

$$(9.6.3) \qquad u \in \mathcal{S}'(\mathbf{R}^n), \qquad u \in H_{(s + m_N)}^{loc} \text{ at } x_0, \qquad u \in H_{(s + m)}^{loc} \text{ at } (x_0, \xi_0),$$

it follows that $a(x, D)u \in H_{(s)}^{loc}$ at (x_0, ξ_0).

Proof. Choose $\varphi \in C_0^\infty(\mathbf{R}^n)$ equal to 1 in a neighborhood of x_0 so that $v = \varphi u \in H_{(s + m_N)}(\mathbf{R}^n)$. Since the kernel of $a(x, D)$ is in C^∞ and rapidly decreasing outside the diagonal, we have $a(x, D)((1 - \varphi)u) \in C^\infty$ in a neighborhood of x_0. Now $v \in H_{(s + m)}^{loc}$ at (x_0, ξ_0), so we can choose $\chi \in S_{1,0}^0$ equal to 1 in a conic neighborhood of (x_0, ξ_0) so that $\chi(x, D)v \in H_{(s + m)}(\mathbf{R}^n)$. This implies

$$a(x, D)\chi(x, D)v \in H_{(s)}(\mathbf{R}^n).$$

If we choose $\chi_0 \in S_{1,0}^\mu$ for some μ to be chosen later, with support in the set where $\chi = 1$, then

$$\chi_0(x, D)a(x, D)(1 - \chi(x, D)) \in \operatorname{Op} S_{1,1}^{\mu+m_N}$$

if we multiply from left to right and use Theorem 9.6.4. In fact, $\operatorname{Op}(\chi_0^{(\alpha)} a_{(\alpha)})(1 - \chi(x, D)) \in \operatorname{Op} S^{-\infty}$ by Theorem 9.6.4, and if $r \in S_{1,1}^{m_N+\mu}$ then $r(x, D)(1 - \chi(x, D)) \in \operatorname{Op} S_{1,1}^{m_N+\mu}$ by Theorem 9.5.2. If we choose $\mu < s$ it follows from Corollary 9.3.6 that

$$\chi_0(x, D)a(x, D)(1 - \chi(x, D))v \in H_{(s-\mu)}(\mathbf{R}^n), \quad \text{hence} \quad \chi_0(x, D)a(x, D)v \in H_{(s-\mu)}(\mathbf{R}^n),$$

which implies that $a(x, D)u \in H_{(s)}^{\mathrm{loc}}$ at (x_0, ξ_0). (The precaution of taking $\mu < s$ is superfluous if we use Theorem 9.6.4', but we have preferred to use only the easier Theorem 9.6.4. A similar remark is true for the proof of Theorem 9.6.7 below.)

Next we prove a converse statement:

Theorem 9.6.7. *If a satisfies the hypotheses of Theorem 9.6.2 and is noncharacteristic at (x_0, ξ_0), and if*

$$(9.6.4) \qquad u \in \mathcal{S}'(\mathbf{R}^n), \qquad u \in H_{(s+m_N)}^{\mathrm{loc}} \text{ at } x_0, \qquad a(x, D)u \in H_{(s)}^{\mathrm{loc}} \text{ at } (x_0, \xi_0),$$

then $u \in H_{(s+m)}^{\mathrm{loc}}$ at (x_0, ξ_0).

For the proof we need a lemma:

Lemma 9.6.8. *If a satisfies the hypotheses of Theorem 9.6.2, $m_N < m$, and $\chi \in S_{1,1}^\mu$,*

$$(9.6.5) \qquad\qquad (1 + |\eta|)^m \leq C|a(x, \eta)|, \qquad \text{if } (x, \eta) \in \operatorname{supp} \chi,$$

then one can find $b \in S_{1,1}^{\mu-m}$ with support contained in that of χ such that

$$(9.6.6) \qquad\qquad b(x, D)a(x, D) - \chi(x, D) \in \operatorname{Op} S_{1,1}^{\mu+m_N-m}.$$

Proof. Set $m_N = m - \gamma N$. By hypothesis $\gamma > 0$. The convexity inequalities $(4.2.17)'$ give

$$a_{(\beta)} \in S_{1,1}^{m+(1-\gamma)|\beta|}, \qquad |\beta| < N.$$

It follows from (9.6.5) that $b_0 = \chi/a \in S_{1,1}^{\mu-m}$. Hence Theorem 9.6.4 gives

$$b_0(x, D)a(x, D) = \chi(x, D) + \chi_1(x, D) + R_1(x, D), \qquad \chi_1 \in S_{1,1}^{\mu-\gamma}, \quad R_1 \in S_{1,1}^{\mu-m+m_N},$$

where $\chi_1(x, \eta) = \sum_{0 < |\alpha| < N} b_0^{(\alpha)}(x, \eta) D_x^\alpha a(x, \xi)/\alpha!$, hence $\operatorname{supp} \chi_1 \subset \operatorname{supp} \chi$. Repeating the argument with χ replaced by χ_1, proves the lemma after N steps.

Proof of Theorem 9.6.7. As in the proof of Theorem 9.6.6 we reduce the proof immediately to the case where $u \in H_{(s+m_N)}(\mathbf{R}^n)$. Choose $\chi_1 \in S_{1,0}^0$ equal to 1 in a conic neighborhood Γ of (x_0, ξ_0) so that $\chi_1(x, D)a(x, D)u \in H_{(s)}(\mathbf{R}^n)$, and choose $\chi \in S_{1,0}^\mu$ with support in Γ so that (9.6.5) is valid. With $b \in S_{1,1}^{\mu-m}$ provided by Lemma 9.6.8 we have

$$\chi(x, D) - b(x, D)a(x, D) \in \operatorname{Op} S_{1,1}^{\mu+m_N-m},$$

hence $\chi(x, D)u - b(x, D)a(x, D)u \in H_{(s+m-\mu)}(\mathbf{R}^n)$ by Corollary 9.3.6 if $\mu < s + m$. Now

$$b(x, D)\chi_1(x, D)a(x, D)u \in H_{(s+m-\mu)}(\mathbf{R}^n),$$

again by Corollary 9.3.6, and

$$b(x, D)(1 - \chi_1(x, D))a(x, D)u \in H_{(\infty)}(\mathbf{R}^n),$$

for $b(x, D)(1-\chi_1(x, D)) \in S^{-\infty}$ by Theorem 9.6.4 since $1-\chi_1$ vanishes in a cone containing supp b. Thus $\chi(x, D)u \in H_{(s+m-\mu)}(\mathbf{R}^n)$ which proves that $u \in H^{loc}_{(s+m)}$ at (x_0, ξ_0).

Remark. Theorems 9.6.6 and 9.6.7 remain valid with $H_{(\cdot)}$ replaced by $C^\cdot_*$ everywhere. The proofs need very little change since Theorems 9.3.10 and 9.5.3 give analogues for Zygmund spaces of Corollary 9.3.6 and Theorem 9.5.2 for Sobolev spaces.

In Theorems 9.6.6 and 9.6.7 we do not really care what N is. It is clear that if $a_{(\beta)} \in \tilde{S}_{1,1}^{\mu+|\beta|}$, when $|\beta| = N$, then this remains true for $|\beta| \geq N$.

Definition 9.6.9. We shall say that $a \in S_{1,1}^m$ has order m and *reduced order* μ if $a_{(\beta)} \in \tilde{S}_{1,1}^{\mu+|\beta|}$ when $|\beta|$ is large enough.

Note that if a has reduced order μ then $a_{(\beta)}^{(\alpha)}$ has reduced order $\mu + |\beta| - |\alpha|$. If $a \in \tilde{S}_{1,1}^{m_a}$ and $b \in \tilde{S}_{1,1}^{m_b}$ have reduced order μ_a, μ_b, then $ab \in \tilde{S}_{1,1}^{m_a+m_b}$ has reduced order $\max(m_a + \mu_b, \mu_a + m_b)$; the same is true for the symbol of the composition $a(x, D)b(x, D)$ and for all terms in the expansion of its symbol.

9.7. The sharp Gårding inequality. In the applications of the paradifferential calculus to the study of the propagation of singularities it is essential to have good lower bounds for operators with nonnegative symbols. The following theorem contains the "sharp Gårding inequality" as stated in Hörmander [4, Theorem 18.1.14], and we shall prove it by a modification of the proof given there.

Theorem 9.7.1. Let $a \in \tilde{S}_{1,1}^{m_0}$ and $a_{(\beta)} \in \tilde{S}_{1,1}^{m_2+|\beta|}$ when $|\beta| = 2$, and assume that $m_2 > m_0 - 4$. If $\operatorname{Re} a \geq 0$ it follows that

$$(9.7.1) \qquad \operatorname{Re}(a(x, D)u, u) \geq -C\|u\|^2_{((m_0+m_2)/4)}, \quad u \in \mathcal{S}(\mathbf{R}^n).$$

Proof. If $m_2 \geq m_0$ then the theorem follows from the fact that $a(x, D)$ is continuous from $H_{(m_0/2)}(\mathbf{R}^n)$ to $H_{(-m_0/2)}(\mathbf{R}^n)$, so we may assume that $m_2 < m_0$. Set

$$(9.7.2) \qquad m_0 + m_2 = 4m, \quad m_0 - m_2 = 4\mu.$$

Then

$$m_0 = 2m + 2\mu, \quad m_2 = 2m - 2\mu; \quad 0 < \mu < 1.$$

By the convexity inequalities (4.2.17)' the hypothesis implies

$$(9.7.3) \qquad a_{(\beta)} \in \tilde{S}_{1,1}^{m_{|\beta|}+|\beta|} \quad \text{for } |\beta| \leq 2; \quad m_j = m_0 - 2\mu j.$$

We split a as in the proof of Lemma 9.4.1. Since $r(x, D)$ is continuous from $H_{(m)}$ to $H_{(-m)}$ it suffices to study $\sum_{\nu \geq 0} a_\nu(x, D)$.

As shown in the proof of Hörmander [4, Theorem 18.1.14] we can choose an even function $\psi \in \mathcal{S}(\mathbf{R}^{2n})$ such that

$$(9.7.4) \qquad \iint \psi(x, \eta)\, dx d\eta = 1; \quad (\psi(x, D)u, u) \geq 0, \quad u \in \mathcal{S}.$$

With $q_\nu = 2^{\nu(1-\mu)}$ we write $a_\nu = b_\nu + c_\nu$ where

(9.7.5)
$$b_\nu(x, \eta) = \iint \psi((x - y)q_\nu, (\eta - \theta)/q_\nu)a_\nu(y, \theta)\, dy\, d\theta$$
$$= \iint \psi(yq_\nu, \theta/q_\nu)a_\nu(x - y, \eta - \theta)\, dy\, d\theta.$$

The operator $\psi((x - y)q_\nu, (D - \theta)/q_\nu)$ is unitarily equivalent to $\psi(x, D)$, hence positive by (9.7.4), and since $\operatorname{Re} a_\nu(y, \theta) \geq 0$ it follows that $b_\nu(x, D) + b_\nu(x, D)^* \geq 0$. The proof will be completed if we show that

(9.7.6)
$$c(x, \eta) = \sum_0^\infty c_\nu(x, \eta) \in \tilde{S}_{1,1}^{2m}.$$

We shall first prove that $c \in S_{1,1}^{2m}$, that is,

(9.7.7)
$$\left| \sum_0^\infty c_{\nu(\beta)}^{(\alpha)}(x, \eta) \right| \leq C_{\alpha\beta}(1 + |\eta|)^{2m + |\beta| - |\alpha|}.$$

Since $b_{\nu(\beta)}^{(\alpha)}$ is obtained by differentiating on a_ν in the second integral of (9.7.5), it suffices to prove (9.7.7) when $\alpha = \beta = 0$, which simplifies the notation. Then (9.7.7) follows if we prove that for any N

(9.7.8)
$$|c_\nu(x, \eta)| \leq C2^{2\nu m},$$

(9.7.9)
$$|c_\nu(x, \eta)| \leq C_N(2^\nu + |\eta|)^{-N} \quad \text{if } |\eta| < 2^{\nu-2} \text{ or } |\eta| > 2^{\nu+2}.$$

In fact, when estimating $\sum_0^\infty c_\nu$ we can use (9.7.9) except for five terms which are estimated by (9.7.8). To prove (9.7.9) we observe that in the support of the first integrand in (9.7.5) we have $2^{\nu-1} \leq |\theta| \leq 2^{\nu+1}$. If $|\eta| < 2^{\nu-2}$ it follows that $|\theta - \eta| > 2^{\nu-2}$, hence $2^\nu + |\eta| \leq 5|\theta - \eta|$; if $|\eta| > 2^{\nu+2}$ then $|\theta - \eta| \geq |\eta|/2$, hence $2^\nu + |\eta| \leq 3|\theta - \eta|$. In both cases we obtain
$$5|\eta - \theta|/q_\nu \geq (2^\nu + |\eta|)/2^{\nu(1-\mu)} \geq (2^\nu + |\eta|)^\mu,$$
for $\mu < 1$. Since $\psi \in \mathcal{S}(\mathbf{R}^{2n})$ and $a_\nu = O(2^{m_0\nu})$, we obtain (9.7.9), for $c_\nu = -b_\nu$ then. To prove (9.7.8) we use the second integral in (9.7.5) where by Taylor's formula and (9.7.3)
$$\left| a_\nu(x - y, \eta - \theta) - \sum_{|\alpha+\beta|<2} a_{\nu(\beta)}^{(\alpha)}(x, \eta)(-y)^\beta(-\theta)^\alpha/\alpha!\beta! \right|$$
$$\leq C \sum_{|\alpha+\beta|=2} 2^{\nu(m_0 - 2\mu|\beta| + |\beta| - |\alpha|)}|y^\beta \theta^\alpha| = C2^{2\nu m} \sum_{|\alpha+\beta|=2} |(yq_\nu)^\beta(\theta/q_\nu)^\alpha|,$$
since $2\mu - 2\mu|\beta| + |\beta| - |\alpha| - (1 - \mu)(|\beta| - |\alpha|) = \mu(2 - |\alpha| - |\beta|) = 0$. Now ψ is even, so the first order terms drop out from the integral (9.7.5) and (9.7.8) follows since $\psi \in \mathcal{S}(\mathbf{R}^{2n})$ and $c_\nu = a_\nu - b_\nu$.

(9.7.7) means that $c(x, \eta) \in S_{1,1}^{2m}$. Write $b = \sum_0^\infty b_\nu$, thus $a = b + c$, and recall that $\psi((x - y)q_\nu, (D - \theta)/q_\nu)$ is self-adjoint. Hence decomposing $\overline{a(x, \eta)}$ in the same way as $a(x, \eta)$ gives
$$\bar{a}(x, D) = b(x, D)^* + c_1(x, D)$$
where $c_1 \in S_{1,1}^{2m}$. Subtracting this from the equation
$$a(x, D)^* = b(x, D)^* + c(x, D)^*$$
we obtain $c(x, D)^* - c_1(x, D) = a(x, D)^* - \bar{a}(x, D)$. With the notation c_2 for the symbol of $c(x, D)^*$ it follows by Theorem 9.6.2 that
$$c_2(x, \eta) - c_1(x, \eta) - \langle iD_x, D_\eta \rangle \overline{a(x, \eta)} \in S_{1,1}^{m_2},$$
which proves that $c_2 \in S_{1,1}^{m_1} = S_{1,1}^{2m}$. Hence $c \in \tilde{S}_{1,1}^{2m}$ which completes the proof.

CHAPTER X

PARADIFFERENTIAL CALCULUS

10.1. Introduction. The results of Sections 9.6 and 9.7 are not yet in the most convenient form for the applications where one is given a symbol which is not even in C^∞ with respect to x. A simple but typical example is multiplication by a function $a \in L^\infty$. Bony's paraproduct is defined by regularizing a to a symbol in $\tilde{S}^0_{1,1}$, with additional properties if a is Hölder continuous. We shall discuss such regularisations in Section 10.2, paying particular attention to how far the resulting operator differs from straight multiplication. This will give the paradifferential calculus of Bony [1] in the original form, with extensions given in Meyer [2]. In Section 10.3 we return to the discussion of composite functions $F(u)$ begun in Section 8.5, comparing the operator L_u defined there with paramultiplication by $F'(u)$. This gives the linearization theorem of Bony [1] (see also Rauch [1]). As an application we also give another proof of the microlocal regularity of products discussed in Section 8.3.

10.2. Regularisation of symbols and paradifferential calculus. Multiplication by a function $a(x)$ can be regarded as a pseudo-differential operator with this symbol. To smooth it to a symbol $a(x, \eta)$ of type $1, 1$, it is natural to multiply the Fourier transform $\hat{a}(\xi)$ by a function $\chi(\xi, \eta)$ which is essentially homogeneous of degree 0 and has support staying away from the horizontal space $\{(\xi, 0)\}$ and the twisted diagonal $\{(\xi, -\xi)\}$ at infinity, for these play a special role in the calculus of operators of type $1, 1$.

Definition 10.2.1. By $\mathcal{B}$ we shall denote the set of all $\chi \in C^\infty(\mathbf{R}^n \times \mathbf{R}^n)$ which are symbols of order 0 (with no space variables),

$$(10.2.1) \qquad |\partial^\alpha_{\xi,\eta}\chi(\xi,\eta)| \leq C_\alpha(1 + |\xi| + |\eta|)^{-|\alpha|}, \quad \xi, \eta \in \mathbf{R}^n,$$

such that for some constant B

$$(10.2.2) \qquad \chi(\xi,\eta) = 0 \quad \text{when } |\xi| > B(|\eta| + 1) \text{ or } |\eta| > B(|\xi + \eta| + 1).$$

By $\mathcal{B}_c$, where $c = 0$ or $c = 1$, we shall denote the subset such that in addition

$$(10.2.3) \qquad \chi(\xi,\eta) = c \quad \text{when } |\eta| > B(|\xi| + 1).$$

We shall regularize by multiplication on the Fourier transform side with functions in $\mathcal{B}_1$; the choice of χ is determined modulo $\mathcal{B}_0$. There are functions in $\mathcal{B}_1$ for every $B > 2$.

Proposition 10.2.2. Assume that all η derivatives of $a(x, \eta)$ are in C^ℓ_*, and that

$$(10.2.4) \qquad |a^{(\alpha)}(\cdot, \eta)|^*_\ell \leq C_\alpha(1 + |\eta|)^{m-|\alpha|}, \quad \forall \alpha.$$

If $\chi \in \mathcal{B}$ and $\hat{a}_\chi(\xi, \eta) = \chi(\xi, \eta)\hat{a}(\xi, \eta)$, where the Fourier transforms are taken with respect to the x variables only, it follows that for every β

$$(10.2.5) \qquad (a_\chi)_{(\beta)} \in \begin{cases} \tilde{S}^m_{1,1}, & \text{if } |\beta| < \varrho \\ \tilde{S}^{m+|\beta|-\varrho}_{1,1}, & \text{if } |\beta| > \varrho. \end{cases}$$

Thus a_χ is of reduced order $m - \varrho$. If $\chi \in \mathcal{B}_0$ then

$$(10.2.6) \qquad\qquad\qquad\qquad a_\chi \in \tilde{S}_{1,1}^{m-\varrho}.$$

Proof. Define $\mathcal{B}^\mu$ as $\mathcal{B}$ but with the exponent $-|\alpha|$ in (10.2.1) replaced by $\mu - |\alpha|$. We shall prove that when $\chi \in \mathcal{B}^\mu$ or $\chi \in \mathcal{B}_0^\mu$ the same result is valid, with m replaced by $m+\mu$. Since differentiation of $\hat{a}_\chi(\xi, \eta)$ with respect to η gives one term where χ is differentiated, so that μ is replaced by $\mu - 1$, and one where a is differentiated, so that m is replaced by $m - 1$, it is enough when proving this statement to estimate a_χ and its derivatives with respect to x. Differentiation of a_χ with respect to x_j corresponds to multiplication of $\chi(\xi, \eta)$ by $i\xi_j$, which increases μ by 1. Hence it is enough to prove that

$$(10.2.7) \qquad |\partial_x^\beta a_\chi(x, \eta)| \le \begin{cases} C_{\beta,\chi}(1 + |\eta|)^{\mu+m}, & \text{if } |\beta| < \varrho, \\ C_{\beta,\chi}(1 + |\eta|)^{\mu+m+|\beta|-\varrho}, & \text{if } |\beta| > \varrho, \end{cases} \quad \text{if } \chi \in \mathcal{B}^\mu,$$

and that

$$(10.2.8) \qquad\qquad |a_\chi(x, \eta)| \le C_\chi(1 + |\eta|)^{\mu+m-\varrho}, \qquad \text{if } \chi \in \mathcal{B}_0^\mu.$$

The estimate (10.2.7) (resp. (10.2.8)) is an immediate consequence of (8.6.8) (resp. (8.6.6)), for the functions

$$\xi \mapsto \chi(\xi(1 + |\eta|), \eta)(1 + |\eta|)^{-\mu}$$

belong to a bounded subset of $C_0^\infty(\mathbf{R}^n)$ (resp. $C_0^\infty(\mathbf{R}^n \setminus 0)$) when $|\eta|$ is large.

Remark. For later reference we observe that in the preceding estimates we only used the part of (10.2.2) which requires that $\chi(\xi, \eta) = 0$ when $|\xi| > B(|\eta| + 1)$. However, if the other part of (10.2.2) is not assumed then (10.2.6) must be changed to $a_\chi \in S_{1,1}^{m-\varrho}$ and similarly for (10.2.5).

From (10.2.6) it follows for symbols satisfying (10.2.4) that modulo $\tilde{S}_{1,1}^{m-\varrho}$ the symbol a_χ is independent of the choice of $\chi \in \mathcal{B}_1$.

Definition 10.2.3. By A_ϱ^m we shall denote the class of symbols satisfying (10.2.4). If $a \in A_\varrho^m$ we shall denote by $a_\mathcal{B}$ (any element in) the residue class of a_χ modulo $\tilde{S}_{1,1}^{m-\varrho}$ when $\chi \in \mathcal{B}_1$. The corresponding operators are called *paradifferential operators* corresponding to a.

If $\varrho < 0$ then $a_\chi \in \tilde{S}_{1,1}^{m-\varrho}$ by (10.2.5), so we obtain the uninteresting residue class of 0. Thus the definition is only interesting when $\varrho \ge 0$. Note that

$$A_\infty^m = \bigcap_\varrho A_\varrho^m = S_{1,0}^m; \qquad a - a_\chi \in S_{1,0}^{-\infty} \qquad \text{if } a \in S_{1,0}^m, \quad \chi \in \mathcal{B}_1.$$

(See the proof of Theorem 10.2.6 below.)

To interpret the calculus results in Section 9.6 it is also important to note that for all α, β

$$(10.2.9) \qquad\qquad (a_{(\beta)}^{(\alpha)})_\chi - (a_\chi)_{(\beta)}^{(\alpha)} \in \tilde{S}_{1,1}^{m-\varrho+|\beta|-|\alpha|}, \qquad \chi \in \mathcal{B}_1.$$

It is enough to prove this for first order derivatives. Since

$$(\partial_{x_j} a)_\chi = \partial_{x_j}(a_\chi), \qquad (\partial_{\eta_j} a)_\chi + a_{\partial_{\eta_j}\chi} = \partial_{\eta_j}(a_\chi),$$

and $(1 + |\eta|^2)^{\frac{1}{2}}\partial_{\eta_j}\chi(\xi, \eta) \in \mathcal{B}_0$ if $\chi \in \mathcal{B}_1$, (10.2.9) follows from (10.2.6). From Theorem 9.6.2 we now obtain:

Theorem 10.2.4. *If $a \in A_\varrho^m$ then the adjoint of $a_\mathcal{B}(x, D)$ is a paradifferential operator corresponding to*

$$\sum_{j \leq \varrho} \langle iD_x, D_\eta \rangle^j \overline{a(x, \eta)}/j!,$$

where the terms are in $A_{\varrho-j}^{m-j}$. It is of order m and reduced order $m - \varrho$.

Proof. Let N be the smallest integer $> \varrho$, and let $\chi \in \mathcal{B}_1$. By (10.2.5)

$$(a_\chi)_{(\beta)} \in \tilde{S}_{1,1}^{m+|\beta|-\varrho}, \qquad |\beta| = N,$$

so we can apply Theorem 9.6.2 with $m_N = m - \varrho$. Thus the symbol of $a_\chi(x, D)^*$ is modulo $\tilde{S}_{1,1}^{m-\varrho}$ equal to

$$\sum_{j \leq \varrho} \langle iD_x, D_\eta \rangle^j \overline{a_\chi(x, \eta)}/j! \equiv \left(\sum_{j \leq \varrho} (\langle iD_x, D_\eta \rangle^j \overline{a(x, \eta)}/j!)_\chi. \right.$$

The proof is complete.

Note that for each term the corresponding paradifferential operators are of reduced order $m - \varrho$. There is an analogous version of Theorem 9.6.4:

Theorem 10.2.5. *If $a \in A_\varrho^m$ and $b \in A_\varrho^\mu$, $\varrho > 0$, then the composition $b_\mathcal{B}(x, D)a_\mathcal{B}(x, D)$ is a paradifferential operator corresponding to*

$$\sum_{|\alpha| \leq \varrho} i^{-|\alpha|} b^{(\alpha)}(x, \eta) a_{(\alpha)}(x, \eta)/\alpha!$$

where the terms are in $A_{\varrho-|\alpha|}^{m+\mu-|\alpha|}$. It is of order $m + \mu$ and reduced order $m + \mu - \varrho$.

Proof. If N is the smallest integer $> \varrho$, then

$$(a_\chi)_{(\alpha)} \in \tilde{S}_{1,1}^{m+|\alpha|-\varrho}, \qquad |\alpha| = N,$$

so the symbol of $b_\chi(x, D)a_\chi(x, D)$ is by Theorem 9.6.4 equal to

$$\sum_{|\alpha| \leq \varrho} i^{-|\alpha|} b_\chi^{(\alpha)}(x, \eta) a_{\chi(\alpha)}(x, \eta)/\alpha! \qquad (\mathrm{mod}\ \tilde{S}_{1,1}^{m+\mu-\varrho}).$$

By (10.2.9) this is equal to

$$\sum_{|\alpha| \leq \varrho} i^{-|\alpha|} (b^{(\alpha)})_\chi(x, \eta)(a_{(\alpha)})_\chi(x, \eta)/\alpha! \qquad (\mathrm{mod}\ \tilde{S}_{1,1}^{m+\mu-\varrho}).$$

Here $b^{(\alpha)} \in A_\varrho^{\mu-|\alpha|}$, $a_{(\alpha)} \in A_{\varrho-|\alpha|}^m$. Since $\mu - |\alpha| + m - (\varrho - |\alpha|) = \mu + m - \varrho$, it follows from Proposition 8.6.9 for a suitable choice of χ and χ_1 that

$$|(b^{(\alpha)})_\chi(x, \eta)(a_{(\alpha)})_\chi(x, \eta) - (b^{(\alpha)}a_{(\alpha)})_{\chi_1}(x, \eta)| \leq C(1 + |\eta|)^{\mu+m-\varrho}.$$

In view of (10.2.6) differentiations with respect to η acting on χ or χ_1 give contributions with order lowered by $1 + \varrho$ the first time and 1 later on, so we obtain

$$|\partial_x^\beta \partial_\eta^\gamma \left((b^{(\alpha)})_\chi(x, \eta)(a_{(\alpha)})_\chi(x, \eta) - (b^{(\alpha)}a_{(\alpha)})_{\chi_1}(x, \eta) \right)| \leq C_{\alpha\beta}(1 + |\eta|)^{\mu+m-\varrho+|\beta|-|\gamma|},$$

if $|\beta| = 0$, hence for all β by Bernstein's inequality.

Theorem 10.2.6. *If $a \in A^m_\varrho$ is noncharacteristic at (x_0, ξ_0) and $\varrho > 0$, then a_B is noncharacteristic at (x_0, ξ_0).*

Proof. If $0 < \sigma < \varrho$ and $\chi \in \mathcal{B}_1$ it follows from (8.6.9) and the proof of Proposition 10.2.2 that

$$|a(x, \eta) - a_\chi(x, \eta)| \leq C(1 + |\eta|)^{m+\sigma-\varrho}.$$

Hence a lower bound $|a(x, \eta)| \geq c(1 + |\eta|)^m$ with $c > 0$ implies a similar bound for a_χ when $|\eta|$ is large, and conversely.

We shall now discuss a paradifferential version of the sharp Gårding inequality. When doing so we shall need conditions on Hölder norms and not only on Zygmund norms.

Theorem 10.2.7. *Let $0 \leq \operatorname{Re} a \in A^m_\varrho$, where $0 < \varrho \leq 2$, and assume that even*

$$|a^{(\alpha)}(\cdot, \eta)|_\varrho \leq C_\alpha (1 + |\eta|)^{m-|\alpha|},$$

with Hölder norm of order ϱ. Then we have for a corresponding paradifferential operator

$$(10.2.10) \qquad \operatorname{Re}(a_B(x, D)u, u) \geq -C\|u\|^2_{(2m-\varrho)/4}, \qquad u \in \mathcal{S}.$$

Proof. Let $\chi \in \mathcal{B}_1$. We shall prove that for some C

$$\operatorname{Re} a_\chi(x, \eta) \geq -C(1 + |\eta|^2)^{(m-\varrho)/2}.$$

Since $a_{\chi(\beta)} \in \tilde{S}^{m+|\beta|-\varrho}$ when $|\beta| = 2$, by (8.6.1), the theorem will then follow if we apply Theorem 9.7.1 to $a_\chi(x, \eta) + C(1 + |\eta|^2)^{(m-\varrho)/2}$. Let ψ_η be the inverse Fourier transform of $\xi \mapsto \chi(\xi(1 + |\eta|), \eta)$. These functions belongs to a bounded set in C^∞_0 so ψ_η belongs to a bounded set in $\mathcal{S}$. Now

$$a_\chi(x, \eta) = \int a(x - y/(1 + |\eta|), \eta)\psi_\eta(y)\, dy,$$

$$\int \psi_\eta(y)\, dy = 1, \qquad \int y^\alpha \psi_\eta(y)\, dy = 0 \quad \text{for } \alpha \neq 0, \text{ if } |\eta| \text{ is large.}$$

Thus we have

$$a_\chi(x, \eta) - a(x, \eta) = \int \left(a(x - y/(1 + |\eta|), \eta) - \sum_{|\beta| < \varrho} a_{(\beta)}(x, \eta)(-y/(1 + |\eta|))^\beta \right) \psi_\eta(y)\, dy.$$

By Taylor's formula the integral can be estimated by

$$C \int |y|^\varrho (1 + |\eta|)^{m-\varrho} |\psi_\eta(y)|\, dy \leq C'(1 + |\eta|)^{m-\varrho},$$

which completes the proof.

In this section we have in particular defined a *paraproduct* $a_B(x, D)u$ if $a \in C^\varrho_*(\mathbf{R}^n)$ for some $\varrho > 0$. If $u \in H_{(s)}(\mathbf{R}^n)$ it belongs to $H_{(s)}(\mathbf{R}^n)$, and modulo $H_{(s+\varrho)}(\mathbf{R}^n)$ it is independent of the choices made. It is bilinear but not symmetric in the two factors. Indeed, if a_B is defined by means of the function $\chi \in \mathcal{B}_1$, then the Fourier transform of $a_B(x, D)u$ is

$$\xi \mapsto (2\pi)^{-n} \int \hat{a}(\xi - \eta)\chi(\xi - \eta, \eta)\hat{u}(\eta)\, d\eta,$$

while that of $u_B(x, D)a$ is

$$\xi \mapsto (2\pi)^{-n} \int \hat{u}(\xi - \eta)\chi(\xi - \eta, \eta)\hat{a}(\eta)\, d\eta = (2\pi)^{-n} \int \hat{a}(\xi - \eta)\chi(\eta, \xi - \eta)\hat{u}(\eta)\, d\eta.$$

if $u \in S(\mathbf{R}^n)$ for example. Hence

$$(10.2.11) \qquad au - a_\chi(x, D)u - u_\chi(x, D)a = a_\Phi(x, D)u, \quad u \in S(\mathbf{R}^n),$$

where $\Phi(\xi, \eta) = 1 - \chi(\xi, \eta) - \chi(\eta, \xi)$ vanishes when $|\eta| > B(|\xi| + 1)$ or $|\xi| > B(|\eta| + 1)$ but $\Phi(\xi, \eta) = 1$ in the neighborhood of infinity of the twisted diagonal where $B(|\xi + \eta| + 1) < \min(|\xi|, |\eta|)$. The proof of Proposition 10.2.2 still gives $a_\Phi \in S_{1,1}^{-\varrho}$, as pointed out in a remark after the proof. In view of Corollary 9.3.6 and Theorem 9.3.10 we have proved:

Theorem 10.2.8. *If $a \in C_*^\varrho(\mathbf{R}^n)$ for some $\varrho > 0$, and $u \in H_{(s)}(\mathbf{R}^n)$ for some $s > -\varrho$, then (10.2.11) is valid with $a_\Phi \in S_{1,1}^{-\varrho}$, hence $a_\Phi(x, D)u \in H_{(s+\varrho)}(\mathbf{R}^n)$; if $u \in C_*^\sigma(\mathbf{R}^n)$ for some $\sigma > -\varrho$ then (10.2.11) is valid and $a_\Phi(x, D)u \in C_*^{\sigma+\varrho}(\mathbf{R}^n)$.*

In particular, it follows from the theorem that if $a \in C_*^\varrho$ for every ϱ, then $au - a_B u \in H_{(+\infty)}$.(This is also clear since $a - a_B \in S_{1,0}^{-\infty}$ then.)

Theorem 10.2.8 contains Proposition 8.6.8. In fact, assume that the hypotheses of Proposition 8.6.8 are fulfilled and let $\varrho_1 > 0$, for example. Then $(u_1)_\chi(x, D)u_2 \in C_*^{\varrho_2}$ and $(u_1)_\Phi(x, D)u_2 \in C_*^{\varrho_1+\varrho_2}$. If $\varrho_2 > 0$ then $(u_2)_\chi(x, D)u_1 \in C_*^{\varrho_1}$. If $\varrho_2 = 0$ then $(u_2)_\chi(x, D)u_1 \in C_*^{\varrho_1-\varepsilon}$ for any $\varepsilon > 0$, in particular for $\varepsilon = \varrho_1$. Finally, if $\varrho_2 < 0$ then $(u_2)_\chi(x, D)u_1 \in C_*^{\varrho_1+\varrho_2}$ since $(u_2)_\chi \in \tilde{S}_{1,1}^{-\varrho_2}$.

However, to prove Theorem 8.3.1 using our new tools we must examine the continuity properties of a_Φ using the full strength of Theorem 9.3.5 when $a \in H_{(s)}(\mathbf{R}^n)$. Recall that $H_{(s)}(\mathbf{R}^n) \subset C_{s-n/2}^*(\mathbf{R}^n)$ by Proposition 8.6.10.

Proposition 10.2.9. *Let χ_0 denote a function satisfying (9.3.3), let $a \in H_{(s)}(\mathbf{R}^n)$, and define $a_{\chi_0,\varepsilon}$ by $\hat{a}_{\chi_0,\varepsilon}(\xi, \eta) = \chi_0(\xi + \eta, \varepsilon\eta)\hat{a}(\xi)$. Then it follows that*

$$(10.2.12) \qquad \left(\int_{R \le |\xi| \le 2R} |\partial_\eta^\alpha a_{\chi_0,\varepsilon}(x, \eta)|^2\, d\eta \right)^{\frac{1}{2}}$$
$$\le C_\alpha \varepsilon^{n/2} R^{n/2-s}(\varepsilon R)^{n/2-|\alpha|}\|a\|_{(s)}, \quad 0 < \varepsilon < 1/2.$$

The operator $a_\Phi(x, D)$ in (10.2.11) is in $\mathrm{Op}\, S_{1,1}^{n/2-s}$ and maps $H_{(t)}(\mathbf{R}^n)$ to $H_{(t+s-n/2)}(\mathbf{R}^n)$ if $t + s > 0$.

Proof. Since $a \in C_{s-n/2}^*(\mathbf{R}^n)$ it follows from the remark after the proof of Proposition 10.2.2 that $a_\Phi \in S_{1,1}^{n/2-s}$. By Cauchy-Schwarz' inequality

$$|\partial_\eta^\alpha a_{\chi_0,\varepsilon}(x, \eta)|^2 \le \left((2\pi)^{-n} \int |\partial_\eta^\alpha \chi_0(\xi + \eta, \varepsilon\eta)\hat{a}(\xi)|\, d\xi \right)^2 \le C_\alpha(\varepsilon|\eta|)^{n-2|\alpha|} \int_{|\xi+\eta|<\varepsilon|\eta|} |\hat{a}(\xi)|^2\, d\xi,$$

for $\partial_\eta^\alpha \chi_0(\xi + \eta, \varepsilon\eta)$ is a sum of terms of the form $\chi_1(\xi + \eta, \varepsilon\eta)$ multiplied by nonnegative powers of ε, where χ_1 is homogeneous of degree $-|\alpha|$ outside a compact set. Hence

$$\int_{R \le |\eta| \le 2R} |\partial_\eta^\alpha a_{\chi_0,\varepsilon}(x, \eta)|^2\, d\eta \le C_\alpha'(\varepsilon R)^{2n-2|\alpha|} \int_{R(1-\varepsilon) \le |\xi| \le 2R(1+\varepsilon)} |\hat{a}(\xi)|^2\, d\xi,$$

for $R \leq |\eta| \leq 2R$ and $|\xi + \eta| < \varepsilon|\eta|$ implies $R(1 - \varepsilon) \leq |\xi| \leq 2R(1 + \varepsilon)$ and $|\xi + \eta| \leq 2\varepsilon R$. The last integral can be estimated by a constant times $R^{-2s}\|a\|^2_{(s)}$, which proves (10.2.12).

Now it follows from Theorem 9.3.5 with $m = n/2 - s$ and $\sigma = n/2$ that $a_\Phi(x, D)$ is continuous from $H_{(t)}(\mathbf{R}^n)$ to $H_{(t+s-n/2)}(\mathbf{R}^n)$ if $t + s > 0$, for $\hat{a}_\Phi = \hat{a}$ at infinity in a conic neighborhood of the twisted diagonal. The proof is complete.

We can now give another proof of Theorem 8.3.1 when $s_1 + s_2 > 0$ using the new tools. When $s_1 + s_2 > 0$ it follows from Proposition 10.2.9 that $(u_1)_\Phi(x, D)u_2 \in H_{(s_1+s_2-n/2)}$. If $s_1 > n/2$ then $(u_1)_\chi \in \tilde{S}^0_{1,1}$ and $(u_1)_\chi u_2 \in H_{(s_2)}$. If $s_1 < n/2$ then $(u_1)_\chi \in \tilde{S}^{n/2-s_1}_{1,1}$, and $(u_1)_\chi u_2 \in H_{(s_1+s_2-n/2)}$; when $s_1 = n/2$ then $(u_1)_\chi u_2 \in H_{(s_2-\varepsilon)}$ for every $\varepsilon > 0$, which is good enough since $s < s_2$ by hypothesis then. The term $(u_2)_\chi(x, D)u_1$ is of course similar so we obtain Theorem 8.3.1 except in the borderline case where $s_1 + s_2 = 0$, for which the combination of Proposition 10.2.9 and Theorem 9.3.5 does not give sufficient control of $a_\Phi(x, D)$.

However, our new techniques improve Theorem 8.3.3:

Theorem 10.2.10. *Let* $u_j \in H_{(s_j)}(\mathbf{R}^n)$, $j = 1, 2$, *where* $s_1 + s_2 > 0$. *Then*

  (i) $u_1 u_2 \in H^{\text{loc}}_{(s_2)}$ *outside* $WF(u_1)$ *if* $s_1 > n/2$.

  (ii) $u_1 u_2 \in H^{\text{loc}}_{(s_1+s_2-n/2)}$ *outside* $WF(u_1)$ *if* $s_1 < n/2$.

  (iii) $u_1 u_2 \in H^{\text{loc}}_{(s_1+s_2-n/2)}$ *outside* $WF(u_1) \cup WF(u_2)$.

Proof. We have just seen that $(u_1)_\Phi(x, D)u_2 \in H_{(s_1+s_2-n/2)}$, and $(u_2)_\chi(x, D)u_1$ is in $H^{\text{loc}}_{(s_1+s_2-n/2)}$ outside $WF(u_1)$ by Theorem 9.6.6 since $(u_2)_\chi(x, D)$ is of reduced order $n/2 - s_2$. In the same way we obtain $(u_1)_\chi(x, D)u_2 \in H^{\text{loc}}_{(s_1+s_2-n/2)}$ outside $WF(u_2)$, which gives (iii). If $s_1 > n/2$ then $(u_1)_\chi \in \tilde{S}^0_{1,1}$, and if $s_1 < n/2$ then $(u_1)_\chi \in S^{n/2-s_1}_{1,1}$, which proves the statements (i) and (ii) for $(u_1)_\chi(x, D)u_2$ also.

Finally we observe that by Theorem 9.6.6 and the remark at the end of Section 9.6

$$a_B(x, D)u \in H^{\text{loc}}_{(t+\varrho-m)} \qquad \text{outside } WF(u), \qquad \text{if } u \in H_{(t)} \text{ and } a \in A^m_\varrho;$$

$$a_B(x, D)u \in C^{\sigma+\varrho-m}_* \qquad \text{outside } WF(u), \qquad \text{if } u \in C^\sigma_* \text{ and } a \in A^m_\varrho.$$

10.3. Bony's linearisation theorem. We shall now return to the discussion of a composite function $F(u)$ begun in Section 8.5. We found there that if $F \in C^\infty(\mathbf{R})$ and $u \in L^\infty(\mathbf{R}^n, \mathbf{R})$, $u_t = \varphi(D/t)u$, where $\varphi \in C^\infty_0(\mathbf{R}^n)$, $\varphi(\xi) = 1$ when $|\xi| < \frac{1}{2}$ and $\varphi(\xi) = 0$ when $|\xi| > 1$, then

$$F(u) = F(u_1) + L_u u$$

where L_u is the pseudo-differential operator with symbol

$$a(x, \xi) = \int_1^\infty F'(u_t(x))\psi(\xi/t)\,dt/t, \quad \psi(\xi) = -\langle \xi, \partial\varphi(\xi)/\partial\xi \rangle.$$

Here $F(u_1) \in C^\infty$ so it is of no interest in regularity questions. Bony's linearisation theorem relates L_u to a paraproduct:

Theorem 10.3.1. *If* $u \in C^\varrho_*$, *where* $\varrho > 0$, *then* L_u *differs from a paradifferential operator corresponding to* $F'(u) \in C^\varrho_*$ *by an element in* Op $S^{-\varrho}_{1,1}$.

Proof. From Proposition 8.6.12 we know that $F'(u) \in C^\varrho_*$. A paradifferential operator corresponding to $F'(u)$ is defined by the symbol

$$b(x, \eta) = \int_1^\infty (\varphi(4D/t)F'(u))\psi(\eta/t)\,dt/t,$$

for

$$\chi(\xi, \eta) = \int_1^\infty \varphi(4\xi/t)\psi(\eta/t)\, dt/t$$

satisfies the conditions (10.2.1), (10.2.2) and (10.2.3) with $c = 1$. In fact, $\chi(\xi, \eta)$ is homogeneous of degree 0 when $|\eta| > 1$, since the integration may be extended to $(0, \infty)$ then, $\chi(\xi, \eta)$ vanishes when $|\eta| < \frac{1}{2}$ or $|\xi| > |\eta|/2$ and is equal to 1 when $|\eta| > 1$ and $|\xi| < |\eta|/8$. Thus the statement means that

$$b(x, \eta) - a(x, \eta) = \int_1^\infty \left(\varphi(4D/t)F'(u) - F'(\varphi(D/t)u) \right) \psi(\eta/t)\, dt/t$$

is in $S_{1,1}^{-\varrho}$. For $\partial_\eta{}^\alpha(b(x, \eta) - a(x, \eta))$ we have a similar formula with $\psi(\eta/t)$ replaced by $\psi^{(\alpha)}(\eta/t)t^{-|\alpha|}$. The integrand vanishes except when $\frac{1}{2} < |\eta|/t < 1$, and $\int_{|\eta|}^{2|\eta|} dt/t = \log 2$, so it follows from Proposition 8.6.13 that

$$|a_{(\beta)}^{(\alpha)}(x, \eta) - b_{(\beta)}^{(\alpha)}(x, \eta)| \le C_{\alpha\beta} |\eta|^{|\beta| - |\alpha| - \varrho}$$

as claimed.

Corollary 10.3.2. Let $F \in C^\infty$. If $u \in C_*^\varrho$ at x_0 and $u \in C_*^\sigma$ at (x_0, ξ_0), where $0 < \varrho \le \sigma \le 2\varrho$, then $F(u) \in C_*^\varrho$ at x_0 and $F(u) \in C_*^\sigma$ at (x_0, ξ_0). If $u \in C_*^\varrho \cap H_{(t)}^{loc}$ at x_0 where $\varrho > 0$, and $u \in H_{(t)}^{loc}$ at (x_0, ξ_0), then $F(u) \in H_{(t)}^{loc}$ at (x_0, ξ_0) if $0 < t \le s + \varrho$. If $s > n/2$ we may take $\varrho = s - n/2$ and omit the condition $u \in C_*^\varrho$ at x_0.

Proof. We may assume that the hypothesis made locally at x_0 is valid globally in $\mathbf{R}^n$, for otherwise we just multiply by a cutoff function. Then

$$F(u) = Pu + Ru$$

where P is the paramultiplication operator corresponding to $F'(u) \in A_\varrho^0$ and $R \in \operatorname{Op} S_{1,1}^{-\varrho}$. Then $Ru \in C_*^{2\varrho}$ (resp. $H_{(s+\varrho)}$), and $Pu \in C_*^\sigma$ (resp. $H_{(t)}^{loc}$) at (x_0, ξ_0) when $\sigma - \varrho \le \varrho$ (resp. $t - \varrho \le s$) by Theorem 9.6.6 and its analogue for C_* where we can take $m = 0$ and $m_N = -\varrho$ by (10.2.5).

The preceding discussion works just as well if $F(u)$ is replaced by $F(x, u)$ where $F \in C^\infty(\mathbf{R}^{n+1})$. In fact, the proofs and statements of Propositions 8.6.12 and 8.6.13 are modified in an obvious way then, if $F(x, 0) = 0$ as we may assume. It is also clear that no essential change is needed if u takes its values in $\mathbf{R}^N$ for some N and $F \in C^\infty(\mathbf{R}^{n+N})$. (We must assume uniform bounds for the derivatives when u is bounded.) In particular, we can apply the result to a nonlinear differential equation of order m in $\mathbf{R}^n$ which we write in the form

$$(10.3.1) \qquad F(x, J_m u(x)) = 0, \qquad J_m u(x) = \{\partial^\alpha u(x)\}_{|\alpha| \le m}.$$

One calls $J_m u(x)$ the m-jet of u at x. The differential equation can be regarded as an equation for the m-jet. We shall write

$$F_\alpha(x, J_m) = \partial F(x, J_m)/\partial u_\alpha, \qquad \text{if } J_m = \{u_\alpha\}_{|\alpha| \le m}.$$

Let $N_{m,n} = \binom{m+n}{n}$ be the dimension of the space of m-jets.

Corollary 10.3.3. *Let $u \in C_*^{m+\varrho}$, $\varrho > 0$, be a solution of the nonlinear differential equation $F(x, J_m u(x)) = 0$, where $F \in C^\infty(\mathbf{R}^{n+N_{m,n}})$ has bounded derivatives of all orders. Then $Pu \in C_*^{2\varrho}$ if P is a paradifferential operator corresponding to*

$$(10.3.2) \qquad \sum_{|\alpha| \le m} F_\alpha(x, J_m u)(i\eta)^\alpha \in A_\varrho^m.$$

If $u \in H_{(m+s)}$ and $s + \varrho > 0$ it follows that

$$(10.3.3) \qquad Pu \in H_{(s+\varrho)}.$$

In particular,

$$(10.3.4) \qquad u \in H_{(s+m)}, \quad s > n/2 \implies Pu \in H_{(2s-n/2)}.$$

Proof. This is a consequence of Corollary 9.3.6, Theorem 9.3.10, and the extension of Theorem 10.3.1 just discussed; if $s > n/2$ we have $u \in C_*^{m+\varrho}$ for $\varrho = s - n/2$ by Sobolev's lemma (Proposition 8.6.10) if $u \in H_{(m+s)}$.

The result, which is due to Bony [1] in a slightly weaker form and was given by Meyer [2] in the perfected form stated here, can also be given a local version. Suppose that the regularity assumptions and the differential equation are just given locally in an open set $X \subset \mathbf{R}^n$. We can then take two functions χ_0, χ_1 in $C_0^\infty(X)$ such that $\chi_1 = 1$ in a neighborhood of supp χ_0 and use that $v = \chi_1 u$ satisfies the regularity assumptions globally and the equation $\chi_0(x)F(x, J_m v) = 0$ in the whole space. In a set where $\chi_0 = 1$ the equation is not much affected.

The differential equation

$$\sum_{|\alpha| \le m} F_\alpha(x, J_m u(x))\partial^\alpha v(x) = 0$$

corresponding to (10.3.2) is a classically well known object, the *linearised equation*. It is obtained by looking for small perturbations εv of u such that $F(x, J_m(u(x) + \varepsilon v(x))) = o(\varepsilon)$. A solution of a nonlinear differential equation is called *elliptic, hyperbolic, noncharacteristic,* ... when the linearisation at the solution is elliptic, hyperbolic, noncharacteristic, ... It is therefore in full agreement with classical terminology to introduce

Definition 10.3.4. *The solution $u \in C_*^{m+\varrho}$, $\varrho > 0$, of the nonlinear differential equation $F(x, J_m u(x)) = 0$ is called noncharacteristic at (x_0, ξ_0) if*

$$\sum_{|\alpha|=m} F_\alpha(x, J_m u(x))\xi^\alpha \ne 0 \text{ at } (x_0, \xi_0).$$

We have now developed all that is needed to prove Bony's noncharacteristic regularity theorem:

Theorem 10.3.5. *Let $u \in C_*^{m+\varrho}$ in a neighborhood of x_0, where $\varrho > 0$, and assume that u satisfies the nonlinear differential equation $F(x, J_m u(x)) = 0$. Then $u \in C_*^{m+2\varrho}$ locally at every noncharacteristic point (x_0, ξ_0) for the solution. If in addition $u \in H_{(m+s)}^{loc}$ at x_0, then $u \in H_{(m+s+\varrho)}^{loc}$ at (x_0, ξ_0); when $s > n/2$ we can drop the condition $u \in C_*^{m+\varrho}$ and conclude that $u \in H_{(m+2s-n/2)}^{loc}$ at (x_0, ξ_0).*

Proof. By Theorem 10.2.6 the paradifferential operator P in Corollary 10.3.3 is also noncharacteristic at (x_0, ξ_0), and $Pu \in H_{(s+\varrho)}$ at x_0 by Corollary 10.3.3. By Proposition

10.2.2 it is of reduced order $m - \varrho$. If $u \in H^{loc}_{(m+s)} = H^{loc}_{(m-\varrho+s+\varrho)}$ at x_0, it follows from Theorem 9.6.7 that $u \in H^{loc}_{(m+s+\varrho)}$ at (x_0, ξ_0). By the remark at the end of Section 9.6 we get the statement on Zygmund spaces in the same way.

Theorem 10.3.5 can be regarded as the case $k = m$ of Theorem 8.4.13 which was not accessible to the elementary methods used in Section 8.4. We shall now show that there is a better result for a quasilinear equation of order m.

Theorem 10.3.6. *Let* $u \in H^{loc}_{(s+m-1/2)}$, $s > (n-1)/2$ *and* $s > n/4$, *and assume that* u *satisfies the quasilinear differential equation*

$$(10.3.5) \qquad \sum_{|\alpha|=m} a_\alpha(x, J_{m-1}u(x))\partial^\alpha u + c(x, J_{m-1}u(x)) = 0,$$

where a_α *and* c *are in* C^∞. *Then it follows that* $u \in H^{loc}_{(2s+m-n/2)}$ *at every noncharacteristic point* (x_0, ξ_0).

Proof. It is no restriction to assume that $u \in \mathcal{E}' \cap H_{(s+m-1/2)}$. Since $J_{m-1}u \in H_{(s+1/2)}$ and $s + 1/2 > n/2$ we have

$$(10.3.6) \qquad a_\alpha(x, J_{m-1}u(x)) \in H_{(s+1/2)}, \quad \partial^\alpha u \in H_{(s-1/2)}, \qquad |\alpha| = m,$$

so the equation is well defined. Let A_α be paramultiplication by $a_\alpha(x, J_{m-1}u)$, let S_α be paramultiplication by $\partial^\alpha u$, and let L_α, L be paradifferential operators associated with the linearizations of a_α and c. (We suppress the argument $(x, J_{m-1}u(x))$ for the sake of brevity.) By the first part of (10.3.6) and Theorem 10.2.8 we have

$$a_\alpha \partial^\alpha u - (A_\alpha \partial^\alpha u + S_\alpha a_\alpha) \in H_{(2s-n/2)},$$

since $2s > n/2$. Furthermore, $a_\alpha - L_\alpha u \in H_{(2s+1-n/2)}$ by Corollary 10.3.3, and since we shall see below that S_α is of order < 1 it follows that

$$(10.3.7) \qquad a_\alpha \partial^\alpha u - (A_\alpha \partial^\alpha u + S_\alpha L_\alpha u) \in H_{(2s-n/2)}.$$

Corollary 10.3.3 also gives

$$c - Lu \in H_{(2s+1-n/2)},$$

and adding this and (10.3.7) for all α with $|\alpha| = m$ we obtain

$$Pu = \Big(\sum_{|\alpha|=m} (A_\alpha \partial^\alpha + S_\alpha L_\alpha) + L \Big) u \in H_{(2s-n/2)}.$$

To conclude using Theorem 9.6.7 that $u \in H^{loc}_{(2s+m-n/2)}$ at noncharacteristic points for the linearized operator we must show

(1) that these are noncharacteristic for the mth order operator P,
(2) that the reduced order of P is $\leq m - s - 1/2 + n/2$.

The operator A_α is of order 0 and reduced order $n/2 - s - 1/2$, so $A_\alpha \partial^\alpha$ is of order m and has the required reduced order. The symbol is equal to

$$(a_\alpha(x, J_{m-1}u(x)) + O(|\eta|^{n/2-s-1/2}))(i\eta)^\alpha.$$

The operator S_α is of order 0 if $s > (n+1)/2$, of any positive order if $s = (n+1)/2$ and of order $(n+1)/2 - s < 1$ if $(n-1)/2 < s < (n+1)/2$; the reduced order is $(n+1)/2 - s$.

The operators L_α and L are of order $m-1$ and reduced order $m-1-(s+1/2-n/2)$. Hence $S_\alpha L_\alpha$ is of order $< m$ and the reduced order is $\le m-s-1/2+n/2$ since

$$(n+1)/2-s+m-1 = m-s-1/2+n/2, \qquad 1+m-1-(s+1/2-n/2) = m-s-1/2+n/2.$$

The symbol of P is

$$\sum_{|\alpha|=m} a_\alpha(x, J_{m-1}u(x))(i\eta)^\alpha + O(|\eta|^{m-\kappa})$$

for some $\kappa > 0$, which proves the condition (1) above and completes the proof of the theorem.

The same proof gives a more general result, also due to Bony [1]:

Theorem 10.3.7. *Let $u \in H^{\text{loc}}_{(s+d)}$, $s > n/4$, be a solution of the nonlinear differential equation*

$$(10.3.8) \qquad \sum_{m_0 < |\alpha| \le m} a_\alpha(x, J_{m(\alpha)}u(x))\partial^\alpha u + c(x, J_{m_0}u(x)) = 0,$$

where a_α and c are in C^∞, and

$$(10.3.9) \qquad \begin{array}{ccc} m(\alpha) < |\alpha|, & m(\alpha) + |\alpha| \le 2d, & m_0 \le d; \\ m_0 < s+d-n/2, & m(\alpha) < s+d-n/2. \end{array}$$

Then it follows that $u \in H^{\text{loc}}_{(2s+m-n/2)}$ at every noncharacteristic point (x_0, ξ_0).

Proof. We follow the proof of Theorem 10.3.6 with the same notation. Since

$$a_\alpha(x, J_{m(\alpha)}u(x)) \in H_{(s+d-m(\alpha))},$$

and $s > n/4$, $m(\alpha) < d$, it follows from Theorem 10.2.8 that

$$a_\alpha \partial^\alpha u - (A_\alpha \partial^\alpha u + S_\alpha a_\alpha) \in H_{(2s-n/2)},$$

for $s+d-|\alpha|+(s+d-m(|\alpha|))-n/2 \ge 2s-n/2 > 0$ by (10.3.9). We have

$$a_\alpha - L_\alpha u \in H_{(2(s+d-m(\alpha))-n/2)}, \quad \partial^\alpha u \in H_{(s+d-|\alpha|)};$$

since $2(d-m(\alpha)) > 0$ and $2(d-m(\alpha)) > n/2-s-d+|\alpha|$ because $s+d-n/2-m(\alpha) > 0 \ge m(\alpha) + |\alpha| - 2d$, it follows from the calculation of the order of S_α below that

$$a_\alpha \partial^\alpha u - (A_\alpha \partial^\alpha u + S_\alpha L_\alpha u) \in H_{(2s-n/2)}.$$

The conditions $s+d-m_0 > n/2$ and $m_0 \le d$ give that $c - Lu \in H_{(2s-n/2)}$, so we have $Pu \in H_{(2s-n/2)}$. Apart from obvious repetitions what remains is to show that P has reduced order $\le -s+d+n/2$. The reduced order of A_α is $n/2 - (s+d-m(\alpha))$, so the reduced order of $A_\alpha \partial^\alpha$ is

$$n/2 - s - d + m(\alpha) + |\alpha| \le n/2 - s + d.$$

The order of S_α is $n/2 - (s+d-|\alpha|)$ if this is positive, 0 if this is negative and any positive number if this is 0; the reduced order is always $n/2 - (s+d-|\alpha|)$. The order of L_α is $m(\alpha)$ and the reduced order is $m(\alpha) + n/2 - (s+d-m(\alpha))$. Now

$$n/2 - (s+d-|\alpha|) + m(\alpha) \le n/2 - s + d,$$

again since $|\alpha| + m(\alpha) \leq 2d$, and

$$0 + m(\alpha) + n/2 - (s + d - m(\alpha)) < n/2 - s + d$$

since $2m(\alpha) < m(\alpha) + |\alpha| \leq 2d$. This proves the required bound for the reduced order of P since the reduced order of L is $m_0 + n/2 - (s + d - m_0) \leq n/2 - s + d$. The proof is complete.

Remark. One should of course choose

(10.3.10) $$d = \max(m_0, \max(m(\alpha) + |\alpha|)/2))$$

in order to satisfy the second and third condition (10.3.9) and optimize the conclusions. If $d = m_0$ then $s > n/2$ by the fourth condition, hence $s + m - n/2 > m_0 = d$; if $d = (m(\alpha) + |\alpha|)/2$ for some α then $s + m - n/2 > m(\alpha) + m - d \geq 2d - d = d$ by the last condition (10.3.9). This shows that the microlocal regularity in the conclusion of Theorem 10.3.7 is always better than the assumed local regularity.

Corollary 10.3.8. *Let the hypotheses of Theorem 10.3.7 be fulfilled and assume in addition that $u \in H^{loc}_{(\sigma+d)}$ at (x_0, ξ) when (x_0, ξ) is characteristic. Then it follows that $u \in H^{loc}_{(\sigma+d)}$ at x_0 and that $u \in H^{loc}_{(2\sigma+m-n/2)}$ at (x_0, ξ) when (x_0, ξ) is not characteristic.*

Proof. The statement becomes stronger if d is decreased while $s + d$ and $\sigma + d$ are kept constant. Hence we may assume that d is given by (10.3.10). By Theorem 10.3.7 and Proposition 8.2.6 we have $u \in H^{loc}_{(t+d)}$ at x_0 if

$$t + d \leq \sigma + d, \qquad t + d \leq 2s + m - n/2.$$

By the preceding remark $s + m - n/2 - d > 0$. Unless $s = \sigma$, which means that we have nothing to prove, it follows that we can take $t > s$. Repeating the argument with s replaced by t and so on, we increase s in larger and larger steps until $t = \sigma$. Then the last statement follows from Theorem 10.3.7. The proof is complete.

Corollary 10.3.8 shows that, as for linear differential equations, the lowest regularity must always occur at the characteristics. The following chapter will be devoted to the study of the regularity in that set.

CHAPTER XI

PROPAGATION OF SINGULARITIES

11.1. Introduction. In Section 10.3 we have proved a general theorem on microlocal regularity of solutions of nonlinear differential equations outside the characteristic set. In the characteristic set one cannot assert that all solutions have a higher degree of smoothness than postulated from the beginning, even in the linear case. However, one can prove that the smoothness is constant along bicharacteristic curves. This is obvious for a solution of the differential equation $\partial u/\partial x_n = 0$; the bicharacteristics are straight lines in the x_n direction and smoothness is of course constant in that direction since u does not depend on x_n. In spite of the triviality of this example we shall devote Section 11.2 to a discussion of the equation $\partial u/\partial x_n = f$, or more generally $Lu = f$ where L is a smooth real vector field. This will give motivations for the later results using a minimum of technicalities. As a byproduct we also give some results of Rauch and Reed [1] on propagation of singularities for second order hyperbolic equations in two variables. The weaker higher dimensional analogue due to Beals [1, 2] is studied in Sections 11.5 and 11.6. After the introductory discussion in Section 11.2 we study the propagation of singularities for solutions of pseudo-differential equations in Section 11.3. The applications to nonlinear differential equations are given in Section 11.4.

11.2. First order scalar differential equations. Let us first consider a solution of the simplest possible differential equation

$$(11.2.1) \qquad \partial u/\partial x_n = f$$

in an open set $X \subset \mathbf{R}^n$ containing the interval $I = [0, a]e_n$, $e_n = (0, \ldots, 0, 1)$.

Proposition 11.2.1. *If $u \in H^{\mathrm{loc}}_{(s)}$ at 0 and $f \in H^{\mathrm{loc}}_{(s)}$ at I, then $u \in H^{\mathrm{loc}}_{(s)}$ at I. If $\xi \in \mathbf{R}^n \setminus 0$ and $u \in H^{\mathrm{loc}}_{(s)}$ at $(0, \xi)$ and $f \in H^{\mathrm{loc}}_{(s)}$ at $I \times \{\xi\}$, then $u \in H^{\mathrm{loc}}_{(s)}$ at $I \times \{\xi\}$. (If $\xi_n \neq 0$ then $u \in H^{\mathrm{loc}}_{(s+1)}$ at $I \times \{\xi\}$ of course.)*

Proof. If $0 \le t \le a$ then

$$(11.2.2) \qquad u(\cdot + te_n) - u = \int_0^t f(\cdot + se_n)\, ds$$

in a neighborhood of 0, for both sides are equal when $t = 0$ and they have the same derivative with respect to t as functions of t with distribution values. This gives the first statement at once. If the hypotheses in the second one are fulfilled then we can choose $\chi \in C_0^\infty$ with support in a small neighborhood of 0 and equal to 1 in another neighborhood and a conic neighborhood Γ of ξ_0 so that the Fourier transforms of χu and of $\chi f(\cdot + se_n)$ for $0 \le s \le t$ are bounded in $L^2(\Gamma)$ with respect to the measure $(1 + |\xi|^2)^s \, d\xi$. Hence (11.2.2) shows that the Fourier transform of $\chi u(\cdot + te_n)$ is also in this L^2 space as claimed.

The proposition remains valid if we replace the equation (11.2.1) by

$$(11.2.1)' \qquad \partial u/\partial x_n + au = f,$$

where $a \in C^\infty$. In fact, if $\partial A/\partial x_n = Aa$ and $A \neq 0$, this is equivalent to $\partial(Au)/\partial x_n = Af$. This reduction works just as well if u and f are vector valued and a is a square matrix. The result also extends to the more general first order equation

$$(11.2.1)'' \qquad \sum_1^n a_j(x)\partial u/\partial x_j + au = f,$$

where $a_j \in C^\infty$ are real valued and $\sum_1^n a_j^2 \neq 0$. For the proof we choose in a neighborhood of an arbitrary point new coordinates $y_1, \ldots, y_n$ such that

$$\sum_1^n a_j(x)\partial/\partial x_j = \partial/\partial y_n.$$

The parallels of the y_n axis then correspond to the orbits of the vector field $\sum_1^n a_j(x)\partial/\partial x_j$. If we accept the fact that $H_{(s)}^{\mathrm{loc}}$ is invariant under diffeomorphisms, which has not been proved here, the first statement in Proposition 11.2.1 shows that $u \in H_{(s)}^{\mathrm{loc}}$ along such an orbit if this is true for f and at some point also for u. To translate the second statement from the y variables to the x variables we need to know that the condition $u \in H_{(s)}^{\mathrm{loc}}$ at (x, ξ) is invariant under a diffeomorphism if (x, ξ) is transformed as a cotangent vector. (See e.g. Hörmander [4, section 18.1].) Accepting this as a fact we observe that if $x(t)$ are solutions of the differential equations

$$dx_j/dt = a_j(x), \quad j = 1, \ldots, n,$$

which depend on a parameter τ, then the derivatives $\dot{x}_j$ with respect to τ satisfy the equation

$$(11.2.3) \qquad d\dot{x}_j/dt = \sum_{k=1}^n \partial a_j/\partial x_k \dot{x}_k, \quad j = 1, \ldots, n.$$

In the y variables such vectors $\dot{x}$ become constant along the orbit. For covectors ξ which are constant along the orbit when expressed in the y variables it follows that $\sum_1^n \dot{x}_j \xi_j$ is constant along the orbit when (11.2.3) holds, that is,

$$\sum_{j,k=1}^n \partial a_j/\partial x_k \dot{x}_k \xi_j + \sum_{j=1}^n \dot{x}_j d\xi_j/dt = 0.$$

Since $\dot{x}_j$ may be arbitrary, this means that

$$d\xi_j/dt = -\sum_{k=1}^n \partial a_k/\partial x_j \xi_k, \quad j = 1, \ldots, n.$$

If $a(x, \xi) = \sum_1^n a_j(x)\xi_j$ is the principal symbol of $(11.2.1)''$ (divided by i) these equations together with the equations of the orbit give the Hamilton equations

$$(11.2.4) \qquad dx_j/dt = \partial a(x, \xi)/\partial \xi_j, \qquad d\xi_j/dt = -\partial a(x, \xi)/\partial x_j, \quad j = 1, \ldots, n,$$

of the *bicharacteristics*. For the equations $(11.2.1)''$ the second statement in Proposition 11.2.1 now takes the following form: If $f \in H_{(s)}^{\mathrm{loc}}$ at every point of an interval I on a bicharacteristic of $(11.2.1)''$ and $u \in H_{(s)}^{\mathrm{loc}}$ at some point on I, then $u \in H_{(s)}^{\mathrm{loc}}$ at every point in I. This is the basic form of the results to be proved in Sections 11.3 and 11.4. A very essential difference in the case of differential operators of higher order is that the first equations (11.2.4) depend on ξ then. This will cause new difficulties in the nonlinear case since bicharacteristics crossing over the same point x can then transmit singularities to each other.

Let us now consider a continuous solution of a nonlinear equation

$$(11.2.5) \qquad\qquad \partial u/\partial x_n = f(x, u)$$

where $f \in C^\infty$. If $u \in H^{\text{loc}}_{(s)}(X)$ for some $s > 0$ then $f(x, u) \in H^{\text{loc}}_{(s)}(X)$ by Theorem 8.5.1. If $\varphi \in C^\infty_0(X)$, then $v = \varphi u \in H_{(s)}$ and $\partial v/\partial x_n \in H_{(s)}$, that is,

$$\int |\hat{v}(\xi)|^2 (1 + |\xi|^2)^s (1 + |\xi_n|^2) d\xi < \infty.$$

Since Cauchy-Schwarz' inequality gives

$$(1 + |\xi'|^2)^s \left(\int |\hat{v}(\xi)| \, d\xi_n \right)^2 \leq \pi \int |\hat{v}(\xi)|^2 (1 + |\xi|^2)^s (1 + \xi_n^2) \, d\xi_n, \quad \xi' = (\xi_1, \ldots, \xi_{n-1}),$$

we see that $v(\cdot, x_n)$ for fixed x_n is also in $H_{(s)}$ as a function of $x' = (x_1, \ldots, x_{n-1})$, for the Fourier transform is $\xi' \mapsto \int e^{ix_n \xi_n} \hat{v}(\xi', \xi_n) \, d\xi_n / 2\pi$. Conversely, assume that u is a continuous solution of (11.2.5) in a neighborhood of the interval I in Proposition 11.2.1, and let $u_0(x') = u(x', 0) \in H^{\text{loc}}_{(s)}$ for some $s > 0$. The solution of the Cauchy problem for (11.2.5) can be written

$$u(x) = \Phi(x, u_0(x')),$$

where Φ is a C^∞ function in a neighborhood of $I \times \{u_0(0)\}$. Hence the assumption that $u_0 \in H^{\text{loc}}_{(s)}$ at 0 implies that $u(\cdot, x_n) \in H^{\text{loc}}_{(s)}$ at 0 if $0 \leq x_n \leq a$, and the differential equation gives infinite differentiability with respect to x_n. This implies that $u \in H^{\text{loc}}_{(s)}$ in a neighborhood of I.

Our results on microlocal regularity of composite functions can also be used to discuss the microlocal regularity of u: If $u_0 \in H^{\text{loc}}_{(s)}$ in a neighborhood of 0, where $s > (n-1)/2$, and $u_0 \in H^{\text{loc}}_{(\sigma)}$ at $(0, \xi')$, then Corollary 10.3.2 gives

$$u(\cdot, x_n) \in H^{\text{loc}}_{(\sigma)} \text{ at } (0, \xi') \text{ if } 0 \leq x_n \leq a \text{ and } \sigma \leq 2s - (n-1)/2.$$

This is uniform even for x_n in a neighborhood of $[0, a]$, and since u is infinitely differentiable in x_n it follows that $u \in H^{\text{loc}}_{(\sigma)}$ at $I \times (\xi', 0)$. Hence we have proved:

Proposition 11.2.2. Let $u \in H^{\text{loc}}_{(s)}$ for some $s > (n-1)/2$ be a continuous solution of (11.2.5) in a neighborhood of $I = [0, a]e_n$. Then $\xi_n = 0$ if $(x, \xi) \in WF(u)$. If $u \in H^{\text{loc}}_{(\sigma)}$ at $(0, (\xi', 0))$ and $s \leq \sigma \leq 2s - (n-1)/2$, it follows that $u \in H^{\text{loc}}_{(\sigma)}$ at $I \times \{(\xi', 0)\}$.

We shall now prove that the bound on σ here cannot be improved. To do so we take $f(x, u) = \psi'(x_n)u^2$ where $\psi \in C^\infty$ vanishes in a neighborhood of 0. Then the solution of the Cauchy problem gives

$$u(\cdot, x_n) = u_0/(1 - \psi(x_n)u_0),$$

and the assertion is a consequence of the following result:

Proposition 11.2.3. *Let $s > \nu/2 \geq 1$ and assume that for all real valued $u \in H_{(s)}(\mathbf{R}^\nu)$ with $|u| < 1/2$ and $WF(u) \subset \{(0, (\xi_1, 0, \ldots, 0))\}$ we have $u/(1 - u) \in H_{(\sigma)}^{\mathrm{loc}}$ at $(0, \eta)$ when $\eta' = (\eta_2, \ldots, \eta_\nu) \neq 0$. Then $\sigma \leq 2s - \nu/2$.*

Proof. The hypothesis implies that

$$u/(1 - u) - u/(1 + u) = 2u^2/(1 - u^2) \in H_{(\sigma)}^{\mathrm{loc}} \quad \text{at } (0, \eta) \text{ when } \eta' \neq 0.$$

To test this we denote by v the function in Example 8.2.4 with n replaced by ν and s replaced by $s + a$ where $a > (1 + \varrho(\nu - 1))/2$. Then $v \in H_{(s)}$ (see (8.2.7)), and we can take $u = c(v + \tilde{v})$, that is, $\hat{u}(\xi) = c(\hat{v}(\xi) + \hat{v}(-\xi))$ with a small positive c. Since $u/(1 - u)$ is in S away from 0 the hypothesis implies that the Fourier transform of $u^2/(1 - u^2) = u^2 + u^4 + \ldots$ is in L^2 with respect to the measure $(1 + |\xi|^2)^\sigma \, d\xi$ in an open cone Γ. The Fourier transform is $\geq \hat{u} * \hat{u}/(2\pi)^n$ since $\hat{u} \geq 0$, hence

$$\int |\hat{u} * \hat{u}(\xi)|^2 (1 + |\xi|^2)^\sigma \, d\xi < \infty.$$

We have

$$\hat{u} * \hat{u}(\xi) \geq 2c^2 \int \hat{v}(\xi + \theta)\hat{v}(\theta) \, d\theta = 2c^2 \int d\theta_1 \int \hat{v}(\xi_1 + \theta_1, \xi' + \theta')\hat{v}(\theta_1, \theta') \, d\theta'.$$

When $\theta_1 \gg |\xi|$ the two factors in the integrand are larger than $\frac{1}{2}\theta_1^{-s-a}$ when $|\xi' + \theta'| < \frac{1}{4}\theta_1^\varrho$ resp. $|\theta'| < \frac{1}{2}\theta_1^\varrho$. If even $\theta_1^\varrho \gg |\xi|$ then the measure of the intersection of these balls is $> c_1 \theta_1^{(\nu-1)\varrho}$ where $c_1 > 0$, so we obtain for some large C and $c', c'' > 0$

$$\hat{u} * \hat{u}(\xi) \geq c' \int_{C|\xi|^{1/\varrho}}^\infty \theta_1^{-2(s+a)}\theta_1^{(\nu-1)\varrho} \, d\theta_1 \geq c'' |\xi|^{((\nu-1)\varrho - 2(s+a)+1)/\varrho}.$$

Since this is in L^2 in Γ with respect to the measure $(1 + |\xi|^2)^\sigma \, d\xi$, we must have

$$2((\nu - 1)\varrho - 2(s + a) + 1)/\varrho + 2\sigma < -\nu.$$

Letting $2a \to 1 + \varrho(\nu - 1)$ we obtain

$$-4s/\varrho + 2\sigma \leq -\nu,$$

which gives $2\sigma \leq 4s - \nu$ when $\varrho \to 1$, as claimed.

Although we have now proved that microlocal irregularity of high order may spread to other frequencies, there are some very strong regularity properties which are propagated. To study them we start with a simple but important lemma.

Lemma 11.2.4. *Let f be a C^∞ function in $\mathbf{R}$ and let $u \in H_{(s)}^{\mathrm{loc}}$ at $x_0 \in \mathbf{R}^n$ for some $s > n/2$. Then $f(u) \in H_{(s)}^{\mathrm{loc}}$ at x_0. If V is a family of C^∞ vector fields and k is a positive integer such that*

$$V_1 \cdots V_j u \in H_{(s)}^{\mathrm{loc}} \quad \text{at } x_0, \text{ if } V_1, \ldots, V_j \in V \text{ and } j \leq k,$$

then $f(u)$ has the same property.

Proof. The first statement is contained in Theorem 8.5.1. To prove the second one we first show that for any smooth vector field V with compact support and any $u \in H_{(s)}$ we have $Vf(u) = f'(u)Vu$. This is clear if $u \in S$. To prove the general statement we choose a sequence $u_\nu \in S$ with $u_\nu \to u$ in $H_{(s)}$ as $\nu \to \infty$. Then $u_\nu \to u$ uniformly so $f(u_\nu) \to f(u)$ uniformly and $Vf(u_\nu) \to Vf(u)$ in $\mathcal{D}'$, while $f'(u_\nu) \to f'(u)$ weakly in $H_{(s)}$ and $Vu_\nu \to Vu$ strongly in $H_{(s-1)}$. Since $s > 1 - s$ it follows that $f'(u_\nu)Vu_\nu \to f'(u)Vu$ in $\mathcal{D}'$ which proves the claim. Now any finite product of factors of the form $f^{(j)}(u)$ and $V_1 \cdots V_j u$, with $j \leq k$, is in $H_{(s)}^{\mathrm{loc}}$ at x_0, so the lemma follows by successive differentiations.

Combining this lemma with the proof of Proposition 11.2.2 we obtain

Proposition 11.2.2′. *Let $u \in H^{loc}_{(s)}$, for some $s > (n-1)/2$, be a continuous solution of (11.2.5) in a neighborhood of $I = [0, a]e_n$, and assume that*

(11.2.6) $\partial^\alpha u \in H^{loc}_{(s)}$ *at 0, if $|\alpha| \leq k$ and $\alpha_{\nu+1} = \cdots = \alpha_n = 0$.*

Then the same regularity property holds at every point in I.

Note that (11.2.6) implies that $u \in H^{loc}_{(s+k)}$ at $(0, \xi)$ unless $\xi_1 = \cdots = \xi_\nu = 0$. However, it is a much more restrictive condition than that which explains why it is possible to propagate it.

We shall now prove a related result due to Rauch and Reed [1] for a semi-linear wave equation

(11.2.7) $Pu = \partial^2 u/\partial x_1 \partial x_2 + a_1(x)\partial u/\partial x_1 + a_2(x)\partial u/\partial x_2 + a_0(x)u = f(x, u)$

with just two independent variables and $a_j, f \in C^\infty$. Note that locally we can bring any linear second order hyperbolic operator to the form of P by a change of variables. We set $I = [0, a]e_2$ as before. First we study a linear equation:

Lemma 11.2.5. *Assume that $u \in H^{loc}_{(s)}$ at I and that $Pu = F$ where for some positive integer k*

(11.2.8) $\partial_1^j F \in H^{loc}_{(s)}$ *at I if $0 \leq j < k$.*

Then it follows that

(11.2.9) $\partial_1^j u \in H^{loc}_{(s)}$ *at I when $0 \leq j \leq k$*

if this is true at 0.

Proof. Assume first that $k = 1$ and write the equation $Pu = F$ in the form

$$(\partial_2 + a_1)(\partial_1 + a_2)u = F - cu, \qquad c = a_0 - a_1 a_2 - \partial_2 a_2.$$

This can be regarded as an equation for $(\partial_1 + a_2)u$ of the form (11.2.1)′, which proves that $(\partial_1 + a_2)u \in H^{loc}_{(s)}$ at I and gives the statement when $k = 1$. Assume now that $k > 1$ and that the lemma has already been proved for lower values of k. If we apply ∂_1^{k-1} to the equation $Pu = F$ the resulting equation can be written

$$(\partial_2 + a_1)v = g, \qquad v = \partial_1^{k-1}(\partial_1 + a_2)u;$$

$g \in H^{loc}_{(s)}$ at I for it is a linear combination with C^∞ coefficients of $\partial_1^{k-1}F$ and of $\partial_1^j u$ with $j < k$ which are already known to be in $H^{loc}_{(s)}$ by the inductive hypothesis. Thus $v \in H^{loc}_{(s)}$ at I, since this is true at 0 , which proves that $\partial_1^k u \in H^{loc}_{(s)}$ at I.

Note that if P is the homogeneous wave operator then any term $w(x_2)$ in $H^{loc}_{(s)}$ could be added to u without changing the equation or the initial condition. Although we have no control of such terms $w(x_2)$ the lemma holds because $w(x_2)$ drops out when we differentiate with respect to x_1.

From Lemma 11.2.5 we shall now deduce a result on propagation of regularity for solutions of the wave equation (11.2.7) with nonlinearity of order 0 only.

Proposition 11.2.6. *Let u be a solution of the equation (11.2.7) in a neighborhood of $I = [0, a]e_2$ which is in $H_{(s)}^{loc}$ at I for some $s > 1$. If $\partial_1^j u \in H_{(s)}^{loc}$ at 0 when $j \leq k$, then this remains true at I.*

Proof. The claim follows at once from Lemma 11.2.5 when $k = 1$, for $f(x, u) \in H_{(s)}^{loc}$ at I. Assuming now that $k > 1$ and that the proposition is proved already for lower values of k, we obtain $\partial_1^j f(x, u) \in H_{(s)}^{loc}$ at I when $j < k$, by Lemma 11.2.4. Hence Lemma 11.2.5 gives that $\partial_1^k u \in H_{(s)}^{loc}$ at I, which completes the proof.

In particular, if $u \in C^\infty$ at 0, then $\partial_1^j u \in H_{(s)}^{loc}$ for every j in a neighborhood of I independent of j; in particular $\xi_1 = 0$ in the wave front set. This means that on I the wave front set is conormal to the other family of characteristics (which means that it should propagate along that).

A reduction of a hyperbolic differential equation to the form in (11.2.7) can be made in a neighborhood of any characteristic curve. Hence Proposition 11.2.6 implies the following result of Rauch and Reed [1]:

Theorem 11.2.7. *Let u be a solution of a nonlinear differential equation $Pu = f(x, u)$ where P is a linear hyperbolic second order operator in two independent variables with C^∞ coefficients near two characteristic curves Γ_1 and Γ_2 intersecting at x_0. Assume that $u \in H_{(s)}^{loc}$ for some $s > 1$. If $u \in C^\infty$ at some point on Γ_1 and some point on Γ_2, then $u \in C^\infty$ at x_0.*

For the solution of a Cauchy problem this means that *the singular support is contained in the union of the characteristic curves starting from the singular support of the Cauchy data.* This is false for higher order equations in two variables (see Rauch and Reed [2]) and for second order equations in more than two variables (see Beals [1, 2, 3] and Lascar [1]). We shall return to the results of Beals [1, 2] in Section 11.5.

11.3. Linear pseudo-differential equations. In this section we shall prove a very general version of the result on propagation of regularity suggested by Proposition 11.2.1 and the subsequent discussion. As pointed out by Bony [1] one can prove propagation along a Hamilton flow (11.2.4) even when a is just in C^1. Then we only have the Peano existence theorem (Theorem 1.2.6) available for the Hamilton flow and no uniqueness theorem, so we must expect a whole funnel of solutions as in Theorem 1.2.8. Mainly following Bony [1] we shall prepare the discussion with giving another characterization of the set X in Theorem 1.2.8. As there we consider a Cauchy problem

$$(11.3.1) \qquad dx(t)/dt = f(t, x), \qquad x(0) = 0,$$

where $x \in \mathbf{R}^N$ and f is a bounded continuous function; we set

$$K = \{(t, x(t)); \ 0 \leq t \leq T, \ x(t) \text{ satisfies } (11.3.1)\},$$

which is a compact set in $\mathbf{R}^{1+N}$.

Lemma 11.3.1. *If U is a neighborhood of K, then one can find $\varphi \in C_0^\infty(U)$ and $\varepsilon > 0$ such that $\varphi \geq 0$, $\varphi > 0$ in K, and*

$$(11.3.2) \qquad a\partial\varphi/\partial t + \langle b, \partial\varphi/\partial x \rangle \geq 0, \qquad \text{if } (t, x) \in U \text{ and } t \leq T,$$

for all $(a, b) \in \mathbf{R}^{1+N}$ with $|a - 1| < \varepsilon$ and $|b - f(t, x)| < \varepsilon$.

Before the proof we observe that if (11.3.2) holds, then $\varphi(t, x) > 0$ on K if $\varphi(0, 0) > 0$, for (11.3.1) gives $d\varphi(t, x(t))/dt = \partial\varphi/\partial t + \langle f, \partial\varphi/\partial x \rangle$.

Proof. Roughly speaking we want to regularize the characteristic function of the funnel K. We do that in two steps, where the first just widens K and the second smooths it out. With a small $\varepsilon > 0$ we let M_ε be the set of piecewise C^1 functions $x : [-\varepsilon, T] \to \mathbf{R}^N$ such that

$$(11.3.1)' \qquad |dx(t)/dt - f(t, x)| < 2\varepsilon, \quad -\varepsilon \le t \le T, \qquad |x(0)| < \varepsilon,$$

and set

$$U_\varepsilon = \{(t, x(t)); -\varepsilon < t < T, x \in M_\varepsilon\}.$$

This is an open set with $(0, 0) \in U_\varepsilon$. For small ε we have $\overline{U}_\varepsilon \subset U$. In fact, otherwise we could find a sequence $u_j \in M_{1/j}$ converging to a solution of (11.3.1) with $(t, x(t)) \notin U$ for some $t \in [0, T]$, which contradicts that U is a neighborhood of K. We can find $\delta > 0$ depending on ε so small that

$$(t, x) \in (s, y) + U_\varepsilon, \ (s, y) \in \mathbf{R}^{1+N}, \ s \ge 0, \ |(s, y)| < \delta, \ t < T, \ |a-1| < \varepsilon, \ |b - f(t, x)| < \varepsilon$$
$$\implies (t + \tau a, x + \tau b) \in (s, y) + U_\varepsilon, \quad \text{if } 0 < \tau < \delta, \text{ and } t + \tau a < T.$$

In fact, if we write $(t, x) = (s, y) + (t', x')$, where $(t', x') \in U_\varepsilon$, this is true if

$$(t', x') \in U_\varepsilon, \ t' + \tau a < T, \ 0 < \tau < \delta, \ |a-1| < \varepsilon, \ |b - f(t'+s, x'+y)| < \varepsilon, \ s \ge 0, \ |(s, y)| < \delta$$
$$\implies (t' + \tau a, x' + \tau b) \in U_\varepsilon.$$

By the definition of U_ε this is true if $|b - f(t' + \tau a, x' + \tau b)| < 2\varepsilon$, hence if

$$|f(t'+s, x'+y) - f(t'+\tau a, x'+\tau b)| < \varepsilon, \text{ when } 0 < \tau < \delta, \ t'+\tau a < T, \ |(s, y)| < \delta, \ (t', x') \in U_\varepsilon$$

which is true by the uniform continuity of f if δ is small enough. If we define φ as the convolution of the characteristic function of U_ε and a function $\chi \in C_0^\infty$ with support in $\{(s, y); s > 0, |(s, y)| < \delta\}$, such that $\chi \ge 0$ and $\int \chi(s, y) \, ds \, dy = 1$, then $\varphi \in C_0^\infty(U)$ and $\varphi(0, 0) > 0$ if δ is small enough. Since φ is an average of functions which do not decrease in the directions (a, b) in Lemma 11.3.1 when $(t, x) \in U$ and $t < T$, it follows that this is true for φ too, which means that (11.3.2) holds.

Remark. Note that given a neighborhood V of the slice $K_T = \{(T, x) \in K\}$ one can choose U in Lemma 11.3.1 so that (11.3.2) holds except in V. We just have to replace U if necessary by

$$\{(t, x) \in U; t < T\} \cup (V \cap U) \subset U.$$

Now assume that $q(x, \xi)$ is a real valued C^1 function, homogeneous of degree 1 in ξ, defined in $X \times (\mathbf{R}^n \setminus 0)$ where X is an open set in $\mathbf{R}^n$. Then the Hamilton vector field introduced in (11.2.4),

$$H_q = \partial q/\partial \xi \partial/\partial x - \partial q/\partial x \partial/\partial \xi$$

can be considered as a vector field on the cosphere bundle $\Sigma = X \times P_{\mathbf{R}}^{n-1}$, for acting on functions homogeneous of degree 0 in ξ it produces functions homogeneous of degree 0. If $H_q = 0$ at some point as a tangent vector of Σ, then $H_q u = 0$ must be a consequence of the Euler equation $\sum_1^n \xi_j \partial u/\partial \xi_j = 0$, that is, H_q is proportional to the *radial vector field* $\langle \xi, \partial/\partial \xi \rangle$, and dq is proportional to the *canonical one form* $\langle \xi, dx \rangle$.

Now let $(x_0, \xi_0) \in X \times (\mathbf{R}^n \setminus 0)$ be a characteristic point, that is, $q(x_0, \xi_0) = 0$, and assume that H_q is not proportional to the radial vector field at (x_0, ξ_0). If we introduce local coordinates (t, z) on Σ near the class $\gamma_0 \in \Sigma$ of (x_0, ξ_0) such that the component of H_q in the direction of the t axis is not 0, we can apply Lemma 11.3.1 and obtain with $\Psi = t$:

Lemma 11.3.2. *There exists a function* $\Psi \in C^\infty(\Sigma)$ *such that*

(i) $H_q \Psi > 0$ *in an open neighborhod* W *of* γ_0.

(ii) *If* T *is sufficiently small then the union of orbits of* H_q *starting at* γ_0 *on which* $\Psi(\gamma_0) \le \Psi \le \Psi(\gamma_0) + T$ *is a compact subset* K *of* W, *and so is the subset* K_T *where* $\Psi = \Psi(\gamma_0) + T$.

(iii) *If* U *and* V *are arbitrary neighborhoods of* K *and of* K_T *contained in* W *then there exists a function* $\Phi \in C_0^\infty(U)$ *such that* $\Phi \ge 0$, $\Phi > 0$ *in* K *and* $H_{\tilde q} \Phi \ge 0$ *in* $U \setminus V$ *for all* $\tilde q$ *in a* C^1 *neighborhood of* q.

In what follows we regard U, V, W, K, K_T as cones in $X \times (\mathbf{R}^n \setminus 0)$ and consider Ψ, Φ as homogeneous functions in W of degree 0. We denote by Φ_0 a C^∞ function which is equal to Φ when $|\xi| > 1$ and vanishes when $|\xi| < 1/2$.

If $q \in C^1(X \times (\mathbf{R}^n \setminus 0))$ is homogeneous of degree m instead, we can apply Lemma 11.3.2 to qk with $k \in C^\infty$ homogeneous of degree $1 - m$ and > 0. When $q(\gamma_0) = 0$ we have $q = 0$ in K, hence $H_{qk} = kH_q$, so H_{qk} and H_q have the same orbits through γ_0. The condition $H_{qk}\Psi > 0$ at γ_0 can be written $H_q\Psi > 0$ at γ_0.

The main step in the study of propagation of singularities is the following lemma which can be proved by combining the proof of Proposition 3.5.1 in Hörmander [5] with the preliminaries above due to Bony [1].

Lemma 11.3.3. *Let* $P \in \tilde{S}_{1,1}^m$, $\operatorname{Im} P \in \tilde{S}_{1,1}^{m-1}$, *assume that* $P_{(\beta)} \in \tilde{S}_{1,1}^m$ *has reduced order* μ *when* $|\beta| = 1$, *and that*

$$(11.3.3) \qquad q(x,\xi) = \lim_{t \to \infty} P(x, t\xi)/t^m$$

exists in $C^1(\mathbf{R}^n \times (\mathbf{R}^n \setminus 0))$. *Let* $\gamma_0 \in \mathbf{R}^n \times (\mathbf{R}^n \setminus 0)$, $q(\gamma_0) = 0$, *assume that* $H_q(\gamma_0)$ *is not radial, and let* Ψ, K, T *and* K_T *be defined as in Lemma 11.3.2. Then*

$$u \in \mathcal{E}', \quad P(x,D)u \in H_{(s)}^{\text{loc}} \text{ at } K, \quad u \in H_{(s+m-1)}^{\text{loc}} \text{ at } K_T \implies u \in H_{(s+m-1)}^{\text{loc}} \text{ at } K$$

provided that $u \in H_{(s+\mu-1)}^{\text{loc}}(X)$ *for some* $X \subset \mathbf{R}^n$ *such that* $K \subset X \times (\mathbf{R}^n \setminus 0)$.

Proof. The statement is trivial unless $\mu < m$, which we assume from now on. A bootstrapping argument shows that we may also assume, with a fixed $\delta \in (0,1)$ to be chosen later, that $u \in H_{(s+m-1-\delta)}^{\text{loc}}$ in a conic neighborhood of K. Choose $\chi \in S_{1,0}^0$ with support in such a neighborhood $\subset X \times (\mathbf{R}^n \setminus 0)$ so that $\chi = 1$ at infinity in another open conic neighborhood Γ of K. Then $\chi(x,D)u \in H_{(s+m-1-\delta)}(\mathbf{R}^n)$ and $P(x,D)(1 - \chi(x,D))u \in H_{(s)}^{\text{loc}}$ in Γ by Theorem 9.6.6, for Γ does not intersect the wave front set of $(1-\chi(x,D))u$ and P is of reduced order $\mu - 1$, $u \in H_{(s+\mu-1)}^{\text{loc}}(X)$. Replacing u by $\chi(x,D)u$ we may thus assume from now on that $u \in H_{(s+m-1-\delta)}$. We may also assume without restriction that $m = 1$ and that $u \in \mathcal{E}'(X)$, for otherwise we just multiply $P(x,D)$ to the right by $(1 + |D|^2)^{(1-m)/2}$, and multiply u first by $(1 + |D|^2)^{(m-1)/2}$ and then by a suitable cutoff function.

Let U be a conic neighborhood of K where $Pu \in H_{(s)}^{\text{loc}}$, let V_1 be a conic neighborhood of K_T where $u \in H_{(s)}^{\text{loc}}$, and choose $\Phi_V \in C^\infty$ nonnegative with support in a closed cone $\subset V_1$ so that, when $|\eta| > 1$, Φ_V is homogeneous of degree 0 and $\Phi_V = 1$ in another conic neighborhood V of K_T. Then $\Phi_V(x,D)u \in H_{(s)}(\mathbf{R}^n)$. With λ to be chosen later we set

$$(11.3.4) \qquad c_{\lambda,\varepsilon}(x,\eta) = \Phi_0(x,\eta)e^{\lambda\Psi(x,\eta)}|\eta|^s(1 + \varepsilon^2|\eta|^2)^{-1},$$

where Φ_0 and Ψ are given by Lemma 11.3.2 and the discussion after its proof, and write $C_{\lambda,\varepsilon} = c_{\lambda,\varepsilon}(x,D)$. When $\varepsilon > 0$ we have $c_{\lambda,\varepsilon} \in S_{1,0}^{s-2}$, hence $C_{\lambda,\varepsilon}u \in H_{(2-\delta)}$, and $c_{\lambda,\varepsilon}$ is bounded in $S^s_{1,0}$ for fixed λ. If we can prove that the L^2 norm $\|C_{\lambda,\varepsilon}u\|$ remains bounded as $\varepsilon \to 0$, it will follow that $C_{\lambda,0}u \in H_{(0)}$, hence that $u \in H^{loc}_{(s)}$ in K as claimed. It will suffice to prove that for large λ and small ε

$$(11.3.5) \quad \lambda\|C_{\lambda,\varepsilon}v\|^2 \leq A\|C_{\lambda,\varepsilon}P(x,D)v\|^2$$
$$+ A_\lambda(\|\Phi_V(x,D)(1+\varepsilon^2|D|^2)^{-1}v\|^2_{(s)} + \|v\|^2_{(s-\delta)}), \quad v \in C^\infty_0(\mathbf{R}^n),$$

where A is independent of λ. In fact, for reasons of continuity this remains true when $v \in H_{(s-\delta)}(\mathbf{R}^n)$, so we may take $v = u$ here. Since $P(x,D)u \in H^{loc}_{(s)}$ in U the norm of $C_{\lambda,\varepsilon}P(x,D)u$ is bounded as $\varepsilon \to 0$. We have

$$\|\Phi_V(x,D)(1+\varepsilon^2|D|^2)^{-1}u\|_{(s)} \leq \|\Phi_V(x,D)u\|_{(s)} + \|[\Phi_V(x,D),(1+\varepsilon^2|D|^2)^{-1}]u\|_{(s)}$$

where the second term is bounded when $\varepsilon \to 0$ since the symbol of the commutator is bounded in $S^{-1}_{1,0}$ and $u \in H_{(s-1)}(\mathbf{R}^n)$, so the boundedness of $\|C_{\lambda,\varepsilon}u\|$ follows at once.

The energy integral method we shall use starts from the identity

$$2\,\mathrm{Im}(C_{\lambda,\varepsilon}Pv, C_{\lambda,\varepsilon}v) = 2\,\mathrm{Im}(PC_{\lambda,\varepsilon}v, C_{\lambda,\varepsilon}v) + 2\,\mathrm{Im}([C_{\lambda,\varepsilon},P]v, C_{\lambda,\varepsilon}v)$$
$$= i((P^* - P)C_{\lambda,\varepsilon}v, C_{\lambda,\varepsilon}v) + 2\,\mathrm{Im}([C_{\lambda,\varepsilon},P]v, C_{\lambda,\varepsilon}v), \quad v \in C^\infty_0(\mathbf{R}^n).$$

Since $P - P^* \in \mathrm{Op}\,\tilde{S}^0_{1,1}$ is bounded in L^2, it follows that

$$(11.3.6) \quad 2\,\mathrm{Im}([C_{\lambda,\varepsilon},P]v, C_{\lambda,\varepsilon}v) \leq 2\,\mathrm{Im}(C_{\lambda,\varepsilon}Pv, C_{\lambda,\varepsilon}v) + M\|C_{\lambda,\varepsilon}v\|^2$$
$$\leq \|C_{\lambda,\varepsilon}Pv\|^2 + (M+1)\|C_{\lambda,\varepsilon}v\|^2.$$

Theorem 9.6.4$'$ gives

$$i[P, C_{\lambda,\varepsilon}] = \mathrm{Op}(H_Pc_{\lambda,\varepsilon}) + R_{\lambda,\varepsilon}(x,D)$$

where $R_{\lambda,\varepsilon}$ is bounded in $\tilde{S}^{s-2\delta}_{1,1}$ for fixed λ if $P_{(\beta)} \in \tilde{S}^{2-2\delta}_{1,1}$ when $|\beta| = 2$. This is true if δ is small, for $\mu < m = 1$. Hence $|(R_{\lambda,\varepsilon}(x,D)v, C_{\lambda,\varepsilon}v)| \leq M(\lambda)\|v\|^2_{(s-\delta)}$. For large $|\eta|$ we have for some A

$$\mathrm{Re}\,H_P\Phi_0 \geq 0 \text{ outside } V, \quad \mathrm{Re}\,H_P\Psi \geq 2/A \text{ in } \mathrm{supp}\,\Phi_0.$$

Furthermore,

$$|\eta|^{-1}H_P|\eta| = -\sum_{j=1}^n \partial P(x,\eta)/\partial x_j\eta_j|\eta|^{-2}, \quad |(1+\varepsilon^2|\eta|^2)^{-1}H_P(1+\varepsilon^2|\eta|^2)| \leq 2|\eta|^{-1}|H_P|\eta||$$

are bounded functions. If λ is large it follows that there is some $M(\lambda)$ such that for large $|\eta|$

$$\mathrm{Re}\,T_{\lambda,\varepsilon} \geq 0, \quad \text{where} \quad T_{\lambda,\varepsilon} = H_Pc_{\lambda,\varepsilon} - \lambda c_{\lambda,\varepsilon}/A + M(\lambda)\Phi_V|\eta|^s(1+\varepsilon^2|\eta|^2)^{-1}.$$

The last term gives the estimate in V, where $\mathrm{Re}\,H_P\Psi$ does not give a large contribution. From (11.3.6) we now obtain

$$(11.3.7) \quad 2\lambda A^{-1}\|C_{\lambda,\varepsilon}v\|^2 + 2\,\mathrm{Re}(T_{\lambda,\varepsilon}(x,D)v, C_{\lambda,\varepsilon}v) \leq \|C_{\lambda,\varepsilon}Pv\|^2$$
$$+ (M+1)\|C_{\lambda,\varepsilon}v\|^2 + 2M(\lambda)(\|\Phi_V(x,D)(1+\varepsilon^2|D|^2)^{-1}v\|_{(s)}\|C_{\lambda,\varepsilon}v\| + \|v\|^2_{(s-\delta)}).$$

Now $C_{\lambda,\varepsilon}(x,D)^*T_{\lambda,\varepsilon}(x,D) - \mathrm{Op}(c_{\lambda,\varepsilon}T_{\lambda,\varepsilon})$ is bounded in $\mathrm{Op}\,\tilde{S}_{1,1}^{2s-2\delta}$, $\mathrm{Re}\,c_{\lambda,\varepsilon}T_{\lambda,\varepsilon} \geq 0$, for fixed λ, $c_{\lambda,\varepsilon}T_{\lambda,\varepsilon}$ is bounded in $\tilde{S}_{1,1}^{2s}$ and $\partial_x^2(c_{\lambda,\varepsilon}T_{\lambda,\varepsilon})$ is bounded in $\tilde{S}_{1,1}^{2s+2-4\delta}$ for fixed λ, if δ is small enough. Since the conclusion of Theorem 9.7.1 is uniform when the hypotheses on the symbol are uniformly satisfied, it follows that

$$\mathrm{Re}(T_{\lambda,\varepsilon}(x,D)v, C_{\lambda,\varepsilon}v) \geq -\|v\|_{(s-\delta)}^2.$$

Combining this estimate with (11.3.7), we obtain (11.3.5) for large λ.

Remark. Proposition 3.5.1 in Hörmander [5] is more general than Lemma 11.3.3 in the sense that it allows $\mathrm{Im}\,P$ to be nonnegative of order m. The proof given there can be carried over without any difficulty to the more general classes of operators considered here, but the details are left for the reader.

We are now ready to state and prove a global result on propagation of singularities. (It could also be stated for a manifold.)

Theorem 11.3.4. *Let* $P \in \tilde{S}_{1,1}^m$, $\mathrm{Im}\,P \in \tilde{S}_{1,1}^{m-1}$, *assume that* $P_{(\beta)} \in \tilde{S}_{1,1}^m$ *has reduced order* μ *when* $|\beta| = 1$, *and that*

$$q(x,\xi) = \lim_{t\to\infty} P(x,t\xi)/t^m$$

exists in $C^1(\mathbf{R}^n \times (\mathbf{R}^n \setminus 0))$. *Let* $u \in \mathcal{S}'(\mathbf{R}^n)$ *and denote by* U *the open set of all* $(x,\eta) \in \mathbf{R}^n \times (\mathbf{R}^n \setminus 0)$ *such that* $H_q(x,\eta) \neq 0$, $P(x,D)u \in H_{(s)}^{\mathrm{loc}}$ *at* (x,η) *and* $u \in H_{(s+\mu-1)}^{\mathrm{loc}}$ *at* x. *Then* q *vanishes in the set* F *of all* $\gamma \in U$ *such that* $u \notin H_{(s+m-1)}^{\mathrm{loc}}$ *at* γ, *and* F *is invariant under the Hamilton flow in the following sense: If* $\gamma_0 \in F$ *then there exists an orbit* $(a,b) \ni t \mapsto \gamma(t)$ *of* H_q *contained in* F *such that* $a < 0 < b$, $\gamma(0) = \gamma_0$, *and* $\gamma(t)$ *does not converge to a limit in* U *when* $t \to a$ *or* $t \to b$.

Proof. By Theorem 9.6.7 we have $u \in H_{(s+m)}^{\mathrm{loc}}$ at γ if $q(\gamma) \neq 0$. We may assume that H_q is not radial, for the invariance under the Hamilton flow is trivial in the radial case. Since F is closed in U the statement follows if we just prove that γ_0 is in the interior of an orbit contained in F. To do so we use Lemma 11.3.3. Let N be a large integer. If we apply the lemma with T replaced by T/N, we obtain an orbit with end points γ_0 and $\gamma_1 \in F$ where $\Psi(\gamma_1) = \Psi(\gamma_0) + T/N$, for $K_{T/N}$ must contain some point in F. We can repeat the argument with γ_0 replaced by γ_1 and so on, which gives us an orbit γ_N with $\gamma_N(j/N) = \gamma_j \in F$ and $\Psi(\gamma_j) = \Psi(\gamma_0) + jT/N$ when $j = 0, \ldots, N$. All the orbits γ_N are contained in K and are equicontinuous, so we can select a convergent subsequence as $N \to \infty$. The limit is an orbit interval contained in F with γ_0 as one end point. Replacing P by $-P$ we obtain an interval in the other direction which completes the proof.

11.4. Nonlinear differential equations. Combination of the results of Section 11.3 and the linearisation techniques of Section 10.3 gives theorems on propagation of singularities for solutions of nonlinear differential equations. First we shall only discuss fully nonlinear equations as in Theorem 10.3.5.

Theorem 11.4.1. *Let* $u \in H_{(s+m)}^{\mathrm{loc}}(X)$, $s > n/2 + 1$, *be a real valued solution of the differential equation* $F(x, J_m u(x)) = 0$ *where* $F \in C^\infty$. *If*

$$(11.4.1) \qquad\qquad \sigma \leq 2s - n/2$$

then the set of $(x,\eta) \in X \times (\mathbf{R}^n \setminus 0)$ *where* $u \notin H_{(\sigma+m-1)}^{\mathrm{loc}}$ *is contained in the characteristic set, and it is invariant under the Hamilton flow defined by the principal symbol of the linearised equation.*

Proof. We can multiply F and u by cutoff functions to make both u and F vanish outside a compact set. The paradifferential operator P in Corollary 10.3.3 is of order m

and reduced order $m + n/2 - s$. Since $s > n/2 + 1$ the derivatives $P_{(\beta)}$ of the symbol are also of order m and they have reduced order $\mu = m + n/2 + 1 - s < m$ if $|\beta| = 1$. By Theorem 10.2.6 the symbol is equal to

$$\sum_{|\alpha|=m} F_\alpha(x, J_m u(x))(i\eta)^\alpha + O(|\eta|^{m-1}),$$

so P has

$$q(x, \eta) = \sum_{|\alpha|=m} F_\alpha(x, J_m u(x))(i\eta)^\alpha$$

as principal symbol in the sense of (11.3.3). We know from Corollary 10.3.3 that $Pu \in H_{(2s-n/2)}$ and can apply Theorem 11.3.4 with s replaced by σ if (11.4.1) holds, for then we have $Pu \in H^{\text{loc}}_{(\sigma)}$ and

$$\sigma + \mu - 1 = \sigma + m + n/2 - s \leq s + m.$$

The proof is now complete. Note that the statement is vacuous if $\sigma \leq s + 1$. When $\sigma > s + 1$ and (11.4.1) holds then the condition $s > n/2 + 1$ follows, so it is in fact no restriction.

We shall now state and prove an analogue of Theorem 10.3.7 concerning propagation of singularities.

Theorem 11.4.2. *Let $u \in H^{\text{loc}}_{(s+d)}(X)$, $s > n/4$, be a real valued solution of the differential equation (10.3.8) where a_α, c are in C^∞ and (10.3.9) is valid. If (11.4.1) is fulfilled then the set F_σ of $(x, \eta) \in X \times (\mathbf{R}^n \setminus 0)$ where $u \notin H^{\text{loc}}_{(\sigma+m-1)}$ is contained in the characteristic set, and it is invariant under the Hamilton flow defined by the principal symbol of the linearised equation.*

Proof. F_σ is empty unless $s + d < \sigma + m - 1$, and combining this condition with (11.4.1) we obtain

$$(11.4.2) \qquad\qquad s > d - m + 1 + n/2,$$

which we shall assume from now on. We shall also assume that $m_0 < m$ since the fully non-linear case $m_0 = m$ was already studied in Theorem 11.4.1. Since $J_{m(\alpha)}u \in H^{\text{loc}}_{(s+d-m(\alpha))}$ and by (11.4.2)

$$s + d - m(\alpha) > 2d - m(\alpha) - m + 1 + n/2 \geq 1 + n/2, \quad \text{if } |\alpha| = m,$$

the coefficients of the principal symbol of the linearised equation have (Hölder) continuous first derivatives then, so the Hamilton flow exists although it may not be unique.

We can multiply a_α, c and u by cutoff functions to make all these functions vanish outside a compact set. For the paradifferential operator P in the proof of Theorem 10.3.7 we know already that $Pu \in H_{(2s-n/2)}$; what remains is to verify that $P_{(\beta)}$ when $|\beta| = 1$ has order m and reduced order μ with $\sigma + \mu - 1 \leq s + d$ when $\sigma \leq 2s - n/2$, that is, $\mu \leq d + 1 + n/2 - s$, and that the principal symbol in the sense of (11.3.3) is equal to the principal symbol of the linarized operator. In the proof of Theorem 10.3.7 we showed that P has order m and reduced order $d + n/2 - s$; the stronger statement on $P_{(\beta)}$ follows in the same way and gives at the same time uniform Hölder continuity of $P_{(\beta)}(x, \eta)/(1 + |\eta|)^m$, which implies that the principal symbol exists in the sense (11.3.3). By Theorem 10.2.6 it is equal to the principal symbol of the linearised operator which completes the proof; the details are left for the reader.

11.5. Second order hyperbolic equations. Following Beals [2] we shall prove in this section that the restriction (11.4.1) can be relaxed for second order semi-linear hyperbolic equations. At first we denote by $P(x, \partial)$ a second order differential operator with coefficients in $C^\infty(X)$, where X is an open set in $\mathbf{R}^n$, and shall discuss formal manipulation of a nonlinear equation $P(x, \partial)u = f(x, J_1 u)$. We keep the notation for jets introduced in Section 10.3.

Lemma 11.5.1. Assume that $u \in H_{(s)}^{\mathrm{loc}}(X)$ where $s > k+1+n/2$ and k is a nonnegative integer. If $g \in C^\infty$ it follows that with $G \in C^\infty$ and $g_\alpha \in C^\infty$

$$(11.5.1) \qquad P(x, \partial)g(x, J_k u) = G(x, J_{k+1} u) + \sum_{|\alpha|=k} g_\alpha(x, J_k u) \partial^\alpha P(x, \partial)u.$$

Proof. By hypothesis and Sobolev's lemma we have $J_k u \in C^1$, so one differentiation is elementary and gives for some $h_j \in C^\infty$ and $g_\alpha \in C^\infty$

$$\partial_j g(x, J_k u) = h_j(x, J_k u) + \sum_{|\alpha|=k} g_\alpha(x, J_k u) \partial_j \partial^\alpha u.$$

Since $\partial_j \partial^\alpha u \in C^0$ we can differentiate again and obtain with $h \in C^\infty$

$$P(x, \partial)g(x, J_k u) = h(x, J_{k+1} u) + \sum_{|\alpha|=k} g_\alpha(x, J_k u) P(x, \partial) \partial^\alpha u.$$

The commutator $[P(x, \partial), \partial^\alpha]$ is of order $k + 1$ which proves (11.5.1).

The first time that the condition $s > k+1+n/2$ fails we can still make a differentiation, with somewhat greater caution, if $n \geq 2$ as we always assume.

Lemma 11.5.2. Assume that $u \in H_{(s)}^{\mathrm{loc}}(X)$ where $k + n/2 < s \leq k+1+n/2$ and k is a nonnegative integer. If $g \in C^\infty$ it follows that with $G, G_\alpha, g_\alpha, G_{\alpha\beta}$ in C^∞

$$(11.5.2) \quad P(x, \partial)g(x, J_k u) = G(x, J_k u) + \sum_{|\alpha|=k+1} G_\alpha(x, J_k u) \partial^\alpha u$$

$$+ \sum_{|\alpha|=|\beta|=k+1} G_{\alpha\beta}(x, J_k u) \partial^\alpha u \partial^\beta u + \sum_{|\alpha|=k} g_\alpha(x, J_k u) \partial^\alpha P(x, \partial)u.$$

Proof. This is clear from the proof of Lemma 11.5.1 if u is smooth. Now C^∞ is dense in $H_{(s)}^{\mathrm{loc}}$, and $G(x, J_k u), \ldots, g_\alpha(x, J_k u)$ are continuous functions of $u \in H_{(s)}^{\mathrm{loc}}$ with values in $H_{(s-k)}^{\mathrm{loc}}$, while $\partial^\alpha u$ is continuous with values in $H_{(s-k-1)}^{\mathrm{loc}}$ when $|\alpha| = k + 1$, $\partial^\alpha u \partial^\beta u$ is continuous with values in $H_{(s-k-2)}^{\mathrm{loc}}$ when $|\alpha| = |\beta| = k + 1$ by Theorem 8.3.1 since $2(s - k - 1) - n/2 > s - k - 2$, and $\partial^\alpha P(x, \partial)u$ is continuous with values in $H_{(s-k-2)}^{\mathrm{loc}}$ when $|\alpha| = k$. Since $s - k + (s - k - 2) = 2(s - k - 1) > n - 2 \geq 0$, the formula (11.5.2) extends by continuity if we use Theorem 8.3.1 once more.

Suppose now that $u \in H_{(s)}^{\mathrm{loc}}(X)$, $s > n/2$, is a solution of a semi-linear differential equation

$$(11.5.3) \qquad P(x, \partial)u = f_0(x, u) + \sum_{1}^{n} f_j(x, u) \partial_j u$$

where $f_j \in C^\infty$, $j = 0, \ldots, n$. By repeated use of Lemma 11.5.1 we then obtain

$$(11.5.4) \qquad P(x, \partial)^k u = F_k(x, J_k u) \in H^{loc}_{(s-k)}, \qquad s > k + n/2,$$

for this is obvious if $k = 0$, follows from (11.5.3) if $k = 1$, and is proved inductively by (11.5.1) for larger k. Using Lemma 11.5.2 with k replaced by $k-1$ gives in view of Theorem 8.3.1

$$(11.5.5) \quad P(x, \partial)^k u \in H^{loc}_{(2s-2k-n/2-0)} = \bigcap_{\varepsilon > 0} H^{loc}_{(2s-2k-n/2-\varepsilon)}, \qquad k - 1 + n/2 < s \le k + n/2.$$

(Here the equality is a definition which will be convenient in this section.) In fact, $G(x, J_{k-1}u), \ldots, g_\alpha(x, J_{k-1}u)$ are in $H^{loc}_{(s-k+1)}$, $\partial^\alpha u \in H^{loc}_{(s-k)}$ when $|\alpha| = k$, $\partial^\alpha u \partial^\beta u \in H^{loc}_{(2(s-k)-n/2-0)}$ when $|\alpha| = |\beta| = k$, and $\partial^\alpha P(x, \partial)u \in H^{loc}_{(s-k)}$ when $|\alpha| = k - 1$. This suggests the following:

Definition 11.5.3. When $s > n/2$ we shall denote by $H^{loc}_{(s)}(X; P)$ the set of all $u \in H^{loc}_{(s)}(X)$ such that $P(x, \partial)^k u \in H^{loc}_{(s-k)}(X)$ when $0 \le k < s - n/2$ and in addition (11.5.5) is valid.

Note that $2s - 2k - n/2 \le s - k$ in (11.5.5), so condition (11.5.5) is weaker than demanding that $P(x, \partial)^k u \in H^{loc}_{(s-k)}$ for this final value of k.

The discussion preceding the definition proves the first part of the following:

Proposition 11.5.4. If $u \in H^{loc}_{(s)}(X)$, $s > n/2$, is a solution of (11.5.3) then u belongs to $H^{loc}_{(s)}(X; P)$. If $s > n/2 + 1$ then $u \in H^{loc}_{(s)}(X; P)$ implies $\partial_j u \in H^{loc}_{(s-1)}(X; P)$ and $P(x, \partial)u \in H^{loc}_{(s-1)}(X; P)$.

Proof. If $0 \le k < s - 1 - n/2$ then $P(x, \partial)^k(P(x, \partial)u) \in H^{loc}_{(s-1-k)}(X)$, and if $k - 1 + n/2 < s - 1 \le k + n/2$ then $P(x, \partial)^k(P(x, \partial)u) \in H^{loc}_{(2s-2(k+1)-n/2-0)}(X)$ by (11.5.5) which proves that $P(x, \partial)u \in H^{loc}_{(s-1)}(X; P)$.

With differential operators q_ν of order $k + 1 - \nu$ we have

$$P(x, \partial)^k \partial_j u = \sum_{\nu=0}^{k} q_\nu(x, \partial) P(x, \partial)^\nu u.$$

This follows inductively since $[P(x, \partial), q_\nu(x, \partial)]$ is of order $k + 2 - \nu$. If $0 \le k < s - n/2$ it follows that $q_\nu(x, \partial)P(x, \partial)^\nu u \in H^{loc}_{(s-\nu-k-1+\nu)}(X) = H^{loc}_{(s-1-k)}(X)$, which proves that $\partial_j u \in H^{loc}_{(s-1)}(X; P)$.

Note that from (11.5.5) we obtain $u \in H^{loc}_{(2s-n/2-0)}$ at noncharacteristic points, which implies (11.5.4) locally at such points and almost gives back the noncharacteristic regularity theorem in a simple case covered by Theorem 8.4.13.

Remark. When the dimension is large we could continue deriving formulas for higher powers of $P(x, \partial)$ acting on u. However, we would lose two derivatives for every such factor, so no new information on regularity is obtained.

The following proposition shows that $H^{loc}_{(s)}(X; P)$ is an algebra, and that it is even closed under composition:

Proposition 11.5.5. *If $u \in H^{\mathrm{loc}}_{(s)}(X; P)$, where $s > n/2$, and if $g \in C^{\infty}$ then $g(x, u) \in H^{\mathrm{loc}}_{(s)}(X; P)$.*

Proof. Assume first that $s \leq 1 + n/2$. Then the hypothesis is that $u \in H^{\mathrm{loc}}_{(s)}(X)$ and that $P(x, \partial)u \in H^{\mathrm{loc}}_{(2s-2-n/2-0)}(X)$. By Lemma 11.5.2, with a change of notation,

$$P(x, \partial)g(x, u) = \sum_{|\alpha| \leq 2} G_\alpha(x, u)(\partial u/\partial x)^\alpha + g_0(x, u)P(x, \partial)u.$$

Here $G_\alpha(x, u)$ and $g_0(x, u)$ are in $H^{\mathrm{loc}}_{(s)}(X)$ and $\partial u/\partial x_j \in H^{\mathrm{loc}}_{(s-1)}(X)$, so the monomials $(\partial u/\partial x)^\alpha$ are in $H^{\mathrm{loc}}_{(2s-2-n/2-0)}(X)$ by Theorem 8.3.1, for $2s - 2 - n/2 \leq s - 1$. Another application of Theorem 8.3.1 proves that $P(x, \partial)g(x, u) \in H^{\mathrm{loc}}_{(2s-2-n/2-0)}(X)$.

Since $g(x, u) \in H^{\mathrm{loc}}_{(s)}(X)$ it suffices if $s > 1 + n/2$ to prove that $P(x, \partial)g(x, u) \in H^{\mathrm{loc}}_{(s-1)}(X; P)$. By (11.5.1)

$$P(x, \partial)g(x, u) = G(x, J_1 u) + g_0(x, u)P(x, \partial)u.$$

Since $J_1 u \in H^{\mathrm{loc}}_{(s-1)}(X; P)$ it follows that $G(x, J_1 u) \in H^{\mathrm{loc}}_{(s-1)}(X; P)$ if the proposition is already proved with s replaced by $s - 1$ and with u allowed to take values in $\mathbf{R}^N$ for any N. Also $g_0(x, u)$ and $P(x, \partial)u$ are in $H^{\mathrm{loc}}_{(s-1)}(X; P)$, so the proof is completed by induction.

Proposition 11.5.6. *The space $H^{\mathrm{loc}}_{(s)}(X; P)$, $s > n/2$, can be microlocalized: If $A \in \Psi^0_{1,0}(X)$ is properly supported then $AH^{\mathrm{loc}}_{(s)}(X; P) \subset H^{\mathrm{loc}}_{(s)}(X; P)$.*

Proof. Let $u \in H^{\mathrm{loc}}_{(s)}(X; P)$. We can write

$$P(x, \partial)^k A u = \sum_{j \leq k} A_{kj} P(x, \partial)^j u$$

where $A_{kj} \in \Psi^{k-j}_{1,0}(X)$ is also properly supported. This follows at once by induction for it is trivial when $k = 0$, and we have

$$[P(x, \partial), A_{kj}] \in \Psi^{k+1-j}_{1,0}(X).$$

If $j < k < s + 1 - n/2$ then $A_{kj} P(x, \partial)^j u$ belongs to $H^{\mathrm{loc}}_{(s-j-(k-j))}(X) = H^{\mathrm{loc}}_{(s-k)}(X)$, and $A_{kk} P(x, \partial)^k u$ is in $H^{\mathrm{loc}}_{(\sigma)}(X)$ whenever $P(x, \partial)^k u \in H^{\mathrm{loc}}_{(\sigma)}(X)$. The proof is complete.

Remark. The same inductive argument shows that $H^{\mathrm{loc}}_{(s)}(X; P)$ does not depend on the lower order terms of $P(x, \partial)$.

By Proposition 11.5.6 it makes sense to say that u belongs to $H^{\mathrm{loc}}_{(s)}(\cdot; P)$ at a point (x_0, ξ_0) in $X \times (\mathbf{R}^n \setminus 0)$. If (x_0, ξ_0) is noncharacteristic with respect to $P(x, \partial)$, we have already observed that this is equivalent to $u \in H^{\mathrm{loc}}_{(2s-n/2-0)}$ at (x_0, ξ_0), so it is only at the characteristic points that we have obtained a new way of describing smoothness. The important gain there is the following improved analogue of Corollary 8.3.4':

Proposition 11.5.7. *Let $u \in H^{\mathrm{loc}}_{(s)}(X; P)$ for some $s > n/2$, and assume that $u \in H^{\mathrm{loc}}_{(\sigma)}$ at (x_0, ξ_0). If $P(x, \partial)$ is hyperbolic and characteristic at (x_0, ξ_0), and if $s \leq \sigma < 3s - n$, it follows that $g(x, u) \in H^{\mathrm{loc}}_{(\sigma)}$ at (x_0, ξ_0) when $g \in C^{\infty}$.*

The proof will be parallel to that of Corollary 8.3.4' once we have established the following:

Lemma 11.5.8. *Let* $u, v \in H^{\mathrm{loc}}_{(s)}(X; P)$ *for some* $s > n/2$ *and assume that* $u, v \in H^{\mathrm{loc}}_{(\sigma)}$ *at* (x_0, ξ_0). *If* $P(x, \partial)$ *is hyperbolic and characteristic at* (x_0, ξ_0), *and if* $s \leq \sigma < 3s - n$, *it follows that* $vu \in H^{\mathrm{loc}}_{(\sigma)}$ *at* (x_0, ξ_0). *Moreover,* $v \partial_j u \in H^{\mathrm{loc}}_{(\sigma-1)}$ *at* (x_0, ξ_0) *even if the microlocal assumption on* v *is replaced by* $v \in H^{\mathrm{loc}}_{(\varrho)}$ *at* (x_0, ξ_0) *where*

$$\varrho > \max(\sigma - 1, \sigma - s + n/2) = \sigma - \min(1, s - n/2).$$

Proof that Lemma 11.5.8 implies Proposition 11.5.7. If $\sigma = s$ there is nothing to prove. Assume now that the conclusion of Proposition 11.5.7 is valid for all $g \in C^\infty$ when σ is replaced by some $\sigma_0 \in [s, \sigma)$. We have

$$\partial_j g(x, u) = g_j(x, u) + g'(x, u) \partial_j u$$

where $g_j \in C^\infty$, hence $g_j(x, u) \in H^{\mathrm{loc}}_{(\sigma_0)}$ at (x_0, ξ_0) by hypothesis. Since $g'(x, u) \in H^{\mathrm{loc}}_{(s)}(X; P)$ by Proposition 11.5.5 and $g'(x, u) \in H^{\mathrm{loc}}_{(\sigma_0)}$ at (x_0, ξ_0) by hypothesis, it follows from Lemma 11.5.8 with $\varrho = \sigma_0$ that $g'(x, u) \partial_j u \in H^{\mathrm{loc}}_{(\sigma-1)}$ at (x_0, ξ_0) if

$$\sigma < \sigma_0 + \min(1, s - n/2).$$

Thus we conclude that $g(x, u) \in H^{\mathrm{loc}}_{(\sigma)}$ at (x_0, ξ_0) if $\sigma \leq \sigma_0 + \frac{1}{2} \min(1, s - n/2)$. Otherwise we may replace σ_0 by the right-hand side. After a finite number of repetitions of the argument, the proposition is proved.

Proof of Lemma 11.5.8. By Proposition 11.5.6 we can write $u = u_0 + u_1$, $v = v_0 + v_1$ where u_0, u_1, v_0, v_1 satisfy the same hypotheses as u and v and in addition

$$(x_0, \xi_0) \notin WF(u_1) \cup WF(v_1), \qquad u_0 \in H_{(\sigma)}, \quad v_0 \in H_{(\sigma)} \text{ resp. } v_0 \in H_{(\varrho)}.$$

Thus $v_0 u_0 \in H_{(\sigma)}$ resp. $v_0 \partial_j u_0 \in H_{(\sigma-1)}$, and by Theorem 10.2.10 we have at (x_0, ξ_0)

$$v_0 u_1 + v_1 u_0 \in H^{\mathrm{loc}}_{(\sigma)} \quad \text{resp.} \quad v_0 \partial_j u_1 + v_1 \partial_j u_0 \in H^{\mathrm{loc}}_{(\sigma-1)},$$

for $\varrho > \sigma - 1$ and $\varrho + s - 1 - n/2 > \sigma - 1$. What remains is to prove the same microlocal regularity for $v_1 u_1$ resp. $v_1 \partial_j u_1$, that is, that the proposition is true when $(x_0, \xi_0) \notin WF(u) \cup WF(v)$.

If $P(x, \partial)$ is noncharacteristic in $WF(u)$ then $u \in H^{\mathrm{loc}}_{(2s-n/2-0)}$, and since $v \in H^{\mathrm{loc}}_{(s)}$ it follows from Theorem 10.2.10 that $vu \in H^{\mathrm{loc}}_{(3s-n-0)}$ and that $v \partial_j u \in H^{\mathrm{loc}}_{(3s-n-1-0)}$ at (x_0, ξ_0). We can argue in the same way if $P(x, \partial)$ is noncharacteristic in $WF(v)$. Again by Proposition 11.5.6 it follows that it suffices to prove the proposition when $WF(u)$ and $WF(v)$ are contained in small conic neighborhoods of characteristic points (x_0, ξ_1) and (x_0, ξ_2). If ξ_1 and ξ_2 do not have opposite directions we can assume that the wave front sets are contained in $X \times \Gamma_1$ resp. $X \times \Gamma_2$ where Γ_1 and Γ_2 are closed cones with $\xi_0 \notin \Gamma_1 + \Gamma_2$. This is clear if ξ_1 and ξ_2 have the same direction since $(x_0, \xi_0) \notin WF(u) \cup WF(v)$. If ξ_1 and ξ_2 are linearly independent, then the linear combinations $(x_0, t_1 \xi_1 + t_2 \xi_2)$ are not characteristic unless $t_1 t_2 = 0$, since the hyperbolicity implies that there is a hyperplane where the characteristic form $p(x_0, \xi)$ only vanishes at the origin, so $p(x_0, t_1 \xi_1 + t_2 \xi_2)$ is a nonzero multiple of $t_1 t_2$. Thus ξ_0 is not in the closed conic hull of ξ_1 and ξ_2. This implies that $(x_0, \xi_0) \notin WF(vu) \cup WF(v \partial_j u)$, for if $u, v \in \mathcal{E}'$, as we may assume, then the Fourier transforms are rapidly decreasing outside Γ_1 resp. Γ_2, which implies that the Fourier transforms of vu and $v \partial_j u$ are rapidly decreasing outside $\Gamma_1 + \Gamma_2$.

What remains is to discuss the case where $\xi_1 = -\xi_2$ while ξ_0 has a different direction. In that case the wave front set of the products can contain arbitrary frequencies. However, we note that the hyperbolicity implies that ξ_0 is not in the tangent plane of the characteristic surface $\{\xi; p(x_0, \xi) = 0\}$ at ξ_1, for it intersects the characteristic surface in the line $\mathbf{R}\xi_1$. In this section we shall only complete the proof in the constant coefficient case, postponing the discussion of the general case to Section 11.6. After completing the proof of Lemma 11.5.8 in the special case we shall also derive a theorem on propagation of singularities by the arguments used to prove Theorem 8.4.13.

We shall begin by discussing the constant coefficient case essentially as in Section 8.3.

Lemma 11.5.9. *Let $P(\xi)$ be a real nondegenerate quadratic form in $\mathbf{R}^n$, let $\xi_1 \in \mathbf{R}^n \backslash 0$, $P(\xi_1) = 0$, and let Γ be a closed cone in $\mathbf{R}^n$ which does not meet the tangent plane of $P^{-1}(0)$ at ξ_1 except at the origin. If $\Gamma_\pm$ is a sufficiently small closed conic neighborhood of $\pm\xi_1$, supp $\hat{u}_\pm \subset \Gamma_\pm$, and k is a nonnegative integer, $s > 0$, then*

$$(11.5.6) \qquad \|\chi(D)(u_+ u_-)\|_{(s+k-n/2)} \le C \sum_{i+j \le k} \|P(D)^i u_+\|_{(s_i)} \|P(D)^j u_-\|_{(t_j)}$$

if $s_i + t_j = s - i - j$, $P(D)^i u_+ \in H_{(s_i)}(\mathbf{R}^n)$ and $P(D)^j u_- \in H_{(t_j)}(\mathbf{R}^n)$ for $i + j \le k$, and $0 \le \chi \le 1$, supp $\chi \subset \Gamma$.

Proof. We may assume in the proof that $\hat{u}_\pm$ has compact support. The Fourier transform of $U = \chi(D)(u_+ u_-)$ is

$$(11.5.7) \qquad \widehat{U}(\xi) = (2\pi)^{-n} \int \chi(\xi) \hat{u}_+(\xi - \eta) \hat{u}_-(\eta) \, d\eta.$$

We shall estimate the integral when $\xi \in \Gamma$. Set $\xi_+ = \xi - \eta$, $\xi_- = \eta$, thus $\xi_\pm \in \Gamma_\pm$ and $\xi_+ + \xi_- = \xi \in \Gamma$ in the support of the integrand. If $\Gamma_\pm \cap \Gamma = \{0\}$ then

$$(11.5.8) \qquad 1/C \le |\xi_+|/|\xi_-| \le C, \quad \text{if } \Gamma \ni \xi = \xi_+ + \xi_-, \, \xi_\pm \in \Gamma_\pm,$$

which implies $|\xi| \le (1 + C)|\xi_\pm|$. The inequality follows for reasons of homogeneity for if $|\xi_\pm| = 1$ and $|\xi_\mp|$ is very small then $\xi_\pm + \xi_\mp$ is so close to $\Gamma_\pm$ that it cannot belong to Γ. The hypothesis on Γ is that

$$|\theta| \le C|\langle \theta, P'(\xi_1)\rangle|/|\xi_1|, \quad \theta \in \Gamma.$$

For reasons of continuity this remains true with ξ_1 replaced by any point in a small conic neighborhood if C is replaced by $2C$. If $\Gamma_- = -\Gamma_+$ and Γ_+ is a sufficiently small convex conic neighborhood of ξ_1, we can take $\theta = \xi_+ + \xi_-$ and replace ξ_1 by $\xi_+ - \xi_-$, noting that $|\xi_\pm| \le |\xi_+ - \xi_-|$. Since $\langle \xi_+ + \xi_-, P'(\xi_+ - \xi_-)\rangle = 2(P(\xi_+) - P(\xi_-))$ we obtain

$$|\xi| \le 4C|P(\xi_+) - P(\xi_-)|/|\xi_+ - \xi_-| \le 4C(|P(\xi_+)|/|\xi_+| + |P(\xi_-)|/|\xi_-|).$$

With another constant and the notation $\langle \xi \rangle = (1 + |\xi|^2)^{\frac{1}{2}}$ it follows that

$$(11.5.9) \quad \langle \xi \rangle \le C(|P(\xi_+)|/\langle \xi_+ \rangle + |P(\xi_-)|/\langle \xi_- \rangle + 1), \quad \text{if } \Gamma \ni \xi = \xi_+ + \xi_-, \, \xi_\pm \in \Gamma_\pm,$$

for this is obvious when $|\xi_\pm|$ is small. Using (11.5.8) we now obtain

$$\langle \xi \rangle^{s+k-n/2} |\hat{u}_+(\xi_+) \hat{u}_-(\xi_-)|$$

$$\le C' \langle \xi \rangle^{s-n/2} \sum_{i+j \le k} |P(\xi_+)|^i \langle \xi_+ \rangle^{-i} |\hat{u}_+(\xi_+)| |P(\xi_-)|^j \langle \xi_- \rangle^{-j} |\hat{u}_-(\xi_-)|$$

$$\le C'' \sum_{i+j \le k} F_{ij}(\xi, \eta) v_i(\xi - \eta) w_j(\eta),$$

where we have used (11.5.8) and the notation

$$v_i(\xi_+) = |P(\xi_+)|^i \langle\xi_+\rangle^{s_i} |\hat{u}_+(\xi_+)|, \quad w_j(\xi_-) = |P(\xi_-)|^j \langle\xi_-\rangle^{t_j} |\hat{u}_-(\xi_-)|,$$
$$F_{ij}(\xi,\eta) = \begin{cases} \langle\xi\rangle^{s-n/2} \langle\eta\rangle^{-s}, & \text{if } |\xi| \le (1+C)\min(|\eta|, |\xi-\eta|), \\ 0, & \text{otherwise.} \end{cases}$$

Taking $\xi/|\eta|$ as a new integration variable we obtain

$$\int |F_{ij}(\xi,\eta)|^2 \, d\xi \le |\eta|^{n-2s} \int_{|\xi|\le 1+C} (1+|\eta|^2|\xi|^2)^{s-n/2} \, d\xi \le 2^{|s-n/2|} \int_{|\xi|\le 1+C} (|\xi|^{2s-n}+1) \, d\xi,$$

if $|\eta| \ge 1$, and it is obvious that there is a bound when $|\eta| \le 1$. Hence it follows from Lemma 8.3.2 that

$$\left(\int |\widehat{U}(\xi)|^2 \langle\xi\rangle^{2s+2k-n} \, d\xi\right)^{\frac{1}{2}} \le C''' \sum_{i+j\le k} \|v_i\| \|w_j\|,$$

which gives (11.5.6) by Parseval's formula.

Proof of Lemma 11.5.8 in a special case. Assuming that $P(x,\partial)$ is independent of x we can now prove Lemma 11.5.8. By a remark after Proposition 11.5.6 it is no additional restriction to assume that $P(\xi)$ is homogeneous of degree 2, which makes Lemma 11.5.9 applicable. Let $u, v \in \mathcal{E}'$ be as in Lemma 11.5.8 with wave front sets in small conic neighborhoods of the characteristics $(x_0, \pm\xi_1)$, and assume that $\xi_0 \notin \mathbf{R}\xi_1$, which implies that ξ_0 is not in the tangent plane of the characteristic surface at ξ_1. Subtracting functions in $H_{(+\infty)}$ from u and v we may assume instead that $u, v \in \mathcal{S}'$ and that supp $\hat{u}$, supp $\hat{v}$ are contained in small conic neighborhoods of $\pm\xi_1$. Then, with k as in (11.5.5), hence $k \ge 1$,

$$P(D)^i u \in H_{(s_i)} \quad \text{if } 0 \le i \le k \text{ and } s_i = s + i(s - 2k - n/2)/k - \varepsilon$$

for every $\varepsilon > 0$, for this is true when $i = 0$ and when $i = k$. We may replace u by v here. Since $s - k - n/2 \le 0$ by (11.5.5) we have

$$s_i + s_j + i + j = 2s + (i+j)(s-k-n/2)/k - 2\varepsilon \ge 3s - k - n/2 - 2\varepsilon, \quad i+j \le k,$$

which means that (11.5.6) can be used with s replaced by $3s-k-n/2-2\varepsilon$, for $3s-k-n/2 > n/2$ since $s > k-1+n/2$ and $2k-3+n \ge n-1 \ge n/2$. Hence $\chi(D)(uv) \in H_{(3s-n-2\varepsilon)}$, and since χ can be chosen equal to 1 in a conic neighborhood of ξ_0 we obtain $uv \in H^{\mathrm{loc}}_{(3s-n-0)}$ at (x_0, ξ_0). The same argument with $t_j = s_j - 1$ shows that $v\partial_j u \in H^{\mathrm{loc}}_{(3s-n-1-0)}$ at (x_0, ξ_0) for $3s - k - n/2 - 1 > n/2 - 1 \ge 0$.

Now we come to the propagation theorem which we have been aiming for.

Theorem 11.5.10. *Let $u \in H^{\mathrm{loc}}_{(s)}(X)$ for some $s > n/2$ and assume that u satisfies the hyperbolic second order semi-linear differential equation*

$$P(x,\partial)u = f(x,u)$$

where $f \in C^\infty$. If $u \in H^{\mathrm{loc}}_{(\sigma)}$ at a characteristic point (x_0, ξ_0) and if $s \le \sigma < 3s - n + 1$, then it follows that $u \in H^{\mathrm{loc}}_{(\sigma)}$ at the bicharacteristic γ through (x_0, ξ_0).

Proof. Let $s \le \tau < \sigma$ and assume we already know that $u \in H^{\mathrm{loc}}_{(\tau)}$ at γ. By Proposition 11.5.4 we know that $u \in H^{\mathrm{loc}}_{(s)}(X; P)$, so it follows from Proposition 11.5.7 that $f(x, u) \in$

$H^{loc}_{(\tau)}$ at γ, if $\tau < 3s - n$. Hence the linear propagation theorem, a special case of Theorem 11.3.4, shows that $u \in H^{loc}_{(\min(\tau+1,\sigma))}$ at γ. Repeating the argument a finite number of times we conclude that $u \in H^{loc}_{(\sigma)}$ at γ. The proof is complete.

We refer to Beals [2] for extensions of the result to more general equations and to Beals [1] for examples with P equal to the wave operator which show that Theorem 11.5.10 is essentially optimal. In fact, Theorem 11.5.10 shows that if P is the wave operator and u is only singular at $x = 0$ when $t = 0$ then u has about $3s - n$ L^2 derivatives inside the light cone, and Beals has constructed examples of "self-spreading" where this conclusion cannot be improved.

The results of this section have been extended by Chemin [1] to nonlinear hyperbolic differential equations of arbitrary order. A brief description of the results and further references can be found in a survey paper of Bony [2], which also covers the propagation of conormal singularities. These behave roughly speaking as homogeneous functions of high (noninteger) degree in the distance to a surface but are smooth along the surface.

11.6. Beals' restriction theorem. To carry out the technically harder but basically similar proof of Lemma 11.5.8 in the variable coefficient case it is helpful to look at a more general question. One can regard a product uv of two functions u and v in $\mathbf{R}^n$ as the restriction to the diagonal of the tensor product $u \otimes v$ defined by

$$(u \otimes v)(x,y) = u(x)v(y), \quad (x,y) \in \mathbf{R}^n \times \mathbf{R}^n.$$

In other words, uv is the pullback of $u \otimes v$ by the diagonal map $\mathbf{R}^n \ni x \mapsto (x,x) \in \mathbf{R}^n \times \mathbf{R}^n$. We shall consider the more general problem of determining the regularity of the restriction to submanifolds of distributions in $H^{loc}_{(s)}$ which remain in $H^{loc}_{(s)}$ when some powers of a first order pseudo-differential operator are applied to them. This extension of the problem has the advantage that the coordinates can be chosen so that the submanifold has a simple form.

Let X be a C^∞ manifold of dimension n, let Y be a C^∞ submanifold of codimension ν, and denote the embedding $Y \to X$ by γ_Y. It is well known that the restriction map $C^\infty(X) \ni u \mapsto \gamma_Y^* u = u \circ \gamma_Y \in C^\infty(Y)$ extends by continuity to a map $\gamma_Y^* : H^{loc}_{(s)}(X) \to H^{loc}_{(s-\nu/2)}(Y)$ if $s > \nu/2$. We shall need a slightly stronger result, where $N^*(Y)$ denotes the conormal bundle of Y.

Lemma 11.6.1. *The map γ_Y^* can be defined by continuity for every $u \in \mathcal{D}'(X)$ such that $u \in H^{loc}_{(s)}$ at $N^*(Y)$ for some $s > \nu/2$, and $\gamma_Y^* u$ is then in $H^{loc}_{(s-\nu/2)}$ at $(y,\eta) \in T^*(Y)\setminus 0$ if $u \in H^{loc}_{(s)}$ at (y,ξ) for every ξ with ${}^t\gamma'(y)\xi = \eta$.*

Proof. We can choose local coordinates with the origin at y so that Y is defined by $x' = 0$ where $x' = (x_1, \ldots, x_\nu)$, $x'' = (x_{\nu+1}, \ldots, x_n)$ is a splitting of the coordinates in $\mathbf{R}^n$. Then $N^*(Y)$ is defined by $x' = 0$, $\xi'' = 0$. If $\varphi \in C_0^\infty(\mathbf{R}^n)$ and supp φ is sufficiently close to the origin, $v = \varphi u$, then $\hat{v}(\xi)$ is in L^2 with respect to the measure $(1 + |\xi|^2)^s d\xi$ when $|\xi''| < \varepsilon|\xi'|$, for some $\varepsilon > 0$. If $v \in C_0^\infty(\mathbf{R}^n)$ and $v_0 = \gamma_Y^* v$, that is, $v_0(x'') = v(0, x'')$, then $\hat{v}_0(\xi'') = (2\pi)^{-\nu} \int \hat{v}(\xi', \xi'') d\xi'$. Now

$$|\hat{v}_0(\xi'')|^2 = \left|(2\pi)^{-\nu} \int \hat{v}(\xi) d\xi'\right|^2 \le (2\pi)^{-2\nu} \left(\int |\hat{v}(\xi)|^2 (1 + |\xi|^2)^s d\xi'\right) \int (1 + |\xi|^2)^{-s} d\xi'$$

$$= C_s (1 + |\xi''|^2)^{\nu/2-s} \int |\hat{v}(\xi)|^2 (1 + |\xi|^2)^s d\xi',$$

which proves that the map $v \mapsto v_0$ is continuous from $H_{(s)}(\mathbf{R}^n)$ to $H_{(s-\nu/2)}(\mathbf{R}^n)$ if $s > \nu/2$, which is the standard result mentioned above. If we write $v_0 = v_1 + v_2$ where

$$\hat{v}_1(\xi'') = (2\pi)^{-\nu} \int_{\varepsilon|\xi'|>|\xi''|} \hat{v}(\xi', \xi'') d\xi', \quad \hat{v}_2(\xi'') = (2\pi)^{-\nu} \int_{\varepsilon|\xi'|<|\xi''|} \hat{v}(\xi', \xi'') d\xi',$$

it follows in the same way that with another constant C_s

$$\|v_1\|_{(s-\nu/2)} \le C_s \Big(\iint_{\epsilon|\xi'|>|\xi''|} |\hat{v}(\xi)|^2 (1+|\xi|^2)\, d\xi \Big)^{\frac{1}{2}}, \quad s > \nu/2,$$

$$\|v_2\|_{(\mu-\nu/2)} \le C_{\mu,\epsilon} \Big(\iint_{\epsilon|\xi'|<|\xi''|} |\hat{v}(\xi)|^2 (1+|\xi|^2)^\mu\, d\xi \Big)^{\frac{1}{2}},$$

where μ is any real number and $C_{\mu,\epsilon}^2 = (2\pi)^{-\nu} \int_{\epsilon|\theta|<1}(1+|\theta|^2)^{-\mu}\, d\theta$. This gives the desired extension of $\gamma_Y^* v$ to the case where $v \in \mathcal{E}'(\mathbf{R}^n)$ is just in $H_{(s)}^{loc}$ at (x,ξ) for all x and $\xi \ne 0$ with $\xi'' = 0$, for some $s > \nu/2$. Since the integration with respect to ξ'' can be restricted to any cone we conclude that $\gamma^* v \in H_{(s-\nu/2)}^{loc}$ at (y,η), $\eta \ne 0$, if $v \in H_{(s)}^{loc}$ at (x,ξ) when $\xi'' = \eta$ or $\xi'' = 0$. This proves the lemma.

The following is a general version of Lemma 1.10 in Beals [2]:

Theorem 11.6.2. *Assume that $P \in \Psi_{1,0}^1(X)$ is properly supported and has a homogeneous principal symbol p such that*

$$(11.6.1) \qquad p = 0 \quad \text{on the conormal bundle } N^*(Y) \setminus 0.$$

This implies that the base projection $p'_\xi(x,\xi)$ of the Hamilton vector is tangent to Y when $(x,\xi) \in N^*(Y) \setminus 0$. Let $u \in H_{(s)}^{loc}(X)$ where $s > \nu/2$, let $(y,\eta) \in T^*(Y) \setminus 0$, and assume that

$$(11.6.2) \qquad P^j u \in H_{(s)}^{loc}(X), \quad 0 \le j \le k, \quad \text{at } (\gamma_Y(y),\xi) \text{ when } {}^t\gamma'_Y(y)\xi \in \mathbf{R}\eta.$$

Then it follows that $\gamma_Y^* u \in H_{(s+k-\nu/2)}^{loc}$ at (y,η) if

$$(11.6.3) \qquad p(y,\xi) \ne 0 \quad \text{when } (y,\xi) \in WF(u) \text{ and } {}^t\gamma'_Y(y)\xi = \eta,$$

$$(11.6.4) \qquad \langle p'_\xi(y,\xi),\eta \rangle \ne 0 \quad \text{when } (y,\xi) \in WF(u) \text{ and } {}^t\gamma'_Y(y)\xi = 0.$$

Proof. If $Q_0,\dots,Q_j \in \Psi_{1,0}^0(X)$ are properly supported, then

$$(11.6.5) \quad Q_0 P Q_1 \cdots P Q_j u \in H_{(s)}^{loc}(X), \quad 0 \le j \le k, \quad \text{at } (\gamma_Y(y),\xi) \text{ when } {}^t\gamma'_Y(y)\xi \in \mathbf{R}\eta.$$

This is obvious if $j = 0$ and follows by induction with respect to j if we note that

$$Q_{j-1}[P,Q_j] \in \Psi_{1,0}^0, \qquad Q_{j-1}Q_j \in \Psi_{1,0}^0.$$

In particular this shows that (11.6.2) remains valid if an operator in $\Psi_{1,0}^0(X)$ is added to P. We also conclude that φu satisfies the same hypotheses as u if $\varphi \in C_0^\infty$, so the statement is a local one. We may therefore assume that $X \subset \mathbf{R}^n$ and $y = 0$, that $P \in \text{Op}\,S_{1,0}^1$ and $u \in \mathcal{E}'(U)$, where U is a small neighborhood of 0, and that Y is defined by $x' = 0$ with a splitting of the coordinates in $\mathbf{R}^n$ as above.

In terms of the chosen coordinates, (11.6.1) means that

$$(11.6.1)' \qquad p((0,y),(\xi',0)) = 0, \quad \xi' \in \mathbf{R}^\nu \setminus 0, \; y \in \mathbf{R}^{n-\nu},$$

and ${}^t\gamma'_Y(0,y)(\xi',\eta) = \eta$. Differentiation of $(11.6.1)'$ with respect to ξ' gives that $\partial p(x,\xi)/\partial\xi$ is tangent to $Y = \{x; x' = 0\}$ when $(x,\xi) \in N^*(Y)\setminus 0$. The conditions (11.6.3) and (11.6.4) are

$$(11.6.3)' \qquad p(0,\xi) \ne 0 \quad \text{if } (0,\xi) \in WF(u), \; \xi'' = \eta,$$

$$(11.6.4)' \qquad \langle \partial p(0,\xi)/\partial\xi'',\eta \rangle \ne 0 \quad \text{if } (0,\xi) \in WF(u), \; \xi'' = 0.$$

From $(11.6.4)'$ it follows that $(11.6.3)'$ is valid when ξ' is large. Moreover, it follows that there is a conic neighborhood Γ of $\{\xi; (0,\xi) \in WF(u), \xi'' = 0\}$ in $\mathbf{R}^n \setminus \{0\}$, a conic neighborhood Γ'' of η in $\mathbf{R}^{n-\nu} \setminus \{0\}$, and a neighborhood U of 0 such that for some C

$$(11.6.6) \qquad |\xi''| \le C|p(x,\xi)|, \quad \text{if } x \in U, \ \xi \in \Gamma, \ \xi'' \in \Gamma'',$$

and $P^j u \in H_{(s)}^{\mathrm{loc}}$ in $U \times \Gamma$ when $j \le k$. In fact, by Taylor's formula and the homogeneity of p we have when $|\xi''| \ll |\xi'|$

$$p(x,\xi) = |\xi'| p(x, (\xi'/|\xi'|, \xi''/|\xi'|)) = \langle p'_{\xi''}(x, (\xi', 0)), \xi'' \rangle + O(|\xi''|^2/|\xi'|).$$

When $(0, (\xi', 0)) \in WF(u)$, ξ'' is proportional to η, and $x = 0$, it follows from $(11.6.4)'$ that

$$3|\xi''| \le C|\langle p'_{\xi''}(x, (\xi', 0)), \xi'' \rangle|,$$

and for reasons of continuity this remains true with 3 replaced by 2 for $x \in U$, $\xi \in \Gamma$, $\xi'' \in \Gamma''$ if these neighborhoods are small enough. If in addition Γ is chosen so that $|\xi''|/|\xi'|$ is sufficiently small when $(\xi', \xi'') \in \Gamma$, then $(11.6.6)$ follows.

By $(11.6.3)'$ and $(11.6.2)$ we have

$$(11.6.7) \qquad u \in H_{(s+k)}^{\mathrm{loc}} \quad \text{at } (0,\xi) \text{ if } \xi'' = \eta.$$

The difficulty in the proof is to establish enough regularity of u when $\xi'' = 0$ using $(11.6.6)$. It is clear that this condition can only give regularity in the x'' variables, and to work with it we must use pseudo-differential operators with symbols in a class adapted to this situation.

Since the lower order terms of P are not important in $(11.6.2)$ we may assume that $P(0, x'', \xi', 0) \equiv 0$ for the full symbol of P. Set $P_0(x,\xi) = P(0, x'', \xi)$. Then $P_{0(\beta)}^{(\alpha)}(x,\xi) = 0$ when $\xi'' = 0$, if $\alpha'' = 0$, so the mean value theorem gives when $|\xi''| \le 1 + |\xi'|$

$$|P_{0(\beta)}^{(\alpha)}(x,\xi)| \le C_{\alpha,\beta} |\xi''| (1 + |\xi|)^{-|\alpha|}, \quad \text{if } \alpha'' = 0;$$

when $|\xi''| > 1 + |\xi'|$ this follows from the fact that $P \in S_{1,0}^1$.

Let g be the metric

$$(11.6.8) \qquad |dx|^2 + |d\xi'|^2/\langle\xi\rangle^2 + |d\xi''|^2/\langle\xi''\rangle^2,$$

with the standard notation $\langle\xi\rangle = (1 + |\xi|^2)^{\frac{1}{2}}$, for example. This is a σ temperate metric (see Hörmander [2, Section 18.5]), and the weights $M(\xi) = \langle\xi\rangle$ and $m(\xi) = \langle\xi''\rangle$ are g, σ temperate. Hence operators in $S(1,g)$ are continuous in all $H_{(s)}$ spaces. We have proved that $(11.6.1)$ implies that

$$P_0 \in S(m,g),$$

for this means that

$$|P_{0(\beta)}^{(\alpha)}(x,\xi)| \le C_{\alpha,\beta} \langle\xi''\rangle^{1-|\alpha''|} \langle\xi\rangle^{-|\alpha'|},$$

which follows from the fact that $P_0 \in S_{1,0}^1$ when $\alpha'' \ne 0$.

By $(11.6.6)$ we have

$$(11.6.9) \qquad \langle\xi''\rangle \le C(|P_0(x,\xi)| + 1), \quad \text{if } x \in U, \ \xi \in \Gamma, \ \xi'' \in \Gamma''.$$

Let $\chi \in S(m^\mu, g)$ and

$$(11.6.10) \qquad \operatorname{supp}\chi \subset U \times \widehat{\Gamma}, \quad \widehat{\Gamma} = \{\xi \in \Gamma, \ \xi'' \in \Gamma'', \ |\xi''| > 2C\},$$

which implies that $|P_0(x,\xi)| \geq 1$, hence $\langle \xi'' \rangle \leq 2C|P_0(x,\xi)|$ if $(\xi',\xi'') \in \widehat{\Gamma}$. (Later on we shall choose χ as the product of a symbol in $S_{1,0}^0$ with support in $U \times \Gamma$ and a symbol in the ξ'' variables with support in Γ''.) Then

$$(11.6.11) \qquad \Psi(x,\xi) = \chi(x,\xi)/P_0(x,\xi) \in S(m^{\mu-1}, g),$$

$$(11.6.12) \qquad \Psi(x,D)P_0(x,D) - \chi(x,D) \in \mathrm{Op}\, S(m^{\mu-1}, g),$$

or more precisely,

$$\Psi(x,D)P_0(x,D) - \chi(x,D) - \mathrm{Op} \sum_{0<|\alpha|<k} i^{-|\alpha|}\Psi^{(\alpha)}(x,D)P_{0(\alpha)}(x,D)/\alpha! \in \mathrm{Op}\, S(m^{\mu-k}, g).$$

The terms in the sum are in $\mathrm{Op}\, S(m^{\mu-|\alpha|}, g)$. Now we can write

$$P(x,D) = P_0(x,D) + \sum_1^\nu x_j Q_j(x,D),$$

where $Q_j \in S_{1,0}^1 \subset S(M,g)$. If we multiply by $\Psi(x,D)$, use (11.6.12), and the fact that

$$[\Psi(x,D), x_j]Q_j(x,D) \in \mathrm{Op}\, S(m^{\mu-1}, g), \qquad j \leq \nu,$$

because $\partial\Psi/\partial\xi_j \in S(m^{\mu-1}M^{-1}, g)$, $j \leq \nu$, we obtain if the composition series is broken off when the remainder is of weight $m^{\mu-k}$

$$\chi(x,D) = \Psi(x,D)P(x,D) + \Psi_1(x,D) + \Psi_2(x,D) - \sum_1^\nu x_j \Psi(x,D)Q_j(x,D).$$

Here $\Psi_1 \in S(m^{\mu-1}, g)$, $\Psi_2 \in S(m^{\mu-k}, g)$ and $\mathrm{supp}\,\Psi_1 \subset \mathrm{supp}\,\chi$. The sum does not contribute to the restriction to Y. We keep the term $\Psi_2(x,D)$ but repeat the argument with χ replaced by Ψ or Ψ_1, which gives that $\chi(x,D)$ is equal to $A_2(x,D)P(x,D)^2 + A_1(x,D)P(x,D) + A_0(x,D)$ apart from terms in $\mathrm{Op}\, S(m^{\mu-k}, g)$ and terms which do not contribute to the restriction to Y. Here $A_j \in S(m^{\mu-2}, g)$ and $\mathrm{supp}\, A_j \subset \mathrm{supp}\,\chi$, so the argument can be repeated again. After k steps it gives with a change of notation,

$$\gamma_Y^* \chi(x,D)u = \gamma_Y^* \sum_0^k \Psi_j(x,D)P(x,D)^j u, \qquad \Psi_j \in S(m^{\mu-k}, g).$$

If $\mu = k$ then $\Psi_j \in S(1,g)$ which implies that $\Psi_j(x,D)$ is continuous in $H_{(s)}$ for every s.

Choose $\chi_0 \in S_{1,0}^0$ with support in $U \times \Gamma$ so that $\chi_0 = 1$ at infinity in a conic neighborhood of $\{(0,\xi) \in WF(u); \xi'' = 0\}$, and choose $q(\xi'') \in C^\infty$ with $\mathrm{supp}\, q \subset \{\xi'' \in \Gamma''; |\xi''| > 2C\}$ so that q is homogeneous of degree k at ∞ and $\neq 0$ in the direction η. Then $q(D'')\chi_0(x,D) - \chi(x,D) \in \mathrm{Op}\, S(1,g)$ if

$$\chi(x,\xi) = \sum_{|\alpha|<k} i^{-|\alpha|}q^{(\alpha)}(\xi'')\chi_{0(\alpha)}(x,\xi)/\alpha! \in S(m^k, g).$$

The proof of (11.6.5) also proves that $P^j\chi_0(x,D)u \in H_{(s)}^{\mathrm{loc}}$ when $j \leq k$, for $P^j u \in H_{(s)}^{\mathrm{loc}}$ in $U \times \Gamma$ when $j \leq k$. Hence we conclude that

$$q(D'')\chi_0(x,D)^2 u - \chi(x,D)\chi_0(x,D)u \in H_{(s)}^{\mathrm{loc}}, \quad \gamma_Y^*\chi(x,D)\chi_0(x,D)u \in H_{(s-\nu/2)}^{\mathrm{loc}},$$

which proves that $q(D'')\gamma_Y^*\chi_0(x,D)^2 u \in H_{(s-\nu/2)}^{\mathrm{loc}}$. This implies that $\gamma_Y^*\chi_0(x,D)^2 u \in H_{(s+k-\nu/2)}^{\mathrm{loc}}$ at $(0,\eta)$.

For $v = u - \chi_0(x,D)^2 u$ we have $(0,\xi) \notin WF(u)$ if $\xi'' = 0$, for $1 - \chi_0 = 0$ in a conic neighborhood. Thus $v \in H_{(s+k)}^{\mathrm{loc}}$ then, and since (11.6.7) remains valid with u replaced by v, it follows from Lemma 11.6.1 that $\gamma_Y^* v \in H_{(s+k-\nu/2)}^{\mathrm{loc}}$ at $(0,\eta)$. The proof is complete.

In our application we shall have a second order operator so we restate Theorem 11.6.2 for that case:

Theorem 11.6.2'. *Assume that $Q \in \Psi_{1,0}^2(X)$ is properly supported and has a homogeneous principal symbol q such that*

$$(11.6.1)' \qquad q = 0 \quad \text{on the conormal bundle } N^*(Y) \setminus 0.$$

This implies that the base projection $q'_\xi(x,\xi)$ of the Hamilton vector is tangent to Y when $(x,\xi) \in N^*(Y) \setminus 0$. Let $u \in H_{(s)}^{\mathrm{loc}}(X)$ where $s > \nu/2$, let $(y,\eta) \in T^*(Y) \setminus 0$, and assume that

$$(11.6.2)' \qquad Q^j u \in H_{(s-j)}^{\mathrm{loc}}, \quad 0 \leq j \leq k, \quad \text{at } (\gamma_Y(y),\xi) \text{ when } {}^t\gamma'_Y(y)\xi \in \mathbf{R}\eta.$$

Then it follows that $\gamma_Y^* u \in H_{(s+k-\nu/2)}^{\mathrm{loc}}$ at (y,η) if

$$(11.6.3)' \qquad q(y,\xi) \neq 0 \quad \text{when } (y,\xi) \in WF(u) \text{ and } {}^t\gamma'_Y(y)\xi = \eta,$$
$$(11.6.4)' \qquad \langle q'_\xi(y,\xi),\eta \rangle \neq 0 \quad \text{when } (y,\xi) \in WF(u) \text{ and } {}^t\gamma'_Y(y)\xi = 0.$$

Proof. Let E be an elliptic properly supported operator in $\Psi_{1,0}^{-1}$ and set $P = EQ$. Then P and u satisfy (11.6.2), for it follows by induction that

$$P^j u = \sum_{l \leq j} A_{jl} Q^l u, \qquad A_{jl} \in \Psi_{1,0}^{-l},$$

for $EA_{jl} \in \Psi_{1,0}^{-l-1}$ and $E[Q, A_{jl}] \in \Psi_{1,0}^{-l}$. If e is the principal symbol of E then $e \neq 0$ and $p = eq$, which implies $p'_\xi = eq'_\xi$ when $p = 0$. Hence (11.6.1), (11.6.3), (11.6.4) are fulfilled, which proves the statement.

When specializing Theorem 11.6.2' to an analogue of Lemma 11.5.9 we need some elementary facts on the regularity of tensor products:

Lemma 11.6.3. *Let $u \in H_{(s)}^{\mathrm{loc}}$ at $x \in \mathbf{R}^n$ and $v \in H_{(t)}^{\mathrm{loc}}$ at $y \in \mathbf{R}^n$. Then it follows that $w = u \otimes v \in H_{(r)}^{\mathrm{loc}}$ at $(x,y) \in \mathbf{R}^{2n}$ if $r \leq s$, $r \leq t$, and $r \leq s + t$. We have $WF(w) \subset (WF(u) \cup 0) \times (WF(v) \cup 0)$, and $u \otimes v \in H_{(s+t)}^{\mathrm{loc}}$ at $((x,y),(\xi,\eta))$ if $\xi \neq 0, \eta \neq 0$.*

Proof. After multiplication of u and v with cutoff functions with support close to x and y we may assume that $u \in H_{(s)}(\mathbf{R}^n)$ and that $v \in H_{(t)}(\mathbf{R}^n)$. Then $\hat{w}(\xi,\eta) = \hat{u}(\xi)\hat{v}(\eta)$ is rapidly decreasing in a conic neighborhood of (ξ,η) if $\xi \neq 0$ (resp. $\eta \neq 0$) and $\hat{u}$ (resp. $\hat{v}$) is rapidly decreasing in a conic neighborhood of ξ (resp. η). This proves the statement on $WF(w)$. Since

$$(2\pi)^{-2n} \iint |\hat{w}(\xi,\eta)|^2 (1 + |\xi|^2)^s (1 + |\eta|^2)^t \, d\xi \, d\eta = \|u\|_{(s)}^2 \|v\|_{(t)}^2 < \infty,$$

it follows that $\iint_\Gamma |\hat{w}(\xi,\eta)|^2 (1 + |\xi|^2 + |\eta|^2)^{s+t} \, d\xi \, d\eta < \infty$ if Γ is a cone where $C^{-1}|\xi| \leq |\eta| \leq C|\xi|$, which proves the last statement. The first follows since $(1 + |\xi|^2 + |\eta|^2)^r \leq (1 + |\xi|^2)^s (1 + |\eta|^2)^t$. This is obvious when $s, t \geq 0$, since $(1 + |\xi|^2)(1 + |\eta|^2) \geq 1 + |\xi|^2 + |\eta|^2$ and when $s < 0$ or $t < 0$ since $1 + |\xi|^2 + |\eta|^2 \geq 1 + \max(|\xi|^2, |\eta|^2)$, which completes the proof.

Theorem 11.6.4. *Let $P(x,\xi)$ be a real quadratic form in $\xi \in \mathbf{R}^n$ with coefficients in $C^\infty(\mathbf{R}^n)$, let $\xi_1 \in \mathbf{R}^n \setminus 0$, $P(0,\xi_1) = 0$, $a = P'_\xi(0,\xi_1) \neq 0$, and let Γ be a closed cone in $\mathbf{R}^n \setminus 0$ such that $\langle a, \theta \rangle \neq 0$, $\theta \in \Gamma$. Let $s > n/2$, and let s_i, t_i be real numbers such that $s_i + t_j = s - i - j$, when $0 \leq i \leq k$, $0 \leq j \leq k$. Then there exist conic neighborhoods $V_\pm$ of $(0, \pm\xi_1)$ such that $u_+ u_- \in H^{\mathrm{loc}}_{(s+k-n/2)}$ at $0 \times \Gamma$ if*

$$ WF(u_\pm) \subset V_\pm, P(x,D)^j u_+ \in H^{\mathrm{loc}}_{(s_j)}, P(x,D)^j u_- \in H^{\mathrm{loc}}_{(t_j)} \text{ for } 0 \leq j \leq k. $$

Proof. We may assume that $u_\pm \in \mathcal{E}'(U)$, where U is a small neighborhood of 0, and that $P(x,\xi)$ is defined for $x \in \mathbf{R}^n$ with constant coefficients outside U. Set $w = u_+ \otimes u_-$. With P_1 and P_2 denoting the differential operators in $\mathbf{R}^n \times \mathbf{R}^n$ such that P_1 acts as P on $f(x,y)$ as a function of x for fixed y while P_2 acts on $f(x,y)$ as a function of y for fixed x, it is clear that

$$ P_1^i P_2^j w = (P^i u_+) \otimes (P^j u_-). $$

Hence it follows from Lemma 11.6.3 that $P_1^i P_2^j u \in H^{\mathrm{loc}}_{(s_i + t_j)} = H^{\mathrm{loc}}_{(s-i-j)}$ at $((x,y),(\xi,\eta))$ if $\xi \neq 0$ and $\eta \neq 0$. If $Q = P_1 - P_2$ this implies that

(11.6.13) $Q^\kappa w \in H^{\mathrm{loc}}_{(s-\kappa)}$ at $((x,y),(\xi,\eta))$ if $0 \leq \kappa \leq k$, $\xi \neq 0$, $\eta \neq 0$.

We identify the diagonal $\Delta = \{(x,x); x \in \mathbf{R}^n\}$ with $\mathbf{R}^n$ through $\gamma_\Delta(x) = (x,x)$, $x \in \mathbf{R}^n$. Then ${}^t\gamma'_\Delta(x)(\xi,\eta) = \xi + \eta$. The conormal bundle N_Δ^* of Δ is defined by $\xi + \eta = 0$, so the principal symbol $q(x,x,\xi,\eta) = p(x,\xi) - p(x,\eta)$ vanishes there, and (11.6.13) is applicable at N_Δ^*.

Let $\overline{V}_\pm \subset U \times \overline{\Gamma}_\pm$, where $\Gamma_\pm$ are small open conic neighborhoods of $\pm\xi_1$ such that $\overline{\Gamma}_\pm \cap \Gamma = \{0\}$. Then we have (cf. (11.5.8))

(11.6.14)
$$ |\xi| \leq C_1 |\eta| \quad \text{if } \xi \in \Gamma_+ \text{ and } \xi + \eta \in \Gamma, $$
$$ |\eta| \leq C_1 |\xi| \quad \text{if } \eta \in \Gamma_- \text{ and } \xi + \eta \in \Gamma. $$

If $(\xi,\eta) \in WF(w)$ and $0 \neq \xi + \eta \in \Gamma$ it follows that $\xi \neq 0$, $\eta \neq 0$, and that $|\eta|/C_1 \leq |\xi| \leq C_1|\eta|$, so (11.6.13) is valid.

If we write $p(x,\xi) = \langle A(x)\xi, \xi \rangle$ with a symmetric matrix $A(x)$ then $q(x,x,\xi,\eta) = \langle A(x)(\xi - \eta), \xi + \eta \rangle$. By hypothesis $2A(0)\xi_1 = a$ where $\langle a, \theta \rangle \neq 0$, if $\theta \in \Gamma$. Hence

$$ \langle A(0)(\xi - \eta), \theta \rangle \neq 0 \quad \text{if } 0 \neq \theta \in \Gamma, \ \xi \in \Gamma_+, \ \eta \in \Gamma_-, $$

provided that $\Gamma_\pm$ are small enough. If $\theta = {}^t\gamma'_\Delta(0)(\xi,\eta) = \xi + \eta$ this means that $q(0,0,\xi,\eta) \neq 0$, which implies (11.6.3)'. If $\xi + \eta = 0$ then

$$ q'_{\xi,\eta}(0,0,\xi,\eta) = (p'_\xi(0,\xi), -p'_\eta(0,\eta)) = \gamma'_\Delta(0)p'_\xi(0,\xi) $$

is a tangent vector of Δ, and

$$ \langle p'_\xi(0,\xi), \theta \rangle = 2\langle A(0)\xi, \theta \rangle \neq 0 \quad \text{if } 0 \neq \theta \in \Gamma, \ \xi \in \Gamma_+, $$

which verifies (11.6.4)' since $\xi \in \Gamma_+$ if $(0,0,\xi,-\xi) \in WF(u_+ \otimes u_-)$. The statement is now a consequence of Theorem 11.6.2'.

End of proof of Lemma 11.5.8. We have stated Theorem 11.6.2 as a qualitative regularity property rather than an estimate as in Lemma 11.5.9. However, it serves the same purpose (and is equivalent by the closed graph theorem), so the proof of Lemma 11.5.8

given in a special case by means of Lemma 11.5.9 extends to the variable coefficient case if we appeal to Theorem 11.6.4 instead. The only problem is that in Theorem 11.6.4 we had to assume that $s > n/2$ whereas Lemma 11.5.9 is applicable when $s > 0$. The reason for that is that the special nature of a tensor product is not used in the proof of Theorem 11.6.4 via Theorem 11.6.2 whereas the proof of Lemma 11.5.9 exploited Lemma 8.3.2. The problem only becomes acute at the end of the proof of Lemma 11.5.8 where $s > k - 1 + n/2$ may not imply $3s - k - n/2 - 1 > n/2$, for this requires that $2k - 4 + n/2 \geq 0$. This is true when $n \geq 4$. If $n = 2$ we can change coordinates to make the coefficients of $P(x, \xi)$ constant. However, we have not proved Lemma 11.5.8 when $n = 3$ and $k = 1$, that is, $s \leq 1 + n/2 = 5/2$, which means that for $n = 3$ we have only proved Theorem 11.5.10 when $s > 5/2$. M. Beals claims (personal communication) that his proof of Lemma 1.10 in [2] can be modified so that this restriction is removed.

APPENDIX ON PSEUDO-RIEMANNIAN GEOMETRY

A.1. Basic definitions. A C^∞ manifold M, of dimension n, is called pseudo-Riemannian if a non-degenerate quadratic form g is defined in the tangent bundle; M is called Riemannian if g is positive definite. Thus it is assumed that in local coordinates we have a quadratic form

$$\sum_{j,k=1}^n g_{jk}(x)dx^j dx^k, \quad g_{jk} = g_{kj},$$

which transforms so that it is invariantly defined on the tangent bundle. The quadratic form gives a bijective map of the fiber $T_x(M)$ of the tangent bundle at x on the fiber $T_x^*(M)$ of the cotangent bundle; in coordinates $(t^1,\ldots,t^n) \mapsto (t_1,\ldots,t_n)$; $t_j = \sum_1^n g_{jk}(x)t^k$. The inverse matrix is denoted by g^{jk}. From the invariance of the quadratic form it follows that if $\tilde{g}$ is g expressed in new coordinates $\tilde{x}$, then $|g|(\det \partial x/\partial \tilde{x})^2 = |\tilde{g}|$, where we have used the abbreviated notation $|g| = |\det g|$. This means that $\sqrt{|g|}\, dx$, where dx is the Lebesgue measure in local coordinates, is an invariant measure on M, which we shall sometimes denote by $\mathrm{vol}(dx)$. An invariantly defined "Laplace" operator Δ is defined in local coordinates by

$$\Delta u = |g|^{-\frac{1}{2}} \sum_{j,k=1}^n \partial_j(|g|^{\frac{1}{2}}g^{jk}\partial_k u),$$

for if $u, v \in C_0^\infty$ then the corresponding bilinear form

$$(u,v) \mapsto \int \sum_{j,k=1}^n g^{jk}\partial_j u \partial_j v \, \mathrm{vol}(dx)$$

is invariantly defined since the integrand and the measure are. In the Riemannian case it is the Laplace-Beltrami operator with respect to the metric, and when g has Lorentz signature it is of course the d'Alembertian with variable coefficients.

A.2. Geodesic coordinates and curvature. Let us assume for a moment that we have a Riemannian manifold and try to find the shortest curve joining two points x_0 and x_1 in the same coordinate patch. This means that we want to find a C^∞ function $[0,1] \ni t \mapsto x(t) \in M$ with $x(0) = x_0$ and $x(1) = x_1$, such that the arc length s of the curve, defined independently of the parametrisation by

$$s = \int_0^1 \Big(\sum_{j,k=1}^n g_{jk}(x(t))dx^j(t)/dt\, dx^k(t)/dt \Big)^{\frac{1}{2}} dt = \int_0^1 \Big(\sum_{j,k=1}^n g_{jk}(x)dx^j dx^k \Big)^{\frac{1}{2}}$$

is minimized. By Schwarz' inequality we have

$$s^2 \le \int_0^1 \sum_{j,k=1}^n g_{jk}(x)dx^j/dt dx^k/dt\, dt$$

with equality if and only if the integrand is constant, that is, the parameter t is proportional to the arc length. This can always be achieved, so the minimum problem is equivalent to minimizing

$$(A.2.1) \qquad I = \int_0^1 \sum_{j,k=1}^n g_{jk}(x) dx^j /dt dx^k /dt \, dt$$

for all smooth curves from x_0 to x_1. This has the advantage that one can expect a unique solution if the points are sufficiently close. If the minimum is attained, then the Euler equations

$$(A.2.2) \qquad 2\frac{d}{dt} \sum_{k=1}^n g_{jk}(x) dx^k /dt = \sum_{i,k=1}^n \partial g_{ik}(x)/\partial x^j \, dx^i /dt dx^k /dt, \quad j = 1, \ldots, n,$$

must be valid for the *geodesics*. These equations make sense also on a pseudo-Riemannian manifold so we can now drop the hypothesis that (g_{jk}) is positive definite.

By the basic existence theorems for ordinary differential equations the Euler equations (A.2.2) have a unique solution for small t with prescribed initial data $x = x_0, dx/dt = v \in T_{x_0}$ when $t = 0$. The solution $x(x_0, t, v)$ is a C^∞ function and depends on tv rather than on t and v, because (A.2.2) is independent of t and homogeneous in dt. This means that $x(x_0, t, v) = X(x_0, tv)$ where X is a C^∞ function from a neighborhood of the zero section in $T(M)$ to M with $X(x_0, 0) = x_0$ and $\partial_v X(x_0, 0)$ equal to the identity. By the inverse function theorem we can therefore introduce X as local coordinates in a neighborhood of x_0. If we do that for fixed x_0, and denote the metric in the new coordinates by $\sum G_{jk}(X) dX^j \, dX^k$, the fact that $t \mapsto tX$ is a solution of A.2.2 for every x means that

$$(A.2.2)' \qquad 2\frac{\partial}{\partial t} \sum_{k=1}^n G_{jk}(tX) X^k = \sum_{i,k=1}^n (\partial_j G_{ik})(tX) X^i X^k, \quad j = 1, \ldots, n.$$

If we multiply by X^j and sum, it follows that $\sum G_{jk}(tX) X^j X^k$ is independent of t, thus $\sum (G_{jk}(tX) - G_{jk}(0)) X^j X^k = 0$. The derivative with respect to X^j is

$$2\sum (G_{jk}(tX) - G_{jk}(0)) X^k + t \sum (\partial_j G_{ik})(tX) X^i X^k$$

which must therefore vanish. Combined with (A.2.2)' this gives

$$2S_j + 2t\partial S_j /\partial t = 0, \qquad S_j = \sum (G_{jk}(tX) - G_{jk}(0)) X^k.$$

Thus tS_j is independent of t, hence equal to 0, so we obtain

$$(A.2.3) \qquad \sum_1^n G_{jk}(X) X^k = \sum_1^n G_{jk}(0) X^k, \quad j = 1, \ldots, n.$$

Conversely, if we have a coordinate system such that (A.2.3) is valid, then reversing the preceding argument shows that the straight lines through the origin (in the parameter space) are geodesics. In the Riemannian case we could make $G_{jk}(0)$ equal to the identity matrix by a linear change of coordinates, and correspondingly we could transform to a normal form in the case of other signatures. However, the notation will be simpler and more uniform if we leave $G_{jk}(0)$ arbitrary. The coordinate systems now obtained are called *geodesic coordinates*; they are uniquely determined up to a linear transformation (up to an orthogonal one if one has insisted on normal forms).

If we expand G in a Taylor series,

$$G = G^0 + G^1 + G^2 + \dots$$

where G^j is homogeneous of degree j in X, it follows from (A.2.3) that $G^1 = 0$, for the equations $\sum_k G^1_{jk}(X)X^k = 0$ imply that if we write $G^1_{jk}(X) = \sum_l G_{jkl}X^l$, then G_{jkl} is symmetric in the first two indices and antisymmetric in the last two. Thus

$$G_{jkl} = -G_{jlk} = -G_{ljk} = G_{lkj} = G_{klj} = -G_{kjl} = -G_{jkl}$$

which proves the claim. Note that since the G_{jk} are stationary at the center of a geodesic coordinate system, the Laplacian is given there by the same expression as when the coefficients are constant.

The first interesting term is therefore G^2. We shall write

$$G^2(X;Y) = \langle G^2(X)Y, Y \rangle,$$

which is a symmetric quadratic form in X as well as in Y and has the fundamental property

(A.2.4) $$\partial G^2(X;Y)/\partial Y = 0, \quad \text{if } Y = X.$$

In particular, $G^2(X;X) = 0$. Since the dimension of the space of quadratic forms in n variables is $n(n+1)/2$ and that of cubic forms is $n(n+1)(n+2)/6$, it is easily seen that the space of forms satisfying (A.2.4) is of dimension

$$(n(n+1)/2)^2 - n^2(n+1)(n+2)/6 = n^2(n^2-1)/12.$$

We can polarize G^2 to a 4-linear form $G^2(X_1, X_2; Y_1, Y_2)$ which is symmetric in X_1, X_2 as well as in Y_1, Y_2 and has the property

$$G^2(X;Y) = G^2(X, X; Y, Y).$$

Then (A.2.4) yields $G^2(X, X; X, Y) = 0$, hence

(A.2.4)' $$G^2(X_1, X_2; X_3, X_4) + G^2(X_3, X_1; X_2, X_4) + G^2(X_2, X_3; X_1, X_4) = 0.$$

From (A.2.4)' we easily obtain $G^2(X, X; Y, Y) = -2G^2(X, Y; X, Y)$, hence the symmetry

(A.2.5) $$G^2(X;Y) = G^2(Y;X).$$

When $n = 2$ the fact that the quadratic form $X \mapsto G^2(X;Y)$ degenerates on the line generated by Y shows that it must be a constant times the square of the determinant of X and Y. This proves again that (A.2.5) holds, for arbitrary n, and it follows that $G^2(X;Y)$ only depends on $X \wedge Y$. This is better expressed by introducing the Riemann-Christoffel curvature tensor, which is the 4-linear form defined by

(A.2.6) $$R(X_1, X_2; X_3, X_4) = 2G^2(X_1, X_4; X_2, X_3) - 2G^2(X_1, X_3; X_2, X_4).$$

Using (A.2.4)' we obtain

$$-6G^2(X_1, X_2; X_3, X_4) = R(X_1, X_4; X_2, X_3) + R(X_1, X_3; X_2, X_4),$$
$$-3G^2(X, X; Y, Y) = R(X, Y; X, Y).$$

We have a one to one correspondence between the two 4-linear forms $G^2(X_1, \ldots, X_4)$ and $R(X_1, \ldots, X_4)$ with the properties:

symmetry for G^2 when $X_1 \leftrightarrow X_2$ or $X_3 \leftrightarrow X_4$ or $(X_1, X_2) \leftrightarrow (X_3, X_4)$; for R when $(X_1, X_2) \leftrightarrow (X_3, X_4)$;

antisymmetry for R when $X_1 \leftrightarrow X_2$ or $X_3 \leftrightarrow X_4$;

circular antisymmetry for both G^2 and R as described in (A.2.4)'. For R this is called the first Bianchi identity.

For R the symmetries mean that $R(X_1, X_2; X_3, X_4)$ is a symmetric bilinear form in $X_1 \wedge X_2$ and $X_3 \wedge X_4$ with an additional symmetry expressed by the Bianchi identity.

We shall now write down the conventional expression for the Riemann curvature tensor in arbitrary local coordinates $x^1, \ldots, x^n$. With the Christoffel symbols of the first kind defined by

$$\Gamma_{ijk} = \tfrac{1}{2}(\partial_j g_{ik} + \partial_i g_{jk} - \partial_k g_{ij})$$

the curvature tensor has the covariant components

$$(A.2.7) \qquad R_{ijkl} = \partial_l \Gamma_{ikj} - \partial_k \Gamma_{ilj} + \sum_{r,s=1}^{n}(\Gamma_{ilr}\Gamma_{kjs} - \Gamma_{ikr}\Gamma_{ljs})g^{rs}.$$

This is true at the center of a geodesic coordinate system, for the non-linear terms vanish there and the linear terms

$$(A.2.7)' \qquad \tfrac{1}{2}(\partial_i \partial_l g_{kj} + \partial_j \partial_k g_{il} - \partial_j \partial_l g_{ik} - \partial_i \partial_k g_{lj})$$

agree with (A.2.6). If we introduce some new coordinates $y = x + O(|x|^2)$ at the origin of the geodesic coordinates x, then the new metric tensor has the components

$$\tilde{g}_{ij}(y) = \sum_{r,s=1}^{n} g_{rs}(x)\partial x_r/\partial y_i \, \partial x_s/\partial y_j.$$

Since $dg_{rs} = 0$ at 0 we obtain there if $(g_{rs}(0))$ has diagonal form

$$\partial \tilde{g}_{ij}/\partial y_k = \partial g_{ij}/\partial x_k + \sum_{r=1}^{n} g_{rr}(0)(\partial^2 x_r/\partial y_i \partial y_k \, \partial x_r/\partial y_j + \partial x_r/\partial y_i \partial^2 x_r/\partial y_j \partial y_k) + O(|y|^2),$$

so the new Christoffel symbols are

$$\tilde{\Gamma}_{ijk} = \Gamma_{ijk} + \sum_{r=1}^{n} g_{rr}(0)\partial^2 x_r/\partial y_i \partial y_j \, \partial x_r/\partial y_k + O(|y|^2) = g_{kk}(0)\partial^2 x_k/\partial y_i \partial y_j + O(|y|).$$

Hence

$$\partial \tilde{\Gamma}_{ikj}/\partial y_l + \sum_{t,s=1}^{n} \tilde{\Gamma}_{ilt}\tilde{\Gamma}_{kjs}g^{st}(0) = \partial \Gamma_{ikj}/\partial x_l + g_{jj}(0)\partial^3 x_j/\partial y_i \partial y_k \partial y_l$$

$$+ \sum_{r=1}^{n} g_{rr}(0)(\partial^2 x_r/\partial y_i \partial y_k \partial^2 x_r/\partial y_j \partial y_l + \partial^2 x_r/\partial y_i \partial y_l \partial^2 x_r/\partial y_k \partial y_j).$$

When we interchange k and l and subtract, it follows that $\tilde{R}_{ijkl} = R_{ijkl}$ if $\tilde{R}_{ijkl}$ is defined by (A.2.7).

Remark. The derivation above is very close to that outlined by Riemann, and derives historically from the Legendre theorem on angular excess for a geodesic triangle on a surface. Note that the curvature tensor is always given by $(A.2.7)'$ at a point where $dg_{jk} = 0$ for all j, k even if the coordinates are not geodesic there. However, $(A.2.6)$ is not always valid then.

From the curvature tensor one obtains the symmetric *Ricci tensor* by contracting with respect to the first and third index,

$$R_{jl} = \sum_{i,k=1}^{n} g^{ik} R_{ijkl},$$

and the contraction of R is the *scalar curvature*

$$S = \sum_{j,l=1}^{n} g^{jl} R_{jl} = \sum_{i,j,k,l=1}^{n} g^{ik} g^{jl} R_{ijkl}.$$

In the Riemannian case with $n = 2$ we have $R_{jk} = K g_{jk}$ where K is the Gaussian curvature, so $S = 2K$. Clearly the Ricci tensor is not interesting then. However, since the set of Ricci tensors which can occur at a point where g is prescribed is linear and invariant under the orthogonal group, it must for any n be either the multiples of the identity or the set of all symmetric tensors or the set of symmetric tensors with zero trace. If we take for M the direct product of $\mathbf{R}$ and S^{n-1}, say, with the natural metrics, and $n \geq 3$, the computation below shows that the Ricci tensor is positive and not a multiple of the identity, so it can be completely arbitrary.

We shall end the section with computing the curvature for the simplest example, the sphere. For the sphere with radius R in $\mathbf{R}^{n+1}$ with center at $(0, \ldots, 0, R)$, the stereographic projection to the tangent plane at 0 gives the parametrisation

$$\mathbf{R}^n \ni x \mapsto (4R^2 x, 2R|x|^2)/(4R^2 + |x|^2) \in \mathbf{R}^{n+1}.$$

With this parametrisation the metric is

$$(A.2.8) \qquad\qquad |dx|^2/(1 + |x/2R|^2)^2.$$

In particular, at $x = 0$ it is $|dx|^2(1 - |x|^2/2R^2) + O(|x|^4))$. At the origin the metric is therefore Euclidean and the coefficients are stationary, so by $(A.2.7)'$ we obtain

$$(A.2.9) \qquad\qquad R_{ijkl} = (g_{ik}g_{jl} - g_{il}g_{jk})/R^2$$

with these coordinates and therefore for arbitrary coordinates in view of the invariance. (We cannot use $(A.2.6)$ since the coordinates are not geodesic.) The Ricci tensor and scalar curvature are

$$(A.2.10) \qquad\qquad R_{jl} = (n-1)g_{jl}/R^2, \qquad S = n(n-1)/R^2.$$

For later reference we finally observe, that it follows from $(A.2.8)$ by a translation of x that also $|dx|^2/(1 + |x + a|^2/4R^2)^2$ is the metric for a parametrisation of the sphere with radius R. When $x = 0$ the metric is $|dx|^2/(1 + |a|^2/4R^2)^2 + O(|x|)$. We therefore take $x/(1 + |a|^2/4R^2)$ as new coordinates and conclude that also the metric

$$(A.2.11) \qquad |dx|^2/(1 + 2\langle \alpha, x \rangle + \beta|x|^2)^2, \qquad \alpha = a/4R^2, \quad \beta = |\alpha|^2 + 1/4R^2,$$

is a metric for the sphere of radius R, so the corresponding curvature is still given by $(A.2.9)$, $(A.2.10)$. There are no other conditions on α and β.

A.3. Conformal changes of metric. The curvature tensor can be written in the form

$$(A.3.1) \qquad R_{ijkl} = W_{ijkl} + \frac{R_{ik}g_{jl} - R_{il}g_{jk} + R_{jl}g_{ik} - R_{jk}g_{il}}{n-2} - \frac{S(g_{ik}g_{jl} - g_{il}g_{jk})}{(n-1)(n-2)};$$

this defines the *Weyl tensor* W. The coefficients in the formula are chosen so that the contraction of W over any pair of indices is equal to 0; note that W also has the symmetries of R. From (A.2.9) and (A.2.10) it follows that the Weyl tensor for a sphere is equal to 0. We shall now prove that if one makes a conformal transformation, replacing g by $\tilde{g} = e^{2\varphi}g$, then the new Ricci tensor, scalar curvature, and Weyl tensor are

$$(A.3.2) \quad \widetilde{R}_{jk} = R_{jk} - (n-2)\partial_j\partial_k\varphi + (n-2)\partial_j\varphi\partial_k\varphi - (\Delta\varphi + (n-2)\Delta_1\varphi)g_{jk},$$

$$(A.3.3) \qquad \widetilde{S} = e^{-2\varphi}(S - 2(n-1)\Delta\varphi - (n-1)(n-2)\Delta_1\varphi),$$

$$(A.3.4) \qquad \widetilde{W}_{ijkl} = e^{-2\varphi}W_{ijkl}.$$

Here the derivatives are computed at the center of the old geodesic coordinates (or $\partial_j\partial_k$ should be understood as covariant differentiation), and $\Delta_1\varphi = \sum_{j,k=1}^{n} g^{jk}\partial_j\varphi\partial_k\varphi$ is "the first Beltrami operator".

The statements (A.3.2), (A.3.3), (A.3.4) are rational identities involving g_{jk} and the derivatives of order ≤ 2. Their validity is therefore independent of the signature of (g_{jk}) so we may assume that we have a Riemannian metric. It suffices to prove the transformation laws at 0 assuming that the coordinates are geodesic with respect to g, that $g_{jk}(0) = \delta_{jk}$, and that $\varphi(0) = 0$. Choose $\alpha \in \mathbf{R}^n$ so that $\varphi(x) = -2\langle x, \alpha\rangle + O(|x|^2)$, and choose β as in (A.2.11) for some $R > 0$. We have

$$\varphi(x) = -\log(1 + 2\langle x, \alpha\rangle + \beta|x|^2) + q(x) + O(|x|^3),$$

where $q(x) = \sum_{i,j=1}^{n} q_{ij}x_i x_j$ is a quadratic form; thus

$$\varphi(x) = -2\langle x, \alpha\rangle - \beta|x|^2 + 2\langle x, \alpha\rangle^2 + q(x) + O(|x|^3).$$

Write $g_{jk}(x) = \delta_{jk} + G_{jk}(x) + O(|x|^3)$ where G_{jk} is a quadratic form. Then

$$\tilde{g}_{jk}(x) = \delta_{jk}/(1 + 2\langle\alpha, x\rangle + \beta|x|^2)^2 + 2q(x)\delta_{jk} + G_{jk}(x) + O(|x|^3).$$

For the metric tensor $\delta_{jk}/(1+2\langle x, \alpha\rangle + \beta|x|^2)^2$ the curvature is given by (A.2.9), (A.2.10). If $y_1, \ldots, y_n$ are geodesic coordinates for this metric, then $\tilde{g}_{jk}$ expressed in the y coordinates will in addition to the geodesic metric of the sphere with radius R consist of the terms $2q(y)\delta_{jk} + G_{jk}(y) + O(|y|^3)$. By (A.2.7)' the curvature tensor will therefore be the sum of that for the sphere, the curvature tensor R_{ijkl} corresponding to the metric tensor $\delta_{jk} + G_{jk}$, and the curvature tensor R^q corresponding to $(1 + 2q)\delta_{jk}$, which by (A.2.7)' is

$$(A.3.5) \qquad R^q_{ijkl} = 2(q_{il}\delta_{jk} + q_{jk}\delta_{il} - q_{jl}\delta_{ik} - q_{ik}\delta_{jl}).$$

Direct calculation gives

$$(A.3.6) \qquad R^q_{jl} = 2((2-n)q_{jl} - \mathrm{Tr}\, q\delta_{jl}), \qquad S^q = 2(2 - 2n)\,\mathrm{Tr}\, q = 2(1-n)\Delta q.$$

Recalling (A.2.10) we obtain

$$\widetilde{S} = n(n-1)/R^2 + S + 2(1-n)(\Delta\varphi + 2n\beta - 4|\alpha|^2),$$

which yields (A.3.3) if we recall that $|2\alpha| = |\partial\varphi(0)|$ and that $\beta = |\alpha|^2 + 1/4R^2$. The auxiliary variable R drops out as it should. For the Ricci tensor we obtain

$$\widetilde{R}_{jl} = (n-1)\delta_{jl}/R^2 + R_{jl} + 2((2-n)q_{jl} - \operatorname{Tr} q\delta_{jl});$$
$$2q_{jl} = \partial_j\,\partial_l\varphi + 2\beta\delta_{jl} - \partial_j\varphi\partial_l\varphi = \partial_j\,\partial_l\varphi + \tfrac{1}{2}|\partial\varphi|^2\delta_{jl} + \delta_{jl}/2R^2 - \partial_j\varphi\partial_l\varphi,$$
$$2\operatorname{Tr} q = \Delta\varphi + (\tfrac{1}{2}n - 1)|\partial\varphi|^2 + n/2R^2.$$

Again the radius R drops out and we obtain

$$\widetilde{R}_{jl} = R_{jl} + (2-n)\partial_j\,\partial_l\varphi + (2-n)|\partial\varphi|^2\delta_{jl} - (2-n)\partial_j\varphi\partial_l\varphi - \Delta\varphi\delta_{jl},$$

which confirms (A.3.2).

It remains to show that the Weyl tensor is unchanged. This requires that we prove that the Weyl tensor is 0 for the sphere and that the Weyl tensor part of R^q is also 0. The statement on the sphere is of course a special case, already recorded, so we just have to check using (A.3.5) and (A.3.6) that R^q is reconstructed correctly from its contractions when $\operatorname{Tr} q = 0$. Then we have

$$W_{ijkl} = 2(q_{il}\delta_{jk} + q_{jk}\delta_{il} - q_{jl}\delta_{ik} - q_{ik}\delta_{jl}) + 2(q_{ik}\delta_{jl} - q_{il}\delta_{jk} + q_{jl}\delta_{ik} - q_{jk}\delta_{il}) = 0,$$

which completes the proof.

The Ricci tensor determines the scalar curvature so the terms in (A.3.1) are not independent of each other. It is therefore more natural to rewrite (A.3.1) using the *traceless Ricci tensor* $B_{ij} = R_{ij} - (S/n)g_{ij}$. (This is always equal to 0 when $n = 2$.) Then (A.3.1) takes the form

$$(\text{A.3.1})' \qquad R_{ijkl} = W_{ijkl} + \frac{B_{ik}g_{jl} - B_{il}g_{jk} + B_{jl}g_{ik} - B_{jk}g_{il}}{n-2} + \frac{S(g_{ik}g_{jl} - g_{il}g_{jk})}{n(n-1)};$$

with $W = 0$ and without the term involving B it is also valid when $n = 2$. Since the Riemann curvature tensor takes its values in a space of dimension $n^2(n^2-1)/12$, it follows that the Weyl tensor takes its values in a space of dimension $n(n+1)/2$ less if $n \geq 3$, that is, dimension $n(n+1)(n+2)(n-3)/12$. In particular, the Weyl tensor always vanishes when $n = 3$. However, when $n > 3$ the Weyl tensor may be different from 0. Then it guarantees that there is no conformal metric which is *flat*, that is, has Riemann curvature tensor equal to 0.

We shall now examine how the Laplace operator is transformed when one passes to a conformal metric, and compare the result with the transformation law for the scalar curvature. With $\widetilde{\Delta}$ denoting the Laplacian for the metric $\tilde{g} = e^{2\varphi}g$, and with a constant a to be chosen later we have

$$\widetilde{\Delta}(e^{a\varphi}u) = e^{-n\varphi}g^{-\frac{1}{2}}\sum_{j,k=1}^{n}\partial_j(e^{(n-2)\varphi}g^{\frac{1}{2}}g^{jk}\partial_k(e^{a\varphi}u))$$

$$= e^{(a-2)\varphi}(\Delta u + (n-2+2a)\sum_{j,k=1}^{n}g^{jk}\partial_j\varphi\partial_k u) + Fu,$$

where F does not depend on u. Now we choose $a = 1 - \frac{n}{2}$ so that the first order terms disappear. If the formula is applied with $u = e^{-a\varphi}$, it follows that

$$-F = e^{2(a-1)\varphi}\Delta(e^{-a\varphi}) = e^{(a-2)\varphi}(a^2\Delta_1\varphi - a\Delta\varphi),$$

where the last equality is justified using geodesic coordinates. By (A.3.3) we have

$$a\Delta_1\varphi - \Delta\varphi = \tfrac{1}{2}((2 - n)\Delta_1\varphi - 2\Delta\varphi) = (e^{2\varphi}\tilde{S} - S)/(2n - 2),$$

and since $-a/(2n - 2) = (n - 2)/(4n - 4)$ it follows that

$$\tilde{\Delta}(e^{a\varphi}u) = e^{(a-2)\varphi}(\Delta u + (n - 2)/(4n - 4)(e^{2\varphi}\tilde{S} - S)u), \text{ or}$$

(A.3.7) $\qquad \left(\tilde{\Delta} - \dfrac{\tilde{S}(n - 2)}{4n - 4}\right)(e^{\varphi(2-n)/2}u) = e^{-\varphi(n+2)/2}\left(\Delta - \dfrac{S(n - 2)}{4n - 4}\right)u.$

The operator $\Delta - S(n-2)/(4n-4)$ is called the *conformal Laplacian*. The transformation law (A.3.7) makes it easy to pass from solutions for one metric to solutions for another conformal one. Note that in a flat space, where $S = 0$, the conformal Laplacian is the standard Laplacian, so we shall be able to study its solutions by studying the conformal Laplacian in a conformally equivalent situation. In the Euclidean case one may for example use the stereographic projection to obtain a compact situation. In the following section we shall discuss a related conformal map of Minkowski space into the Einstein universe.

A.4. A conformal imbedding of Minkowski space. The $1+n$ dimensional Einstein universe is the product $\mathbf{R}\times S^n$ with the pseudo-Riemannian metric $dT^2 - dX^2$, where $T \in \mathbf{R}$ and $X \in S^n$ with dX^2 defined by the imbedding of S^n as the unit sphere in $\mathbf{R}^{n+1}$. Apart from two antipodal points we can parametrize S^n by

(A.4.1) $\qquad (0, \pi) \times S^{n-1} \ni (\alpha, \omega) \mapsto (\cos\alpha, \sin\alpha\,\omega) = X = (X_0, \vec{X}) \in \mathbf{R}^{n+1},$

the exceptional points being $(\pm 1, 0, \ldots, 0)$. Then

$$dX^2 = d\alpha^2 + \sin^2\alpha|d\omega|^2,$$

where $|d\omega|^2$ is the metric in S^{n-1} (a recursive definition). Now consider the map

(A.4.2) $\qquad \Psi : \mathbf{R} \times S^n \ni (T, X) \mapsto \dfrac{1}{\cos T + X_0}(\sin T, \vec{X}) \in \mathbf{R}^{1+n}.$

It is clear that Ψ is smooth (analytic) when $\cos T + X_0 > 0$. With the notation in (A.4.1) we can write

$$\Psi(T, X) = \dfrac{1}{\cos T + \cos\alpha}(\sin T, \sin\alpha\,\omega) = (t, r\omega), \text{ where}$$
$$r = \tfrac{1}{2}\left(\tan\dfrac{T + \alpha}{2} - \tan\dfrac{T - \alpha}{2}\right), \qquad t = \tfrac{1}{2}\left(\tan\dfrac{T + \alpha}{2} + \tan\dfrac{T - \alpha}{2}\right).$$

When $T = 0$ the map is the stereographic projection of S^n on the plane $X_0 = 0$ from the pole $(-1, 0, \ldots, 0)$. For general T, it is the projection from $(-\cos T, 0, \ldots, 0)$ defined when $X_0 + \cos T > 0$, that is, X is to the right of the projection point, and with range in the plane where $X_0 + \cos T = 1$.

To calculate an inverse of Ψ we set $\Omega = \cos T + \cos\alpha$ and use that

$$\sin T = \Omega t, \quad \sin\alpha = \Omega r.$$

Hence $1 = (\Omega - \cos T)^2 + \Omega^2 r^2 = \Omega^2(1 + r^2 - t^2) - 2\Omega\cos T + 1$; using the symmetry in α and T we therefore have

$$\cos T = \tfrac{1}{2}\Omega(1 + r^2 - t^2), \quad \cos\alpha = \tfrac{1}{2}\Omega(1 + t^2 - r^2).$$

Squaring one of these relations we obtain

$$\frac{2}{\Omega} = \sqrt{(1+r^2-t^2)^2 + 4t^2} = \sqrt{(1+t^2-r^2)^2 + 4r^2} = \sqrt{(1+(r-t)^2)(1+(r+t)^2)},$$

and $\alpha \in [0,\pi)$, $T \in (-\pi,\pi)$ are now uniquely determined. The inverse we have found is well defined and analytic in the whole of Minkowski space, which proves that Ψ is a bijection from $\{(T,X) \in \mathbf{R} \times S^n; |T| < \pi, X_0 + \cos T > 0\}$ to the whole Minkowski space. The condition $X_0 + \cos T > 0$ means that $\alpha + |T| < \pi$ if $X_0 = \cos \alpha$ with $\alpha \in [0,\pi)$.

It remains to verify that Ψ is conformal and compute the conformality factor; the relation to stereographic projection is just a promising start. Write $f = 1/\Omega$, thus

$$f = 1/(\cos T + \cos \alpha), \quad t = f \sin T, \quad \vec{x} = f \sin \alpha\, \omega,$$
$$(dt)^2 = (\sin T df + f \cos T dT)^2, \quad |d\vec{x}|^2 = (\sin \alpha df + f \cos \alpha d\alpha)^2 + f^2 \sin^2 \alpha |d\omega|^2,$$
$$(dT)^2 - |dX|^2 = dT^2 - d\alpha^2 - \sin^2 \alpha |d\omega|^2.$$

To eliminate $|d\omega|^2$ we must try the conformal factor f^2, so we form

$$(dt)^2 - |d\vec{x}|^2 - f^2((dT)^2 - |dX|^2) = (\sin^2 T - \sin^2 \alpha)(df)^2 - f^2 \sin^2 T (dT)^2$$
$$+ f^2 \sin^2 \alpha (d\alpha)^2 + 2f df(\sin T \cos T dT - \sin \alpha \cos \alpha d\alpha).$$

We can regard the vanishing of this as a quadratic equation in df/f and see that it is equivalent to

$$(df/f)(\sin^2 T - \sin^2 \alpha) + \sin T \cos T dT - \sin \alpha \cos \alpha d\alpha = \pm(\cos \alpha \sin T dT - \sin \alpha \cos T d\alpha).$$

Since $df/f = (\sin T dT + \sin \alpha d\alpha)/(\cos T + \cos \alpha)$, the left-hand side is

$$(\sin T dT + \sin \alpha d\alpha)(\cos \alpha - \cos T) + \sin T \cos T dT - \sin \alpha \cos \alpha d\alpha$$
$$= \sin T \cos \alpha dT - \sin \alpha \cos T d\alpha,$$

which verifies the conformality.

Theorem A.4.1. *The map* Ψ *defined by* (A.4.2) *when* $\cos T + X_0 > 0$, $-\pi < T < \pi$, *is a conformal bijection on Minkowski space, with*

(A.4.3) $$dT^2 - dX^2 = (\cos T + X_0)^2(dt^2 - |d\vec{x}|^2).$$

We shall also need to compute the pushforward of the vector fields in $\mathbf{R}^{1+n}$ by the inverse of Ψ. To avoid lengthy computations we begin with the simple case $n = 1$. Then S^n is parametrized by $\mathbf{R}/2\pi\mathbf{Z} \ni \alpha \mapsto (\cos \alpha, \sin \alpha) \in S^1$, and with these coordinates Ψ becomes the map

$$(T,\alpha) \mapsto (t,x) = (\sin T, \sin \alpha)/\Omega, \quad \Omega = \cos T + \cos \alpha.$$

We have

$$\frac{\partial}{\partial T} = \Omega^{-2}\left((1 + \cos T \cos \alpha)\frac{\partial}{\partial t} + \sin \alpha \sin T \frac{\partial}{\partial x}\right),$$
$$\frac{\partial}{\partial \alpha} = \Omega^{-2}\left(\sin \alpha \sin T \frac{\partial}{\partial t} + (1 + \cos T \cos \alpha)\frac{\partial}{\partial x}\right),$$

which gives

$$\frac{\partial}{\partial t} = (1 + \cos T \cos \alpha)\frac{\partial}{\partial T} - \sin \alpha \sin T \frac{\partial}{\partial \alpha}$$

$$\frac{\partial}{\partial x} = -\sin \alpha \sin T \frac{\partial}{\partial T} + (1 + \cos T \cos \alpha)\frac{\partial}{\partial \alpha}.$$

Thus the pushforwards of $\partial/\partial t$ and of $\partial/\partial x$ extend to analytic vector fields in the entire Einstein universe. This remains true for any space dimension n, but the proofs require somewhat more work since we do not have convenient global coordinates to work with then. We shall therefore convert the preceding result using the stereographic projection from $(-1,0)$, that is, $Y = \psi(X) = \sin \alpha/(1 + \cos \alpha) = \tan(\frac{1}{2}\alpha)$. Then

$$\sin \alpha = 2Y/(1+Y^2), \quad \cos \alpha = (1-Y^2)/(1+Y^2), \partial/\partial Y = 2(1+Y^2)^{-1}\partial/\partial \alpha.$$

With the variables Y instead of α on S^1 we thus obtain

(A.4.4)
$$\frac{\partial}{\partial t} = \left(1 + \frac{1-Y^2}{1+Y^2}\cos T\right)\frac{\partial}{\partial T} - Y \sin T \frac{\partial}{\partial Y},$$

$$\frac{\partial}{\partial x} = -\frac{2Y}{1+Y^2}\sin T \frac{\partial}{\partial T} + \tfrac{1}{2}(1+Y^2 + (1-Y^2)\cos T)\frac{\partial}{\partial Y}.$$

Let us now pass to the n dimensional case, parametrizing with the stereographic projection

$$Y = \psi(X) = \frac{\sin \alpha}{1 + \cos \alpha}\omega = \tan(\tfrac{1}{2}\alpha)\omega$$

where the notation is as in (A.4.1). Then

$$\omega = \frac{Y}{|Y|}, \ \tan(\tfrac{1}{2}\alpha) = |Y|, \ \sin \alpha = \frac{2|Y|}{1+|Y|^2}, \ \cos \alpha = \frac{1-|Y|^2}{1+|Y|^2}, \ \sin \alpha \omega = \frac{2}{1+|Y|^2}Y.$$

With the notation $a = |Y|^2$ we can write $\Psi(T,X) = (f(T,a), g(T,a)Y) = (t,x)$ where

$$f(T,a) = \frac{(1+a)\sin T}{(1+a)\cos T + 1 - a}, \ g(T,a) = \frac{2}{(1+a)\cos T + 1 - a}.$$

Thus

(A.4.5)
$$\frac{\partial}{\partial T} = f_T'\frac{\partial}{\partial t} + g_T'\langle Y, \frac{\partial}{\partial x}\rangle,$$

$$\frac{\partial}{\partial Y_j} = 2f_a'Y_j\frac{\partial}{\partial t} + g\frac{\partial}{\partial x_j} + 2g_a'Y_j\sum_1^n Y_k\frac{\partial}{\partial x_k},$$

where the second equation can be written

$$\frac{\partial}{\partial x_j} = \frac{1}{g}\frac{\partial}{\partial Y_j} - Y_j V, \quad gV = 2f_a'\frac{\partial}{\partial t} + 2g_a'\langle Y, \frac{\partial}{\partial x}\rangle.$$

Now the equations

$$\frac{\partial}{\partial T} = f_T'\frac{\partial}{\partial t} + g_T'\langle Y, \frac{\partial}{\partial x}\rangle$$

$$0 = 2f_a'\frac{\partial}{\partial t} + 2g_a'\langle Y, \frac{\partial}{\partial x}\rangle - gV$$

$$\langle Y, \frac{\partial}{\partial Y}\rangle = \qquad g\langle Y, \frac{\partial}{\partial x}\rangle + gaV$$

do not at all depend on the dimension. They are uniquely solvable for $\partial/\partial t$, $\langle Y, \partial/\partial x \rangle$ and V, for elimination of V between the last two equations gives in the right-hand side a matrix which is essentially the Jacobian of Ψ in the one dimensional case. In that case we have

$$V = \frac{1}{Y}\left(\frac{1}{g}\frac{\partial}{\partial Y} - \frac{\partial}{\partial x}\right) = \frac{1}{Y}\left(\tfrac{1}{2}((1+Y^2)\cos T + 1 - Y^2)\right)\frac{\partial}{\partial Y} + \frac{2Y}{1+Y^2}\sin T \frac{\partial}{\partial T}$$
$$- \tfrac{1}{2}(1+Y^2 + (1-Y^2)\cos T)\frac{\partial}{\partial Y}) = \frac{2}{1+Y^2}\sin T \frac{\partial}{\partial T} + (\cos T - 1)\langle Y, \frac{\partial}{\partial Y}\rangle.$$

Hence we obtain
(A.4.6)
$$\frac{\partial}{\partial t} = (1 + \frac{1-|Y|^2}{1+|Y|^2}\cos T)\frac{\partial}{\partial T} - \sin T \langle Y, \frac{\partial}{\partial Y}\rangle,$$
$$\frac{\partial}{\partial x_j} = -\frac{2Y_j}{1+|Y|^2}\sin T \frac{\partial}{\partial T} + \tfrac{1}{2}((1+|Y|^2)\cos T + 1 - |Y|^2)\frac{\partial}{\partial Y_j} + (1 - \cos T)Y_j\langle Y, \frac{\partial}{\partial Y}\rangle.$$

The vector fields on the right are clearly analytic vector fields on $\mathbf{R} \times (S^n \setminus (-1, 0, \ldots, 0))$. To prove analyticity also at the other pole we switch to the coordinates $Z_j = Y_j/|Y|^2$ corresponding to stereographic projection from the other pole. Then $Y_j = Z_j/|Z|^2$ and $\langle Y, \partial/\partial Y \rangle = -\langle Z, \partial/\partial Z \rangle$, $\partial/\partial Y = |Z|^2\partial/\partial Z - 2Z\langle Z, \partial Z \rangle$, and we obtain
(A.4.6)'
$$\frac{\partial}{\partial t} = (1 + \frac{|Z|^2-1}{|Z|^2+1}\cos T)\frac{\partial}{\partial T} + \sin T\langle Z, \frac{\partial}{\partial Z}\rangle,$$
$$\frac{\partial}{\partial x_j} = -\frac{2Z_j}{|Z|^2+1}\sin T\frac{\partial}{\partial T} + \tfrac{1}{2}((|Z|^2+1)\cos T + |Z|^2 - 1)\frac{\partial}{\partial Z_j} - (1 + \cos T)Z_j\langle Z, \frac{\partial}{\partial Z}\rangle.$$

Thus we have proved:

Theorem A.4.2. *The pushforwards of the vector fields $\partial/\partial t$ and $\partial/\partial x_j$ in $\mathbf{R}^{1+n}$ to the Einstein universe by the inverse of the map Ψ defined by (A.4.2) are restrictions of analytic vector fields on the entire Einstein universe defined by (A.4.6) and (A.4.6)' in terms of the stereographic projections of S^n.*

To simplify notation we sometimes use the notation $\partial/\partial t$ and $\partial/\partial x_j$ for the vector fields on the Einstein universe defined by (A.4.6), (A.4.6)' when no confusion seems possible. With that notation we shall prove that

(A.4.7) $$\frac{\partial\Omega}{\partial t} = -\sin T\frac{1-|Y|^2}{1+|Y|^2}\Omega, \quad \frac{\partial\Omega}{\partial x_j} = -\frac{2\cos T}{1+|Y|^2}Y_j\Omega; \qquad \Omega = \cos T + X_0.$$

With the coordinates Y of the stereographic projection the fact that Ω only depends on T and on $a = |Y|^2$ gives as above that it suffices to verify (A.4.7) when $n = 1$. Then it follows at once if we use the coordinate α on S^1:

$$\frac{\partial\Omega}{\partial t} = (1 + \cos T \cos\alpha)(-\sin T) + \sin\alpha\sin T\sin\alpha = -\sin T\cos\alpha(\cos T + \cos\alpha),$$
$$\frac{\partial\Omega}{\partial x} = \sin\alpha\sin T\sin T - (1 + \cos T\cos\alpha)\sin\alpha = -\sin\alpha\cos T(\cos T + \cos\alpha).$$

Note that the factors in front of Ω in the right-hand sides of (A.4.7) are also analytic in terms of the coordinates Z, so they are analytic functions on the entire Einstein universe.

We must also compute the pushforward of vector fields on $\mathbf{R} \times S^n$ to $\mathbf{R}^{1+n}$. A basis for these vector fields is

$$\frac{\partial}{\partial T}, \quad X_j \frac{\partial}{\partial X_k} - X_k \frac{\partial}{\partial X_j}; \quad j, k = 0, \ldots, n.$$

When $j, k = 1, \ldots, n$ these vector fields are infinitesimal generators of rotations in $\mathbf{R}^n$ which operate in $\mathbf{R}^{1+n}$ in the same way, so the corresponding vector fields in $\mathbf{R}^{1+n}$ are $x_j \partial/\partial x_k - x_k \partial/\partial x_j$. What remains is to study the vector fields

$$\frac{\partial}{\partial T}, \quad X_0 \frac{\partial}{\partial X_k} - X_k \frac{\partial}{\partial X_0}; \quad k = 1, \ldots, n.$$

When $n = 1$ and $\alpha \in \mathbf{R}/2\pi\mathbf{Z}$ is taken as parameter on S^1, $X_0 = \cos\alpha$, $X_1 = \sin\alpha$, then

$$\frac{\partial}{\partial \alpha} = -\sin\alpha \frac{\partial}{\partial X_0} + \cos\alpha \frac{\partial}{\partial X_1} = X_0 \frac{\partial}{\partial X_1} - X_1 \frac{\partial}{\partial X_0},$$

so the vector fields to study are $\partial/\partial T$ and $\partial/\partial \alpha$. Since $\Omega^{-2} \sin\alpha \sin T = tx$ and

$$\Omega^{-2}(1 + \cos T \cos\alpha) = \tfrac{1}{4}(4\Omega^{-2} + (1 + r^2 - t^2)(1 + t^2 - r^2)) = \tfrac{1}{2}(1 + r^2 + t^2),$$

our earlier formulas yield

(A.4.8)
$$\frac{\partial}{\partial T} = \tfrac{1}{2}(1 + t^2 + |x|^2)\frac{\partial}{\partial t} + tx\frac{\partial}{\partial x}$$
$$\frac{\partial}{\partial \alpha} = tx\frac{\partial}{\partial t} + \tfrac{1}{2}(1 + t^2 + |x|^2)\frac{\partial}{\partial x}.$$

In the n dimensional case we know from (A.4.5) that $\partial/\partial T$ is a linear combination of $\partial/\partial t$ and $\langle x, \partial/\partial x \rangle$ with coefficients depending only on t and $|x|^2$ and not on the dimension. (Note that $Y = x/(1 + \cos\alpha)$.) These coefficients can be read off from (A.4.8), so we have in general

(A.4.9)
$$\frac{\partial}{\partial T} = \tfrac{1}{2}(1 + t^2 + |x|^2)\frac{\partial}{\partial t} + t\langle x, \frac{\partial}{\partial x} \rangle.$$

In terms of the stereographic projection coordinates Y used above, we have

$$X_0 \frac{\partial}{\partial X_k} - X_k \frac{\partial}{\partial X_0} = \tfrac{1}{2}(1 - |Y|^2)\frac{\partial}{\partial Y_k} + Y_k \langle Y, \frac{\partial}{\partial Y} \rangle.$$

It suffices to verify the equality when applied to $X_0 = (1 - |Y|^2)/(1 + |Y|^2)$ or to $X_j = 2Y_j/(1 + |Y|^2)$, $j \neq 0$, which is completely straightforward. If we multiply the second set of equations (A.4.5) by Y_j and sum, we obtain

$$\langle Y, \frac{\partial}{\partial Y} \rangle = 2af_a'\frac{\partial}{\partial t} + (g + 2ag_a')\langle Y, \frac{\partial}{\partial x} \rangle.$$

Hence $\langle Y, \partial/\partial Y \rangle$ is a linear combination of $\partial/\partial t$ and $\langle x, \partial/\partial x \rangle$ with coefficients depending on t and $|x|^2$ but not on the dimension. Since

$$\tfrac{1}{2}(1 - |Y|^2)g = (1 - |Y|^2)/(\Omega(1 + |Y|^2)) = \cos\alpha/\Omega = \tfrac{1}{2}(1 + t^2 - r^2),$$

it follows that

$$X_0\frac{\partial}{\partial X_k} - X_k\frac{\partial}{\partial X_0} = \tfrac{1}{2}(1+t^2-|x|^2)\frac{\partial}{\partial x_k} + x_k(c_1\frac{\partial}{\partial t} + c_2\langle x,\frac{\partial}{\partial x}\rangle),$$

where c_1 and c_2 are functions of t and $|x|^2$ independent of the dimension. From the case $n=1$ in (A.4.8) we conclude that $c_1 = t$ and that $c_2 = 1$, so we have

(A.4.10) $$X_0\frac{\partial}{\partial X_k} - X_k\frac{\partial}{\partial X_0} = \tfrac{1}{2}(1+t^2-|x|^2)\frac{\partial}{\partial x_k} + x_k(t\frac{\partial}{\partial t} + \langle x,\frac{\partial}{\partial x}\rangle)).$$

The vector field on the right (the pushforward by Ψ of that on the left, to be quite correct) is of the form (6.3.8) with $\theta_k = -1/2$ and all other coordinates of θ equal to 0, in addition to the constant vector field $2^{-1}\partial/\partial x_k$. The vector field in (A.4.9) corresponds to $\theta_0 = 1/2$ and all other coordinates equal to 0, in addition to $2^{-1}\partial/\partial t$. We have therefore obtained all the conformal vector fields with quadratic coefficients which occur in (6.3.8).

To be able to transfer L^2 estimates between Minkowski space and the Einstein universe we must compute the surface measure in S^n in terms of the stereographic projection $Y = \psi(X)$. The conformality of ψ implies

$$|dX|^2 = (1+X_0)^2|dY|^2; \quad \text{here } 1+X_0 = 1+\cos\alpha = 2/(1+|Y|^2).$$

With the coordinates Y the metric tensor of S^n is thus equal to $4(1+|Y|^2)^{-2}$ times the identity matrix, so the square root of the determinant is $2^n(1+|Y|^2)^{-n}$, which proves that if u is a measurable function in $\mathbf{R}^n$, then

(A.4.11) $$\int_{S^n} |u(\psi(X)|^2\, dS(X) = 2^n \int_{\mathbf{R}^n} |u(x)|^2 (1+|x|^2)^{-n}\, dx.$$

References

Alinhac, S. [1], *Blowup for nonlinear hyperbolic equations*, Birkhäuser, Boston, Basel, Berlin, 1995.

Alinhac, S. [2], *Blowup of classical solutions of nonlinear hyperbolic equations: A survey of recent results*, Partial differential equations and mathematical physics (L. Hörmander and A. Melin, ed.), Birkhäuser, Boston, Basel, Berlin, 1996, pp. 15–24.

Alinhac, S. [3], *Blowup of small data solutions for a class of quasilinear wave equations in two space dimensions: an ouline of the proof*, Geometrical optics and related topics (F. Colombini and N. Lerner, eds.), Birkhäuser, Boston, Basel, Berlin, 1997.

Aubin, T. [1], *Non-linear analysis on manifolds. Monge-Ampère equations*, Springer Verlag, New York, Heidelberg, Berlin, 1982.

Beals, M. [1], *Self-spreading and strength of singularities for solutions to semilinear wave equations*, Ann. of Math. **118** (1983), 187–214.

Beals, M. [2], *Propagation of smoothness for nonlinear second order strictly hyperbolic differential equations*, A.M.S. Proc. Symp. Pure Math. **43** (1985), 21–44.

Beals, M. [3], *Spreading of singularities for a semilinear wave equation*, Duke Math. J. **49** (1982), 275–286.

Bony, J.M. [1], *Calcul symbolique et propagation des singularités pour les équations aux dérivées partielles non linéaires*, Ann. Sc. Ec. Norm. Sup. **14** (1981), 209–246.

Bony, J. M. [2], *Analyse microlocale des équations aux dérivées partielles non linéaires*, Springer Lecture Notes in Math. **1495** (1991), 1–45.

Bressan, A. [1], *The semigroup approach to systems of conservation laws*, Mathematica Contemporanea (to appear).

Burgers, J.M. [1], *A mathematical model illustrating the theory of turbulence*, Adv. in Appl. Mech. **1** (1948), 171–179.

Bourdaud, G. [1], *Sur les opérateurs pseudo-différentiels à coefficients peu réguliers*, Thèse, Univ. de Paris-Sud (1983), 1–154.

Bourdaud, G. [2], *Une algèbre maximale d'opérateurs pseudo-différentiels*, Comm. Partial Diff. Eq. **13** (1988), 1059–1083.

Chemin, J.-Y. [1], *Interaction contrôlée dans les E. D. P. non linéaires strictement hyperboliques*, Bull. Soc. Math. France **116** (1988), 341–383.

Chemin, J.-Y. and Lerner, N. [1], *Flot de champs de vecteurs non Lipschitziens et équations de Navier-Stokes*, J. Differential Equations **121** (1995), 314–328.

Ching, Chin-Hung [1], *Pseudodifferential operators with non regular symbols*, J. Differential Equations **11** (1972), 436–447.

Christodoulou, D. [1], *Global solutions of nonlinear hyperbolic equations for small initial data*, Comm. Pure App. Math. **39** (1986), 267–282.

Chueh, K.N., Conley, C.C. and Smoller, J.A. [1], *Positively invariant regions for systems of nonlinear diffusion equations*, Indiana Univ. Math. J. **26** (1977), 373–392.

Coifman, R., Lions, P. L., Meyer, Y. and Semmes, S. [1], *Compensated compactness and Hardy spaces*, Jour. Math. Pures Appl. **72** (1993), 247–286.

Dafermos, C. [1], *Generalized characteristics and the structure of solutions of hyperbolic conservation laws*, Indiana Univ. Math. J. **26** (1977), 1097–1119.

Dafermos, C. [2], *Regularity and large time behaviour of solutions of a conservation law without convexity*, Proc. Roy. Soc. Edinburgh **99A** (1985), 201–239.

Delort, J.-M. [1], *Sur le temps d'existence pour l'équation de Klein-Gordon semi-linéaire en dimension 1*, Prépublications Math. de l'Univ. Paris-Nord 96-33.

DiPerna, R. [1], *Hyperbolic conservation laws and shock waves*, Series of lectures during the summer term 1981, 1–59.

DiPerna, R.J. [2], *Convergence of approximate solutions to conservation laws*, Arch. Rat. Mech. Anal. **82** (1983), 27–70.

DiPerna, R.J. [3], *Compensated compactness and general systems of conservation laws*, Trans. Amer. Math. Soc. **292** (1985), 383–420.

Federer, H. [1], *Geometric measure theory*, Springer Verlag, Berlin, Heidelberg, New York, 1969.

Filippov, A.F. [1], *Differential equations with discontinuous right-hand side*, Amer.Math.Soc.Transl. Ser. 2 **42** (1960), 199–231.

Friedlander, G. [1], *On the radiation field of pulse solutions of the wave equation*, Proc. Roy. Soc. A **269** (1962), 53–65.

Friedlander, G. [2], *Ons the radiation field of pulse solutions of the wave equation. II*, Proc. Roy. Soc. A. **279** (1964), 386-394.

Friedlander, G. [3], *On the radiation field of pulse solutions of the wave equation. III*, Proc. Roy. Soc. A. **299** (1967), 264–278.

Friedrichs, K. and Lax, P.D. [1], *Systems of conservation laws with a convex extension*, Proc. Nat. Acad. Sci. U.S.A. **68** (1971), 1686–1688.

Gagliardo, E. [1], *Ulteriori proprieta di alcune classi di funzioni in piu variabili*, Ric. Mat. **8** (1959), 24–51.

Georgiev, V. [1], *Decay estimates for the Klein-Gordon equations*, Comm. Partial Diff. Eq. **17** (1992), 1111–1139.

Glimm, J. [1], *Solutions in the large for nonlinear systems of conservation laws*, Comm. Pure Appl. Math. **18** (1965), 685–715.

Hopf, E. [1], *The partial differential equation $u_t + uu_x = \mu u_{xx}$*, Comm. Pure Appl. Math. **3** (1950), 201–230.

Hörmander, L. [1], *The lifespan of classical solutions of non-linear hyperbolic equations*, Springer Lecture Notes in Math. **1256** (1987), 214–280.

Hörmander, L. [2], *Remarks on the Klein-Gordon equation*, Journées Equations aux dérivées partielles Saint-Jeande-Monts Juin 1987, pp. I-1–I-9.

Hörmander, L. [3], *L^1, L^∞ estimates for the wave operator*, Analyse Mathématique et Applications, Gauthier-Villars, Paris, 1988, pp. 211–234.

Hörmander, L. [4], *The analysis of linear partial differential operators I-IV*, Springer Verlag, Berlin, Heidelberg, New York, Tokyo, 1983–1985.

Hörmander, L. [5], *On the existence and the regularity of solutions of pseudo-differential equations*, L'Ens. Math. **17** (1971), 99–163.

Hörmander, L. [6], *On global existence of solutions of non-linear hyperbolic equations in $\mathbf{R}^{1+3}$*, Institut Mittag-Leffler report **9** (1985), 1–22.

Hörmander, L. [7], *The fully non-linear Cauchy problem with small data*, Bol. Soc. Bras. Mat. **20** (1989), 1–27.

Hörmander, L. [8], *On the fully non-linear Cauchy problem with small data. II*, Microlocal analysis and nonlinear waves (M. Beals, R. Melrose, J. Rauch, eds.), Springer Verlag, 1991, pp. 51–81, IMA Volumes in Mathematics and its Applications vol. 30.

Hörmander, L. [9], *Pseudo-differential operators of type 1,1*, Comm. Partial Diff. Eq. **13** (1988), 1085–1111.

Hörmander, L. [10], *Continuity of pseudo-differential operators of type 1,1*, Comm. Partial Diff. Eq. **14** (1989), 231–243.

John, F. [1], *Formation of singularities in one-dimensional non-linear wave propagation*, Comm. Pure Appl. Math. **27** (1974), 377–405.

John, F. [2], *Blowup of radial solutions of $u_{tt} = c^2(u_t)\Delta u$ in three space dimensions*, Mat. Apl. Comput. **4** (1985), 3–18.

John, F. [3], *Existence for large times of strict solutions of nonlinear wave equations in three space dimensions*, Comm. Pure Appl. Math. **40** (1987), 79–109.

John, F. [4], *Blow-up of solutions of nonlinear wave equations in three space dimensions*, Manuscripta Math. **28** (1979), 235–265.

John F. and Klainerman, S. [1], *Almost global existence to nonlinear wave equations in three space dimensions*, Comm. Pure Appl. Math. **37** (1984), 443–455.

Klainerman, S. [1], *Global existence for nonlinear wave equations*, Comm. Pure Appl. Math. **33** (1980), 43–101.

Klainerman, S. [2], *Long time behaviour of solutions to nonlinear wave equations*, Proc. Int. Congr. Math., Warszawa., 1983, pp. 1209–1215.

Klainerman, S. [3], *Uniform decay estimates and the Lorentz invariance of the classical wave equation*, Comm. Pure Appl. Math. **38** (1985), 321–332.

Klainerman, S. [4], *The null condition and global existence to nonlinear wave equations*, Lectures in Applied Mathematics **23** (1986), 293–326.

Klainerman, S. [5], *Global existence of small amplitude solutions to nonlinear Klein-Gordon equations in four space-time dimensions*, Comm. Pure Appl. Math. **38** (1985), 631-641.

Kružkov, S. [1], *First order quasilinear equations with several space variables*, Math. USSR Sbornik **10** (1970), 217–273.

Lascar, B. [1], *Singularités des solutions d'équations aux dérivées partielles nonlinéaires*, C. R. Acad. Sci. Paris **287A** (1978), 521–529.

Lax, P.D.[1], *Hyperbolic systems of conservation laws and the mathematical theory of shock waves*, Regional Conf. in Appl. Math. **11** (1973), 1–48.

Lax, P.D. [2], *Hyperbolic systems of conservation laws II*, Comm. Pure Appl. Math. **10** (1957), 537–566.

Lax, P.D. [3], *Weak solutions of nonlinear hyperbolic equations in two independent variables*, Comm. Pure Appl. Math. **7** (1954), 159–193.

Lax, P.D. [4], *Shock waves and entropy*, Contributions to nonlinear functional analysis, Academic Press, New York, 1971, pp. 603–634.

Lindblad, H. [1], *Blow up for solutions of $\Box u = |u|^p$ with small initial data*, Comm. Partial Diff. Eq. **15** (1990), 757–821.

Lindblad, H. [2], *On the lifespan of solutions of nonlinear wave equations with small initial data*, Comm. Pure Appl. Math. **43** (1990), 445–472.

Li Ta Tsien and Chen Yun-Mei [1], *Initial value problems for nonlinear wave equations*, Comm. Partial Diff. Eq. **13** (1988), 383–422.

Li Ta Tsien and Zhou Yi [1], *A note on the life-span of classical solutions to nonlinear wave equations in four space dimensions*, Indiana Univ. Math. J. (1996) (to appear).

Liu, Tai-Ping [1], *The deterministic version of the Glimm scheme*, Comm. math. Phys. **57** (1977), 135–148.

Meyer, Y. [1], *Régularité des solutions des équations aux dérivées partielles non linéaires*, Springer Lecture Notes in Math. **842** (1980), 293–302.

Meyer, Y. [2], *Remarques sur un théorème de J.M. Bony*, Suppl. ai Rend. del Circolo mat. di Palermo **II:1** (1981), 1-20.

Moriyama, K., Tonegawa, S. and Tsutsumi, Y. [1], *Almost global existence of solutions for the quadratic semilinear Klein-Gordon equation in one space dimension*, Preprint 1996.

Nirenberg, L. [1], *On elliptic partial differential equations*, Ann. Scu. Norm. Sup. Pisa **13(3)** (1959), 115-162.

Olejnik, O.A. [1], *Discontinuous solutions of non-linear differential equations*, Amer. Math. Soc. Transl. Ser. 2 **26** (1957), 95–172.

Olejnik, O.A. [2], *Uniqueness and stability of the generalized solution of the Cauchy problem for a quasilinear equation*, Amer. Math. Soc. Transl. Ser. 2 **33** (1964), 285–290.

Ozawa, T., Tsutaya, K. and Tsutsumi, Y. [1], *Global existence and asymptotic behavior of solutions for the Klein-Gordon equations with quadratic nonlinearity in two space dimensions*, Math. Z. **222** (1996), 341–362.

Rauch, J. [1], *Singularities of solutions to semilinear wave equations*, J. Math. Pures et Appl. **58** (1979), 299–308.

Rauch, J. and M. Reed [1], *Propagation of singularities for semilinear hyperbolic equations in one space variable*, Ann. of Math. **111** (1980), 531–552.

Rauch, J. and M. Reed [1], *Nonlinear microlocal analysis of semilinear hyperbolic systems in one space dimension*, Duke Math. J. **49** (1982), 397–475.

Schaeffer, D.G. [1], *A regularity theorem for conservation laws*, Adv. in Math. **11** (1973), 368–386.

Schatzman, M. [1], *Continuous Glimm functionals and uniqueness of solutions of the Riemann problem*, Indiana Univ. Math. J. **34** (1985), 533–589.

Shatah, J. [1], *Normal forms and quadratic nonlinear Klein- Gordon equations*, Comm. Pure Appl. Math. **38** (1985), 685–696.

Simon, J. C. H. and Taflin, E. [1], *The Cauchy problem for non-linear Klein-Gordon equations*, Comm. Math. Phys. **152** (1993), 433–478.

Smoller, J. [1], *Shock waves and reaction-diffusion equations*, Springer Verlag, New York, Heidelberg, Berlin, 1983.

Stokes, G.G. [1], *On a difficulty in the theory of sound*, London & Edinburgh Philosophical Magazine **33** (1848), 349–356.

Tartar, L. [1], *Compensated compactness and applications to partial differential equations*, Nonlinear Analysis and mechanics: Heriot-Watt Symposium IV, Pitman, San Fransisco, London, Melbourne, 1979, pp. 136–212.

Vol'pert, A. [1], *The spaces BV and quasilinear equations*, Math.USSR Sbornik **2** (1967), 225–267.

von Wahl, W. [1], *L^p decay rates for homogeneous wave equations*, Math. Z. **120** (1971), 93–106.

Weinberger, H. [1], *Invariant sets for weakly coupled parabolic and elliptic systems*, Rend. Mat. Univ. Roma **8** (1975), 295-310.

Yordanov, B. [1], *Blow-up for the one-dimensional Klein-Gordon equation with a cubic nonlinearity*, Preprint 1995, 2 pp.

Young, L.C. [1], *Lectures on the calculus of variations and optimal control theory*, W.B. Saunders, Philadelphia, Pa, 1969.

$\mathbf{R}$	The real numbers.
$\mathbf{R}_+$	The strictly positive real numbers.
$t_\pm$	The positive (negative) part $\max(0, \pm t)$ when $t \in \mathbf{R}$.
$\mathbf{Z}$	The integers.
$\mathbf{C}$	The complex numbers.
$C^k(X)$	Functions in X with continuous derivatives of order $\leq k$.
$C_0^k(X)$	Functions in $C^k(X)$ with compact support.
$\mathcal{D}'(X)$	Schwartz distributions in X.
χ_+^a	A homogeneous distribution on $\mathbf{R}$ defined on p. 91.
$\mathcal{E}'(X)$	Schwartz distributions in X with compact support.
$\mathcal{S}(\mathbf{R}^n)$	Schwartz space of rapidly decreasing C^∞ functions.
$\mathcal{S}'(\mathbf{R}^n)$	The dual space of temperate distributions.
$H_{(s)}$	Sobolev space of distributions with L^2 derivatives of order s (p. 188).
$H_{(-\infty)}$	The union of all $H_{(s)}$.
$H_{(+\infty)}$	The intersection of all $H_{(s)}$.
$H_{(s)}^{\text{loc}}$	See p. 188 for definition in the cotangent bundle.
$H_{(s)}^{\text{loc}}(X; P)$	See Definition 11.5.3, p. 258.
C^ϱ	Hölder class if $\varrho > 0$ is not an integer.
$|\cdot|_\varrho$	The norm in C^ϱ.
C_*^ϱ	Zygmund class of order ϱ, arbitrary $\varrho \in \mathbf{R}$.
$|\cdot|_\varrho^*$	A norm in C_*^ϱ.
$f * g$	The convolution of f and g.
supp u	The support of u.
sing supp u	The singular support of u.
$X \Subset Y$	The closure of X is a compact subset of Y.
$\complement X$	The complement of X (in some larger set).
∂X	The boundary of X.
α	Usually a multiindex $\alpha = (\alpha_1, \ldots, \alpha_n)$.
$|\alpha|$	The length $\alpha_1 + \cdots + \alpha_n$ of α.
$\alpha!$	The multifactorial $\alpha_1! \cdots \alpha_n!$.
x^α	The monomial $x_1^{\alpha_1} \cdots x_n^{\alpha_n}$ in $\mathbf{R}^n$.
$\langle \xi \rangle$	$(1 + |\xi|^2)^{\frac{1}{2}}$ when $\xi \in \mathbf{R}^n$.
∂^α	The partial derivative $\partial_1^{\alpha_1} \cdots \partial_n^{\alpha_n}$ where $\partial_j = \partial/\partial x_j$.
$a_{(\beta)}^{(\alpha)}(x, \xi)$	Short notation for $\partial_\xi^\alpha \partial_x^\beta a(x, \xi)$.
$J_m u(x)$	The m-jet of u at x, the array $(\partial^\alpha u(x))_{|\alpha| \leq m}$.
D^α	The partial derivative $D_1^{\alpha_1} \cdots D_n^{\alpha_n} = i^{-|\alpha|}\partial^\alpha$, $D_j = -i\partial_j$.
Δ	The Laplace operator.
$\square$	The wave operator.
$S_{\varrho,\delta}^m$	Symbols of order m and type ϱ, δ (see pp. 200, 211).
S^m	Symbols of order m and type $1, 0$; same as $S_{1,0}^m$ (see p. 193).
S^n	Occasionally this denotes the Euclidean unit sphere in $\mathbf{R}^{n+1}$.

$S^{-\infty}$	The intersection $\cap_m S^m = \cap_m S^m_{\varrho,\delta}$ for arbitrary ϱ, δ.
$\tilde{S}^m_{1,1}$	A subset of $S^m_{1,1}$ defined on p. 226.
A^m_ϱ	A class of irregular symbols (see p. 236).
$a(x, D)$	The pseudo-differential operator with symbol $a(x, \xi)$ (see p. 193).
$\mathrm{Op}\, a$	Another notation for $a(x, D)$.
$\Psi^m(X)$	The pseudo-differential operators in X of order m and type $1, 0$ (see p. 198).
$\Psi^m_{\varrho,\delta}$	Pseudo-differential operators with symbol in $S^m_{\varrho,\delta}$.
$\tilde{\Psi}^m_{1,1}$	Pseudo-differential operators with symbol in $\tilde{S}^m_{1,1}$ (see p. 226).
WF	The wave front set (see Definition 8.2.2, p. 193).
Id	The identity (matrix).
Tr	The trace of a matrix .
$[A, B]$	The commutator $AB - BA$ of two operators.
${}^tT$	The transpose of a linear map (matrix) T.
δ_{ij}	The Kronecker symbol $\delta_{ij} = 0$ when $i \neq j$, $\delta_{jj} = 1$, the unit matrix.

Printing and Binding: AZ Druck und Datentechnik GmbH, Kempten